Foundations and Societal Considerations of Gene-Edited Plants

Sana Basharat
https://orcid.org/0000-0002-6737-5611
Hainan University, China

Muhammad Waseem
http://orcid.org/0000-0001-7947-7722
Hainan University, China

IGI Global
Scientific Publishing
Publishing Tomorrow's Research Today

Vice President of Editorial Melissa Wagner
Director of Acquisitions Mikaela Felty
Director of Book Development Jocelynn Hessler
Production Manager Mike Brehm
Cover Design Jose Rosado

Published in the United States of America by
IGI Global Scientific Publishing
701 East Chocolate Avenue
Hershey, PA, 17033, USA
Tel: 717-533-8845 | Fax: 717-533-7115
Website: https://www.igi-global.com E-mail: cust@igi-global.com

Library of Congress Cataloging-in-Publication Data

LCCN: 2026037915 (CIP Data Pending)
ISBN13: 9798260035757
Isbn13Softcover: 9798260035764
EISBN13: 9798260035771
British Cataloguing in Publication Data
A Cataloguing in Publication record for this book is available from the British Library.

Table of Contents

Detailed Table of Contents

TALENs offer high specificity, their complex, labor-intensive design limits scalability; CRISPR enables faster editing and multiplexing. Newer base and prime editors further increase precision by introducing targeted changes without double-strand breaks, reducing unintended effects. Coupled with synthetic biology and AI for guide design and off-target prediction, gene editing is accelerating plant biotechnology. Remaining hurdles—off-target mutations, ethics, and regulation—demand continued research and international coordination to maximize benefits for sustainable agriculture and food security.

Crop development underpins agricultural progress and food security. Traditional breeding—selection, hybridization, and mutation—has delivered durable, high-yield varieties but is slow and limited by available genetic diversity. Gene-editing tools such as CRISPR-Cas9, TALENs, and ZFNs now enable precise genetic changes to improve yield, nutrition, disease resistance, and stress tolerance. This chapter reviews gene editing in modern agriculture, highlighting biofortification, extended shelf life, and enhanced resistance in major crops, as well as applications for climate-resilient crops and livestock. It also discusses barriers to adoption, including public perception, ethical concerns, and regulatory complexity. Case studies on climate-resilient rice, heat-tolerant cattle, and pest-resistant maize illustrate integrated breeding strategies. Future gains will depend on advances such as pangenomics and genomic selection, alongside clearer regulation and improved public engagement.

Gene editing is a groundbreaking technology in sustainable agriculture, whereby plant genomes can be strictly edited to produce quality, high-yielding, nutritious, and stress-tolerant crops. The CRISPR/Cas and other technologies enhance resistance to pests, diseases, drought, salinity, and heat, as well as resource-use efficiency and decrease the use of chemical inputs. Gene editing, unlike traditional genetic

drivers of consumer attitudes. Misinformation and sensational reporting can amplify perceived risks, while cultural and ethical concerns—such as religious views and "naturalness"—shape regional differences in acceptance. Survey evidence and market trends suggest that transparent labeling and confidence in regulators strongly influence purchasing decisions. The chapter reviews case studies of public engagement to identify effective practices, including clear science communication, educational outreach, and policy transparency. Overall, it argues for a balanced strategy that integrates scientific evidence with ethical reflection and sustained public dialogue, as improved transparency and misconception reduction are essential for wider adoption in support of food security and sustainable agriculture.

Muhammad Rizwan Khan, Quaid-i-Azam University, Islamabad, Pakistan
Sameel Hassan, Department of Agriculture and Forest Sciences, University of Tuscia, Viterbo, Italy
Tauqeer Ahmaed Qadri, Department of Biosciences, COMSATS University, Islamabad, Pakistan
Muhammad Waseem, Hainan University, Haikou, China

Site-specific plant gene editing is transforming agricultural biotechnology. Programmable nucleases, especially CRISPR/Cas systems, enable crops with higher yield, enhanced stress tolerance, and improved nutritional quality. However, broad deployment remains constrained by technical and translational barriers. Key limitations include off-target mutations, inefficient delivery and transformation, and low homology-directed repair (HDR) rates, which reduce reproducibility and scalability. Adoption is further slowed by complex tissue-culture requirements, specialized infrastructure, and skilled personnel, increasing development costs. Regulatory uncertainty, ethical debate, and divergent public attitudes also create region-specific hurdles. This chapter traces the evolution of plant gene editing from early tools to base and prime editors, reviews current constraints, and highlights emerging solutions such as AI-guided target design, nanoparticle-enabled delivery, and in planta transformation. Case studies in wheat and leafy greens illustrate practical applications.

Recent advances in the CRISPR-Cas system of genome editing have led to a breakthrough in the ability to precisely modify the plant microbiome and interactions between plants and microorganisms. Modifying plant and microbe communities using the CRISPR technology has the potential to enhance agricultural production, thus minimizing the excessive application of chemicals as fertilizers and pesticides. This is a powerful strategy that has shown promise in increasing the ability of beneficial microorganisms to fix atmospheric nitrogen and dissolve insoluble phosphates. Moreover, the integration of CRISPR technology with synthetic biology, metagenomics, and multi-omics has provided a new perspective on developing microbiome engineering for agriculture.

Preface

ABOUT THIS BOOK

Genome editing has emerged as one of the most transformative technologies in plant science, offering unprecedented opportunities to improve crops with greater precision, speed, and efficiency than conventional breeding approaches. Since the development of CRISPR/Cas systems and other programmable genome editing tools, plant biotechnology has entered a new era in which targeted genetic modifications can address challenges related to food security, climate change, sustainable agriculture, and environmental conservation. Alongside these scientific advances, however, important questions have emerged regarding regulatory oversight, ethical considerations, public acceptance, and the responsible implementation of genome editing technologies.

Foundations and Societal Considerations of Gene-Edited Plants provides a comprehensive exploration of the scientific principles, technological developments, and broader societal dimensions of plant genome editing. The volume introduces the molecular basis of genome editing, including CRISPR/Cas systems and other genome engineering platforms, while presenting comparative perspectives on the strengths and limitations of different technologies. It further examines how genome editing complements conventional breeding strategies and contributes to sustainable agricultural production.

Beyond the scientific foundations, this book addresses critical issues surrounding the governance and adoption of gene-edited crops. Topics including regulatory frameworks, biosafety, public perception, ethical considerations, and future policy development are discussed in the context of rapidly evolving international regulations and agricultural innovation. The book also explores emerging research frontiers, including microbiome engineering and integrated approaches that combine genome editing with systems biology and precision agriculture.

By bringing together contributions from experts in molecular biology, genetics, plant biotechnology, breeding, microbiology, regulatory science, and agricultural policy, this volume provides readers with a balanced understanding of both the scientific advances and the broader implications of genome editing technologies.

We hope this book will serve as an authoritative reference for researchers, students, educators, policymakers, regulators, and biotechnology professionals seeking to understand the scientific foundations, practical applications, and societal implications of plant genome editing. As the field continues to evolve rapidly, we anticipate that this volume will contribute to informed discussion, responsible innovation, and future advances in sustainable agriculture.

Sana Basharat
Hainan University, China

Muhammad Waseem
Hainan University, China

Chapter 1

Introduction to Gene Editing in Plants:
Gene Editing in Plants

Tauqeer Ahmaed Qadri
http://orcid.org/0009-0000-9617-4512
COMSATS University, Islamabad, Pakistan

Muhammad Anas
http://orcid.org/0000-0001-5127-8043
University of Selcuk, Konya, Turkey

Amjid Khan
http://orcid.org/0009-0001-5475-5156
University of South Africa, South Africa

Muhammad Rizwan Khan
Quaid-i-Azam University, Islamabad, Pakistan

Shabbar Abbas
http://orcid.org/0000-0003-3435-1266
COMSATS University, Islamabad, Pakistan

Javed Iqbal
http://orcid.org/0000-0002-9032-2622
Bacha Khan University, Charsadda, Pakistan

Farishta Zarshan
http://orcid.org/0009-0007-8445-2950
Bacha Khan University, Charsadda, Pakistan

ABSTRACT

Plant genome editing—especially CRISPR/Cas—enables precise DNA changes for faster, sustainable crop improvement. This chapter covers tools from ZFNs/TALENs to Cas12/13, base/prime editing and dCas9, plus applications in stress tolerance, disease resistance, yield, nutrition and shelf life. It also reviews epigenome/RNA editing, synthetic biology, AI-guided design, single-cell insights, global regulation, ethics, equity, and key technical challenges.

DOI: 10.4018/979-8-2600-3575-7.ch001

INTRODUCTION

Plant genome editing refers to a suite of molecular techniques that enable precise targeted modifications to the DNA of plant species. The CRISPR/Cas9 system has become a revolutionary tool for inducing specific changes at specific genomic loci. Unlike traditional genetic modification, which usually involves insertion of foreign DNA into the plant genome, genome editing results in alteration of the plant's native genome to the desired traits without the introduction of exogenous genetic material. This precision prevents the development of unintended effects and accelerates the development of better plant varieties (Bhatt et al., 2022). Genome editing in plants has a wide scope of applications, including high yield, nutritional content improvement, resistance to biotic and abiotic stresses, and shelf-life extension. In recent years, advances have been made in developing gene-edited crops, including non-browning bananas, long-lasting strawberries, and vitamin D-enriched tomatoes, which could soon hit supermarket shelves if regulatory approvals are forthcoming (Anas et al., 2025; Sharma et al., 2023).

IMPORTANCE OF GENOME EDITING IN AGRICULTURE

Climate change, population growth, and the need for sustainable food production are putting increasing pressure on the global agricultural sector. These issues can be addressed by rapid development of crop varieties with better traits through genome editing (Ahmed et al., 2023; Kumar et al., 2022). For instance, gene-edited crops have been developed to improve resistance to diseases, pests, and environmental stresses, such as drought and salinity. Furthermore, genome editing has enabled the improvement of the nutritional content, shelf life, and yield of crops, which is good for food security and minimizes post-harvest losses (Shelake et al., 2022). An obvious example of CRISPR/Cas9 technology is that it can be used to create wheat varieties immune to powdery mildew, a common fungal disease that has the potential to affect wheat production worldwide. Scientists can precisely switch on or off the genes that make the plant susceptible, so the plant can strengthen its natural defense, so that chemical pesticides are not needed, and a more sustainable and resilient agricultural system is ensured (Choudhary et al., 2022; Raza et al., 2025). In addition, genome editing plays a significant role in the creation of crops with improved tolerance to environmental stress. For instance, CRISPR/Cas9 has been used to modify genes for drought and salinity tolerance in rice and tomatoes, allowing them to survive in unfavorable environments. To enhance resistance and tolerance, genome editing has been used to improve the nutritional quality of crops.

Researchers have used CRISPR/Cas9 to enhance the iron and zinc contents of rice to eliminate micronutrient deficiencies in many developing countries (Li et al., 2022).

COMPARISON WITH TRADITIONAL BREEDING AND TRANSGENICS

Traditional crop breeding entails slower selection, crossing, and cultivation of specific plants with desirable traits, as well as the intersection of desirable species. Breeding this method is difficult due to the lack of variation within a species' genetics. Attempts to speed up the process can be made using transgenic methods, in addition to public disapproval (Ahmar et al., 2020). However, these issues are avoided when genome editing is used. This method eliminates the need to insert DNA, enabling a faster, smoother, and smoother regulation flow. Modifying it in such a way as to not affect exterior DNA increases precision, reduces outside DNA adjustments, maintains genetic control, and makes the process more efficient. Reliance on procedures such as this leads to a stronger outcome (Van Eck, 2020). Genome editing offers superior precision and efficiency compared with traditional breeding methods, which take decades to create new traits in crops. Genome editing has the potential to reduce the development time of new crop traits to several years. As genome editing does not require foreign DNA introduction, unlike transgenic methods, it could help address public concerns regarding genetically modified organisms (GMOs) (Nawaz, 2021).

SOCIO-ECONOMIC AND ENVIRONMENTAL IMPLICATIONS

The use of genome editing technology in agriculture has a major impact on socio-economic structures and environmental conditions. The economic advantages of genome editing include cost reduction in crop development while reducing chemical input dependency, which benefits producers and consumers alike (Bullock et al., 2021). The use of genome-edited crops in agriculture helps to create a more sustainable farming system by decreasing pesticide and fertilizer requirements and reducing environmental impacts. This technology offers social benefits that could help overcome nutritional deficiencies and strengthen food security in developing areas (Husaini & Sohail, 2024).

The responsible implementation of genome-editing technologies depends critically on addressing equitable access issues, ethical questions, and regulatory structures. The Union's tough regulations concerning genome-edited crops have generated discussions about intellectual property rights and how they could hinder innovation (Menz et al., 2020). Genome-edited crops have different levels of public acceptance across various regions. Genome-edited crops are available in markets in the United States and Canada, but Europe maintains a cautious approach because of regulatory restrictions and public apprehension (Spök et al., 2022).

This chapter provides a complete survey of plant genome editing technologies while explaining their scientific foundations, agricultural applications, and socio-economic and environmental implications. The objectives are to:

- Describe the basic procedures and instruments used to edit plant genomes.
- Evaluate how genome editing differs from traditional plant breeding and transgenic methods.
- Exploring how genome editing techniques enhance crop development through practical applications and advantages.
- Research on both the economic and ecological consequences resulting from genome editing technology implementation.

EVOLUTION OF GENE EDITING TECHNOLOGIES

Timeline: From Zinc Finger Nucleases (ZFNs) and TALENs to CRISPR

The exploration of gene editing technologies has been associated with critical points that have improved the precision of genome modifications in plants.

Zinc Finger Nucleases (ZFNs)

ZFNs, introduced at the very end of the 1990s, constituted the first "tool" that allowed targeted genome manipulation. These constructed proteins are represented by a zinc finger DNA-binding domain linked to a nuclease domain that is typically obtained from the FokI restriction enzyme (Shamshirgaran et al., 2022). Through the production of specific zinc finger arrays, ZFNs can promote their delineation to target specific DNA sequences and cause double strand breaks (DSBs) that trigger the natural patching up of the cell (the Heidelberg process) and hence the desired mutations. Interestingly, although ZFNs were pioneers in this area, they are hard to design and may have off-target effects that limit their use (Shamshirgaran et al., 2022).

Transcription Activator-Like Effector Nucleases (TALENs)

TALENs, a more versatile means of editing a genome, emerged in the early 2010s, following the development of ZFNs. TALENs are made from transcription activator-like effectors (TALEs) of *Xanthomonas* bacteria and comprise customizable DNA-binding domains that are fused to FokI nuclease (Shamshirgaran et al., 2022). Each TALE repeat identifies a single nucleotide, making it easier and more accurate to target than ZFNs. TALENs have been successfully used in different plant species, leading to the development of improved crops. Their large size and need for protein engineering for every target site are problematic (Bhardwaj & Nain, 2021).

CRISPR-Cas Systems

CRISPR-Cas system enables genome editing. Derived from the adaptive immune system of bacteria, the CRISPR-Cas9 system employs a gRNA to act as a sort of address tag to help direct Cas9 nuclease to a defined location in the genome and induce DSBs. This RNA-guided mechanism streamlines and eases the design process, making genome editing much easier and convenient (Wang et al., 2022). After its microbiological application adaptation for eukaryotic systems in 2012, CRISPR-Cas9 has been extensively used in plant studies, making crop improvement involving trait improvement an efficient process (Erdoğan et al., 2023). There are others such as Cas12 and Cas13 that exit beyond Cas9 and have expanded the genome editing tool kit, while Cas12 possesses various cleaving patterns, Cas13 uses RNA for confinement, offering a way to perform transcriptome engineering. This has expanded the domain of genome editing applications in plants (Azubuike et al., 2022).

Table 1. Comparative Overview of Major Gene Editing Tools in Plants: Mechanisms, Advantages, Limitations and Applications.

Gene Editing Tool	Discovery & Milestones	Mechanism of Action	Targeting Specificity	Advantages	Limitations	References
Zinc Finger Nucleases (ZFNs)	Introduced in late 1990s	Custom-designed zinc finger domains bind to specific DNA sequences; FokI nuclease induces double-strand breaks (DSBs)	Moderate; specificity depends on engineered protein-DNA interaction	First-generation programmable nuclease; enabled early genome editing	Complex design; high cost; off-target effects; limited flexibility	(Cathomen & Joung, 2008)
Transcription Activator-Like Effector Nucleases (TALENs)	Developed in 2009	TALE domains bind specific DNA sequences; FokI nuclease creates DSBs	High, due to longer recognition sequences	Improved specificity over ZFNs; easier to design; efficient in plants	Labor-intensive construction; delivery remains a challenge	(Ochiai & Yamamoto, 2015)
CRISPR/Cas9	Revolutionized gene editing in 2012–2013	RNA-guided endonuclease Cas9 cleaves target DNA at complementary sites via guide RNA (gRNA)	High; easily programmable via gRNA	Simple, cost-effective, scalable; multiplexing possible	Potential off-targets; PAM site required; delivery and tissue culture issues	(Ochiai & Yamamoto, 2015; Wang & Doudna, 2023)
CRISPR/ Cas12 (Cpf1)	Identified in 2015	Similar to Cas9, but recognizes T-rich PAMs; produces staggered DSBs	High; flexible targeting due to T-rich PAM	Smaller size; different PAM; no tracrRNA required	Less characterized than Cas9; PAM restriction	(Bandyopadhyay et al., 2020)
CRISPR/ Cas13	Discovered in 2016	Targets and cleaves RNA instead of DNA; useful for post-transcriptional gene silencing	Very high; RNA-targeting allows fine-tuning gene expression	RNA editing; virus resistance; transient effects	Limited use in stable trait modification; delivery challenges	(Pramanik et al., 2021)
dCas9 (Dead Cas9)	Modified form of Cas9 (nuclease-dead)	Binds DNA without cutting; used for gene regulation (CRISPRi/a)	High; retains target specificity without cleavage	Epigenome editing, transcriptional regulation	No DNA modification; requires co-activators or repressors	(Backes, 2020)

continued on following page

Table 1. Continued

Gene Editing Tool	Discovery & Milestones	Mechanism of Action	Targeting Specificity	Advantages	Limitations	References
Base Editing	Developed in 2016–2017	Fuses dCas9 or nCas9 to cytidine or adenine deaminase to create precise point mutations without DSBs	Very high; enables single nucleotide substitutions	Precise edits; no DSBs; fewer off-targets	Limited to transition mutations; editing window constraints	(Saber Sichani et al., 2023)
Prime Editing	Introduced in 2019	Combines nCas9 and reverse transcriptase with pegRNA to write new sequences into genome	Extremely high; highly programmable	Enables insertions, deletions, and all base changes without DSBs	Lower efficiency in plants (still under optimization); complex design	(Mikhaylova et al., 2024)

MILESTONES IN PLANT GENOME EDITING

The application of genome-editing technologies in plants has led to numerous breakthroughs.

- **Disease Resistance**: CRISPR-Cas9 has been taken as an application used to promote the resistance of rice varieties to bacterial blight by conjugating susceptibility genes, thus reducing the use of chemical pesticides (Zafar et al., 2020).
- **Abiotic stress tolerance**: Raising the editing of genes related to drought and salinity resistance has led to the development of crops that can survive in extreme environmental conditions vital for agriculture when confronted with climate change (KhokharVoytas et al., 2023).
- **Nutritional Enhancement**: Genome editing has been applied to biofortify crops; for example, provitamin A in rice has been enhanced and micronutrient deficiencies in populations dependent on staple crops (Shoeb et al., 2024).
- **Yield Improvement**: Changes in the genes that control plant architecture and flowering time have considerably increased the yield of crops such as maize and wheat (Huang et al., 2021).

ADVANCES IN DELIVERY METHODS

The effective delivery of genome-editing components into plant cells is vital for successful genome modifications. A series of techniques was created and perfected:

Agrobacterium-Mediated Transformation

Agrobacterium tumefaciens, a soil bacterium, is a naturally occurring DNA transgene in the plant genome. Scientists have used this mechanism to transfer the genes or genome-editing components of choice into plants. This method is widely applied because it is efficient and can incorporate DNA into the plant genome. However, its application is limited to specific plant species and creates transgenic plants, which attracts regulatory concerns (Thomson et al., 2024).

Particle Bombardment (Biolistics)

This physical approach entails shooting plant tissues with high-velocity equipment and tiny DNA-coated particles. Particle bombardment is flexible and applicable to a large category of plants, which cannot be easily transformed using Agrobacterium-mediated approaches. However, it may have tissue-damaging effects and typically causes random integration of DNA, which can cause unintended effects (Ozyigit & Yucebilgili Kurtoglu, 2020).

Protoplast Transfection

Protoplasts are plant cells without cell walls that are capable of the direct uptake of DNA, RNA, or protein complexes. Techniques, such as polyethylene glycol (PEG)-mediated transfection, are used to introduce genome-editing components into protoplasts. This approach allows for transient expression and decreases the likelihood of stable DNA integration. Nevertheless, regeneration of complete plants from protoplasts remains difficult for many species (Chen et al., 2023).

Nanoparticle-Mediated Delivery

New nanotechnology methodologies use nanoparticles to deliver genome-editing reagents to plant cells. The plant cell wall and membrane can be crossed by nanoparticles which provides noninvasive delivery. This strategy may hold promise for DNA-free genome editing and may overcome the regulatory difficulties of transgenic plants (Lv et al., 2020). Nanoparticle-mediated delivery is a frontier in the plant genome, whether or not it is in a developmental stage. The progress in gene editing technologies beginning with ZFNs and TALENs and finishing with CRISPR-Cas systems has significantly enhanced plant biotechnology. In synergy with innovations in delivery modalities, these tools have transformed our ability to engineer plant genomes with precision stage the way for sustainable agriculture and food security (Ahmar et al., 2021).

CRISPR/CAS SYSTEM IN PLANT GENE EDITING

Structure and Mechanism of CRISPR/Cas9

Genome editing has become much easier, simpler, and versatile because of the CRISPR/Cas9 system, as is the case with the adaptive immune system of bacteria. In a natural setting, the CRISPR array holds viral DNA fragments and helps bacteria detect and prevent subsequent infections. The protein Cas9, an integral part of the RNA-guided endonuclease, plays an important role in this defensive mechanism by creating double-strand breaks (DSBs) in target DNA sequences that match the guide RNA (gRNA) (Tripathi et al., 2022). In plant genome editing, the CRISPR/Cas9 system is engineered to target specific genomic loci and gRNA guides Cas9 to the target DNA to produce a DSB. The plant cell's natural repair mechanisms will repair the break through NHEJ or HDR, which will provide insertions, deletions, or exact corrections. Our technology has proven to be useful not only in rice, wheat, maize, tomato plants, and other plants, thereby enabling the production of plants with good characteristics, including resistance to disease, stress, and high nutritional content (Fig 1) (Wada et al., 2020).

Figure 1. Schematic representation of the CRISPR/Cas9 mechanism for targeted genome editing in plants.

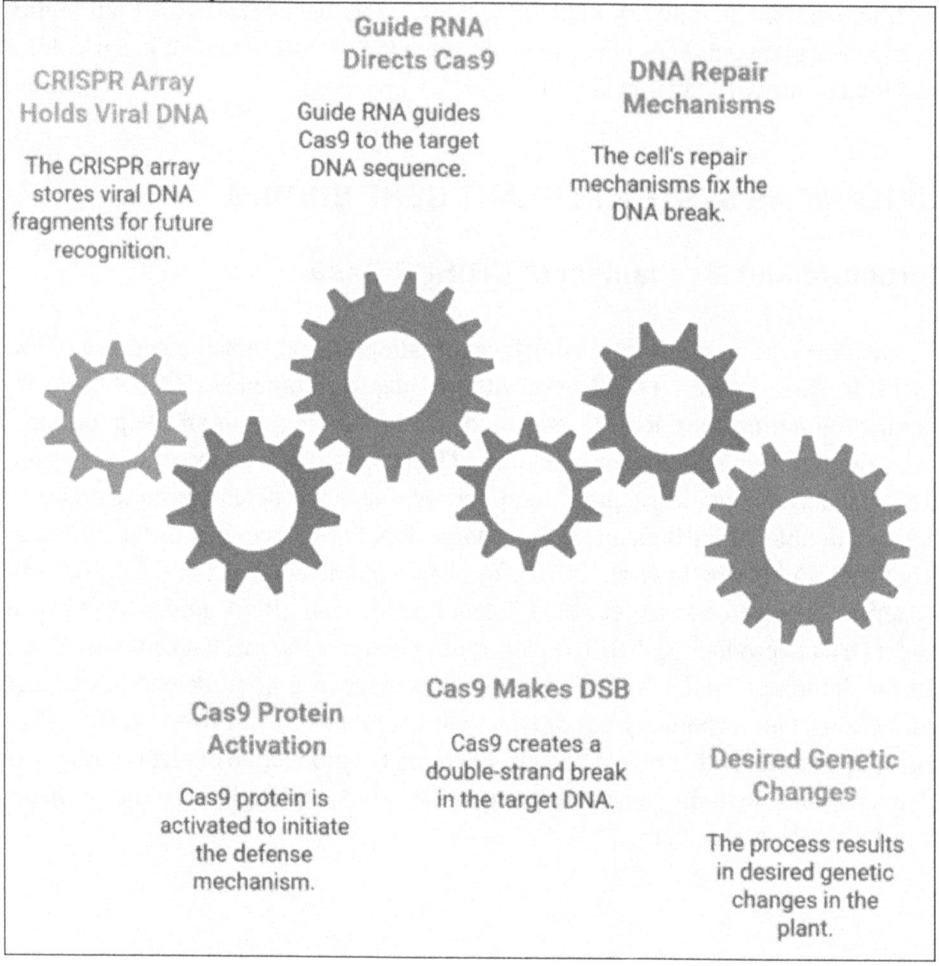

Advances: Cas12, Cas13, dCas9, and Multiplex Editing

Although Cas9 is perhaps the most well-known, Cas proteins have also been identified and modified in a variety of ways with distinct advantages.

- **Cas12 (Cpf1):** Cas12a is a single RNA-guided endonuclease that preferentially recognizes a T-rich protospacer adjacent motif (PAM) and generates staggered DSBs. Unlike Cas9, Cas12a can cleave its own CRISPR RNA (crRNA) independently of a trans-activating crRNA (tracrRNA), making multiplex editing easier. Cas12a can be successfully applied for editing the plant genome, providing effective and precise modifications (Bandyopadhyay et al., 2020).

- **Cas13:** Unlike Cas9 and Cas12, Cas13 specifically targets RNA rather than DNA; this manner of operation provides transient and reversible regulation of genes. In particular, this property is useful for describing gene function and resistance to RNA viruses in plants (Hillary & Ceasar, 2023).
- **dCas9 (dead Cas9):** A catalytically inactive Cas9 (dCas9) does not have endonuclease activity but still has DNA-binding potential. When combined with transcriptional repressors or activators, dCas9 can regulate gene expression without changing the DNA sequence and is applicable to gene regulation and functional genomics (Li et al., 2020).
- **Multiplex Editing**: It based on multi-locus targeting, enabling the modification of complex traits that are quantified by genes in multiple ways. Such an approach has been demonstrated in different plant species, such as Arabidopsis, rice, and maize, using both Cas9 and Cas12 systems. Multiplex editing speeds up the phasing of crops with improved yield, tolerance, and nutrition (Lorenzo et al., 2023).

CRISPRI AND CRISPRA (INTERFERENCE/ ACTIVATION) IN PLANTS

Gene regulation through CRISPR interference (CRISPRi) and CRISPR activation (CRISPRa) is a novel application of the CRISPR/dCas9 system.

- **CRISPRi:** The introduction of dCas9 to transcriptional repressors; for example, the SRDX domain silences gene expression. This technique enables the investigation of gene function and creation of crops with specified traits by repressing given genes (Xu et al., 2023).
- **CRISPRa:** In contrast, the fusion of dCas9 with transcriptional activators allows the upregulation of target genes. CRISPRa has been used to increase the expression of genes that regulate stress responses, metabolic pathways, and developmental processes in plants (Cai et al., 2023).

These systems provide a highly precise and reversible modulation of gene expression, which extends the repertoire to plant functional genomics and crop improvement.

BASE EDITING AND PRIME EDITING: DETAILED MECHANISMS AND EXAMPLES

Instead of DSBs, repair mechanisms of the cell are used by traditional CRISPR/Cas9 editing, which may result in unintended mutations. To overcome this drawback, two alternative techniques have been developed: base and prime editing.

- **Base Editing:** This technique allows for the direct irreversible conversion of one DNA base without the content of DSBs. Base editors are composed of catalytically imperfect cas9 fused to a deaminase enzyme. For instance, cytosine base editors (CBEs) change pairs C•G to T•A, and adenine base editors (ABEs) change pairs A•T to G•C. Base editing has already been used in plants to produce herbicide-resistant rice and enhance drought tolerance in wheat (Gu et al., 2021; Neugebauer et al., 2023).
- **Prime Editing:** Prime editing provides flexible strategies for making targeted insertions, deletions, and all 12 possible base-to-base conversions, without the need for DSBs or donor DNA templates. The system uses a combination of Cas9 nickase and reverse transcriptase guided by a prime editing guide RNA (pegRNA), which specifies the target site and desired edit. Prime editing has been shown in rice and wheat, which allows precise genetic modification of these crops for improvement (Mikhaylova et al., 2024; Zhao et al., 2023).

These advanced editing techniques improve the precision and magnitude of genome editing in plants, along the path to the creation of crops with superior traits and lower off-target effects.

APPLICATIONS IN CROP IMPROVEMENT

Abiotic Stress Tolerance: Drought, Salinity, and Heat

Abiotic stresses, such as drought, salinity, and heat, are serious challenges to agriculture worldwide, diminishing crop yield and food security. The effects of climate change are expected to exacerbate these stresses; therefore, there is a need for crops that can cope with these conditions (Devi et al., 2022). Conventional breeding techniques have been frustrated to a certain extent in breeding stress-tolerant varieties because of the complexity of traits and the long timespan involved. The arrival of CRISPR/Cas technology has increased the efficiency and accuracy of genetic modification to increase crop tolerance to abiotic stresses(Fig 2) (Hafeez et al., 2023).

Figure 2. Enhancing crop tolerance to abiotic stresses.

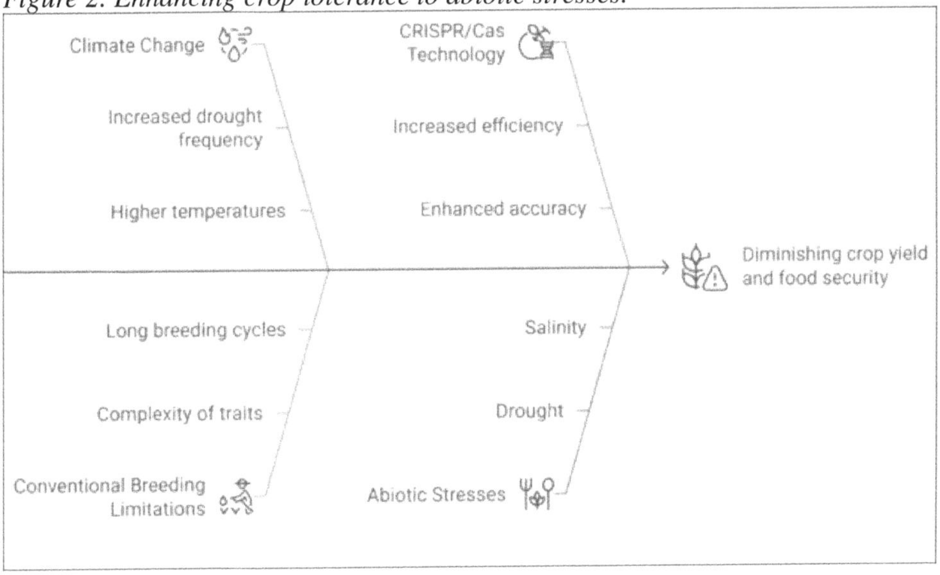

Drought Tolerance

Drought stress affects plant growth and productivity owing to a lack of water. For instance, genes such as OsPYL in rice, which encodes an abscisic acid receptor critical in stress signaling, can be targeted using CRISPR/Cas9 by working researchers (Usman et al., 2020). Mutation of this gene has led to improved drought tolerance by making the plant more capable of closing stomata and running them out of water. Similarly, in maize, the ARGOS8 gene (negative regulator of ethylene signaling) targeted editing has led to higher grain yield under drought stress by improving root growth and water-use efficiency. These genetic enhancements support the maintenance of crop yields during water scarcity, which is a major advantage for areas prone to drought (Singh et al., 2023).

Salinity Tolerance

Therefore, ion imbalance in plants is induced by salinity stress, which retards plant growth and development. The OsRR22 gene, a negative regulator of salt tolerance, was knocked out using CRISPR/Cas9 in rice. Mutations in this gene lead to enhanced plant salt tolerance due to enhanced ion homeostasis (Alam et al., 2022). In wheat, specific modifications in the gene HKT1;5, which controls the uptake of sodium in plants, have decreased shoot accumulation of sodium, thereby increasing salt tolerance. Such changes allow crops to grow in saline soils, which are increasingly appearing as a result of irrigation and sea level rise (Shabala et al., 2025).

Heat Tolerance

High temperatures disturb plant metabolism, inhibit reproductive processes, and lower yields, particularly during flowering. Thermotolerance has been enhanced in tomatoes using CRISPR/Cas9-mediated mutations in the SlMAPK3 gene, because heat shock proteins that can protect cellular components have been upregulated (Singh et al., 2022). Similarly, in rice, the edited OsHTAS gene, which encodes a heat stress-associated protein, has been edited to increase heat tolerance during flowering, such that yield failure from heat stress during critical growth stages is averted. Such improvements are especially applicable for crops cultivated in regions with growing temperatures caused by climate change (Xie et al., 2023).

BIOTIC STRESS RESISTANCE: PEST AND PATHOGEN RESISTANCE

At least one-third of the global crop yield losses are attributed to biotic stresses caused by pests and pathogens. Traditional breeding approaches for enhancing pest and disease resistance usually encounter difficulties such as complicated genetically based interactions between plants and pathogens (Begna, 2020). CRISPR/Cas-based genome editing has provided a sound and precise method to develop crops resistant to pests and pathogens by targeting susceptibility (S) genes or upregulation of resistance (R) genes (Khan et al., 2021).

Fungal Resistance

Among the fungal pathogens found are powdery mildew and blast, which affect wheat and rice crops and cause severe damage to crops (Tembo et al., 2020). CRISPR/Cas9 mediated knockout of the TaMLO gene in wheat, a susceptibility gene for powdery mildew, has provided plants that are highly resistant to this fungal disease (Aroge et al., 2024). Similarly, editing the OsERF922 gene, which negatively regulates resistance to *Magnaporthe oryzae* (the causative agent of rice blast) in rice, has increased resistance to this pathogen, thus increasing yield stability in regions that are affected (Mengchao et al., 2023).

Bacterial Resistance

These are bacterial diseases, including the spread of citrus canker from *Xanthomonas citri*, which can destroy crops. Researchers have used CRISPR/Cas9 to edit the CsLOB1 promoter recognized by the pathogen TAL effector in citrus. Through this change, pathogen recognition is prevented, and hence fungal resistance is induced in citrus canker. This strategy may be further applied to other crops that suffer from bacterial pathogens and may offer bicycles a durable and sustainable way of controlling bacterial diseases (NETO, 2022).

Viral Resistance

Viral diseases, such as tobacco and cassava, pose serious health risks. One of the new applications of CRISPR as a viral resistance strategy is through the CRISPR/Cas13 system, which targets RNA (Robertson et al., 2022). The CRISPR / Cas13 technology has been used to target the coat protein gene of the tobacco mosaic virus in tobacco, with a dramatic reduction in viral accumulation and manifestation of symptoms. This technology promises an innovative control scheme for RNA viruses in many crops that will reduce dependence on chemical treatments and contribute to better disease management in a more general setting (Robertson, 2021).

Insect Resistance

Insects, such as aphids and caterpillars, spread diseases and kill crops. Although still very early, there is potential for the use of CRISPR/Cas9 to make insect-resistant crops an idea. For instance, in melons, researchers have targeted eIF4E, which is responsible for viral replication and insect resistance. When this gene is modified, crops are less susceptible to aphid-transmitted viruses. Such early successes suggest that CRISPR-based approaches may transform pest resistance by creating plants with reduced carcinogenic impacts through the development of pest resistant crops (Zahoor et al., 2021).

YIELD AND QUALITY ENHANCEMENT

Improving yield and quality is one of the major objectives of crop breeding to facilitate the ever-increasing food needs of humanity. CRISPR/Cas systems are potent tools for improving yield, size, shape, and nutritional content in crops (Ahmad et al., 2021).

Yield Enhancement

CRISPR/Cas9 has allowed for target modification of several yield-related components. In the case of rice, scientists have simultaneously edited genes such as Gn1a (associated with the number of grains), GS3 (grain size), and DEP1 (panicle architecture) to enhance the overall yield (Sen et al., 2024). These changes have increased the plant's capacity to produce more grains in each panicle and improved the size of the grains, consequently increasing the yield. Editing the negative regulator grain weight gene TaGW2 in wheat has led to the development of larger, heavier grains, thus contributing to increased yields. These yield improvements are fundamental in maintaining food security in regions in which their population is constantly rising and there is a shortage of arable land (Jian et al., 2024).

Figure 3. Enhancement of crop yield using CRISPR/Cas 9.

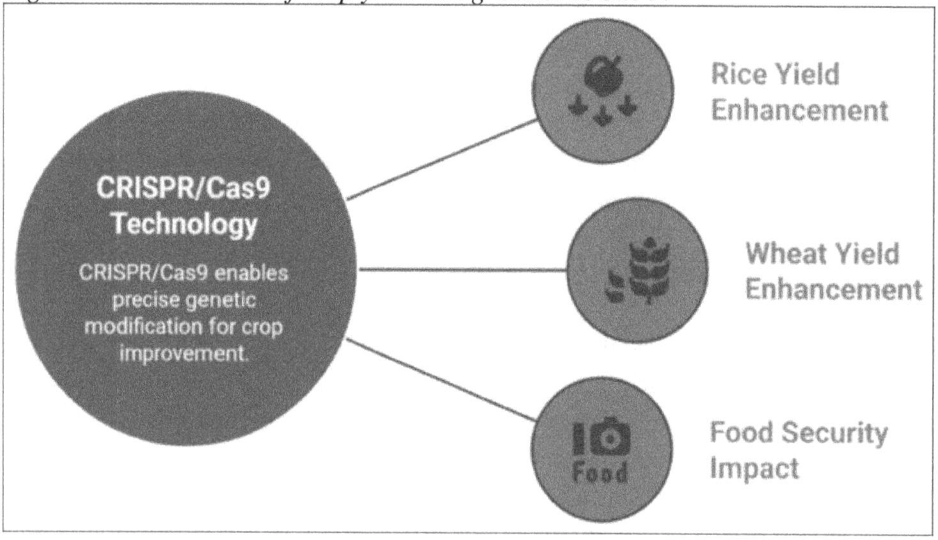

Quality Improvement

Quality traits such as taste, texture, nutritional value, and yield are crucial for crop commercialization. CRISPR/Cas9 has been used to improve these aspects in crops. SlMYB12 has been edited in tomato to increase flavonoid content, thereby enhancing the nutritional quality and antioxidant properties of the fruit (Liu et al., 2021). Moreover, in rice, modification of the Wx gene, which controls the amount of amylose, has resulted in improved eating qualities, a more pleasing texture, and less stickiness. CRISPR/Cas-based editing improves yield and yield traits, allowing the production of crops that are more attractive to consumers and closer to market requirements (Verma et al., 2023).

NUTRITIONAL BIOFORTIFICATION

Hunger for nutrients, especially essential micronutrients, vitamins, and minerals, is prevalent worldwide. Biofortification increases the nutrient content of crops through breeding or genetic manipulation. CRISPR/Cas technology offers an extremely specific and useful way to produce biofortified crops (Shahzad et al., 2021).

Vitamin A Enrichment

Vitamin A deficiency remains a serious public health problem, especially in developing countries, where rice is a staple food. In "Golden Rice," CRISPR/Cas9 has been applied to modify the expression of the biosynthetic pathway for the precursor – β -carotene of vitamin A (Hossain & Bakhsh, 2020). By attacking genes such as OsPSY (phytoene synthase) and OsORANGE (regulator of carotenoid accumulation), researchers have enhanced β-carotene levels in rice grains. This biofortification strategy has the potential to address vitamin A deficiency and improve public health outcomes (Yu et al., 2021).

Iron and Zinc Biofortification

Iron and zinc deficiencies are widespread, especially in areas where regional breads, such as rice and wheat, are the main dietary sources. CRISPR/Cas9-mediated alterations of genes, including OsNAS2 (nicotianamine synthase), in rice have resulted in increased rice grain levels of iron and zinc (Mishra et al., 2024). The iron content of the endosperm is enhanced in wheat when the TaVIT2 gene is targeted, which encodes a transporter for iron storage. These biofortified crops provide an affordable and sustainable way to overcome micronutrient deficiencies in at-risk populations (Kenzhebayeva et al., 2022).

Enhancement

Folate is another important micronutrient, and its deficiency is associated with birth defects and other health difficulties. In tomato, CRISPR/Cas9 has been used to edit the GGH gene, which causes folate degradation. As a result, fruits with high folate levels were obtained. This bio-fortification strategy helps to make these ordinarily consumed crops more nutritious, thus making people healthier and well (Khan et al., 2024).

SHELF-LIFE IMPROVEMENT AND POST-HARVEST TRAITS

Agricultural postharvest losses are a major concern in various tropical and developing regions. By prolonging the shelf life of fruits and vegetables, as well as by enhancing their post-harvest properties, waste will be reduced and food security will be enhanced. CRISPR/Cas technology has been used to improve these traits in several crops (Enyiukwu et al., 2020).

Delayed Ripening

Fruit ripening involves a complex set of genes that regulate cell wall breakdown, aroma, and flavor development. In tomatoes, the RIN gene, which is a major regulator of fruit ripening, was knocked out using CRISPR/ Cas9 (Li et al., 2021). This mutation slows down the ripening process, which means that the fruits produced in this plant are stored for longer periods after harvesting. These tomatoes have a prolonged shelf life; therefore, they also maintain nutrition and taste, which reduces food waste during transportation and storage (Umeohia & Olapade, 2024).

Reduced Browning

Browning due to PPO activity is a major problem for crops such as potatoes and apples. In potatoes, CRISPR / Cas9 has been used on the PPO gene to remove browning, make the tubers more visually attractive, and extend shelf life. In addition to marketability, this mode of cultivation reduces the necessity for chemical preservatives, thus making the product visually appealing for consumption and longer shelf life (González et al., 2020).

Improved Firmness

Fruit firmness is an important trait in consumer preference and shelf life. The SlPG gene, responsible for cell wall degradation, has been modified through the application of CRISPR/Cas9 in plants. The resultant fruits are firmer and retailed less bruising, thus elongating their shelf life, resulting in reduced post-chemical losses (Mathiazhagan et al., 2021).

Improved Storage Quality

Senescence-and chlorophyll degradation-related genes in leafy vegetables, such as lettuce, have been altered using CRISPR/Cas9. These interventions will delay leaf yellowing, making leaves fresher for storage, and therefore, improve shelf-life and reduce wastage (Asrey et al., 2021).

EMERGING FRONTIERS IN PLANT GENE

Epigenome Editing

Epigenetic regulation is essential for regulating gene expression (without recourse to changing the genetic sequence). Modifications such as DNA methylation, histone modification, and non-coding RNA modulation are mechanisms of gene activity regulation. Epigenome editing, that is, specific alteration of these epigenetic marks, has made recent progress and shows great promise for fine-tuning gene expression in plants without permanent genetic alterations (Bure et al., 2022).The advent of epigenetic editing tools has been a significant step forward since researchers can regulate the expression of genes that may be difficult to edit using conventional methods of genome editing. One such approach is the use of CRISPR/Cas9-based epigenetic tools. dCas9 (the dead catalytic form of Cas9), linked to DNA methyltransferases or histone modifiers allows the targeted addition/ removal of epigenetic 'marks' at specific genomic territory. This technique CRISPR/Cas9-based epigenetic modulation has been used to modulate gene expression without changing the underlying DNA sequence (Rahman & Tollefsbol, 2021). In plants, this technology has been proven to work well in controlling important traits such as stress tolerance, growth rate, and metabolic pathways. For instance, in the case of rice, dCas9-based DNA methylation has been used to silence genes associated with lignin biosynthesis, thereby leading to increased production of biomass and better exploitation of cell wall components. Researchers have used epigenetic editing to regulate the expression of genes associated with flowering time, which enables increased synchronization of flowering with the surrounding environment, which is an important attribute for optimizing crops (Nangammbi, 2021). Epigenome editing provides for a temporary and reversible genetic modification of plant traits, which is especially beneficial for concerns that long-term ecological impacts may arise from sustainable gene manipulation (Fig 4). In addition, epigenetic modifications are heritable and can thus become a new path for crop advancement, while preserving gene diversity. This is a very young field, but its potential in the future in terms of agriculture is immense, allowing for fine-tuning of plant traits in response to altered environmental conditions (KhokharVoytas et al., 2023).

Figure 4. Unveiling epigenetic editing in plants.

RNA Editing

RNA editing, which modifies RNA rather than genomic DNA, is another frontline in plant gene editing. This approach can be used to edit certain nucleotides in RNA sequences, thereby influencing gene expression without altering the DNA itself. RNA editing has the advantage of temporary changes in gene expression, which may be useful in cases where permanent genetic changes are not desirable (Van Eck, 2020). The most extensively researched RNA-editing tools are represented by the use of CRISPR/Cas13, which targets RNA, not DNA. CRISPR/Cas13 is an RNA-targeting iteration of the CRISPR system that can be programmed to attach and edit RNA molecules and serves as a powerful tool for silencing/gene modulation in plants. Using this system, researchers can accurately manipulate RNA sequences, fix mutations in messenger RNA (mRNA), or control gene expression at the RNA level (Kordyś et al., 2022). RNA editing in plants has been proven to be of great value in increasing disease and stress tolerance. For example, in Arabidopsis thaliana, researchers have used CRISPR/Cas13 to edit mRNA transcripts involved in immune responses, which has increased the plant's chances of resisting fungal and bacterial invasions. Similarly, RNA editing has been used to change the expression of genes involved in abiotic stress responses, for example, drought and heat tolerance, by changing the stability or possibly the translation of certain mRNA's (Kavuri et al., 2022). Another significant aspect of RNA editing is its capacity to avoid unwanted off-target effects that are characteristic of DNA-based edits. Editing RNA is a way for researchers to obtain permanent genetic changes and therefore provide a more reversible mode of gene regulation. As RNA editing technology continues to develop, it is likely to become an invaluable instrument for precision agriculture, allowing the regulation of gene expression levels in crops with a lower risk of unexpected genetic effects (Manghwar et al., 2020).

Synthetic Biology and Gene Cirçuits

Synthetic biology is a fast-expanding frontier in biology, where biology, engineering, and technology are integrated to design and build novel biological systems and devices. Synthetic biology presents the option of developing new pathways, circuits, or even synthetic organisms that can do new things in plant gene editing. Gene circuits consisting of regulatory elements and genes can be designed to regulate the expression of several genes in response to environmental stimuli (Rizzo et al., 2023).Using gene circuits and implanting them into plant genomes, researchers can make crops that have a notably increased functionality, such as crops that can produce new types of metabolites, resist particular pests or pathogens, or respond to environmental signals in a regulated way. For instance, synthetic gene circuits have been engineered in plants to allow the synthesis of high-value secondary metabolites (pharmaceuticals or biofuels) upon activation by environmental cues. These circuits are based on promoters, regulatory elements, and feedback loops and control the expression of genes in a dynamic responsive system within the plant (D'Amelia et al., 2021). The most promising synthetic biology in plant gene editing is that which provides enhanced biosynthetic capabilities in crops. For instance, synthetic gene circuits have been designed in Arabidopsis to increase the production of terpenoids, which are essential for the defense of plants and have different industrial uses. In addition, synthetic biology has been employed to create plants that take in carbon more efficiently, and this goes a long way to stabilize the world in fighting climate change. With the construction of synthetic biosynthetic pathways, researchers can improve carbon fixation and biomass production, providing a sustainable solution for food security (Mani et al., 2021).

AI and Machine Learning in Guide RNA Design

The design of guide RNAs (gRNAs) is one of the most important steps in CRISPR-based gene editing. The accuracy of the CRISPR system is associated with the correct identification of target sequences, and optimal selection of gRNAs is mandatory for lowering off-target effects and increasing efficiency at the level of gene editing. Conventional gRNA design depends on bioinformatics tools that predict possible target sites, but the complexity of plant genomes demands more intricate approaches (Liu et al., 2020). The art and science of artificial intelligence (AI) and its brilliance in terms of machine learning (ML) have proven to be powerful tools for the optimization of gRNA design. When training algorithms using a large volume rate of genomic sequences and experimental results, it is possible, with high accuracy, to predict the best gRNA sequences using AI models. These models can be used to analyze various parameters, including sequence specificity, likely off-target effects, and binding affinity, together with creating optimized gRNAs for plant gene editing (Dixit et al., 2024; Goshisht, 2024).

Other algorithms that can also be useful are those of machine learning systems, which can predict the outcome of gene editing, likely effects of some edits on off-target mutations, and the nature of plant phenotypes after some particular edits. This effectiveness can help minimize the trial-balancing-in-error character of gene editing, making it faster, more affordable, and less dependent on experimental resources (Bhardwaj et al., 2024). As AI and machine learning technologies develop this aspect, they will continue to be incorporated into plant gene-editing workflows. These tools will not only boost the precision of the design of guide RNA, but will also make gene editing in crops scalable and easier to do even on multiple genes at a time or create the most complex of plant genetic modifications (Liu et al., 2024).

Single-Cell Genomics in Editing Precision

Single-cell genomics is a newly evolving discipline that makes it possible to analyze the genomic and transcriptomic profiles of single cells. This technology has drastically changed our perception of cellular heterogeneity and the precision of gene editing in single cells. In plant gene editing, single-cell genomics enables researchers to follow the results of cell, tissue, or organ modifications, thus facilitating a detailed analysis of how gene editing affects plant development (Wang et al., 2023). Classical gene-editing approaches usually scrutinize cell populations, thus obscuring individual cell results in terms of genetic modification. Single-cell genomics provides the possibility of detecting off-target edits and measuring the efficacy and precision of cell manipulation in specific plant tissues (Bode et al., 2021). For instance, using single-cell RNA sequencing (scRNA-seq) following

CRISPR/Cas9 editing, the research group can analyze transcriptomic alterations that result from purposeful alterations in certain cells, such as leaf or root cells (Lv et al., 2024). Such precision is indispensable at the level of complex traits that must be altered in certain tissues or developmental steps. Single-cell genomics can also assist in recognizing rare or unexpected incidents induced by gene editing, which enhances the safety and accuracy of genome editing in plants. With the opportunity to observe more details about the gene editing process, single-cell genomics will play an important role in contributing to the precision and efficiency of plant genome editing (Gao, 2021).

REGULATORY LANDSCAPE AND GLOBAL STATUS

The regulatory environment that characterizes gene editing in plants is complex and varies greatly worldwide. As plant genome editing technologies, especially CRISPR/Cas-based systems, are becoming increasingly advanced, there is a need for countries to develop frameworks to regulate their usage while balancing the potential benefits of their use with safety concerns (Zhang et al., 2020). In this section, we highlight the regulatory approaches prevalent in various regions, such as the USA, China, European Union, India, and Africa, along with some interesting case studies in our respective countries, aspects related to biosafety, issues around intellectual property (IP), and public perception related to gene-edited crops (Menz et al., 2020).

Regulation in Different Countries

USA

In America, the regulatory regime for gene-edited crops has major inputs. Department of Agriculture (USDA), Environmental Protection Agency (EPA), Food and Drug Administration (FDA). USDA's treatment of gene-edited plants has tended to be permissive by nature in that unless foreign genetic material is present, it is possible to deregulate crops. For instance, crops modified via CRISPR/Cas9 to interfere with plant genes with no foreign DNA invasion generally do not require as much attention as transgenic crops (Turnbull et al., 2021).

However, gene-edited plants are subject to environmental risk assessment to confirm that they do not risk biodiversity or the environment. The FDA provides a direction for the safety of gene-edited crops for human consumption and is especially related to their allergenicity and toxicity. For instance, gene-edited plants, such as CRISPR-modified soybeans, have received safety reviews from the FDA to ensure that they are suitable for food use (Munawar et al., 2024). Although the regulatory approach is rather simple, there is a certain ambivalence, at least within the U.S., over whether the current structure is sufficiently sophisticated to handle the intricacies of new gene editing technologies. Many critics would agree that more transparency is required with respect to labeling and public information to guarantee consumer confidence in gene-edited foods (Mukherjee et al., 2023).

China

China is a leading global user of gene editing in agriculture. The regulatory framework has been accommodating more for genetic adjustments in China than in many other areas (Liang et al., 2022). In recent years, China has drastically simplified the regulatory regime for gene-edited crops. In the same year (2020), China sanctioned the first commercial release of a CRISPR-edited crop, rice that was genetically modified to be resistant to fungal disease. This validation was a major breakthrough for the biotechnology industry in the country, making China a major player in the global market for agricultural biotechnology (Molinari et al., 2024). Regulations in China are mainly handled at the level of the Ministry of Agriculture and Rural Affairs (MARA), where the genetic safety of GMOs is controlled. Although it has remained open to gene-edited crops, the approval procedure for the latter still requires thorough safety tests. Evaluations for environmental safety, health impacts, and possible ecological risks are made on gene-edited crops, but this oversight is somewhat less strict than is true of those GMOs that have a more established procedure towards approval (Liu, 2022). Heaps of money have been expended by China to conduct research on gene editing for crop improvement and disease resistance. These efforts form part of the national planning push to promote food security, increase productivity on farms, and reduce dependency on foreign imports (Kang et al., 2022).

European Union (EU)

Genetically modified organisms (GMOs) are among the most rigid regulatory systems in the European Union. The EU has debated heavily about its regulatory position in gene editing, and individual EU countries have followed different approaches (Jiang, 2020). In 2018, the CJEU, the highest court in the EU, determined that even rapes engineered without the insertion of foreign DNA were subject to the same strong legal regulations traditionally applied to GMOs. This decision has hindered the general adoption of CRISPR-edited crops in the EU because the approval of GMO crops is time- and money-consuming (Faltus, 2023). The EU strategy of gene editing is biased towards the precautionary principle, which guarantees that gene-edited crops are subject to a wide range of possible risks prior to being given a go ahead. This comprises assessing the effects on biodiversity, environment, and food safety. Notwithstanding this pragmatic attitude, some European states, such as the UK, have begun to depart from the EU's hardline, with implications for greater flexibility in the regulation of gene-edited crops post-Brexit in Britain (Brookes & Smyth, 2024).

Africa

The regulatory world for gene editing in Africa differs greatly from one country to another, with some countries rationalizing the role of gene editing more than others who have a more conservative approach to gene editing. Generally, the cultivation of gene-edited crops in Africa remains a blend of the international framework of gene-edited regulations and the national political and social context (Rock et al., 2023). Genetic modification technologies, including gene editing, are the most widely adopted in South Africa as the leader country. Multiple GM crops have already been approved in the country, and the regulating system is rather developed. South Africa is also considering the use of gene editing to enhance maize, soybeans, and potatoes. Other countries across Africa especially in both East and West Africa are still assessing the safety and benefits that gene editing can still be of to the people. There is immense interest in the use of gene-editing technologies to enhance cassava, banana, and sorghum as staple foods in large parts of the continent. However, in these countries, regulatory approval is frequently hindered by anxiety regarding biosafety, food security, and public acceptability (Akinbo et al., 2021).

Case Studies

UK's Precision Breeding Act

In 2023, the UK passed the Precision Breeding Act, which derives the legal cover for the use of gene editing in agriculture. According to this law, gene-edited crops that would have been achieved using traditional breeding methods can be deregulated, meaning that there is no need for a long drawn-out process of approval of genetically modified crops (Watson & Hayta, 2024).This new approach is a dramatic change from what the EU has been thinking, putting the UK in the same slot as countries such as the USA and China in terms of gene-editing technologies. The Precision Breeding Act allows for quicker innovation in the field of crop improvement, including, yet not limited to, disease resistance, drought tolerance, and nutritional upgrade. This novel method has been proven to ignite a lot of interest in gene editing among British scientists and farmers, but the public debate about it is split (Asquer & Krachkovskaya, 2021).

Approval of Edited Rice

China approved its first CRISPR-edited crop, a strain of rice whose genome has been edited to resist fungal disease in 2020. This approval is indicative of China's progressive issues with gene editing in agriculture (Zhou et al., 2021). Gene editing, created by Chinese scientists within the CRISPR/Cas9 system, was considered a breakthrough in the country's biotechnology sector. This approval is an expression of China's commitment to gene editing to enhance food security, enhance crop yield, and reduce pesticide use (Wei et al., 2022).

Labelling, Biosafety, and Public Perception

Labelling of gene-edited foods remains a vital concern for regulators and consumers. Genetically modified foods, as well as those resulting from gene editing technologies, are subject to labelling in many countries, especially in Europe. However, whether gene-edited crops should be labelled separately from conventional crops remains debatable. In the US, for instance, gene-edited foods do not have to be labelled unless their contents are allergens or pose safety concerns (Flanagan, 2020).Biosafety is an essential aspect of gene editing for crop regulation. Before gene-edited crops can be commercially sold, most regulatory arms of each country carry out far-reaching evaluations to determine potential rollout effects, such as gene flow, biodiversity, and off-target effects (Goodman, 2023). The public is impacted by different factors that influence their view of modified crops, such as the cultural attitude toward genetic modification and the perception of the technology's safety to human beings given the environmental challenges it may cause. In areas where GMOs have been strongly opposed, such as the EU, gene-editing technologies present massive barriers to public acceptance (Delhove et al., 2020).

Intellectual Property and Patent Issues

The problem of IP and patents is critical for the development and commercialization of gene-edited crops. The CRISPR/Cas9 patent landscape is complicated by numerous patent holders and by different patent holders for different aspects of the technology. Firms and research bodies that develop gene-edited crops frequently must traverse this complex patent landscape to avoid the latter and enter into license agreements (Jiang, 2020). In many cases, IP concerns can influence the cost and availability of gene-edited crops, especially in developing countries. This ownership of gene-editing patents has also resulted in sparks about patenting natural genetic sequences and the possibility of monopolies over key agricultural technologies (Borg & Policante, 2024).

ETHICAL, SOCIAL, AND ECONOMIC CONSIDERATIONS

The adoption of gene-editing technologies in agriculture has generated a volume of moral, social, and economic debate. Such technologies especially CRISPR /Cas9 present vast opportunities for increasing crop reliability, ensuring food security, and solving climate change problems (Selfa et al., 2023). However, their implementation also brings to the fore important questions surrounding nature manipulation, equality at the level of economic accessibility, and social consequences for smallholder farmers, consumers, and the international agricultural system (Clapp & Ruder, 2020). This section discusses these debates, emphasizing on ethical concern, public acceptance, affordability in economics, and the larger argument of risk-benefit approach of gene-edited crops

Ethical Debates: Playing with Nature vs. Food Security

Playing with nature is one of the main ethical issues associated with gene editing in plants. Critics contend that even gene editing with no foreign DNA changes the natural genome of plants and thus can be viewed as an improbable intervention in the evolution of species. Some opponents wonder whether man should be given the right to such radical modifications over organisms, particularly in agricultural systems comprising billions of people. There is a worry that the outcome of gene editing is not well known in the long term; and intervention could cause unforeseen ecological/environmental impacts which can interrupt natural biodiversity (Idris et al., 2023). However, proponents believe that gene editing can cure problems, such as food security, climate change, and malnutrition. Genetically edited crops could help address the increasing needs of the world's expanding population if they are more resilient to abiotic stresses, such as drought and salinity, more disease resistant, and richer in nutrition. The ethical point here is the question of whether gene editing benefits from improving food security justify the associated risks. In view of their urgency, proponents argue that gene editing could be a critical tool for providing a constant and secure food supply in the future (Anyshchenko, 2022).

Public Acceptance and Misinformation

The acceptance of gene-edited crops by the public is of utmost importance in making such crops a facility and product of commercial bias. Although the safety and efficacy of gene editing are well supported by scientific communities and regulators, the technology is met with skepticism by many consumers. Dissemination of information is a major threat to public acceptance. Sometimes, due to the deterministic nature of perception, media outlets and advocacy groups tend to frame gene-editing technologies as dangerous or unnatural; as a result, the public may be afraid and resistant (Cummings et al., 2024). This skepticism is intensified by the larger social background of genetic modification. Even in most parts of the world, particularly in Europe, it has been a contentious issue for many years, and negativity towards GMOs will spill over to gene-edited crops. By all means, gene-edited crops are very different from standard GMO's (they are not always based on foreign genes), but for the uneducated and mistrustful population, they are the same and indistinguishable. This creates confusion and mistrust (Turnbull et al., 2021). Initiating public awareness regarding the safety, possible benefits, and scientific background of gene editing is key to overcoming these challenges. Openness on the part of the developer and regulator of gene-edited crops and openness in the dissemination of information on scientific workings will help win public confidence. Furthermore, interacting with local communities to arrange for consumers to be heard in making decisions about gene-edited foods may create acceptance and build trust (Piergentili et al., 2021).

Economic Accessibility for Smallholder Farmers

The economic implications of gene editing technologies for smallholder farmers have been a growing issue. In many developing nations, in which smallholder farmers form the major employment segment in the agricultural sector, the ability to access new technologies is hampered by limited access to capital, inadequate infrastructure, and problems with market access (de Roij, 2020). It is a fact that gene editing promises to make crops more productive and resistant, but there are concerns that these benefits will not go to those farmers that need them; the cost for researching and releasing gene-edited crops is substantial, and unless these technologies can be seen to be more accessible on a monetary level, there is a risk that only large-scale commercial farms or wealthy nations will benefit. This can further worsen existing inequalities in the agricultural field and prevent smallholder farmers from competing in the global factor of markets (Bansal & Kaur, 2024).

To obtain gene editing technologies that have a positive economic effect on smallholder farmers, there is a need to mitigate the costs of research and development, ensure affordable access to gene-edited seeds, and ensure that these smallholder farmers are not left out of the benefits of these advancements (de Roij, 2020). Governments, Non-Governmental Organizations, and international development agencies can play a critical role in enabling the uptake of gene-editing technologies by smallholder farmers through subsidizing, training, and technical support (Mukherji et al., 2023). Furthermore, collaboration between private firms, governments, and local communities will be vital to ensure that the benefits of gene editing are not lopsided, such that capable areas are left with little or no gains, whereas food security and agricultural productivity are a problem in resource-poor areas (Barbosa et al., 2021).

RISK-BENEFIT ANALYSIS

One of the major ethical issues in gene editing is identified as a potential risk to the individual being edited. Although the benefits derived from such gene-editing technologies include higher crop yield, better resistance to disease, and better nutritional content, there are latent risks that need to be carefully addressed. Such risks include unintended consequences such as off-target mutations, ecological impacts, and even potential for gene flow from edited crops to wild relatives (Piergentili et al., 2021). The use of gene editing to eradicate issues such as global food security, climate change, and malnutrition represents an opportunity to make a major breakthrough in global agriculture. However, the benefits that may accrue must be balanced with environmental, ecological, and social risks. For instance, genetically edited crops may present drought-resistant plant varieties, but there might be fears that these would have an effect on the local environment or biodiversity. Just as gene-edited crops might lessen the need for chemical pesticides, their effect on non-target species and soil health must be fully considered (Ghimire et al., 2023). To address these challenges, strong regulatory regimes are required to ensure the safety of gene-edited crops before their commercialization. This involves intensive tests on environmental effects, observation for unintended effects, and ensuring that gene-edited crops do not unravel the current agricultural system. In addition, we need constant monitoring after the crop is on the market to check for any side effects after the same are in a large-scale distribution (Jenkins et al., 2021). The analysis of risk-benefit should also address the implications in the larger context, as far as the impacts on farmers, consumers, and communities are concerned. For example, gene editing might improve food security in the future while causing job loss in industries dependent on classic agricultural strategies. Therefore, it is essential to carry out a comprehensive approach to gene editing's possible benefits and hazards from the perspectives of both the environment and socioeconomic activities (Lindberg et al., 2023).

CHALLENGES AND LIMITATIONS

As gene-editing technologies in plant science have transformative capabilities, multiple scientific and technical challenges limit their large-scale use in agriculture. These challenges pertain to precision, delivery mechanisms, limitations of plant tissue cultures, genotype alternatives, and aspects of their complex nature. The success of the sweeping promise of genome editing for sustainable and inclusive crop improvement lies in its ability to overcome these obstacles (Silva et al., 2025).

Off-Target Effects

CRISPR and other gene editing tools have one major disadvantage: they increase the probability of off-target mutations and unintended changes at genomic locations that are similar to the target site. While being highly specific, CRISPR/Cas9 may create off-target mismatches between guide RNA and non-target DNA sequences, leading to unintended edits that may result in illustrative or deleterious gene disruptions and consequently to transformative phenotypes or detrimental traits (Manghwar et al., 2020). Such off-target effects are safety and regulatory issues, particularly when edited plants are used for food production. Therefore, ensuring the accuracy of gene editing is important. New developments including high-fidelity Cas9 variants, such as SpCas9-HF1 and eSpCas9, advanced guide RNA design algorithms, and utilization of orthologous types of Cas proteins with different PAM requirements have greatly minimized off-target activity. Nevertheless, full off-target profiling is still required prior to the commercialization of gene-edited lines, and high-throughput whole-genome sequencing remains a state-of-the-art method for identifying unintended edits (Naeem et al., 2020).

Delivery Efficiency and Tissue Culture Dependency

One of the main technical limitations of plant genome editing is the effective transport of gene-editing elements into plant cells. Conventional methods of delivery, including Agrobacterium tumefaciens-mediated transformation and biolistic (gene gun) methods, are commonly used but are limited by their efficiency and species specificity (Munir et al., 2023; Tripathi & Shukla, 2024). Furthermore, these methods introduce tissue culture, which is labor-intensive and time-consuming, in which whole plants are regenerated from transformed cells or tissues. This tissue culture dependency is a bottleneck, particularly for recalcitrant species or genotypes that are difficult to regenerate (Bennur et al., 2025). Regeneration protocols can be very species- and genotype-specific, which makes it difficult to set up universal workflows for manipulating various crops. In addition, prolonged tissue culture may result in clonal variations (unintended genetic changes) during regeneration, compromising the integrity of edited lines (Emaru & Rwotonen). Therefore, innovative delivery systems, such as nanoparticles, viral vectors, and protoplasts, have been investigated to solve the above-mentioned challenges. Nanotechnology provides intriguing, non-invasive, and DNA-free options for targeting plant cells with CRISPR parts directly into the plant cell while avoiding the incorporation of foreign DNA into the genome. However, scalability, regulatory approval, and efficiency of our field crops are yet to be developed (Gogoi et al., 2024).

Genotype Dependence

Genotype dependence is another major issue in plant gene editing. Transformation and regeneration protocols are not universal for all cultivars. For instance, while other model varieties such as rice or tomato may be readily amenable to being altered in a CRISPR fashion, elite commercial or regionally adapted cultivars manifest a poor performance in that regard (Maren et al., 2022). This limitation limits the usefulness of gene editing tools for a small range of genotypes, hindering their widespread use in different germplasm pools. The need to overcome genotype dependence necessitates high-intensity optimization of protocols for every target genotype, which requires time and resources (Kockum et al., 2023). There is further research on the development of genotype-independent transformation technologies, such as the application of morphogenic regulators (such as the baby boom and Wuschel2 genes) that will increase regeneration efficiency in various genotypes (Lee & Wang, 2023).

Trait Complexity and Polygenic Traits

Many valuable agronomic traits, including yield, drought tolerance, and disease resistance, are quantitative, gene-controlled (polygenic traits), and environmental interactions. Such complex traits may not be given the desired phenotype from the editing of a single gene. Such trait complexity poses a significant obstacle for gene-editing applications. Neat manipulation of several genomic loci is required for effective manipulation of polygenic traits (Hajheidari et al., 2024). Despite the potential promise of multiplex genome editing (editing multiple genes simultaneously using several guide RNAs), it has delivery challenges, expression balance, and off-target risks. Further compounds include the design of effective editing strategies and our limited knowledge of gene networks, regulatory elements, and epistatic interactions (Doudna, 2020). Even if several candidate genes for a trait are identified, editing them can result in unforeseeable phenotypes because they are interdependently functional (Strauss et al., 2022). The progress made in systems biology, genomics, and machine learning-based predictive modeling is supporting efforts to solve these challenges by identifying critical gene networks and targets. However, editing complex traits is currently a significant frontier in plant genome editing (Goshisht, 2024).

FUTURE PERSPECTIVES AND CONCLUSION

With gene editing technologies rapidly advancing at an unprecedented rate, their future in plant science appears to have transformative effects on agriculture, food security, and environmental sustainability. New innovations such as prime editing, epigenome editing, and RNA-guided DNA integration systems are reaching precision and versatility limits. Moreover, by combining artificial intelligence (AI) and machine learning to guide RNA design and off-target prediction, the process of designing more effective and reliable editing strategies is becoming faster. When added to single-cell genomics, synthetic biology, and high-throughput phenotyping, future platforms could allow for the targeted modification of complex polygenic traits with limited unintended consequences. These innovations also open the door to next-generation crops that will be able to survive climate change, be resistant to new pathogens, and provide outstanding nutritional traits, which will be essential for feeding an expanding global population living in an increasingly stressed environment.

In summary, although gene editing has changed the shape of the landscape of plant biology and cultivation, full-scale reappearance of gene editing demands multidisciplinary and global inclusiveness. It is important to develop solutions to overcome existing limitations, such as regulatory disharmonies, ethical problems, and bottlenecks in the technical sciences, to achieve equitable adoption. Policymakers, scientists, farmers, and the public must consult each other on a transparent and well-informed basis for genome editing to be used in a responsible and inclusive manner. With time, gene editing is likely to become a mainstream agricultural tool as increasingly regulatory frameworks mature and societies embrace it from laboratories to fields, farms, and food systems across the globe. This chapter helps build a foundation for understanding the science, scope, challenges, and implications of gene editing in plants to encourage further research and responsible progress on its run discipline.

REFERENCES

Ahmad, S., Tang, L., Shahzad, R., Mawia, A. M., Rao, G. S., Jamil, S., & Wei, X. (2021). CRISPR-based crop improvements: A way forward to achieve zero hunger. *Journal of Agricultural and Food Chemistry*, *69*(30), 8307–8323. DOI: 10.1021/acs.jafc.1c02653 PMID: 34288688

Ahmar, S., Gill, R. A., Jung, K.-H., Faheem, A., Qasim, M. U., Mubeen, M., & Zhou, W. (2020). Conventional and molecular techniques from simple breeding to speed breeding in crop plants: Recent advances and future outlook. *International Journal of Molecular Sciences*, *21*(7), 2590. DOI: 10.3390/ijms21072590 PMID: 32276445

Ahmar, S., Mahmood, T., Fiaz, S., Mora-Poblete, F., Shafique, M. S., Chattha, M. S., & Jung, K.-H. (2021). Advantage of nanotechnology-based genome editing system and its application in crop improvement. *Frontiers in Plant Science*, *12*, 663849. DOI: 10.3389/fpls.2021.663849 PMID: 34122485

Ahmed, M., Asim, M., Ahmad, S., & Aslam, M. (2023). Climate change, agricultural productivity, and food security. In *Global agricultural production: Resilience to climate change* (pp. 31–72). Springer.

Akinbo, O., Obukosia, S., Ouedraogo, J., Sinebo, W., Savadogo, M., Timpo, S., Mbabazi, R., Maredia, K., Makinde, D., & Ambali, A. (2021). Commercial release of genetically modified crops in Africa: Interface between biosafety regulatory systems and varietal release systems. *Frontiers in Plant Science*, *12*, 605937. DOI: 10.3389/fpls.2021.605937 PMID: 33828569

Alam, M. S., Kong, J., Tao, R., Ahmed, T., Alamin, M., Alotaibi, S. S., Abdelsalam, N. R., & Xu, J.-H. (2022). CRISPR/Cas9 mediated knockout of the OsbHLH024 transcription factor improves salt stress resistance in rice (Oryza sativa L.). *Plants*, *11*(9), 1184. DOI: 10.3390/plants11091184 PMID: 35567185

Anas, M., Khattak, W. A., Fahad, S., Alrawiq, N., Alrawiq, H. S., Abdelsalam, N. R., Jaremko, M., & Quraishi, U. M. (2025). Mitigating heavy metal pollution in agriculture: A multi-omics and nanotechnology approach to safeguard global wheat production. *Journal of Hazardous Materials Advances*, *17*, 100584. https://doi.org/https://doi.org/10.1016/j.hazadv.2024.100584. DOI: 10.1016/j.hazadv.2024.100584

Anyshchenko, A. (2022). Aligning policy design with science to achieve food security: The contribution of genome editing to sustainable agriculture. *Frontiers in Sustainable Food Systems*, *6*, 897643. DOI: 10.3389/fsufs.2022.897643

Aroge, T., Zhu, Y., Jin, D.-N., Dara, M. Z. N., Feng, J., Olajuyin, A. M., Abbas, A., & Liu, S.-Y. (2024). Omics and CRISPR-Cas9 molecular perception: A progressive review approach for powdery mildew disease management. *Physiological and Molecular Plant Pathology*, *130*, 102217. DOI: 10.1016/j.pmpp.2024.102217

Asquer, A., & Krachkovskaya, I. (2021). Uncertainty, institutions and regulatory responses to emerging technologies: CRISPR Gene editing in the US and the EU (2012–2019). *Regulation & Governance*, *15*(4), 1111–1127. DOI: 10.1111/rego.12335

Asrey, R., Barman, K., Prajapati, U., Sharma, S., & Yadav, A. (2021). Genetically modified fruit and vegetable-An overview on senescence regulation, postharvest nutraceutical quality preservation and shelf life extension. *The Journal of Horticultural Science & Biotechnology*, *96*(3), 271–287. DOI: 10.1080/14620316.2020.1845986

Azubuike, C. C., Couceiro, F., Robson, S. C., Piccinni, M. Z., Watts, J. E., Williams, J. B., Callaghan, A. J., & Howard, T. P. (2022). Developing biosensors for SARS-CoV-2 wastewater-based epidemiology: A systematic review of trends, limitations and future perspectives. *Sustainability (Basel)*, *14*(24), 16761. DOI: 10.3390/su142416761

Backes, N. (2020). *CRISPR interference strategies for studies in essential gene function and live cell fluorescent imaging of DNA elements in Escherichia coli*. Iowa State University. DOI: 10.31274/etd-20200902-8

Bandyopadhyay, A., Kancharla, N., Javalkote, V. S., Dasgupta, S., & Brutnell, T. P. (2020). CRISPR-Cas12a (Cpf1): A versatile tool in the plant genome editing tool box for agricultural advancement. *Frontiers in Plant Science*, *11*, 584151. DOI: 10.3389/fpls.2020.584151 PMID: 33214794

Bansal, P., & Kaur, N. (2024). Assessing risks associated with large-scale adoption of CRISPR gene-edited crops. *Journal of Crop Science and Biotechnology*, •••, 1–11.

Barbosa, S., Toe, L. P., Thizy, D., Vaz, M., & Carter, L. (2021). Engagement and social acceptance in genome editing for human benefit: Reflections on research and practice in a global context. *Wellcome Open Research*, *5*, 244. DOI: 10.12688/wellcomeopenres.16260.2 PMID: 34095505

Begna, T. (2020). Major challenging constraints to crop production farming system and possible breeding to overcome the constraints. [IJRSAS]. *International Journal of Research Studies in Agricultural Sciences*, *6*(7), 27–46.

Bennur, P. L., O'Brien, M., Fernando, S. C., & Doblin, M. S. (2025). Improving transformation and regeneration efficiency in medicinal plants: Insights from other recalcitrant species. *Journal of Experimental Botany*, *76*(1), 52–75. DOI: 10.1093/jxb/erae189 PMID: 38652155

Bhardwaj, A., & Nain, V. (2021). TALENs—an indispensable tool in the era of CRISPR: A mini review. *Journal of Genetic Engineering and Biotechnology*, *19*(1), 125. DOI: 10.1186/s43141-021-00225-z PMID: 34420096

Bhardwaj, A., Tomar, P., & Nain, V. (2024). Machine Learning-Driven Prediction of CRISPR-Cas9 Off-Target Effects and Mechanistic Insights. *environment, 12*, 13.

Bhatt, P., Singh, S., Alfuraiji, N., & Al-Snafi, A. E. (2022). CRISPR CAS9: A new technology to modify genome-A review. *Open J. Syst. Demonstr. J*, *8*(4), 208–215.

Bode, D., Cull, A. H., Rubio-Lara, J. A., & Kent, D. G. (2021). Exploiting single-cell tools in gene and cell therapy. *Frontiers in Immunology*, *12*, 702636. DOI: 10.3389/fimmu.2021.702636 PMID: 34322133

Borg, E., & Policante, A. (2024). The Gene Editing Business: Rent Extraction in the Biotech Industry. *Review of Political Economy*, •••, 1–36.

Brookes, G., & Smyth, S. J. (2024). Risk-appropriate regulations for gene-editing technologies. *GM Crops & Food*, *15*(1), 1–14. DOI: 10.1080/21645698.2023.2293510 PMID: 38215017

Bullock, D. W., Wilson, W. W., & Neadeau, J. (2021). Gene editing versus genetic modification in the research and development of new crop traits: An economic comparison. *American Journal of Agricultural Economics*, *103*(5), 1700–1719. DOI: 10.1111/ajae.12201

Bure, I. V., Nemtsova, M. V., & Kuznetsova, E. B. (2022). Histone modifications and non-coding RNAs: Mutual epigenetic regulation and role in pathogenesis. *International Journal of Molecular Sciences*, *23*(10), 5801. DOI: 10.3390/ijms23105801 PMID: 35628612

Cai, R., Lv, R., Shi, X., Yang, G., & Jin, J. (2023). CRISPR/dCas9 tools: Epigenetic mechanism and application in gene transcriptional regulation. *International Journal of Molecular Sciences*, *24*(19), 14865. DOI: 10.3390/ijms241914865 PMID: 37834313

Cathomen, T., & Joung, J. K. (2008). Zinc-finger nucleases: The next generation emerges. *Molecular Therapy : The Journal of the American Society of Gene Therapy*, *16*(7), 1200–1207. DOI: 10.1038/mt.2008.114

Chen, K., Chen, J., Pi, X., Huang, L.-J., & Li, N. (2023). Isolation, purification, and application of protoplasts and transient expression systems in plants. *International Journal of Molecular Sciences, 24*(23), 16892. DOI: 10.3390/ijms242316892 PMID: 38069215

Choudhary, A., Kumar, A., Kaur, H., Pandey, V., Singh, B., & Mehta, S. (2022). Breeding strategies for developing disease-resistant wheat: present, past, and future. In *Cereal diseases: nanobiotechnological approaches for diagnosis and management* (pp. 137–161). Springer. DOI: 10.1007/978-981-19-3120-8_8

Clapp, J., & Ruder, S.-L. (2020). Precision technologies for agriculture: Digital farming, gene-edited crops, and the politics of sustainability. *Global Environmental Politics, 20*(3), 49–69. DOI: 10.1162/glep_a_00566

Cummings, C., Selfa, T., Lindberg, S., & Bain, C. (2024). Identifying public trust building priorities of gene editing in agriculture and food. *Agriculture and Human Values, 41*(1), 47–60. DOI: 10.1007/s10460-023-10465-z

D'Amelia, V., Docimo, T., Crocoll, C., & Rigano, M. M. (2021). Specialized metabolites and valuable molecules in crop and medicinal plants: The evolution of their use and strategies for their production. *Genes, 12*(6), 936. DOI: 10.3390/genes12060936 PMID: 34207427

de Roij, S. (2020). *Gene-editing crops to benefit smallholder farmers in developing countries: Realised potential or empty claims of inclusive innovation?* Delhove, J., Osenk, I., Prichard, I., & Donnelley, M. (2020). Public acceptability of gene therapy and gene editing for human use: a systematic review. *Human Gene Therapy, 31*(1-2), 20–46. PMID: 31802714

Devi, B., Kumar, M. N., Chutia, M., & Bhattacharyya, N. (2022). Abiotic and biotic stress challenges of Cassava in changing climate and strategies to overcome: A review. *Scientia Horticulturae, 305*, 111432. DOI: 10.1016/j.scienta.2022.111432

Dixit, S., Kumar, A., Srinivasan, K., Vincent, P. D. R., & Ramu Krishnan, N. (2024). Advancing genome editing with artificial intelligence: Opportunities, challenges, and future directions. *Frontiers in Bioengineering and Biotechnology, 11*, 1335901. DOI: 10.3389/fbioe.2023.1335901 PMID: 38260726

Doudna, J. A. (2020). The promise and challenge of therapeutic genome editing. *Nature, 578*(7794), 229–236. DOI: 10.1038/s41586-020-1978-5 PMID: 32051598

Emaru, A., & Rwotonen, I. Implications of Somaclonal variation in vitro in crop management: A review.

Enyiukwu, D., Bassey, I., Nwaogu, G., Chukwu, L., & Maranzu, J. (2020). Post-harvest spoilage and management of fruits and vegetables: A perspective on small-holder agricultural systems of the tropics. *Greener Trends in Plant Pathology and Entomology, 3*(1), 01-17.

Erdoğan, İ., Cevher-Keskin, B., Bilir, Ö., Hong, Y., & Tör, M. (2023). Recent Developments in CRISPR/Cas9 Genome-Editing Technology Related to Plant Disease Resistance and Abiotic Stress Tolerance. *Biology (Basel), 12*(7), 1037–1037. DOI: 10.3390/biology12071037 PMID: 37508466

Faltus, T. (2023). The applicability of the European GMO legislation to epigenetically modified organisms. *Frontiers in Bioengineering and Biotechnology, 11*, 1124131. DOI: 10.3389/fbioe.2023.1124131 PMID: 36923460

Flanagan, K. (2020). *Utilizing Science Fiction to Visualize Ethical Issues Surrounding Human Genome Editing, Influence Ethical Reasoning Skills in Education, and Guide Policy Implementation*. Drew University.

Gao, C. (2021). Genome engineering for crop improvement and future agriculture. *Cell, 184*(6), 1621–1635. DOI: 10.1016/j.cell.2021.01.005 PMID: 33581057

Ghimire, B. K., Yu, C. Y., Kim, W.-R., Moon, H.-S., Lee, J., Kim, S. H., & Chung, I. M. (2023). Assessment of benefits and risk of genetically modified plants and products: Current controversies and perspective. *Sustainability (Basel), 15*(2), 1722. DOI: 10.3390/su15021722

Gogoi, N., Susila, H., Leach, J., Müllner, M., Jones, B., & Pogson, B. J. (2024). Developing frameworks for nanotechnology-driven DNA-free plant genome-editing. *Trends in Plant Science*. PMID: 39477773

González, M. N., Massa, G. A., Andersson, M., Turesson, H., Olsson, N., Fält, A.-S., Storani, L., Décima Oneto, C. A., Hofvander, P., & Feingold, S. E. (2020). Reduced enzymatic browning in potato tubers by specific editing of a polyphenol oxidase gene via ribonucleoprotein complexes delivery of the CRISPR/Cas9 system. *Frontiers in Plant Science, 10*, 497481. DOI: 10.3389/fpls.2019.01649 PMID: 31998338

Goodman, D. (2023). Agri-Biotechnology and the Failed Promises of the Seed-Chemical Complex, CRISPR and Gene Editing, and Regulatory Capture. In *Transforming Agriculture and Foodways* (pp. 53–70). Bristol University Press. DOI: 10.1332/policypress/9781529231465.003.0005

Goshisht, M. K. (2024). Machine learning and deep learning in synthetic biology: Key architectures, applications, and challenges. *ACS Omega, 9*(9), 9921–9945. DOI: 10.1021/acsomega.3c05913 PMID: 38463314

Gu, S., Bodai, Z., Cowan, Q. T., & Komor, A. C. (2021). Base editors: expanding the types of DNA damage products harnessed for genome editing. *Gene and Genome Editing, 1*, 100005.

Hafeez, A., Ali, B., Javed, M. A., Saleem, A., Fatima, M., Fathi, A., Afridi, M. S., Aydin, V., Oral, M. A., & Soudy, F. A. (2023). Plant breeding for harmony between sustainable agriculture, the environment, and global food security: An era of genomics-assisted breeding. *Planta, 258*(5), 97. DOI: 10.1007/s00425-023-04252-7 PMID: 37823963

Hajheidari, M., Sunyaev, S., & de Meaux, J. (2024). Are complex traits underpinned by polygenic molecular traits? A reflection on the complexity of gene expression. *Plant & Cell Physiology*, •••, pcae140. PMID: 39626022

Hillary, V. E., & Ceasar, S. A. (2023). A review on the mechanism and applications of CRISPR/Cas9/Cas12/Cas13/Cas14 proteins utilized for genome engineering. *Molecular Biotechnology, 65*(3), 311–325. DOI: 10.1007/s12033-022-00567-0 PMID: 36163606

Hossain, M. J., & Bakhsh, A. (2020). Development and applications of transplastomic plants; a way towards eco-friendly agriculture. *Environment, climate, plant and vegetation growth*, 285-322.

Huang, X., Hilscher, J., Stoger, E., Christou, P., & Zhu, C. (2021). Modification of cereal plant architecture by genome editing to improve yields. *Plant Cell Reports, 40*(6), 953–978. DOI: 10.1007/s00299-021-02668-7 PMID: 33559722

Husaini, A. M., & Sohail, M. (2024). Agrochemical-free genetically modified and genome-edited crops–towards a 'greener' green revolution and achieving the United Nation's Sustainable Development Goals. *Journal of Biotechnology*. Advance online publication. DOI: 10.1016/j.jbiotec.2024.04.015

Idris, S. H., Mat Jalaluddin, N. S., & Chang, L. W. (2023). Ethical and legal implications of gene editing in plant breeding: A systematic literature review. *Journal of Zhejiang University. Science. B, 24*(12), 1093–1105. DOI: 10.1631/jzus.B2200601 PMID: 38057267

Jenkins, D., Dobert, R., Atanassova, A., & Pavely, C. (2021). Impacts of the regulatory environment for gene editing on delivering beneficial products. *In Vitro Cellular & Developmental Biology. Plant : Journal of the Tissue Culture Association, 57*(4), 609–626. DOI: 10.1007/s11627-021-10201-4 PMID: 34429575

Jian, C., Pan, Y., Liu, S., Guo, M., Huang, Y., Cao, L., & Hou, J. (2024). The TaGW2-TaSPL14 module regulates the trade-off between tiller number and grain weight in wheat. *Journal of Integrative Plant Biology*, *66*(9), 1953–1965. DOI: 10.1111/jipb.13723 PMID: 38924348

Jiang, L. (2020). Commercialization of the gene-edited crop and morality: Challenges from the liberal patent law and the strict GMO law in the EU. *New Genetics and Society*, *39*(2), 191–218. DOI: 10.1080/14636778.2019.1686968

Kang, Y., Deng, H., Pray, C., & Hu, R. (2022). Managers' attitudes toward gene-editing technology and companies' R&D investment in gene-editing: The case of Chinese seed companies. *GM Crops & Food*, *13*(1), 309–326. DOI: 10.1080/21645698.2022.2140567 PMID: 36382611

Kavuri, N. R., Ramasamy, M., Qi, Y., & Mandadi, K. (2022). Applications of CRISPR/Cas13-based RNA editing in plants. *Cells*, *11*(17), 2665. DOI: 10.3390/cells11172665 PMID: 36078073

Kenzhebayeva, S., Atabayeva, S., Sarsu, F., Abekova, A., Shoinbekova, S., Omirbekova, N., Doktyrbay, G., Beisenova, A., & Shavrukov, Y. (2022). Organ-specific expression of genes involved in iron homeostasis in wheat mutant lines with increased grain iron and zinc content. *PeerJ*, *10*, e13515. DOI: 10.7717/peerj.13515 PMID: 35707120

Khan, A., Pudhuvai, B., Shrestha, A., Mishra, A. K., Shah, M. P., Koul, B., & Dey, N. (2024). CRISPR-mediated iron and folate biofortification in crops: Advances and perspectives. *Biotechnology & Genetic Engineering Reviews*, *40*(4), 4138–4168. DOI: 10.1080/02648725.2023.2205202 PMID: 37092872

Khan, Z., Saboor, T., Ashfaq, M., Saddique, A., & Khanum, P. (2021). Disease resistance in crops through CRISPR/Cas. *CRISPR Crops: The Future of Food Security*, 151-175.

KhokharVoytas, A., Shahbaz, M., Maqsood, M. F., Zulfiqar, U., Naz, N., Iqbal, U. Z., Sara, M., Aqeel, M., Khalid, N., Noman, A., Zulfiqar, F., Al Syaad, K. M., & AlShaqhaa, M. A.KhokharVoytas. (2023). Genetic modification strategies for enhancing plant resilience to abiotic stresses in the context of climate change. *Functional & Integrative Genomics*, *23*(3), 283. DOI: 10.1007/s10142-023-01202-0 PMID: 37642792

Kockum, I., Huang, J., & Stridh, P. (2023). Overview of genotyping technologies and methods. *Current Protocols*, *3*(4), e727. DOI: 10.1002/cpz1.727 PMID: 37026777

Kordyś, M., Sen, R., & Warkocki, Z. (2022). Applications of the versatile CRISPR-Cas13 RNA targeting system. *Wiley Interdisciplinary Reviews. RNA, 13*(3), e1694. DOI: 10.1002/wrna.1694 PMID: 34553495

Kumar, L., Chhogyel, N., Gopalakrishnan, T., Hasan, M. K., Jayasinghe, S. L., Kariyawasam, C. S., & Ratnayake, S. (2022). Climate change and future of agri-food production. In *Future foods* (pp. 49–79). Elsevier. DOI: 10.1016/B978-0-323-91001-9.00009-8

Lee, K., & Wang, K. (2023). Strategies for genotype-flexible plant transformation. *Current Opinion in Biotechnology, 79*, 102848. DOI: 10.1016/j.copbio.2022.102848 PMID: 36463838

Li, S., Chen, K., & Grierson, D. (2021). Molecular and hormonal mechanisms regulating fleshy fruit ripening. *Cells, 10*(5), 1136. DOI: 10.3390/cells10051136 PMID: 34066675

Li, X., Xu, S., Fuhrmann-Aoyagi, M. B., Yuan, S., Iwama, T., Kobayashi, M., & Miura, K. (2022). CRISPR/Cas9 technique for temperature, drought, and salinity stress responses. *Current Issues in Molecular Biology, 44*(6), 2664–2682. DOI: 10.3390/cimb44060182 PMID: 35735623

Li, Z., Xiong, X., & Li, J.-F. (2020). The working dead: Repurposing inactive CRISPR-associated nucleases as programmable transcriptional regulators in plants. *aBIOTECH, 1*(1), 32–40. DOI: 10.1007/s42994-019-00003-z PMID: 36305006

Liang, J., Yang, X., Jiao, Y., Wang, D., Zhao, Q., Sun, Y., Li, Y., & Wu, K. (2022). The evolution of China's regulation of agricultural biotechnology. *aBIOTECH, 3*(4), 237–249. DOI: 10.1007/s42994-022-00086-1 PMID: 36533267

Lindberg, S. A., Peters, D. J., & Cummings, C. L. (2023). Gene-Edited Food Adoption Intentions and Institutional Trust in the United States: Benefits, Acceptance, and Labeling☆. *Rural Sociology, 88*(2), 392-425.

Liu, A. (2022). Legal Measures Addressing Environmental Risks That May Be Caused by GM Crops in China. In *Regulating Genetically Modified Crops in View of Environmental Risks: China's Implementation of International Obligations* (pp. 143-222). Springer. DOI: 10.1007/978-981-19-2480-4_5

Liu, B., Jiang, H., & Li, L. (2024). The innovative applications and prospects of artificial intelligence in agricultural research: Gene editing, crop improvement, and intelligent data analysis. *Advances in Resources Research, 4*(4), 624–653.

Liu, G., Zhang, Y., & Zhang, T. (2020). Computational approaches for effective CRISPR guide RNA design and evaluation. *Computational and Structural Biotechnology Journal*, *18*, 35–44. DOI: 10.1016/j.csbj.2019.11.006 PMID: 31890142

Liu, Q., Yang, F., Zhang, J., Liu, H., Rahman, S., Islam, S., Ma, W., & She, M. (2021). Application of CRISPR/Cas9 in crop quality improvement. *International Journal of Molecular Sciences*, *22*(8), 4206. DOI: 10.3390/ijms22084206 PMID: 33921600

Lorenzo, C. D., Debray, K., Herwegh, D., Develtere, W., Impens, L., Schaumont, D., & De Boe, Y. (2023). BREEDIT: A multiplex genome editing strategy to improve complex quantitative traits in maize. *The Plant Cell*, *35*(1), 218–238. DOI: 10.1093/plcell/koac243 PMID: 36066192

Lv, Z., Jiang, R., Chen, J., & Chen, W. (2020). Nanoparticle-mediated gene transformation strategies for plant genetic engineering. *The Plant Journal : For Cell and Molecular Biology*, *104*(4), 880–891. DOI: 10.1111/tpj.14973 PMID: 32860436

Lv, Z., Jiang, S., Kong, S., Zhang, X., Yue, J., Zhao, W., Li, L., & Lin, S. (2024). Advances in Single-Cell Transcriptome Sequencing and Spatial Transcriptome Sequencing in Plants. *Plants*, *13*(12), 1679. DOI: 10.3390/plants13121679 PMID: 38931111

Manghwar, H., Li, B., Ding, X., Hussain, A., Lindsey, K., Zhang, X., & Jin, S. (2020). CRISPR/Cas systems in genome editing: Methodologies and tools for sgRNA design, off-target evaluation, and strategies to mitigate off-target effects. *Advanced Science (Weinheim, Baden-Württemberg, Germany)*, *7*(6), 1902312. DOI: 10.1002/advs.201902312 PMID: 32195078

Mani, V., Park, S., Kim, J. A., Lee, S. I., & Lee, K. (2021). Metabolic perturbation and synthetic biology strategies for plant terpenoid production—An updated overview. *Plants*, *10*(10), 2179. DOI: 10.3390/plants10102179 PMID: 34685985

Maren, N. A., Duan, H., Da, K., Yencho, G. C., Ranney, T. G., & Liu, W. (2022). Genotype-independent plant transformation. *Horticulture Research*, *9*, uhac047. DOI: 10.1093/hr/uhac047 PMID: 35531314

Mathiazhagan, M., Chidambara, B., Hunashikatti, L. R., & Ravishankar, K. V. (2021). Genomic approaches for improvement of tropical fruits: Fruit quality, shelf life and nutrient content. *Genes*, *12*(12), 1881. DOI: 10.3390/genes12121881 PMID: 34946829

Mengchao, Q., Hui, T., Xuetao, S., Chongyang, Z., Feng, H., Min, W., & Shutong, W. (2023). Transcriptome Analysis of oserf922 Mutants Reveals New Insights into Rice Blast Resistance. *Rice Science*, *30*(5), 374–378. DOI: 10.1016/j.rsci.2023.05.002

Menz, J., Modrzejewski, D., Hartung, F., Wilhelm, R., & Sprink, T. (2020). Genome edited crops touch the market: A view on the global development and regulatory environment. *Frontiers in Plant Science*, *11*, 586027. DOI: 10.3389/fpls.2020.586027 PMID: 33163013

Mikhaylova, E., Kuluev, B., Gerashchenkov, G., Chemeris, D., Garafutdinov, R., Kuluev, A., Baymiev, A. K., Baymiev, A. K., & Chemeris, A. (2024). Prime-editing methods and pegRNA design programs. *Molecular Biology*, *58*(1), 17–32. DOI: 10.1134/S0026893324010084 PMID: 39852132

Mishra, M., Afzal, S., & Singh, N. K. (2024). CRISPR/Cas9-Mediated Genome Engineering for Production of Biofortified Crops. In *Phytoremediation and Biofortification* (pp. 303-333). Apple Academic Press.

Molinari, M. D. C., Pagliarini, R. F., Florentino, L. H., Lima, R. N., Arraes, F. B. M., Abbad, S. V.,…Nepomuceno, A. L. (2024). Navigating the Path from Lab to Market: Regulatory Challenges and Opportunities for Genome Editing Technologies for Agriculture. *Plant Genome Editing Technologies: Speed Breeding, Crop Improvement and Sustainable Agriculture*, 25-63.

Mukherjee, M., Posch, K., Molina, S. J., Taymor, K., & Keller, A. (2023). Comparing regulatory processes in genome editing and autonomous vehicles: How institutional environments shape sociotechnical imaginaries. *The Review of Policy Research*, *40*(3), 433–457. DOI: 10.1111/ropr.12532

Mukherji, A., Arndt, C., Arango, J., Flintan, F. E., Derera, J., Francesconi, W.,… Mockshell, J. Y. (2023). Achieving agricultural breakthrough: A deep dive into seven technological areas.

Munawar, N., Faheem, M., Niamat, A., Munir, A., Khan, S. H., Zahoor, M. K., & Ahmad, A. (2024). Regulatory, ethical, social, and biosafety concerns in genome-edited horticultural crops. In *CRISPRized Horticulture Crops* (pp. 421–438). Elsevier. DOI: 10.1016/B978-0-443-13229-2.00026-0

Munir, F., Azeem, S., Amir, R., Gillani, M., Gul, A., Mustafa, A., & Fatima, N. (2023). Genetic Transformation Methods in Cereal Crops. In *Cereal Crops* (pp. 273–294). CRC Press.

Naeem, M., Majeed, S., Hoque, M. Z., & Ahmad, I. (2020). Latest developed strategies to minimize the off-target effects in CRISPR-Cas-mediated genome editing. *Cells*, *9*(7), 1608. DOI: 10.3390/cells9071608 PMID: 32630835

Nangammbi, N. (2021). *Functional Testing of Xylan-Associated Genes in Arabidopsis and Application of CRISPR/Cas9 Gene Editing in Hybrid Poplar Trees*. University of Pretoria.

Nawaz, S. (2021). *Beyond naturalness?: social dimensions of gene editing in agriculture*. University of British Columbia.

NETO, R. R. D. S. (2022). Characterization of CsLIEXP1 as a new susceptibility gene to Xanthomonas citri subsp. citri and gene editing by CRISPR/Cas9 for citrus canker tolerance.

Neugebauer, M. E., Hsu, A., Arbab, M., Krasnow, N. A., McElroy, A. N., Pandey, S., & Banskota, S. (2023). Evolution of an adenine base editor into a small, efficient cytosine base editor with low off-target activity. *Nature Biotechnology*, *41*(5), 673–685. DOI: 10.1038/s41587-022-01533-6 PMID: 36357719

Ochiai, H., & Yamamoto, T. (2015). Genome editing using zinc-finger nucleases (ZFNs) and transcription activator-like effector nucleases (TALENs). *Targeted genome editing using site-specific nucleases: ZFNs, TALENs, and the CRISPR/Cas9 system*, 3-24.

Ozyigit, I. I., & Yucebilgili Kurtoglu, K. (2020). Particle bombardment technology and its applications in plants. *Molecular Biology Reports*, *47*(12), 9831–9847. DOI: 10.1007/s11033-020-06001-5 PMID: 33222118

Piergentili, R., Del Rio, A., Signore, F., Umani Ronchi, F., Marinelli, E., & Zaami, S. (2021). CRISPR-Cas and its wide-ranging applications: From human genome editing to environmental implications, technical limitations, hazards and bioethical issues. *Cells*, *10*(5), 969. DOI: 10.3390/cells10050969 PMID: 33919194

Pramanik, D., Shelake, R. M., Kim, M. J., & Kim, J.-Y. (2021). CRISPR-mediated engineering across the central dogma in plant biology for basic research and crop improvement. *Molecular Plant*, *14*(1), 127–150. DOI: 10.1016/j.molp.2020.11.002 PMID: 33152519

Rahman, M. M., & Tollefsbol, T. O. (2021). Targeting cancer epigenetics with CRISPR-dCAS9: Principles and prospects. *Methods : A Companion to Methods in Enzymology*, *187*, 77–91. DOI: 10.1016/j.ymeth.2020.04.006 PMID: 32315755

Raza, A., Anas, M., Bhardwaj, S., Mir, R. A., Charagh, S., Elahi, M., Zhang, X., Mir, R. R., Weckwerth, W., Fernie, A. R., Siddique, K. H. M., Hu, Z., & Varshney, R. K. (2025). Harnessing metabolomics for enhanced crop drought tolerance. *The Crop Journal*, *13*(2), 311–327. https://doi.org/https://doi.org/10.1016/j.cj.2025.01.001. DOI: 10.1016/j.cj.2025.01.001

Rizzo, P., Chavez, B. G., Dias, S. L., & D'Auria, J. C. (2023). Plant synthetic biology: From inspiration to augmentation. *Current Opinion in Biotechnology*, *79*, 102857. DOI: 10.1016/j.copbio.2022.102857 PMID: 36502769

Robertson, G. (2021). *Establishing the CRISPR/Cas13a genome editing system in Nicotiana benthamiana for RNA targeting applications*. Stellenbosch University.

Robertson, G., Burger, J., & Campa, M. (2022). CRISPR/Cas-based tools for the targeted control of plant viruses. *Molecular Plant Pathology*, *23*(11), 1701–1718. DOI: 10.1111/mpp.13252 PMID: 35920132

Rock, J. S., Schnurr, M. A., Kingiri, A., Glover, D., Stone, G. D., Ely, A., & Fischer, K. (2023). Beyond the Genome: Genetically modified crops in Africa and the implications for Genome Editing. *Development and Change*, *54*(1), 117–142. DOI: 10.1111/dech.12750

Saber Sichani, A., Ranjbar, M., Baneshi, M., Torabi Zadeh, F., & Fallahi, J. (2023). A review on advanced CRISPR-based genome-editing tools: Base editing and prime editing. *Molecular Biotechnology*, *65*(6), 849–860. DOI: 10.1007/s12033-022-00639-1 PMID: 36547823

Selfa, T., Cummings, C., Dahlstrom, M., Lindberg, S., Peters, D., Wolf, C., & Bain, C. (2023). *Gene editing in agriculture: Social concerns, public engagement, and governance* (Vol. 48). Sage Publications Sage CA.

Sen, P., Chatterjee, A., Kumar, D., Bhattacharyya, S., Bandyopadhyay, S., & Sarkar, A. (2024). Genetic and functional mechanisms of yield-related genes in rice. *Acta Physiologiae Plantarum*, *46*(3), 29. DOI: 10.1007/s11738-024-03667-3

Shabala, S., Chen, X., Yun, P., & Zhou, M. (2025). Salinity tolerance in wheat: Rethinking the targets. *Journal of Experimental Botany*, •••, eraf152. PMID: 40201956

Shahzad, R., Jamil, S., Ahmad, S., Nisar, A., Khan, S., Amina, Z., Kanwal, S., Aslam, H. M. U., Gill, R. A., & Zhou, W. (2021). Biofortification of cereals and pulses using new breeding techniques: Current and future perspectives. *Frontiers in Nutrition*, *8*, 721728. DOI: 10.3389/fnut.2021.721728 PMID: 34692743

Shamshirgaran, Y., Liu, J., Sumer, H., Verma, P. J., & Taheri-Ghahfarokhi, A. (2022). Tools for efficient genome editing; ZFN, TALEN, and CRISPR. *Applications of genome modulation and editing*, 29-46.

Sharma, P., Pandey, A., Malviya, R., Dey, S., Karmakar, S., & Gayen, D. (2023). Genome editing for improving nutritional quality, post-harvest shelf life and stress tolerance of fruits, vegetables, and ornamentals. *Frontiers in Genome Editing*, *5*, 1094965. DOI: 10.3389/fgeed.2023.1094965 PMID: 36911238

Shelake, R. M., Kadam, U. S., Kumar, R., Pramanik, D., Singh, A. K., & Kim, J.-Y. (2022). Engineering drought and salinity tolerance traits in crops through CRISPR-mediated genome editing: Targets, tools, challenges, and perspectives. *Plant Communications*, *3*(6), 100417. DOI: 10.1016/j.xplc.2022.100417 PMID: 35927945

Shoeb, E., Venkataraman, S., Badar, U., & Hefferon, K. (2024). Genome editing for crop biofortification. *Applications of Genome Engineering in Plants*, 91-121.

Silva, J. C. F., de Godoy Machado, K. L., de Souza Silva, A. F., Dias, R., Bodnar, V. R., Vieira, W. O.,...Pauli, I. (2025). Challenges and Opportunities for New Frontiers and Technologies to Guarantee Food Production: A Broad Systematic Perspective.

Singh, A., Pandey, H., Pandey, S., Lal, D., Chauhan, D., Aparna, , Antre, S. H., B, S., & Kumar, A. (2023). Drought stress in maize: Stress perception to molecular response and strategies for its improvement. *Functional & Integrative Genomics*, *23*(4), 296. DOI: 10.1007/s10142-023-01226-6 PMID: 37697159

Singh, A. K., Mishra, P., Kashyap, S. P., Karkute, S. G., Singh, P. M., Rai, N., Bahadur, A., & Behera, T. K. (2022). Molecular insights into mechanisms underlying thermo-tolerance in tomato. *Frontiers in Plant Science*, *13*, 1040532. DOI: 10.3389/fpls.2022.1040532 PMID: 36388532

Spök, A., Sprink, T., Allan, A. C., Yamaguchi, T., & Dayé, C. (2022). Towards social acceptability of genome-edited plants in industrialised countries? Emerging evidence from Europe, United States, Canada, Australia, New Zealand, and Japan. *Frontiers in Genome Editing*, *4*, 899331. DOI: 10.3389/fgeed.2022.899331 PMID: 36120531

Strauss, S. H., Slavov, G. T., & DiFazio, S. P. (2022). Gene-editing for production traits in forest trees: Challenges to integration and gene target identification. *Forests*, *13*(11), 1887. DOI: 10.3390/f13111887

Tembo, B., Mulenga, R. M., Sichilima, S., M'siska, K. K., Mwale, M., Chikoti, P. C., & Peterson, G. L. (2020). Detection and characterization of fungus (Magnaporthe oryzae pathotype Triticum) causing wheat blast disease on rain-fed grown wheat (Triticum aestivum L.) in Zambia. *PLoS One*, *15*(9), e0238724. DOI: 10.1371/journal.pone.0238724 PMID: 32956369

Thomson, G., Dickinson, L., & Jacob, Y. (2024). Genomic consequences associated with Agrobacterium-mediated transformation of plants. *The Plant Journal : For Cell and Molecular Biology*, *117*(2), 342–363. DOI: 10.1111/tpj.16496 PMID: 37831618

Tripathi, A., & Shukla, S. (2024). Methods of genetic transformation: Major emphasis to crop plants. *Journal of Microbiology, Biotechnology and Food Sciences*, *13*(4), e10276–e10276.

Tripathi, S., Khatri, P., Fatima, Z., Pandey, R. P., & Hameed, S. (2022). A landscape of CRISPR/Cas technique for emerging viral disease diagnostics and therapeutics: Progress and prospects. *Pathogens (Basel, Switzerland)*, *12*(1), 56. DOI: 10.3390/pathogens12010056 PMID: 36678404

Turnbull, C., Lillemo, M., & Hvoslef-Eide, T. A. (2021). Global regulation of genetically modified crops amid the gene edited crop boom–a review. *Frontiers in Plant Science*, *12*, 630396. DOI: 10.3389/fpls.2021.630396 PMID: 33719302

Umeohia, U. E., & Olapade, A. A. (2024). Quality attributes, physiology, and Post-harvest Technologies of Tomatoes (Lycopersicum esculentum)–A review. *American Journal of Food Science and Technology*, *12*(2), 42–64. DOI: 10.12691/ajfst-12-2-1

Usman, B., Nawaz, G., Zhao, N., Liao, S., Liu, Y., & Li, R. (2020). Precise editing of the OsPYL9 gene by RNA-guided Cas9 nuclease confers enhanced drought tolerance and grain yield in rice (Oryza sativa L.) by regulating circadian rhythm and abiotic stress responsive proteins. *International Journal of Molecular Sciences*, *21*(21), 7854. DOI: 10.3390/ijms21217854 PMID: 33113937

Van Eck, J. (2020). Applying gene editing to tailor precise genetic modifications in plants. *The Journal of Biological Chemistry*, *295*(38), 13267–13276. DOI: 10.1074/jbc.REV120.010850 PMID: 32723863

Verma, V., Kumar, A., Partap, M., Thakur, M., & Bhargava, B. (2023). CRISPR-Cas: A robust technology for enhancing consumer-preferred commercial traits in crops. *Frontiers in Plant Science*, *14*, 1122940. DOI: 10.3389/fpls.2023.1122940 PMID: 36824195

Wada, N., Ueta, R., Osakabe, Y., & Osakabe, K. (2020). Precision genome editing in plants: State-of-the-art in CRISPR/Cas9-based genome engineering. *BMC Plant Biology*, *20*(1), 1–12. DOI: 10.1186/s12870-020-02385-5 PMID: 32450802

Wang, J. Y., & Doudna, J. A. (2023). CRISPR technology: A decade of genome editing is only the beginning. *Science*, *379*(6629), eadd8643. DOI: 10.1126/science.add8643 PMID: 36656942

Wang, J. Y., Pausch, P., & Doudna, J. A. (2022). Structural biology of CRISPR–Cas immunity and genome editing enzymes. *Nature Reviews. Microbiology*, *20*(11), 641–656. DOI: 10.1038/s41579-022-00739-4 PMID: 35562427

Wang, S., Sun, S.-T., Zhang, X.-Y., Ding, H.-R., Yuan, Y., He, J.-J., Wang, M.-S., Yang, B., & Li, Y.-B. (2023). The evolution of single-cell RNA sequencing technology and application: Progress and perspectives. *International Journal of Molecular Sciences*, *24*(3), 2943. DOI: 10.3390/ijms24032943 PMID: 36769267

Watson, O., & Hayta, S. (2024). Precision breeding in agriculture and food systems in the United Kingdom. *Transgenic Research*, *33*(6), 1–6. DOI: 10.1007/s11248-024-00397-7 PMID: 39105945

Wei, X., Pu, A., Liu, Q., Hou, Q., Zhang, Y., An, X., & Wu, S. (2022). The bibliometric landscape of gene editing innovation and regulation in the worldwide. *Cells*, *11*(17), 2682. DOI: 10.3390/cells11172682 PMID: 36078090

Xie, Y., Shen, Q., Li, F., Ni, S., & Yu, J. (2023). Temperature response of plants and heat tolerance in rice: A review. *Advances in Agronomy*, *179*, 135–203. DOI: 10.1016/bs.agron.2023.01.003

Xu, L., Sun, B., Liu, S., Gao, X., Zhou, H., Li, F., & Li, Y. (2023). The evaluation of active transcriptional repressor domain for CRISPRi in plants. *Gene*, *851*, 146967. DOI: 10.1016/j.gene.2022.146967 PMID: 36261092

Yu, Y., Yu, J., Wang, Q., Wang, J., Zhao, G., Wu, H., Zhu, Y., Chu, C., & Fang, J. (2021). Overexpression of the rice ORANGE gene OsOR negatively regulates carotenoid accumulation, leads to higher tiller numbers and decreases stress tolerance in Nipponbare rice. *Plant Science : An International Journal of Experimental Plant Biology*, *310*, 110962. DOI: 10.1016/j.plantsci.2021.110962 PMID: 34315587

Zafar, K., Khan, M. Z., Amin, I., Mukhtar, Z., Yasmin, S., Arif, M., Ejaz, K., & Mansoor, S. (2020). Precise CRISPR-Cas9 mediated genome editing in super basmati rice for resistance against bacterial blight by targeting the major susceptibility gene. *Frontiers in Plant Science*, *11*, 575. DOI: 10.3389/fpls.2020.00575 PMID: 32595655

Zahoor, M. K., Ahmad, A., Zahoor, M. A., Majeed, H. N., Zulhussnain, M., & Ranian, K. (2021). CRISPR/Cas-based insect resistance in crops. *CRISPR Crops: The Future of Food Security*, 117-149.

Zhang, D., Hussain, A., Manghwar, H., Xie, K., Xie, S., Zhao, S., Larkin, R. M., Qing, P., Jin, S., & Ding, F. (2020). Genome editing with the CRISPR-Cas system: An art, ethics and global regulatory perspective. *Plant Biotechnology Journal*, *18*(8), 1651–1669. DOI: 10.1111/pbi.13383 PMID: 32271968

Zhao, Z., Shang, P., Mohanraju, P., & Geijsen, N. (2023). Prime editing: Advances and therapeutic applications. *Trends in Biotechnology*, *41*(8), 1000–1012. DOI: 10.1016/j.tibtech.2023.03.004 PMID: 37002157

Zhou, W., Yuan, Y., Zhang, Y., & Chen, D. (2021). A decade of CRISPR gene editing in China and beyond: A scientometric landscape. *The CRISPR Journal*, *4*(3), 313–320. DOI: 10.1089/crispr.2020.0148 PMID: 34152220

Chapter 2
Molecular Mechanism of CRISPR/Cas9 in Plants

Sakina Abbas
http://orcid.org/0000-0002-5149-1925
State Key Laboratory of Seed Innovation, Institute of Genetics and Developmental Biology, Chinese Academy of Sciences, Beijing, China

Syeda Anber Zahra
http://orcid.org/0009-0009-2933-9001
Shanghai Zhongqiao Vocational and Technical University, Shanghai, China

Faiza Zahra
University of Punjab, Pakistan

Haider Ali
National Agricultural Research Center, Islamabad, Pakistan

Ghulam Rubab Soomro
University of Sindh, Jamshoro, Pakistan

ABSTRACT

CRISPR/Cas9 has revolutionized crop improvement by enabling precise, efficient genome editing. This chapter explains CRISPR/Cas9 molecular mechanisms in plants, tracing the history and significance of genome editing and CRISPR's impact on disease resistance and trait enhancement. It details core components—PAM, Cas9, and guide RNA—and their roles in target recognition, DNA binding, and cleavage, followed by repair via NHEJ and HDR. Plant-specific challenges (cell walls, polyploidy, and regeneration limits) are discussed alongside key applications. Limitations such as editing efficiency, off-target effects, and ethical concerns are

DOI: 10.4018/979-8-2600-3575-7.ch002

addressed, and future directions include improved CRISPR variants, integration with omics, and contributions to sustainable agriculture, climate resilience, and global food security.

1. INTRODUCTION

CRISPR/Cas9 is a genome-editing technique derived from the bacterial adaptive immune system, consisting of clustered regularly interspaced short palindromic repeats (CRISPR) and the associated Cas9 protein. It enables rapid, cost-effective, and precise modification of genomic DNA sequences (Asmamaw & Zawdie, 2021). It has a variety of laboratory applications, including agriculture. The need to develop reliable, cost-effective methods to ensure sufficient nutritional availability for people worldwide is at an all-time high due to escalating demand for food security, driven by population growth and the growing threat of climate change (Pandian et al., 2024). So, plant biotechnology is now needed more than ever to change plants to meet the needs of global agriculture. Plant biotechnology (PBT) is a set of scientific methods and techniques used to identify and modify plant genes to produce specific products or desired traits. The methods available in PBT can be used to express and enhance beneficial traits in crops, while undesirable traits, including susceptibility to pathogens, can be eliminated (Cong Gan & P.K. Ling, 2022).

The development of "Sequence-Specific Nucleases" (SSNs) technology, such as Zinc Finger Nucleases (ZFNs), Transcriptional Activator-like Effector Nucleases (TALENs), and the CRISPR/Cas9 system, has greatly improved the precision of plant genome engineering (Sun et al., 2016; Zhang et al., 2017). Sun et al. (2016) conducted a comprehensive comparison of the three methods mentioned above. The CRISPR/Cas9 system is often chosen over other methods because it is easy to use, inexpensive, quick, and accurate. It has drawn the curiosity of the scientific community regarding genome editing. The CRISPR/Cas system has been the subject of extensive research for genome editing in many species since Ishino et al. (1987) identified the first CRISPR/Cas9 locus and Jansen et al. (2002) conducted the first thorough study.

2. HISTORICAL BACKGROUND AND DISCOVERY

The key advancement in the history of CRISPR/Cas started in 1987, when the gene was discovered. Despite not knowing their functions, there are distinct, repetitive DNA sequences in *E. coli.* Later, it was found that CRISPR functions as an adaptive system in bacteria and archaea, characterized by a unique sequence. The spacer regions used are based on viral DNA (Hille & Charpentier, 2016; Ishino et al., 2018). In 2012, Jennifer Doudna and Emmanuelle Charpentier made an important discovery that was published today. The article details how synthetic RNA can be used to target Cas9, a CRISPR-associated protein. To cut DNA sequences in vitro, to revolutionize genome editing (Garneau et al., 2010; Gostimskaya, 2022). In 2013, Feng et al. modified CRISPR/Cas9 to work in eukaryotic cells, enabling its extensive use in genetic studies, medicine, and agriculture. (Hsu et al., 2014). Since then, CRISPR/Cas9 has been used to modify the genomes of plants, animals and humans, contributing to crop improvement, disease modelling and therapeutic development. The same year, Doudna and Charpentier were awarded the Nobel Prize in Chemistry. Charpentier for having made a major contribution. The CRISPR/Cas technique is advancing, with novel variants such as Cas13 and Cas12 extending its applications to base editing, RNA editing, and more, providing ample opportunities to tackle global challenges (Tavakoli et al., 2021). As shown in Figure 1, in agriculture, it is possible to have a lower share of this total.

Figure 1. History and discovery of the components of the CRISPR-Cas system.

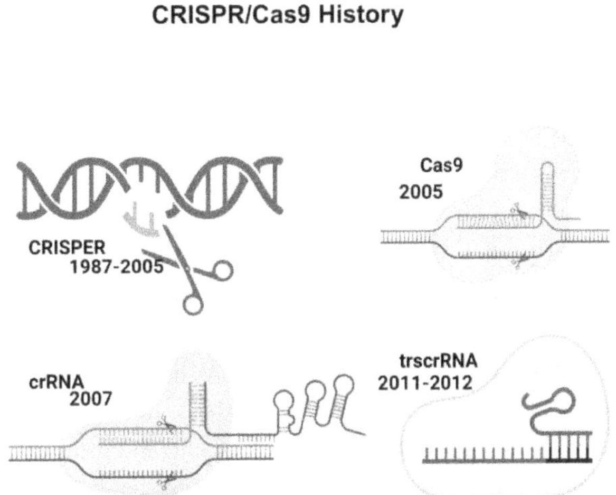

In 1987, CRISPR was first discovered in bacterial genomes and later in archaeal genomes in 1995 (Ishino, Krupovic, & Forterre, 2018). Initial information on proteins associated with the first occurrence of the CRISPR locus and the Cas9 protein was published in 2002. In 2005, the role of CRISPR Loci in prokaryotic immunity to foreign genetic elements was explored. In 2007 and 2011, it was discovered that the complex contained two RNA molecules in addition to the protein: tracrRNA and crRNA. In 2012, research that won the Nobel Prize described how the system was built in a laboratory by assembling all the components, thereby fusing two RNA strands.

3. SCOPE OF THE CHAPTER

The molecular mechanism of the CRISPR/Cas9 complex, particularly its use in plants, is covered in this chapter. We will investigate the recognition and cleavage of target DNA. The complex (CRISPR/Cas9) is designed and delivered into plant cells, and the plant's cellular machinery breaks the DNA. Plant-specific elements will also be discussed, including the challenges of delivering CRISPR/Cas9 components into plant cells, how editing efficacy relates to plant chromatin architecture, and the potential for off-target effects on genomes. To make the best use of CRISPR/Cas9 in plants, it is crucial to understand the underlying molecular mechanisms and processes. The purpose of this chapter is to cover the entire CRISPR/Cas9 process in Plants and its potential applications in solving global agricultural problems.

4. CRISPR/CAS9 SYSTEM: BASIC COMPONENTS

Based on function and structure, the CRISPR/Cas system is divided into two main classes: Class I (types I, III, and IV) and Class II (types II, V, and VI). Class I systems have multi-subunit complexes of Cas proteins. Class II systems depend on a single Cas protein. Over the past few years, the Type II CRISPR/1Cas9 system has been rigorously analyzed and is widely employed in genetic engineering due to its relatively simple structure (Arora & Narula, 2017). In the CRISPR/Cas9 system, there are two essential components. Cas9 protein and the guide RNA (gRNA) are the major components. The Cas9 protein, which is found in *Streptococcus pyogenes* (SpCas9), an important multi-domain DNA endonuclease, consists of 1368 amino acids and is known as "genetic scissors" due to its ability to cut double-stranded DNA (Asmamaw & Zawdie, 2021). CRISPR/Cas9 protects against infections caused by phages, viruses, and other foreign genetic material (Asmamaw & Zawdie, 2021). A set of CRISPR/Cas9-associated (Cas) genes encodes Cas proteins that can cut DNA. The system also includes CRISPR/Cas9 "repeat-spacer arrays" that are transcribed into "CRISPR/Cas9 RNA" (crRNA) and "trans-activating CRISPR/Cas9 RNA" (tracrRNA), as shown in Fig. 2 (Hille & Charpentier, 2016). When prokaryotes encounter foreign genetic material, Cas proteins cleave it into shorter fragments, which are subsequently incorporated into the "CRISPR array" as novel spacers. Upon subsequent invasions by the identical foreign entity, the crRNA identifies and associates with the targeted DNA fragments, directing Cas proteins to split the foreign genes and safeguard the host (Asmamaw & Zawdie, 2021). The two structural lobes that make up Cas9 are the nuclease (NUC) lobe and the globular recognition (REC) lobe. The REC lobe, comprising REC1 and REC2 domains, binds the guide RNA. In contrast, the NUC lobe comprises the RuvC and HNH domains, which cleave single-stranded DNA, as well as the domain that interacts with the Protospacer Adjacent Motif (PAM), verifies PAM specificity, and facilitates binding to the targeted DNA (Nishimasu et al., 2014). The guide RNA contains two components: crRNA and tracrRNA. The crRNA is an 18–20-nucleotide sequence that recognizes the intended DNA through base pairing, while the tracrRNA serves as a scaffold for Cas9 attachment.

Figure 2. Components of the CRISPR-Cas9 system.

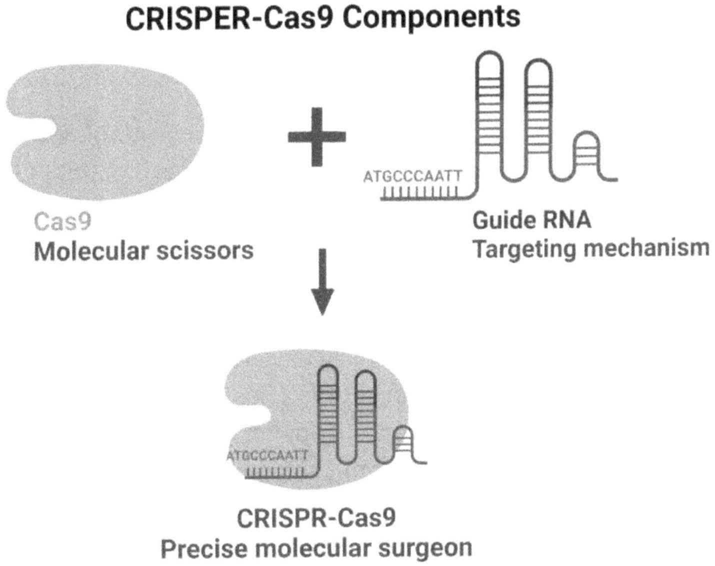

In prokaryotes, the guide RNA targets viral DNA; however, in genetic engineering, tracRNA and crRNA can be combined to form a single guide RNA (sgRNA) (Asmamaw & Zawdie, 2021). This sgRNA can be targeted at almost any gene sequence, making the CRISPR/Cas9 system a powerful tool for genome editing.

5. MECHANISMS OF CRISPR/CAS-9 GENOME EDITING

The CRISPR/Cas9 genome editing mechanism consists of three primary stages: recognition, cleavage, and repair. The intended sgRNA recognizes its target sequence in the coding region of interest and directs Cas9 through its 5ʹcrRNA supplementary base pairs. The absence of sgRNA results in Cas9 inactivation. The Cas9 nuclease induces double-strand breaks at a site located three pairs of bases upstream of the protospacer adjacent motif. The length of the shorter conserved DNA sequence, commonly referred to as the PAM sequence, varies among bacteria, generally ranging from 2 to 5 base pairs and positioned downstream of the deletion site. Cas9, a widely used DNA SeqR, a nuclease for genome editing, recognizes the PAM sequence 5'-NGG-3', where N is any nucleotide. Although the Cas9 enzyme helps open DNA at target sites, the mechanisms of action of these sequences remain poorly understood. However, when a target site was found in the presence of an appropriate PAM, it led to localized DNA melting, which in turn enabled the formation of an RNA-DNA hybrid. Later, the Cas9 protein will be activated to initiate DNA cleavage. The most common type of DSB is blunt-ended, created by the complementary cleavage of the non-complementary segment by RuvC and the segment by the HNH domain. Domain that is to be targeted in the intended target DNA sequence. Ultimately, the host cell's machinery repairs the double-strand break.

The crystal structures of various Cas9 endonucleases reveal a bilobed architecture. It has two nucleic acid-binding loops and has active sites in proximity. The enzymes show a conserved core structure. It is composed of a large NUC lobe and a small REC lobe. Each REC lobe is divided into two parts, known as REC1 and REC2, and there is a large α-helical section, a unique region of Cas9, and the arginine-rich Bridge Helix, as shown in Figure 3a. The NUC lobe contains the RuvC and HNH nuclease domains, along with the PI domain that engages with PAM. The REC lobe features two distinct grooves that bind nucleic acids, whereas the NUC lobe is characterized by one narrow groove and one broader major groove. Figure 3b illustrates how the separately functioning Cas9, a pliable protein, binds and cleaves the DNA targets in a sequence-dependent manner. Single-particle electron microscopy reveals that Cas9 remains in an auto-inhibited state in the absence of nucleic acid ligands, transitioning to an active form upon the loading of guide RNA. The lobes of the REC and NUC undergo conformational remodeling to form a central channel. This facilitates the following arrangement of the RNA-DNA heteroduplex. The REC lobe is the primary site of guide RNA interactions within the dual Cas9-sgRNA complex. Following that, the integrated structure searches for canonical PAM motifs in the non-complementary half of the DNA sequence. It has been suggested that, because Cas9 fails to demonstrate energy-dependent helicase activity, PAM recognition destabilizes the adjacent sequence, leading to the formation of an R-loop (Figure 3c).

Figure 3. Mechanisms of CRISPR/Cas9 Genome Editing.

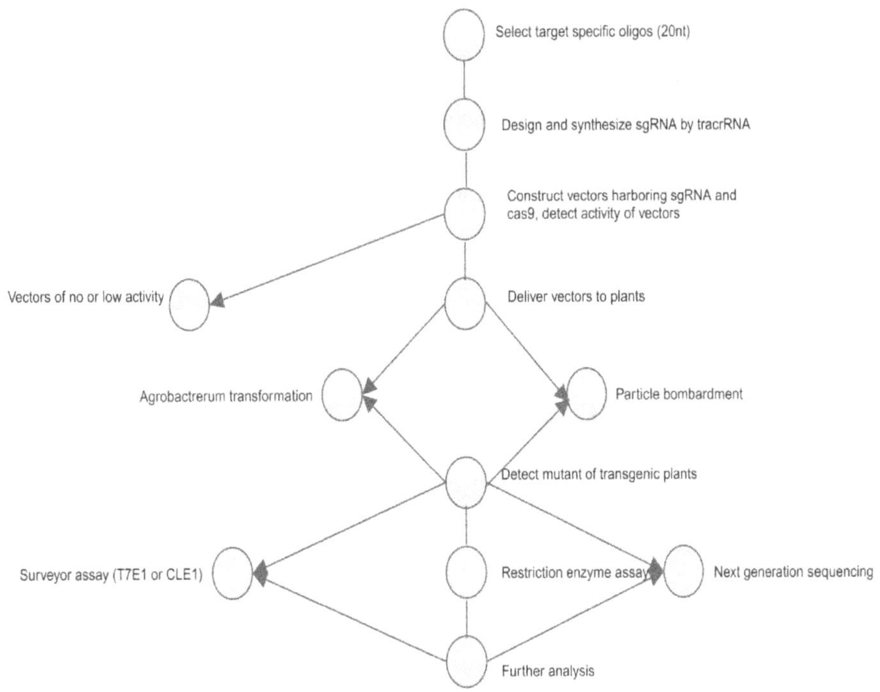

A currently uncovered process that regulates DNA recognition and unwinding is PAM-dependent. Initial crosslinking experiments indicated that the two tryptophan-containing loops within the PI domain contribute to PAM recognition. We now have an improved understanding of the molecular processes that underlie PAM recognition and DNA strand splitting by Cas9, when combined with a substantially duplexed target DNA containing the canonical PAM motif and a gRNA.

The researchers have clarified a crucial base-recognition code for PAM binding that determines the GG dinucleotides at the PAM on the non-complementary strand through the involvement of two separate arginine residues (R1333 and R1335). Additionally, the PAM duplex is connected to the phosphate cluster of the PAM sequence at +1 in the complementary strand via a K1107-S1109 loop. This loop is formed through associations with two supplementary minor-groove residues, serine and lysine (K1107 and S1109) (Figure 3d). The phosphate group helps ensure proper positioning of the complementary DNA strand for base pairing and hybridization with the guiding RNA, thereby splitting the DNA segments (Figure 3e). Following that, the Cas9-sgRNA complex assesses the neighboring DNA for any possible similarity with the gRNA. In the starting region (8–12 bp), complementary pairing of nucleotides gradually destabilizes the targeted DNA, promoting the formation of an RNA–DNA heteroduplex (Figure 3f). Hidden inside the NUC and REC lobes of Cas9, this molecule forms a four-way intersection with the arginine-rich Bridge Helix. In response, the NUC lobe's two nuclease domains prepare to cleave targets. Cloaking occurs on the complementary strand as the tertiary complex gets closer, and nicking occurs on the non-complementary strand by RuvC. The process by which Cas9 regenerates after dissociating from sgRNA remains unclear. The examination of Cas9's structure has revealed potential avenues for creating Cas9 variants with unique PAM specificities.

6. CRISPR/CAS9: TOOLS FOR CRAFTING THE GENOME

The significant aspect of effectively editing the target gene is the design of sgRNA using CRISPR/Cas9. To assist with sgRNA design and evaluate off-target effects, Zhang and colleagues developed an online tool called “CRISPR Design” (http://www.genome-engineering.org/). This tool offers two designs: Single Sequence mode designs sgRNAs between 23 and 500 nucleotides in length, while Batch mode can predict multiple sgRNAs in a single analysis (Li, Chu, et al., 2022). This program facilitates the assessment of off-target effects and mismatches during the sgRNA design phase. The CRISPR Design tool classifies sgRNAs by colour: red, blue, and green, which represent their specificity levels. Additionally, various tools are available to assist in the design of sgRNA, including Cas-OFFinder95, CRISPR-P94, SSFinder98, Cas OT97, Cas-Designer96, and E-CRISPR93 (Nishimasu et al., 2014). There are various advantages and disadvantages to each approach. Technical constraints on transformation efficiency, tissue specificity and regenerative abilities, transgene integration, and genotype dependence, despite their widespread use (Ricroch et al., 2023). So, it really does matter to consider the species to be studied, the scientific questions to be asked, and the type of editing to be achieved the desired outcomes prior to selecting a delivery system. Also requires further optimization, such as enhancing editing efficiency; novel vector design, delivery protocols, and plant regeneration systems are needed with fewer unwanted side effects. If you have a firm understanding of the pros and cons of each approach, you can better integrate the CRISPR/Cas toolbox into a variety of research and plant breeding applications.

7. APPLICATION OF CRISPR/CAS9 FOR EDITING PLANT GENOMES

CRISPR/Cas9 accelerates crop improvement and functional genomic research through precise modifications. It seeks to reveal the activity of developmental and stress-response genes. Gene knock-in approaches introduce beneficial alleles by inserting the gene at a particular locus, thereby improving yields, nutrition, and disease resistance (El-Mounadi et al., 2020). Haque et al. (2018) found that any association of a faulty Cas9 with "transcriptional activators or repressors" can impact transcription. This change allows gene expression to be precisely modulated without changing DNA. Prime and base editing can be used to make single-nucleotide changes without "double-stranded" DNA, making it more efficient. This will enable modifications or deletions. Off-target effects are reduced, and editing precision is increased (Deb et al., 2022). These changes accelerate the adoption of climate-smart, healthier crops and promote sustainable agriculture by tackling Climate change-induced global food security challenges (Yang et al., 2017). The development of the CRISPR/Cas9 system is likely to transform the genetic engineering of crops.

7.1 Gene Knockout

The inactivation of a gene in vivo or in vitro is called a gene knockout. The gene knockout mechanism for telomere shortening was first used in embryonic stem (ES) cells via homologous recombination, targeting an essential gene with a reporter gene. Disruption of the target gene's function is achieved by inserting an exon of the target gene and a selectable marker cassette (Tang & Zhang, 2023). This revolutionary technique was time-consuming and often low-productive. It was intended to enhance targeting efficiency, but in practice, it was not widely used for this purpose. Site-specific nucleases, such as zinc finger nucleases (ZFNs) and transcription activator-like effector nucleases (TALENs), significantly enhanced the efficiency and precision of targeted gene disruption. However, their Complexity, technical issues, and high cost were the factors that prevented their widespread use in protein engineering (Collonnier et al., 2017). By enabling quick, efficient, and cost-effective gene-target disruption across different cell types and species, the discovery of CRISPR/Cas9 has transformed gene knockout methods. This technique is based on the Cas9 nuclease's ability to guide a "single-guide RNA" (sgRNA) to target a 20-nucleotide sequence near a PAM. The double-stranded break is repaired by error-prone non-homologous end joining (NHEJ). Cas9-mediated DNA changes at its target site (DSB). Usually, the NHEJ process introduces insertions or deletions (indels) in the open reading frame, essentially "knocking out" the gene (Li et al., 2002). A recent study has optimized the CRISPR/Cas9 technique by enhancing guide RNA structure, reducing off-target effects, and employing high-fidelity Cas9 types. Enhanced-specificity-designed Cas9 nucleases, for instance, have greatly reduced off-target cleavage, thereby boosting confidence in the resulting knockout phenotypes (Thorneycroft et al., 2001). Furthermore, multiplex gene knockout, in which numerous sgRNAs are delivered concurrently, enables single-experiment analysis of gene families and redundant pathways. Investigating complex biological networks and disorders with many genetic components depends especially on this skill. While these advances have been made, there are some drawbacks to gene knockout techniques. Genetic redundancy and compensation are major challenges, in which the knock-out of a target gene may be masked by compensation (upregulation or functional replacement) of related genes, thereby masking the phenotype (Li et al., 2022). Furthermore, permanent gene disruption in multicellular animals can cause developmental lethality, so conditional knockout techniques are used to investigate gene function in specific tissues or developmental stages. Still, the ongoing improvement of gene knockout techniques has fundamentally changed our knowledge of gene function and made possible discoveries in developmental biology, neurology, immunology, and cancer (Bouché & Bouchez, 2001)

7.2 Gene Knock-In

Precise insertion of the desired DNA sequences helps to introduce specific genetic modifications right at predetermined genomic locations, without much fuss. Gene knock-in approaches are widely used for a variety of reasons, such as mapping protein domain functions, inserting reporter genes to monitor gene expression patterns, and crafting models that match naturally occurring, disease-associated mutations (Rozov et al., 2022). Using lengthy homology arms, traditional knock-in techniques replaced endogenous sequences with modified ones by homologous recombination. These techniques were time and resource-intensive, as their efficiency in mammalian cells had previously been poor. The arrival of CRISPR/Cas9 has greatly enhanced gene knock-in efficiency. CRISPR/Cas9 induces a DSB at a specific locus, facilitating homology-directed repair (HDR) when an exogenous donor template with the intended sequence and adjacent homology arms is present. Recent studies show that knock-in efficiency, which enables biallelic changes in one generation, has achieved highly effective knock-in of a reporter gene at the H11 locus in pigs, thereby generating useful models for investigating human disorders (Puchta & Fauser, 2013). Effective gene knock-in depends much on the relative balance between the HDR and NHEJ repair systems. Whereas NHEJ predominates in the G1 phase, HDR is more active in the S and G2 phases.

Small compounds to prevent NHEJ, synchronizing cells in the S/G2 phase, or utilizing single-stranded oligodeoxynucleotides (ssODNs) as donor templates are novel approaches to improve HDR efficiency. Furthermore, the donor's template design is essential; although at the expense of more complex vector construction, longer homology arms (usually 500–1000 bp) increase the probability of successful integration (Mao et al., 2019). Prime and base editing have recently become alternatives to DSBs for introducing minor changes. However, these techniques are limited in the size of the insertable sequence; hence, for integrating larger DNA fragments, conventional knock-in techniques are preferred. Moreover, while off-target integration of the donor sequence remains an issue, developments in donor design and delivery techniques have greatly reduced these hazards (Mittler et al., 2004). Studying gene regulation and modeling disease depends on gene knock-in. Inserting reporter constructs or epitope tags at endogenous sites provides insight into the temporal and spatial dynamics of gene expression. For instance, real-time visualization of expression in live organisms, made possible by the knock-in of fluorescent proteins under the control of endogenous promoters, has become a valuable resource in developmental biology and neurology (Rozov et al., 2019).

There are still difficulties, even if great progress has been made. Further improvement of knock-in techniques is required, given the intrinsic inefficiencies of HDR, especially in non-dividing cells, and the potential for undesired genomic rearrangements. Still, the integration of CRISPR/Cas9 into knock-in techniques marks a significant advance in precision genome engineering, hence extending the spectrum of conceivable uses in both fundamental research and therapeutic environments (Collonnier et al., 2017).

7.3 Gene Regulation

Strategies for modulating endogenous gene expression, apart from lasting genetic changes, unraveling dynamic regulatory networks, and applying therapeutics, depend on reversible control of gene expression. Strategies for gene control aim to modulate transcription levels while preserving the original DNA sequence (Urquidi Camacho et al., 2020). Although they provide temporary gene silencing, conventional techniques, such as antisense oligonucleotides or RNA interference (RNAi), can suffer from partial suppression and off-target consequences. Powerful replacements have come from current advances in CRISPR-based gene control. Use catalytically inactive Cas9 (dCas9), which maintains the ability to attach to target DNA sequences specified by sgRNAs while requiring endonuclease activity, CRISPR interference (CRISPRi) and CRISPR activation (CRISPRa) employ dCas9 connected with transcriptional repressors, such as the "Kruppel-associated box "(KRAB) domain in CRISPRi, to inhibit transcription initiation or extension at a target locus (Hunt et al., 2004; Sternberg, Redding, Jinek, Greene, & Doudna, 2014). Such complexes include VP64 and the synergistic activation mediator (SAM), which increase gene expression. These CRISPR-based control methods have several advantages over RNAi. Yang et al. (2018) suggest that they could change transcription at the genome level to achieve more complete and accurate silencing or activation of the target genes. In addition, dCas9 is permanently bound to DNA; when bound, it has a much more limited regulatory role, thereby decreasing the probability of undesirable side effects on other genes. Recent research shows that multiplexed CRISPR/Cas9 can target multiple genes simultaneously, making it easier to systematically knock out large numbers of genes. Epigenome editing with CRISPR/Cas9 has been argued by Mahmoudieh et al. (2025) to have evolved into a technology capable of modifying chromatin states and controlling transcription. This incorporation of dCas9 into complexes that include epigenetic modifiers (e.g., histone acetyltransferases, p300) or histone deacetylases (HDACs) can locally remodel chromatin to promote either active or repressed gene expression. This can lead to permanent changes in gene expression. Although the editing is transient, epigenetic restrictions have been shown to persist even after the transient complex forms. This outlines their potential use in long-term therapeutic interventions (Zhakypbek et al., 2025). Additionally, non-coding RNAs can be regulated, and their interactions can be altered. Methods of gene control include enhancers and promoters. Enhancer targeting of dCas9 may alter the expression of other genes downstream and inhibit transcription factor binding. This Genome regulatory architecture is an important consequence of capacity and is fundamental to the discovery of new targets for developing therapeutic strategies in diseases where enhancer activity is dysregulated. Despite these positive developments, there are still some challenges (Lauria & Rossi, 2011). Primary cells might be hard to transfect with large dCas9 fusion proteins in vivo. Also, off-target interactions typically do not pose as great a risk as on-target interactions. Even if it does not directly affect cellular activity, it might impact it through other mechanisms (Zhou et al., 2025). Over time, anticipated Ongoing developments in the field of more compact dCas9 variants, guide RNA design, and delivery technology are solutions to these problems (Le et al., 2025). CRISPR/Cas genome editing of the target gene is shown in Figure 4.

Figure 4. CRISPR/Cas9 editing of target genes.

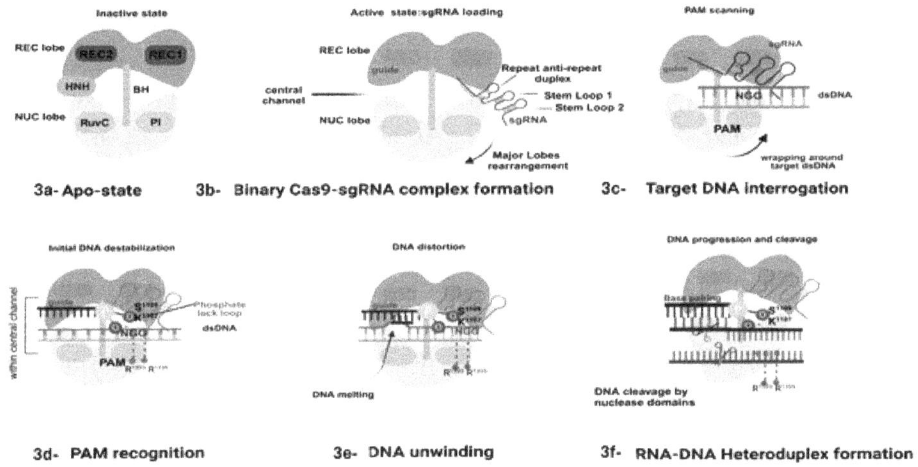

7.4 CRISPR Interference and Activation for Gene Regulation and Multiplex Editing

Recent developments in gene control using CRISPR have allowed scientists not only to alter gene expression in cells but also to regulate gene function without affecting the genomic DNA. Two key strategies, "CRISPRi" and "CRISPRa," are effective methods for precisely regulating genes. By controlling the expression levels, reversible, dynamic control of target genes can be achieved (McCarty et al., 2020). Simultaneous changes to the three components of a multiplex assembly have been achieved because of parallel developments in multiplex editing. Polygenic traits and the dissection of complex genes can be achieved by considering multiple genes that contribute to the networks, as in Abdelrahman et al. (2021).

Using catalytically inactive dCas9, CRISPRi preserves the ability to bind target DNA in a guide RNA (gRNA)-dependent manner without producing DSB. By inhibiting RNA polymerase binding or elongation and by being joined to repressor domains (such as KRAB), dCas9 essentially reduces transcription. As a 20-nucleotide sequence can accurately direct dCas9 binding, this targeted gene knockdown offers an alternative to RNA interference with higher specificity and fewer off-target effects (Singh et al., 2022). Conversely, CRISPR employs the same dCas9 platform but fuses it to transcriptional activator domains (such as p65, Rta, and VP64, in the VPR system) to regulate gene expression. By targeting enhancer or promoter sites,

CRISPR recruits the cell's transcriptional machinery, thereby increasing transcription of the endogenous gene (Pan & Qi, 2023; Li et al., 2026)

CRISPRa can be used to activate multiple genes concurrently with different gRNAs, thereby offering a useful method for clarifying gene function in pathways where overexpression is advantageous. Many species have adopted these systems, including plants and mammalian cells. For Arabidopsis, for example, dCas9-based activators have been effectively employed to control developmental pathways, thereby mimicking conventional transgenic overexpression without permanently altering the genome (Rajput et al., 2021). Similarly, CRISPRi has been employed to reduce gene expression in human cells with minimal off-target transcriptional modifications.

CRISPRi has been used to knock down gene expression in human cells with few off-target transcriptional alterations, thereby underscoring its relevance for functional genomic screens and therapeutic applications. One advantage of CRISPR-i and CRISPR-a over permanent genome editing is their reversible character. To stop the inducing signal, e.g., by inducing blocking gene expression, thus validating temporal control, this is accomplished by the addition of a promoter or a light/chemical activator of gene activity (Cui et al., 2023). Especially when developmental research is conducted, this important control can be beneficial, since continual gene alteration can have negative consequences. Multiplex is a technique in which genome editing can alter multiple genomic regions in a single cell. Since they depend on short guide RNAs, CRISPR/Cas9 systems have illustrated great flexibility for multiplexed applications. A tRNA-processing system or a self-cleaving transcript containing a ribozyme could be used to achieve this. Researchers can use a single vector to design and synthesize many gRNAs. This allows us to target Cas9 at multiple loci simultaneously, especially for polygenic traits, in which many genes contribute to the phenotype (Kasi et al., 2023).

For agricultural development, for example, it is workable to coordinate the removal of redundant gene families in hexaploid wheat. Targeting three homoalleles of a susceptibility gene helped to produce disease resistance. Similarly, multiplex editing has enabled the rapid generation of higher-order mutants in model plants such as Arabidopsis, facilitating functional genomics studies and the discovery of new pathways (Pan et al., 2021). Multiplex editing, which has recently been developed, not only speeds up editing but also reduces the time required. To reduce the risk of off-target effects, all targets should be altered in a synchronized manner. High-throughput techniques such as deep sequencing of pooled editing events, when appropriately designed, have revealed that most targeted loci can be modified simultaneously with minimal mosaicism. A multiplex editing program, in conjunction with CRISPR/Cas9, expands the range of tools available for decomposing gene networks. By simultaneously inhibiting a set of genes and switching on another group, scientists might trace complex regulatory circuits that govern cell fate, stress response, and metabolism (Hunt et al., 2004; Li et al., 2026). For instance, several transcription factors in human cells have recently been discovered that may respond to multiplex CRISPR/Cas9, thereby altering cell recognition. Regarding regenerative medicine and cell treatment, this result has great potential (Aksoy et al., 2022). While impressive achievements in multiplex editing and CRISPR/Cas9 have been made, there are still hurdles to overcome. Haque et al. (2018) stated that maintaining the constant expression of several gRNAs is challenging; this is particularly the case in plants, as transformation efficiencies can vary. In addition, dCas9 fusion proteins are not as permanent an approach as nuclease-active Cas9, and their high concentrations can make it difficult to efficiently deliver them to the appropriate target cells, especially in the clinic. The latest attempts to address these problems have focused on developing more effective, smaller activator or repressor domains and improved delivery vehicles, such as viral vectors and nanoparticles. According to a 2020 study by El-Mounadi et al., another challenge is the possibility that different gRNAs interfere with one another. Optimizing the distance and level of expression for some gRNAs will help mitigate competition for dCas9 binding. This, in turn, affects the efficiency of gene regulation. Moreover, although off-target effects in CRISPR/Cas9 are frequently disappointing compared to nuclease-based editing, comprehensive genome-wide characterization can be performed. Specificity, particularly in therapeutics, remains essential in studies. Integrating inducible systems with multiplex CRISPR/Cas9 is likely to increase the temporal resolution and controllability of gene regulation. These systems employ light, small molecules, or temperature fluctuations to regulate gene expression. Moreover, advances in computational tools for guide RNA design will continue to improve the efficiency and specificity of these systems, thus paving the way for greater acceptance in both fundamental research and therapeutic applications (Singh et al., 2022).

The difficulties and restrictions of using CRISPR/Cas9 to change a plant genome. Although CRISPR/Cas9 is transforming plant biotechnology, numerous technological, biological, and societal issues hinder its widespread use (Robertson et al., 2022). These difficulties include ethical and legal issues that affect public opinion and commercial acceptability, varying efficiency and specificity due to plant-specific conditions, and off-target effects arising from these variables. Addressing these issues will help ensure the reliability and societal acceptability of CRISPR/Cas9-mediated genetic changes (Pan et al., 2021).

8. APPLICATIONS FOR CRISPR/CAS9 IN PLANTS

The manipulation of genes allows for the adjustment and tweaking of specific phenotypes or characteristics expressed by plants, specifically crops. The expected result would be an improved crop, which could have positive effects on society. Genome editing with CRISPR/Cas9 technology is highly dependable because it does not cause random or unexpected changes in the DNA sequence (Murray et al., 2023). Since its discovery, CRISPR/Cas9 editing of plant genomes has been the subject of in-depth study of its potential applications. Much of the research on genome-editing technology has been conducted in *Arabidopsis thaliana*. This model plant was convenient and useful for genetic experiments before it was applied to crops (Cong Gan & P.K. Ling, 2022). As an example, a successive transformation technique was developed and applied in *A. thaliana*, a model plant, to enhance CRISPR-based gene targeting, and in a study of genes such as PDS3, DUO1, and AG, Tsutsui and Higashiyama first examined the performance of the pKAMA-ITACHI Red vector in CRISPR/Cas9 in *A. thaliana* (Tsutsui & Higashiyama, 2016). Expanding the technology's reach to other plant species follows its initial validation on *A. thaliana*. Table 1: The catalog of plant and crop species that have been successfully altered by means of CRISPR/Cas9 technology during the last several years is listed in Table 1.

Table 1. List some of the crops and plants that have been efficiently modified using CRISPR/Cas9 technology in the past few years.

Plant	Target gene	sgRNA promoter	Cas9 version	Delivery method	Cas9 promoter	Mutation frequency (%)	Editing method	References
Arabidopsis thaliana	BRI1, GAI, JAZ1	AtU6-26	Human codon-optimized Cas9	Agrobacterium-mediated transformation	2×35S	30—84	NHEJ	(Castel, et al., 2019)
	A non-functional GFP	AtU6-26	Chlamydomons	—	CaMV 35S	N/A	—	(Feng et al., 2013)
	TRY, ETC2, CPC	U6-26, -29	Maize codon-optimized Cas9	—	2×35S	42—90	—	(Xing et al., 2014)
	SPL4, FT	AtU6	Human codon-optimized Cas9	—	AtICU2	10.00—84.78	—	(Hyun et al., 2014)
	ADH1	AtU6-26	Streptococcus thermophilus and Staphylococcus aureus codon-optimized Cas9	—	PcUbi4-2	6.1—98.5	—	(Schiml, et al., 2014)
	CHLI1, TT4, CHLI2,	OsU3	Human codon-optimized Cas9	Agrobacterium-mediated transformation	OsUBQ1	10—89	—	(Mao et al., 2013)
	ADH1	AtU6-26	Arabidopsis codon-optimized Cas9	—	PcUbi4-2	42.8	—	(Schiml, et al., 2014)
	TRY, ETC2, CPC	U6-26, -29	Maize codon-optimized Cas9	—	2×35S	42—90	—	(Xing et al., 2014)

continued on following page

Table 1. Continued

Plant	Target gene	sgRNA promoter	Cas9 version	Delivery method	Cas9 promoter	Mutation frequency (%)	Editing method	References
	ADH1	AtU6-26	Streptococcus thermophilus and Staphylococcus aureus codon-optimized Cas9	—	PcUbi4-2	6.1—98.5	—	(Schiml, et al., 2014)
	SPL4, FT	AtU6	Human codon-optimized Cas9	—	AtICU2	10.00—84.78	—	(Hyun et al., 2014)
	PYL1, PYR1, PYL2	AtU6-26, AtU3b, At7SL-2	Human codon-optimized Cas9	—	AtUBQ1	13—93	—	(Zhang et al., 2016)
	ETC2, CPC, TRY, etc.	U6-26p, U6-29p	Zea mays codon-optimized Cas9	—	EC1.2	N/A	—	(Wang et al., 2015)
Brassica oleracea	BolC.GA4.a	U6-26	Streptococcus pyogenes Cas9	—	35S	10	—	(Lawrenson et al., 2015)
Cucumis sativus	eIF4E	AtU6	Plant codon-optimized Cas9	Agrobacterium-mediated transformation	35S	N/A	—	(Chandrase-karan et al., 2016)
Medicago truncatula	GUS	U6	Soybean codon-optimized Cas9	—	2×35S	N/A	—	(Michno et al., 2015)
Marchantia polymorpha	MpARF1	MpU6-1	Human codon-optimized Cas9	—	CaMV 35S andMpEF1α	N/A	—	(Sugano et al., 2014)
Nicotiana benthamiana	GFP	AtU6-26	Chlamydomons	Agrobacterium infiltration	CaMV 35S	N/A	—	(Jiang et al., 2013)
	NbPDS	AtU6	Human codon-optimized Cas9	—	35S	1—3	NHEJ	(Nekrasov et al., 2013)
	NbPDS	AtU6	Plant codon-optimized Cas9	—	35S PPDK	9—39	NHEJ, HDR	(Li et al., 2013)
Nicotiana tabacum	NtPDR6, NtPDS	AtU6-26	Plant codon-optimized Cas9	Agrobacterium-mediated transformation	2×35S	81.8—87.5	—	(Gao et al., 2014)
Oryza sativa	YSA, ROC5, SPP	OsU6-2	Human codon-optimized Cas9	—	CaMV 35S	4.8—75	—	(Feng et al., 2013)
	LAZY1, CAO1	OsU3	Rice codon-optimized Cas9	—	OsUbi	83—92	—	(Miao et al., 2013)
	OsMYB1	OsU3	Human codon-optimized Cas9	—	OsUBQ1	50—89	—	(Mao et al., 2013)
	OsBEL	AtU6-26	Plant codon-optimized Sp Cas9	—	2×35S	2—16	—	(Xu et al., 2014)

continued on following page

Table 1. Continued

Plant	Target gene	sgRNA promoter	Cas9 version	Delivery method	Cas9 promoter	Mutation frequency (%)	Editing method	References
	OsMPK5	OsU6	Human codon-optimized Cas9	—	CaMV 35S	3—8	—	(Xie & Yang, 2013)
	OsPDS, OsEPSPS, OsPMS3	OsU6, OsU3	Human codon-optimized Cas9	—	35S, OsUBQ1	21.1—66.7	—	(Shan et al., 2013)
	ALS	OsU6	Rice codon-optimized SpCas9	—	2×P35S	0.147—1	HDR	(Endo et al., 2015)
Solanum lycopersicum	SlAGO7	AtU6	Codon-optimized Cas9	—	35S	N/A	NHEJ	(Brooks et al., 2014)
	SlPIF4, SlPDS	AtU6-26	Human codon-optimized Cas 9	—	CaMV 35S, AtUBQ	72.7—100	—	(Pan et al., 2016)
Sorghum bicolor	DsRED2	OsU6	Monocot codon-optimized synthetic Cas9	—	Rice Actin 1	N/A	—	(Feng et al., 2013)
Solanum tuberosum	StALS1	AtU6	Arabidopsis codon-optimized Cas9	—	35S	3—60	—	(Butler et al., 2015)
Triticum aestivum	Tapds, Tainox	CaMVE35S	Human codon-optimized Cas9	—	CaMVE35S	18—22	—	(Upadhyay et al., 2013)
	TaMLOA1, TaMLOD1, TaMLOB1	TaU6	Plant codon-optimized Cas9	—	Ub1	23—38	—	(Wang et al., 2014)
	TaMLO	TaU6	Rice codon-optimized Cas9	Protoplast transformation	2×35S	26.5—38	—	(Shan et al., 2013)
	TaLOX2	TaU6	Rice codon-optimized Cas9	Particle bombardment	2×35S	45	—	(Shan, et al., 2014)
Zea mays	ZmIPK	ZmU3	Plant codon-optimized Cas9	Agrobacterium-mediated transformation	2×35S	16.4—19.1	—	(Liang, et al., 2014)
	MS45, MS26, LIG,	ZmU6	Maize codon-optimized Cas9	—	Ubi	0.13—3.9	NHEJ, HDR	(Svitashev et al., 2015)
	ZmHKT1	AtU6-26, OsU3, TaU3	Human and Maize codon-optimized Cas9	—	2×35S, Ubi1	N/A	—	(Xing et al., 2014)

9. CHALLENGES AND LIMITATIONS

9.1 Off-Target Effects

Off-target activity" is the term used by Manghwar et al. (2020) for unintended cleavage at genomic sites of low affinity to the target. CRISPR/Cas9 can cause DSBs at non-target loci due to the inherent tolerance of sgRNA, which is especially pronounced in the PAM-distal region (Sturme et al., 2022). Several factors affect the observed promiscuity: the sgRNA's secondary structure, the overall sequence, and the chromatin state at the target site (Battaglia-de Wilde, 1822). Researchers strive to employ a wide range of approaches to reduce off-target effects. This is because today's bioinformatics methods enable us to choose optimized small guide RNAs (sgRNAs) containing more GCs and less homology to non-target sites. This helps minimize the risk of accidental interactions. Currently, there is no uniformity in the field for the identification and reporting of off-target mutations in crop genome editing. This is because of the extensive use of diverse in situ resources. Different prediction tools based on in-silico and sequencing-based methods for the detection of deficiency. The application of different thresholds and control strategies is investigated. This makes the detection methods used to estimate the rate of reported off-target mutations highly variable and difficult to assess. to be used in every study (Liu et al., 2020).

9.2 Designed High-Fidelity Cas9 Variations

With certain variants showing decreases in off-target activity by over 95% (Pedrazzoli et al., 2023), eSpCas9 and SpCas9-HF1 have been developed to enhance the enzyme's fidelity by reducing nonspecific DNA contacts. Cas9 may be transformed into a nickase that cleaves only one DNA strand by altering one of the nuclease domains, e.g., H840A or D10A. Two nickases used together produce a DSB only if both sgRNAs bind in proximity, hence enhancing specificity (Xie et al., 2020).

9.3 Plant Specificity and Efficiency Affecting Performance

Many factors control editing productivity in plants directly affecting the binding and activity of the "Cas9/sgRNA complex," including the local chromatin state, which includes histone modifications and DNA methylation (Banakar et al., 2022). There is a decrease in editing efficiency in areas of reduction. Compacted chromatin or high methylation may be due to the presence of multiple homologous gene copies in polyploid species (e.g., wheat or cotton), which pose further problems. Careful design: The sgRNAs should target conserved sequences and avoid paralogous ones. To avoid variations in the results of editing (Vu et al., 2021). Whether via Agrobacterium-mediated or other methods, gene delivery can occur via transformation, particle bombardment, or ribonucleoprotein (RNP) complexes. method used will greatly influence both the efficiency and accuracy of gene editing. Because the shorter the exposure time of the Cas9, the lower the off-target effects and the higher the on-target effects. Most of the time, transient delivery methods such as RNP electroporation produce these results. (Zhu et al., 2020). A major limitation on the accuracy of sgRNA design and off-target prediction is the lack of high-quality reference genome sequences for species with only limited genomic information. To address this high genetic diversity, a genotype-specific, high-quality reference genome is needed (Zhang et al., 2023). Although multiplexing may speed up trait stacking and functional genomics studies, to avoid competition and unintended sgRNA expression, sgRNA expression levels must be carefully controlled. The goal is to maintain good on-target cleavage with minimal off-target effects; The application of high-fidelity Cas9 variants in plant systems is an emerging trend (Naim et al., 2020). High efficiency and specificity require that these features be continually altered across many plant species to enable the production of new crop varieties (de Koning et al., 2023).

9.4 Knowledge Gaps

Despite the remarkable advances in the genome-editing technique CRISPR/Cas9 in plants, several problems remain unsolved. The major challenge is that HDR is often inefficient, making the addition of precise genes or allelic swaps difficult in many plant and animal species. In addition, concerns exist about the risk of off-target mutations, plant genotype dependency of transformation, and the straightforward hassle of consistent plant editing regeneration. Bringing CRISPR components to hard-to-transform crops has yet to be a breeze, and the durability of the edited traits across generations is still further down the road. Overcoming these problems is crucial for boosting editing accuracy, speed, and the range of applications for CRISPR technologies in crop breeding.

Most of the reviews cited to date have discussed crop improvement, applications, and the success of CRISPR/Cas9. However, less attention has been paid to the ongoing technical challenges that hinder its adoption. In particular, the low efficiency of HDR, unwanted off-target effects, and transformation bottlenecks are under-researched and warrant further investigation.

10. ADVANCES IN CRISPR TECHNOLOGY

CRISPR/Cas9 has transformed the field of plant genome engineering; however, several technical and regulatory hurdles remain. Future studies are needed to make editing more efficient in recalcitrant species, improve delivery methods, reduce off-target mutations, and develop transgene-free editing methods. New genome-editing tools such as base editors, prime editors, and RNA-targeting systems are likely to expand the range of genome-editing applications in plants. Complementarily, the combination of artificial intelligence, high-throughput phenotyping, and multi-omics will enhance the discovery of target genes and the prediction of traits. This progress will enhance the use of genome editing to produce stress-resistant crops for sustainable agriculture in the context of climate change and will be combined.

10.1 Base Editing

Base editing is very different from the usual CRISPR/Cas9 system, which changes genes by creating DSBs. A base editor, which can directly and permanently replace one nucleotide with another, can help prevent DSBs or the need to use donor DNA templates (Li et al., 2021). Base editors are available in two forms: "adenine base editors" (ABEs) and "cytosine base editors" (CBEs). Both types of base editors have been widely acknowledged. The changing of cytosine (C) to CBEs changes adenine (A), guanine (G), or thymine (T) on the other strand. These are made using a different dCas9 construct that is not as effective, in combination with a cytidine deaminase enzyme. According to Bharat et al. (2020), point mutations have been successfully induced in several plant species. In this way, desirable characteristics such as disease resistance and nutritional value can be imparted. For rice and wheat, for example, CBEs have been used to modify genes involved in pathogen detection. and hormone regulation. This renders the plants more resistant to some bacteria and fungi. diseases (Anzalone et al., 2019).

10.2 Prime Editing

Prime editing is even more adaptable than CRISPR/Cas9 editing and addresses some of its drawbacks. The prime editing approach was introduced by Anzalone et al. (2019). The "pegRNA", a changed "reverse transcriptase", and a "Cas9 nickase" are used in this process (Li et al., 2026). Not only does the pegRNA direct the complex to the targeted locus, but it also serves as a guide for the addition of specific base changes, as well as deletions or insertions of tiny amounts. However, prime editors can perform all 12 possible base-to-base conversions and short indels with higher fidelity and a lower rate of byproducts (Li et al., 2023) than base editors. Initial studies on plants have shown that prime editing can generate precise modifications to the genome. Still, many crop species have editing efficiencies less than those seen in mammalian cells. Researchers are abruptly improving pegRNA designs and delivery techniques, as well as exploring ways to modify the cellular DNA repair system to improve the efficiency of plants' primary editing. Recent changes to pegRNAs, including the addition of protection motifs to lower degradation, have shown promise in improving the general efficiency and stability of primary editing systems (Molla et al., 2021; Li et al., 2026).

10.3 Epigenetic Editing in Polyploid Species

Epigenetic editing is the modification of gene activity without altering the DNA sequence, and it might be a future extension of CRISPR technology's capabilities in plants. This has been particularly useful in polyploids where there are many copies of each gene, which can contribute differently to a particular trait. Therefore, all copies may have an unexpected impact on development, while one copy may have little impact. Epigenetic editing enhances studies of complex plant genome regulation and sub-genome gene expression (Song & Chen, 2015).

Epigenetic editing involves the use of catalytically inactive Cas9 (dCas9) that can bind to DNA without making cuts. A guide RNA (gRNA) can guide dCas9 to a genomic region to either activate or repress transcription by targeting it to chromatin via the effector domain of dCas9 (Cardi et al., 2023). The effector domains can regulate DNA methylation, histone marks, transcription, and repression. The method provides precise control of gene expression without double-strand breaks or permanent sequence changes (Moradpour & Abdulah, 2019).

This approach is well-suited to polyploids such as wheat, cotton, canola, and potato. They contain numerous gene copies that are expressed redundantly or at different levels across tissues, developmental stages, or environments. Epigenetic editing can increase the activity of the positive homeolog, reduce the activity of the negative regulator, or rebalance copy activity. Dose effects are more important than gene disruption for the following traits; therefore, such control is important: flowering time, stress tolerance, biomass buildup, and yield (Jogam et al., 2022).

Another benefit may be that epigenetic editing is reversible. The effect could be mediated by the dCas9-effector complex or by the stability of the epigenetic mark, since the DNA sequence remains intact. Gene regulation is flexible and adaptable in plant biotechnology, rather than irreversible mutation. It could also be used to investigate the role of chromatin status in gene function in polyploids with complex, dynamic subgenomic relationships. Epigenetic editing relies on three key components: guide RNA design, targeting DNA, and selection of an effector domain. The direct effect on transcription initiation makes promoters and other regulatory regions the most likely targets to be considered. For highly redundant genomes, several guide RNAs can target several members of the same gene family. To further localize the edits to a specific organ or developmental time point, tissue-specific or inducible promoters can be used (Chen, 2025; Prasad & Shetty, 2026).

10.4 Integration With Other Omics Technologies

The use of CRISPR/Cas systems in conjunction with comprehensive omics platforms such as genomics, transcriptomics, and metabolomics results in an even greater enhancement of their success. This integrated plan allows researchers to connect genetic changes to subsequent molecular, biological, and physiological responses by providing a broad-based understanding of the biological systems being studied and of cellular and metabolic disruptions (Borah et al., 2024).

10.4.1 Genomics Integrating

A simple method to further enhance the success of CRISPR/Cas systems would be to use complete omics. Platforms such as genomics, transcriptomics, and metabolomics. According to Modrzejewski et al. (2020) and Nayak (2021), certain genes can be correlated with each other using this integrative technique; mutations lead to subsequent cellular, molecular, and metabolic alterations. Genomic selection and two more tools in the toolbox for finding genes are "genome-wide association studies" (GWAS), which correlate with certain characteristics. GWAS data and CRISPR-based knockout or knock-in humanized experiments will enable scientists to dissect complex traits and to test genes of interest functionally. On the molecular level, for example, in rice and wheat crops, the combination of CRISPR editing with whole-genome sequencing has simplified the elucidation of the functions of key regulatory genes that control grain yield and stress responses. (Kumar et al., 2021).

10.4.2 Transcriptomics

A transcriptomic study offers insight into how gene expression networks respond to CRISPR-driven changes. RNA sequencing (RNA-seq) and other approaches allow one to measure changes in the transcriptome after gene editing, therefore exposing compensatory pathways or unforeseen regulating effects (Wei et al., 2020). By combining single-cell RNA-seq techniques (e.g., CRISPR-seq, CROP-seq, and Perturb-seq) with CRISPR screens, researchers have mapped gene regulatory networks with hitherto unheard-of precision (Razzaq et al., 2021). Transcriptomic profiling after CRISPR editing can help to clarify stress-responsive networks or developmental pathways altered by focused gene interference in plant systems. Comparing the transcriptomes of altered and wild-type plants helps scientists identify differentially expressed genes that serve as biomarkers for desirable traits, including drought tolerance or more efficient fertilizer use (Zhou et al., 2021).

10.4.3 Metabolomics

Metabolomics, the comprehensive analysis of small molecules and metabolic intermediates within biological systems, complements genomic and transcriptomic data by linking genetic variation to observable phenotypic outcomes (Drapal et al., 2023). Significant alterations in metabolite profiles resulting from CRISPR-induced mutations will influence plant development, yield, and stress tolerance. Metabolite Profiling is enabled on a comprehensive scale for gene-edited plants using advanced techniques of nuclear magnetic resonance (NMR) and mass spectrometry. Using CRISPR to alter the activity of key enzymes in carotenoid biosynthesis or fatty acid metabolism, for example, has been shown to change metabolite levels and improve the nutritional value/taste profile (Enfissi et al., 2021). When metabolomic data are combined with genomic and transcriptomic studies, this could facilitate the validation of CRISPR modifications and guide the selection of the most suitable targeted genes for crop improvement (Sabzehzari et al., 2020; Fraser et al., 2020).

10.5 Potential for Sustainable Agriculture

10.5.1 Addressing Global Food Security

By 2050, the world population is expected to surpass 10 billion, further driving the need for enhanced agricultural production and food security. CRISPR/Cas technology has unique potential to generate varieties in agriculture with high yield, nutritional value, and resistance to biotic and abiotic stresses (Animasaun et al., 2023). Precision genome editing now makes it possible to precisely modify traits that would have been hard to alter by traditional breeding techniques. For example, CRISPR has been used to inactivate genes that can adversely affect the productivity of vital crops such as maize, wheat, and rice. Studies by Yang et al. (2017) and Munaweera et al. (2022) indicate that altering the cytokinin oxidase/dehydrogenase (CKX) genes can increase the amounts of cytokinins but can also increase the number of grains and the yield. Changes in regulatory components of stress-responsive pathways have also enabled the development of varieties that can withstand salt and drought, which is important given uncertain weather (Kumar et al., 2022).

10.5.2 Enhancing Nutritional Quality and Crop Resilience

CRISPR/Cas9 could be used to improve the nutritional components of agricultural products and boost productivity. By fine-tuning metabolic pathways using prime and base-editing approaches, one could improve the production of beneficial fatty acids, antioxidants, and vitamins. CRISPR/Cas9 technology has been used to modify fatty acid desaturase genes to increase the content of monounsaturated fatty acids. Notably, these fatty acids are associated with better heart health. Additionally, gene editing can confer resistance to diseases and pests in crops, helping reduce reliance on chemical pesticides (Kaul et al., 2022). Researchers have developed rice and tomato plants that are more resistant to fungal diseases and bacterial blight by enhancing the activity of genes responsible for plant defense or by targeting genes known to confer susceptibility to these diseases. Technology has advanced, enabling the use of fewer chemicals and better crop management under harsh conditions, thereby supporting environmental sustainability (Kaya, 2025).

10.5.3 Mitigating Climate Change

Climate change directly affects agriculture and is an important factor in farmers' lives. To ensure food security in a changing climate, it is important to develop crops that are better able to cope with climate impacts. Advanced genome-editing tools can improve water-use efficiency, produce carbon-storing plants, and even create drought-resistant plants, says KhokharVoytas et al. (2023). For example, researchers are trying to create drought-resistant crops by manipulating genes that control root structure and stomatal opening and closing. In the same way, changing the genes involved in photosynthetic efficiency and biomass accumulation could produce crops that sequester more atmospheric carbon; therefore, this could be a valuable method to combat climate change. All these actions contribute to broader global efforts to mitigate greenhouse gas emissions and strengthen food production systems (Francisco et al., 2020).

11. CONCLUSION

The discovery of CRISPR/Cas9 has had a significant impact on plant biology and biotechnology, revolutionizing genome editing with a more efficient and precise method for modifying plant DNA. Using this approach, scientists could introduce certain genetic changes through cellular repair systems. To do this, researchers take advantage of bacteria's natural immunity to detect specific gene sequences, induce double-strand breaks, and perform analogous functions. The application of this chemical to plants has led to breakthroughs in several fields, including but not limited to agricultural production, resistance to abiotic and biotic stresses, and nutritional improvement. CRISPR/Cas9 is a scientific and industrial method of genetic modification that is more widely used and more versatile than its precursor. This is because it allows for certain modifications without introducing foreign DNA from any other source. It has made plant breeding more efficient, allowing novel traits to be developed in a noticeably shorter period than was once believed possible. For instance, CRISPR has been used to create varieties of crops with improved disease resistance and to bio-fortify them (make them more nutritious) with higher levels of essential nutrients. In basic research, CRISPR/Cas9 has been an indispensable tool for exploring gene and regulatory network functions. This has allowed researchers to delve much further into the investigation than into agriculture. This phenomenon has led to a better understanding of plant development, energy, and environmental responses. The potential of CRISPR/Cas9 in plant biology is really encouraging. Now, researchers are trying to create new Cas proteins that can target more PAMs, have fewer off-target effects, and be better editors. As CRISPR technology advances, its applications will expand, enabling modification of previously inaccessible portions of the genome. Given the changing climate and rising global population, research is underway into the potential of CRISPR and other cutting-edge technologies, such as synthetic biology, high-throughput sequencing, and artificial intelligence, to modify crop properties. However, there are still obstacles to overcome, including societal and regulatory issues, as well as the need to obtain approval to use CRISPR-based technology. To address these problems, scientists, lawmakers, and other stakeholders need to come together to develop solutions and ensure the responsible use of CRISPR/Cas9 for the good of the world. To conclude, CRISPR/Cas9 is a game-changer in the fields of plant science and biotechnology. It offers a robust toolbox to tackle major challenges in sustainable agriculture and the environment. This improved technology could have a profound impact on how plants are cultivated, studied, and utilized in the service of humankind and the environment.

REFERENCES

Abdelrahman, M., Wei, Z., Rohila, J. S., & Zhao, K. (2021). Multiplex genome-editing technologies for revolutionizing plant biology and crop improvement—. *Frontiers in Plant Science*, *12*, 721203. DOI: 10.3389/fpls.2021.721203 PMID: 34691102

Aksoy, E., Yildirim, K., Kavas, M., Kayihan, C., Yerlikaya, B. A., Çalik, I., & Demirel, U. (2022). General guidelines for CRISPR/Cas-based genome editing in plants. *Molecular Biology Reports*, *49*(12), 12151–12164. DOI: 10.1007/s11033-022-07773-8 PMID: 36107373

Animasaun, D. A., Adedibu, P. A., Shkryl, Y., Emmanuel, F. O., Tekutyeva, L., & Balabanova, L. (2023). Modern plant biotechnology: An antidote against global food insecurity. *Agronomy (Basel)*, *13*(8), 2038. DOI: 10.3390/agronomy13082038

Anzalone, A. V., Randolph, P. B., Davis, J. R., Sousa, A. A., Koblan, L. W., Levy, J. M., Chen, P. J., Wilson, C., Newby, G. A., Raguram, A., & Liu, D. R. (2019). Search-and-replace genome editing without double-strand breaks or donor DNA. *Nature*, *576*(7785), 149–157. DOI: 10.1038/s41586-019-1711-4 PMID: 31634902

Arora, L., & Narula, A. (2017). Gene Editing and Crop Improvement Using the CRISPR-Cas9 System. *Frontiers in Plant Science*, *8*, 1932. DOI: 10.3389/fpls.2017.01932 PMID: 29167680

Asmamaw, M., & Zawdie, B. (2021). Mechanism and Applications of CRISPR/Cas-9-Mediated Genome Editing. *Biologics*, *15*(1), 353–361. PMID: 34456559

Banakar, R., Schubert, M., Kurgan, G., Rai, K. M., Beaudoin, S. F., Collingwood, M. A., & Zhang, F. (2022). Efficiency, specificity, and temperature sensitivity of Cas9 and Cas12a RNPs for DNA-free genome editing in plants. *Frontiers in Genome Editing*, *3*, 760820. DOI: 10.3389/fgeed.2021.760820 PMID: 35098208

Bharat, S. S., Li, S., Li, J., Yan, L., & Xia, L. (2020). Base editing in plants: Current status and challenges. *The Crop Journal*, *8*(3), 384–395. DOI: 10.1016/j.cj.2019.10.002

Borah, A., Singh, S., Chattopadhyay, R., Kaur, J., & Bari, V. K. (2024). Integration of CRISPR/Cas9 with multi-omics technologies to engineer secondary metabolite production in medicinal plants: Challenges and Prospects. *Functional & Integrative Genomics*, *24*(6), 207. DOI: 10.1007/s10142-024-01486-w PMID: 39496976

Bouché, N., & Bouchez, D. (2001). Arabidopsis gene knockout: Phenotypes wanted. *Current Opinion in Plant Biology*, *4*(2), 111–117. DOI: 10.1016/S1369-5266(00)00145-X PMID: 11228432

Brooks, C., Nekrasov, V., Lippman, Z. B., & Van Eck, J. (2014). Efficient Gene Editing in Tomato in the First Generation Using the Clustered Regularly Interspaced Short Palindromic Repeats/CRISPR-Associated9 System. *Plant Physiology*, *166*(3), 1292–1297. DOI: 10.1104/pp.114.247577 PMID: 25225186

Butler, N. M., Atkins, P. A., Voytas, D. F., & Douches, D. S. (2015). Generation and Inheritance of Targeted Mutations in Potato (*Solanum tuberosum L.*) Using the CRISPR/Cas System. *PLoS One*, *10*(12), e0144591. DOI: 10.1371/journal.pone.0144591 PMID: 26657719

Cardi, T., Murovec, J., Bakhsh, A., Justyna Boniecka, B., Bull, T., & Milica Zlatković, S. E. (2023). CRISPR/Cas-mediated plant genome editing: Outstanding challenges a decade after implementation. *Trends in Plant Science*, *28*(10), 1144–1165. DOI: 10.1016/j.tplants.2023.05.012 PMID: 37331842

Castel, B., Tomlinson, L., Locci, F., Yang, Y., & Jones, J. D. G. (2019). Optimization of T-DNA architecture for Cas9-mediated mutagenesis in Arabidopsis. *PLoS One*, *14*(1), e0204778. DOI: 10.1371/journal.pone.0204778 PMID: 30625150

Chandrasekaran, J., Brumin, M., Wolf, D., Leibman, D., Klap, C., Pearlsman, M., & Gal-On, A. (2016). Development of broad virus resistance in non-transgenic cucumber using CRISPR/Cas9 technology. *Molecular Plant Pathology*, *17*(7), 1140–1153. DOI: 10.1111/mpp.12375 PMID: 26808139

Chen, Z. J. (2025). Empowering plant epigenetics to breed resilience of crops: From nucleolar dominance to transgenerational epigenetic inheritance. *The Plant Genome*, *18*(2), e70064. DOI: 10.1002/tpg2.70064 PMID: 40556420

Collonnier, C., Guyon-Debast, A., Maclot, F., Mara, K., Charlot, F., & Nogué, F. (2017). Towards mastering CRISPR-induced gene knock-in in plants: Survey of key features and focus on the model *Physcomitrella patens. Methods : A Companion to Methods in Enzymology*, *121*, 103–117. DOI: 10.1016/j.ymeth.2017.04.024 PMID: 28478103

Cui, Y., Cao, Q., Li, Y., He, M., & Liu, X. (2023). Advances in cis-element-and natural variation-mediated transcriptional regulation and applications in gene editing of major crops. *Journal of Experimental Botany*, *74*(18), 5441–5457. DOI: 10.1093/jxb/erad248 PMID: 37402253

De Koning, R., Daryanavard, H., Garmyn, J., Kiekens, R., Toili, M. E. M., & Angenon, G. (2023). Fine-tuning CRISPR/Cas9 gene editing in common bean (*Phaseolus vulgaris L.)* using a hairy root transformation system and in silico prediction models. *Frontiers in Plant Science*, *14*, 1233418. DOI: 10.3389/fpls.2023.1233418 PMID: 37929181

Deb, S., Amrita, C. B. K., & Rama, R. S. (2022). Applications of CRISPR/Cas9 technology for modification of the plant genome. *Genetica*, *150*(1), 1–12. DOI: 10.1007/s10709-021-00146-2 PMID: 35018532

Drapal, M., Enfissi, E. M., Almeida, J., Rapacz, E., Nogueira, M., & Fraser, P. D. (2023). The potential of metabolomics in assessing global compositional changes resulting from the application of CRISPR/Cas9 technologies. *Transgenic Research*, *32*(4), 265–278. DOI: 10.1007/s11248-023-00347-9 PMID: 37166587

El-Mounadi, K., Morales-Floriano, M. L., & García-Ruiz, H. (2020). Principles, applications, and biosafety of plant genome editing using CRISPR-Cas9. *Frontiers in Plant Science*, *11*(56), 1–14. DOI: 10.3389/fpls.2020.00056 PMID: 32117392

Endo, M., Mikami, M., & Toki, S. (2015). Biallelic Gene Targeting in Rice. *Plant Physiology*, *170*(2), 667–677. DOI: 10.1104/pp.15.01663 PMID: 26668334

Enfissi, E. M., Drapal, M., Perez-Fons, L., Nogueira, M., Berry, H. M., Almeida, J., & Fraser, P. D. (2021). New plant breeding techniques and their regulatory implications: An opportunity to advance metabolomics approaches. *Journal of Plant Physiology*, *258*, 153378. DOI: 10.1016/j.jplph.2021.153378 PMID: 33631493

Feng, Z., Zhang, B., Ding, W., Liu, X., Yang, D.-L., Wei, P., & Zhu, J.-K. (2013). Efficient genome editing in plants using a CRISPR/Cas system. *Cell Research*, *23*(10), 1229–1232. DOI: 10.1038/cr.2013.114 PMID: 23958582

Francisco, R. P., & Camargo, R. (2020). Emerging advanced technologies to mitigate the impact of climate change in Africa. *Plants*, *9*(3), 381. DOI: 10.3390/plants9030381 PMID: 32204576

Fraser, P. D., Aharoni, A., Hall, R. D., Huang, S., Giovannoni, J. J., Sonnewald, U., & Fernie, A. R. (2020). Metabolomics should be used to identify and characterize gene-edited crops. *The Plant Journal : For Cell and Molecular Biology*, *102*(5), 897–902. DOI: 10.1111/tpj.14679 PMID: 31923321

Gan, Cong, W., & P.K. Ling, A. (2022). CRISPR/Cas9 in plant biotechnology: Applications and challenges. *Biotechnologia*, *103*(1), 81–93. DOI: 10.5114/bta.2022.113919 PMID: 36605382

Gao, J., Wang, G., Ma, S., Xie, X., Wu, X., Zhang, X., & Xia, Q. (2014). CRISPR/Cas9-mediated targeted mutagenesis in *Nicotiana tabacum*. *Plant Molecular Biology*, *87*(2), 99–110. PMID: 25344637

Garneau, J. E., Dupuis, M. È., Villion, M., Romero, D. A., Barrangou, R., Boyaval, P., & Moineau, S. (2010). The CRISPR/Cas bacterial immune system cleaves bacteriophage and plasmid DNA. *Nature*, *468*(7320), 67–71. DOI: 10.1038/nature09523 PMID: 21048762

Gostimskaya, I. (2022). CRISPR–Cas9: A History of Its Discovery and Ethical Considerations of Its Use in Genome Editing. *Biochemistry. Biokhimiia*, *87*(8), 777–788. DOI: 10.1134/S0006297922080090 PMID: 36171658

Haque, E., Taniguchi, H., Hassan, M. M., Bhowmik, P., Karim, M. R., Śmiech, M., & Islam, T. (2018). Application of CRISPR/Cas9 genome editing technology for the improvement of crops cultivated in tropical climates: Recent progress, prospects, and challenges. *Frontiers in Plant Science*, *9*(617), 1–15. DOI: 10.3389/fpls.2018.00617 PMID: 29868073

Hille, F., Charpentier, E. (2016). CRISPR-Cas: biology, mechanisms and relevance. Philosophical Transactions of the Royal Society. *Biological Sciences,* 371,1707. 20150496.

Hsu, P. D., Lander, E. S., & Zhang, F. (2014). Development and Applications of CRISPR-Cas9 for Genome Engineering. *Cell*, *157*(6), 1262–1278. DOI: 10.1016/j.cell.2014.05.010 PMID: 24906146

Hunt, L., Lerner, F., & Ziegler, M. (2004). NAD–new roles in signaling and gene regulation in plants. *The New Phytologist*, *163*(1), 31–44. DOI: 10.1111/j.1469-8137.2004.01087.x PMID: 33873776

Hyun, Y., Kim, J., Cho, S. W., Choi, Y., Kim, J. S., & Coupland, G. (2014). Site-directed mutagenesis in *Arabidopsis thaliana* using dividing tissue-targeted RGEN of the CRISPR/Cas system to generate heritable null alleles. *Planta*, *241*(1), 271–284. DOI: 10.1007/s00425-014-2180-5 PMID: 25269397

Ishino, Y., Krupovic, M., & Forterre, P. (2018). History of CRISPR-Cas from Encounter with a Mysterious Repeated Sequence to Genome Editing Technology. *Journal of Bacteriology*, *200*(7), e00580–e17. DOI: 10.1128/JB.00580-17 PMID: 29358495

Jansen, R., Embden, J. D. A., & van, . (2002). Identification of genes that are associated with DNA repeats in prokaryotes. *Molecular Microbiology*, *43*(6), 1565–1575. DOI: 10.1046/j.1365-2958.2002.02839.x PMID: 11952905

Jiang, W., Zhou, H., Bi, H., Fromm, M., Yang, B., & Weeks, D. P. (2013). Demonstration of CRISPR/Cas9/sgRNA-mediated targeted gene modification in Arabidopsis, tobacco, sorghum, and rice. *Nucleic Acids Research*, *41*(20), e188. DOI: 10.1093/nar/gkt780 PMID: 23999092

Jogam, P., Sandhya, D., Alok, A., Peddaboina, V., Allini, V. R., & Zhang, B. (2022). A review on CRISPR/Cas-based epigenetic regulation in plants. *International Journal of Biological Macromolecules*, *219*(31), 1261–1271. DOI: 10.1016/j.ijbiomac.2022.08.182 PMID: 36057300

Kasi Viswanath, K., Hamid, A., Ateka, E., & Pappu, H. R. (2023). CRISPR/Cas, multiomics, and RNA interference in virus disease management. *Phytopathology*, *113*(9), 1661–1676. DOI: 10.1094/PHYTO-01-23-0002-V PMID: 37486077

Kaul, T., Sony, S. K., Bharti, J., Motelb, K. F. A., Verma, R., Thangaraj, A., & Eswaran, M. (2022). CRISPR genome editing brings global food security into the first Lane: Enhancing nutrition and stress resilience in crops. In *Next-generation plant breeding approaches for stress resilience in cereal crops:285–344*. Springer Nature Singapore. DOI: 10.1007/978-981-19-1445-4_9

Kaya, C. (2025). Plant Metabolic Engineering for Enhanced Nutrition and Food Security During Climate Uncertainty. *Food and Energy Security*, *14*(1), e70060. DOI: 10.1002/fes3.70060

KhokharVoytas, A., Shahbaz, M., Maqsood, M. F., Zulfiqar, U., Naz, N., Iqbal, U. Z., Sara, M., Aqeel, M., Khalid, N., Noman, A., Zulfiqar, F., Al Syaad, K. M., & AlShaqhaa, M. A.KhokharVoytas. (2023). Genetic modification strategies for enhancing plant resilience to abiotic stresses in the context of climate change. *Functional & Integrative Genomics*, *23*(3), 283. DOI: 10.1007/s10142-023-01202-0 PMID: 37642792

Kumar, A., Anju, T., Kumar, S., Chhapekar, S. S., Sreedharan, S., Singh, S., Choi, S. R., Ramchiary, N., & Lim, Y. P. (2021). Integrating omics and gene editing tools for rapid improvement of traditional food plants for diversified and sustainable food security. *International Journal of Molecular Sciences*, *22*(15), 8093. DOI: 10.3390/ijms22158093 PMID: 34360856

Kumar, D., Yadav, A., Ahmad, R., Dwivedi, U. N., & Yadav, K. (2022). CRISPR-based genome editing for nutrient enrichment in crops: A promising approach toward global food security—. *Frontiers in Genetics*, *13*, 932859. DOI: 10.3389/fgene.2022.932859 PMID: 35910203

Lauria, M., & Rossi, V. (2011). Epigenetic control of gene regulation in plants. *Biochimica et Biophysica Acta (BBA)-. Gene Regulatory Mechanisms*, *1809*(8), 369–378.

Lawrenson, T., Shorinola, O., Stacey, N., Li, C., Østergaard, L., Patron, N., & Harwood, W. (2015). Induction of targeted, heritable mutations in barley and Brassica oleracea using RNA-guided Cas9 nuclease. *Genome Biology*, *16*(1), 258. DOI: 10.1186/s13059-015-0826-7 PMID: 26616834

Le, H., Simmons, C. H., & Zhong, X. (2025). Functions and Mechanisms of Histone Modifications in Plants. *Annual Review of Plant Biology*, *76*(1), 1–16. DOI: 10.1146/annurev-arplant-083123-070919 PMID: 39952674

Li, C., Chu, W., Gill, R. A., Sang, S., Shi, Y., Hu, X., & Zhang, B. (2022). Computational tools and resources for CRISPR/Cas genome editing. *Genomics, Proteomics & Bioinformatics*, *21*(1), 1–12. PMID: 35341983

Li, H., Chai, Z., Shi, X., Sun, C., Zhang, R., Zhang, Q., & Gao, C. (2026). Multiplexed, precise genome engineering in monocots with twin prime editing systems. *Nature Biotechnology*. Advance online publication. DOI: 10.1038/s41587-026-03174-5 PMID: 42249131

Li, J. F., Norville, J. E., Aach, J., McCormack, M., Zhang, D., Bush, J., & Sheen, J. (2013). Multiplex and homologous recombination–mediated genome editing in Arabidopsis and *Nicotiana benthamiana* using guide RNA and Cas9. *Nature Biotechnology*, *31*(8), 688–691. DOI: 10.1038/nbt.2654 PMID: 23929339

Li, X., Lassner, M., & Zhang, Y. (2002). Deleteagene: A fast neutron deletion mutagenesis-based gene knockout system for plants. *Comparative and Functional Genomics*, *3*(2), 158–160. DOI: 10.1002/cfg.148 PMID: 18628888

Li, Y., Li, W., & Li, J. (2021). The CRISPR/Cas9 revolution continues: From base editing to prime editing in plant science. *Journal of Genetics and Genomics*, *48*(8), 661–670. DOI: 10.1016/j.jgg.2021.05.001 PMID: 34362681

Li, Y., Liang, J., Deng, B., Jiang, Y., Zhu, J., Chen, L., Li, J., & Li, J. (2023). Applications and prospects of CRISPR/Cas9-mediated base editing in plant breeding. *Current Issues in Molecular Biology*, *45*(2), 918–935. DOI: 10.3390/cimb45020059 PMID: 36826004

Li, Y., Wu, X., Zhang, Y., & Zhang, Q. (2022). CRISPR/Cas genome editing improves abiotic and biotic stress tolerance of crops. *Frontiers in Genome Editing*, *4*, 1–17. DOI: 10.3389/fgeed.2022.987817 PMID: 36188128

Liang, Z., Zhang, K., Chen, K., & Gao, C. (2014). Targeted Mutagenesis in *Zea mays* Using TALENs and the CRISPR/Cas System. *Journal of Genetics and Genomics*, *41*(2), 63–68. DOI: 10.1016/j.jgg.2013.12.001 PMID: 24576457

Liu, M. S., Gong, S., Yu, H. H., Jung, K., Johnson, K. A., & Taylor, D. W. (2020). Engineered CRISPR/Cas9 enzymes improve discrimination by slowing DNA cleavage, thereby allowing the release of off-target DNA. *Nature Communications*, *11*(1), 3576. DOI: 10.1038/s41467-020-17411-1 PMID: 32681021

Mahmoudieh, M., Jariani, P., Dehghan, A. K., Karimi, M. R., Vadipour, F., Jahani, M., & Naghavi, M. R. (2025). The future of rubber production: A review of genetic regulation of biosynthetic pathways and genome editing technologies in natural rubber-producing plants. *Tree Genetics & Genomes*, *21*(2), 6. DOI: 10.1007/s11295-025-01690-0

Manghwar, H., Li, B., Ding, X., Hussain, A., Lindsey, K., Zhang, X., & Jin, S. (2020). CRISPR/Cas systems in genome editing: Methodologies and tools for sgRNA design, off-target evaluation, and strategies to mitigate off-target effects. *Advanced Science (Weinheim, Baden-Württemberg, Germany)*, *7*(6), 1902312. DOI: 10.1002/advs.201902312 PMID: 32195078

Mao, Y., Botella, J. R., Liu, Y., & Zhu, J. K. (2019). Gene editing in plants: Progress and challenges. *National Science Review*, *6*(3), 421–437. DOI: 10.1093/nsr/nwz005 PMID: 34691892

Mao, Y., Zhang, H., Xu, N., Zhang, B., Gou, F., & Zhu, J. K. (2013). Application of the CRISPR–Cas System for Efficient Genome Engineering in Plants. *Molecular Plant*, *6*(6), 2008–2011. DOI: 10.1093/mp/sst121 PMID: 23963532

McCarty, N. S., Graham, A. E., Studená, L., & Ledesma-Amaro, R. (2020). Multiplexed CRISPR technologies for gene editing and transcriptional regulation. *Nature Communications*, *11*(1), 1281. DOI: 10.1038/s41467-020-15053-x PMID: 32152313

Miao, J., Guo, D., Zhang, J., Huang, Q., Qin, G., Zhang, X., & Qu, L. J. (2013). Targeted mutagenesis in rice using the CRISPR-Cas system. *Cell Research*, *23*(10), 1233–1236. DOI: 10.1038/cr.2013.123 PMID: 23999856

Michno, J. M., Wang, X., Liu, J., Curtin, S. J., Kono, T. J., & Stupar, R. M. (2015). CRISPR/Cas mutagenesis of soybean and *Medicago truncatula* using a new web tool and a modified Cas9 enzyme. *GM Crops & Food*, *6*(4), 243–252. DOI: 10.1080/21645698.2015.1106063 PMID: 26479970

Mittler, R., Vanderauwera, S., Gollery, M., & Van Breusegem, F. (2004). Reactive oxygen gene network of plants. *Trends in Plant Science*, *9*(10), 490–498. DOI: 10.1016/j.tplants.2004.08.009 PMID: 15465684

Modrzejewski, D., Hartung, F., Lehnert, H., Sprink, T., Kohl, C., Keilwagen, J., & Wilhelm, R. (2020). Which factors affect the occurrence of off-target effects caused by the use of CRISPR/Cas: A systematic review in plants. *Frontiers in Plant Science*, *11*, 574959. DOI: 10.3389/fpls.2020.574959 PMID: 33329634

Molla, K. A., Sretenovic, S., Bansal, K. C., & Qi, Y. (2021). Precise plant genome editing using base editors and prime editors. *Nature Plants*, *7*(9), 1166–1187. DOI: 10.1038/s41477-021-00991-1 PMID: 34518669

Moradpour, M., & Abdulah, S. N. (2019). CRISPR / dC as9 platforms in plants: Strategies and applications beyond genome editing. *Plant Biotechnology Journal*, *18*(1), 1–12. PMID: 31392820

Munaweera, T. I. K., Jayawardana, N. U., Rajaratnam, R., & Dissanayake, N. (2022). Modern plant biotechnology as a strategy in addressing climate change and attaining food security. *Agriculture & Food Security*, *11*(1), 1–28. DOI: 10.1186/s40066-022-00369-2

Murray, J., Samson, C., du Rand, A., & Sheppard, H. (2023). Unintended CRISPR-Cas9 editing outcomes: A review of the detection and prevalence of structural variants generated by gene editing in human cells. *Human Genetics*, *142*(6), 705–720. DOI: 10.1007/s00439-023-02561-1 PMID: 37093294

Naim, F., Shand, K., Hayashi, S., O'Brien, M., McGree, J., Johnson, A. A., & Waterhouse, P. M. (2020). Are the current gRNA ranking prediction algorithms useful for genome editing in plants? *PLoS One*, *15*(1), e0227994. DOI: 10.1371/journal.pone.0227994 PMID: 31978124

Nayak, S. N., Aravind, B., Malavalli, S. S., Sukanth, B. S., Poornima, R., Bharati, P., & Puppala, N. (2021). Omics technologies to enhance plant-based functional foods: An overview. *Frontiers in Genetics*, *12*, 742095. DOI: 10.3389/fgene.2021.742095 PMID: 34858472

Nekrasov, V., Staskawicz, B., Weigel, D., Jones, J. D. G., & Kamoun, S. (2013). Targeted mutagenesis in the model plant *Nicotiana benthamiana* using Cas9 RNA-guided endonuclease. *Nature Biotechnology*, *31*(8), 691–693. DOI: 10.1038/nbt.2655 PMID: 23929340

Nishimasu, H., Ran, F., Hsu, P. D., Konermann, S., Shehata, S. I., Dohmae, N., Ishitani, R., Zhang, F., & Nureki, O. (2014). Crystal Structure of Cas9 in Complex with Guide RNA and Target DNA. *Cell*, *156*(5), 935–949. DOI: 10.1016/j.cell.2014.02.001 PMID: 24529477

Pan, C., & Qi, Y. (2023). CRISPR-Combo–mediated orthogonal genome editing and transcriptional activation for plant breeding. *Nature Protocols*, *18*(6), 1760–1794. DOI: 10.1038/s41596-023-00823-w PMID: 37085666

Pan, C., Sretenovic, S., & Qi, Y. (2021). CRISPR/dCas-mediated transcriptional and epigenetic regulation in plants. *Current Opinion in Plant Biology*, *60*, 101980. DOI: 10.1016/j.pbi.2020.101980 PMID: 33401227

Pan, C., Ye, L., Qin, L., Liu, X., He, Y., Wang, J., & Lu, G. (2016). CRISPR/Cas9-mediated efficient and heritable targeted mutagenesis in tomato plants in the first and later generations. *Scientific Reports*, *6*(1), 24765. DOI: 10.1038/srep24765 PMID: 27097775

Pandian, K., Mustaffa, M. R. A. F., Mahalingam, G., Paramasivam, A., John Prince, A., Gajendiren, M., & Varanasi, S. T. (2024). Synergistic conservation approaches for nurturing soil, food security, and human health towards sustainable development goals. *Journal of Hazardous Materials Advances*, *16*, 100479. DOI: 10.1016/j.hazadv.2024.100479

Pedrazzoli, E., Bianchi, A., Umbach, A., Amistadi, S., Brusson, M., Frati, G., & Casini, A. (2023). An optimized SpCas9 high-fidelity variant for direct protein delivery. *Molecular Therapy : The Journal of the American Society of Gene Therapy*, *31*(7), 2257–2265. DOI: 10.1016/j.ymthe.2023.03.007 PMID: 36905119

Prasad, M., & Shetty, R. (2026). Advancing crop resilience to environmental stresses: Insights from epigenetic modification and CRISPR/dCas9-based editing. *Biochimica et Biophysica Acta (BBA) -. Gene Regulatory Mechanisms*, *1869*(1), 195139. PMID: 41651287

Puchta, H., & Fauser, F. (2013). Gene targeting in plants: 25 years later. *International Journal of Developmental Biology, 57*(6-7-8): 629–637.

Rajput, M., Choudhary, K., Kumar, M., Vivekanand, V., Chawade, A., Ortiz, R., & Pareek, N. (2021). RNA interference and CRISPR/Cas gene editing for crop improvement: Paradigm shift towards sustainable agriculture. *Plants*, *10*(9), 1914. DOI: 10.3390/plants10091914 PMID: 34579446

Razzaq, M. K., Aleem, M., Mansoor, S., Khan, M. A., Rauf, S., Iqbal, S., & Siddique, K. H. (2021). Omics and CRISPR-Cas9 approaches for molecular insight, functional gene analysis, and stress tolerance development in crops. *International Journal of Molecular Sciences*, *22*(3), 1292. DOI: 10.3390/ijms22031292 PMID: 33525517

Ricroch, A. E., Eriksson, D., Dragana Miladinović, S., J., K. van Laere, Wozniak-Gientka, E. (2023). A roadmap for plant genome editing. Europe and beyond. Springer Nature. DOI: 10.1007/978-3-031-46150-7

Robertson, G., Burger, J., & Campa, M. (2022). CRISPR/Cas-based tools for the targeted control of plant viruses. *Molecular Plant Pathology*, *23*(11), 1701–1718. DOI: 10.1111/mpp.13252 PMID: 35920132

Rozov, S. M., Permyakova, N. V., & Deineko, E. V. (2019). The problem of the low rates of CRISPR/Cas9-mediated knock-ins in plants: Approaches and solutions. *International Journal of Molecular Sciences*, *20*(13), 3371. DOI: 10.3390/ijms20133371 PMID: 31323994

Rozov, S. M., Permyakova, N. V., Sidorchuk, Y. V., & Deineko, E. V. (2022). Optimization of genome knock-in methods: Identifying the most efficient genomic regions for transgene expression in plants. *International Journal of Molecular Sciences*, *23*(8), 4416. DOI: 10.3390/ijms23084416 PMID: 35457234

Sabzehzari, M., Zeinali, M., & Naghavi, M. R. (2020). CRISPR-based metabolic editing: Next-generation metabolic engineering in plants. *Gene*, *759*, 144993. DOI: 10.1016/j.gene.2020.144993 PMID: 32717311

Schiml, S., Fauser, F., & Puchta, H. (2014). The CRISPR/Cas system can be used as a nuclease for plant gene targeting and as paired nickases for directed mutagenesis in Arabidopsis, resulting in inheritable progeny. *The Plant Journal : For Cell and Molecular Biology*, *80*(6), 1139–1150. DOI: 10.1111/tpj.12704 PMID: 25327456

Shahid, M. S., Sattar, M. N., Iqbal, Z., Raza, A., & Al-Sadi, A. M. (2021). Next-generation sequencing and the CRISPR-Cas nexus: A molecular plant virology perspective. *Frontiers in Microbiology*, *11*, 609376. DOI: 10.3389/fmicb.2020.609376 PMID: 33584572

Shan, Q., Wang, Y., Li, J., & Gao, C. (2014). Genome editing in rice and wheat using the CRISPR/Cas system. *Nature Protocols*, *9*(10), 2395–2410. DOI: 10.1038/nprot.2014.157 PMID: 25232936

Shan, Q., Wang, Y., Li, J., Zhang, Y., Chen, K., Liang, Z., & Gao, C. (2013). Targeted genome modification of crop plants using a CRISPR-Cas system. *Nature Biotechnology*, *31*(8), 686–688. DOI: 10.1038/nbt.2650 PMID: 23929338

Singh, J., Sharma, D., Brar, G. S., Sandhu, K. S., Wani, S. H., Kashyap, R., & Singh, S. (2022). CRISPR/Cas tool designs for multiplex genome editing and their applications in developing biotic and abiotic-stress-resistant crop plants. *Molecular Biology Reports*, *49*(12), 11443–11467. DOI: 10.1007/s11033-022-07741-2 PMID: 36002653

Song, Q., & Chen, Z. J. (2015). Epigenetic and developmental regulation in plant polyploids. *Current Opinion in Plant Biology*, *24*(10), 101–109. DOI: 10.1016/j.pbi.2015.02.007 PMID: 25765928

Sternberg, S. H., Redding, S., Jinek, M., Greene, E. C., & Doudna, J. A. (2014). DNA interrogation by the CRISPR RNA-guided endonuclease Cas9. *Nature*, *507*(7490), 62–67. DOI: 10.1038/nature13011 PMID: 24476820

Sturme, M. H., van der Berg, J. P., Bouwman, L. M., De Schrijver, A., de Maagd, R. A., Kleter, G. A., & Battaglia-de Wilde, E. (2022). Occurrence and nature of off-target modifications by CRISPR-Cas genome editing in plants. *ACS Agricultural Science & Technology*, *2*(2), 192–201. DOI: 10.1021/acsagscitech.1c00270 PMID: 35548699

Sugano, S. S., Shirakawa, M., Takagi, J., Matsuda, Y., Shimada, T., Hara-Nishimura, I., & Kohchi, T. (2014). CRISPR/Cas9-Mediated Targeted Mutagenesis in the Liverwort *Marchantia polymorpha L. Plant & Cell Physiology*, *55*(3), 475–481. DOI: 10.1093/pcp/pcu014 PMID: 24443494

Sun, Y., Li, J., & Xia, L. (2016). Precise Genome Modification via Sequence-Specific Nucleases-Mediated Gene Targeting for Crop Improvement. *Frontiers in Plant Science*, *7*, 1–15. DOI: 10.3389/fpls.2016.01928 PMID: 28066481

Svitashev, S., Young, J. K., Schwartz, C., Gao, H., Falco, S. C., & Cigan, A. M. (2015). Targeted Mutagenesis, Precise Gene Editing, and Site-Specific Gene Insertion in Maize Using Cas9 and Guide RNA. *Plant Physiology*, *169*(2), 931–945. DOI: 10.1104/pp.15.00793 PMID: 26269544

Tang, X., & Zhang, Y. (2023). Beyond knockouts: Fine-tuning regulation of gene expression in plants with CRISPR-Cas-based promoter editing. *The New Phytologist*, *239*(3), 868–874. DOI: 10.1111/nph.19020 PMID: 37282668

Tavakoli, K., Pour-Aboughadareh, A., Kianersi, F., Poczai, P., Etminan, A., & Shooshtari, L. (2021). Applications of CRISPR-Cas9 as an Advanced Genome Editing System in Life Sciences. *Biotech*, *10*(3), 14–27. DOI: 10.3390/biotech10030014 PMID: 35822768

Thorneycroft, D., Sherson, S. M., & Smith, S. M. (2001). Using gene knockouts to investigate plant metabolism. *Journal of Experimental Botany*, *52*(361), 1593–1601. DOI: 10.1093/jexbot/52.361.1593 PMID: 11479324

Tsutsui, H., & Higashiyama, T. (2016). pKAMA-ITACHI Vectors for Highly Efficient CRISPR/Cas9-Mediated Gene Knockout in *Arabidopsis thaliana. Plant & Cell Physiology*, *58*(1), pcw191. DOI: 10.1093/pcp/pcw191 PMID: 27856772

Upadhyay, S. K., Kumar, J., Alok, A., Tuli, R. (2013). RNA-Guided Genome Editing for Target Gene Mutations in Wheat. *G3: Genes|Genomes|Genetics*, 3(12): 2233–2238.

Urquidi Camacho, R. A., Lokdarshi, A., & von Arnim, A. G. (2020). Translational gene regulation in plants: A green new deal. *Wiley Interdisciplinary Reviews. RNA*, *11*(6), e1597. DOI: 10.1002/wrna.1597 PMID: 32367681

Vu, T. V., Doan, D. T. H., Tran, M. T., Sung, Y. W., Song, Y. J., & Kim, J. Y. (2021). Improvement of the LbCas12a-crRNA system for efficient gene targeting in tomato. *Frontiers in Plant Science*, *12*, 722552. DOI: 10.3389/fpls.2021.722552 PMID: 34447405

Wang, Y., Cheng, X., Shan, Q., Zhang, Y., Liu, J., Gao, C., & Qiu, J. L. (2014). Simultaneous editing of three homoeoalleles in hexaploid bread wheat confers heritable resistance to powdery mildew. *Nature Biotechnology*, *32*(9), 947–951. DOI: 10.1038/nbt.2969 PMID: 25038773

Wang, Y., Zafar, N., Ali, Q., Manghwar, H., Wang, G., Yu, L., & Jin, S. (2022). CRISPR/Cas genome editing technologies for plant improvement against biotic and abiotic stresses: Advances, limitations, and future perspectives. *Cells*, *11*(23), 3928. DOI: 10.3390/cells11233928 PMID: 36497186

Wang, Z. P., Xing, H. L., Dong, L., Zhang, H. Y., Han, C. Y., Wang, X. C., & Chen, Q. J. (2015). Egg cell-specific promoter-controlled CRISPR/Cas9 efficiently generates homozygous mutants for multiple target genes in Arabidopsis in a single generation. *Genome Biology*, *16*(1), 144. DOI: 10.1186/s13059-015-0715-0 PMID: 26193878

Wei, Z., Arazi, T., Hod, N., Zohar, M., Isaacson, T., Doron-Faigenboim, A., & Yedidia, I. (2020). Transcriptome profiling of *Ornithogalum dubium* leaves and flowers to identify key carotenoid genes. *Plants*, *9*(4), 540. DOI: 10.3390/plants9040540 PMID: 32326260

Xie, H., Ge, X., Yang, F., Wang, B., Li, S., Duan, J., & Gu, F. (2020). High-fidelity SaCas9 identified by directional screening in human cells. *PLoS Biology*, *18*(7), e3000747. DOI: 10.1371/journal.pbio.3000747 PMID: 32644995

Xie, K., Minkenberg, B., & Yang, Y. (2015). Boosting CRISPR/Cas9 multiplex editing capability with the endogenous tRNA-processing system. *Proceedings of the National Academy of Sciences of the United States of America*, *112*(11), 3570–3575. DOI: 10.1073/pnas.1420294112 PMID: 25733849

Xie, K., & Yang, Y. (2013). RNA-Guided Genome Editing in Plants Using a CRISPR–Cas System. *Molecular Plant*, *6*(6), 1975–1983. DOI: 10.1093/mp/sst119 PMID: 23956122

Xing, H. L., Dong, L., Wang, Z. P., Zhang, H. Y., Han, C. Y., Liu, B., & Chen, Q. J. (2014). A CRISPR/Cas9 toolkit for multiplex genome editing in plants. *BMC Plant Biology*, *14*(1), 327. DOI: 10.1186/s12870-014-0327-y PMID: 25432517

Xu, R., Li, H., Qin, R., Wang, L., Li, L., Wei, P., & Yang, J. (2014). Gene targeting using the Agrobacterium tumefaciens-mediated CRISPR-Cas system in rice. *Rice (New York, N.Y.)*, *7*(1), 5. DOI: 10.1186/s12284-014-0005-6 PMID: 24920971

Yang, X., Yang, M., Deng, H., & Ding, Y. (2018). A new era in studying RNA secondary structure and its influence on gene regulation in plants. *Frontiers in Plant Science*, *9*, 671. DOI: 10.3389/fpls.2018.00671 PMID: 29872445

Yang, Y. (2017). CRISPR goes global: A snapshot of rules, policies, and attitudes. *The Bulletin.*

Yang, Y., Saand, M. A., Huang, L., Abdelaal, W. B., Zhang, J., Wu, Y., & Wang, F. (2021). Applications of multi-omics technologies for crop improvement. *Frontiers in Plant Science*, *12*, 563953. DOI: 10.3389/fpls.2021.563953 PMID: 34539683

Zhakypbek, Y., Belkozhayev, A. M., Kerimkulova, A., Kossalbayev, B. D., Murat, T., Tursbekov, S., & Allakhverdiev, S. I. (2025). MicroRNAs in Plant Genetic Regulation of Drought Tolerance and Their Function in Enhancing Stress Adaptation. *Plants*, *14*(3), 410. DOI: 10.3390/plants14030410 PMID: 39942972

Zhang, L., Li, G., Zhang, Y., Cheng, Y., Roberts, N., Glenn, S. E., & Qi, Y. (2023). Boosting genome editing efficiency in human cells and plants with novel LbCas12a variants. *Genome Biology*, *24*(1), 102. DOI: 10.1186/s13059-023-02929-6 PMID: 37122009

Zhang, Y., Iaffaldano, B., & Qi, Y. (2021). CRISPR ribonucleoprotein-mediated genetic engineering in plants. *Plant Communications*, *2*(2), 100168. DOI: 10.1016/j.xplc.2021.100168 PMID: 33898980

Zhang, Y., Ma, X., Xie, X., & Liu, Y. G. (2017). CRISPR/Cas9-Based Genome Editing in Plants. *Progress in Molecular Biology and Translational Science*, *149*, 133–150. DOI: 10.1016/bs.pmbts.2017.03.008 PMID: 28712494

Zhang, Z., Mao, Y., Ha, S., Liu, W., Botella, J. R., & Zhu, J. K. (2016). A multiplex CRISPR/Cas9 platform for fast and efficient editing of multiple genes in Arabidopsis. *Plant Cell Reports, 35*(7), 1519–1533. DOI: 10.1007/s00299-015-1900-z PMID: 26661595

Zhou, L., Iqbal, A., Yang, M., & Yang, Y. (2025). Research Progress on Gene Regulation of Plant Floral Organogenesis. *Genes, 16*(1), 79. DOI: 10.3390/genes16010079 PMID: 39858626

Zhou, Q., Galindo-González, L., Hwang, S. F., & Strelkov, S. E. (2021). Application of genomics and transcriptomics to accelerate development of clubroot-resistant canola. *Canadian Journal of Plant Pathology, 43*(2), 189–208. DOI: 10.1080/07060661.2020.1794541

Zhu, H., Li, C., & Gao, C. (2020). Applications of CRISPR–Cas in agriculture and plant biotechnology. *Nature Reviews. Molecular Cell Biology, 21*(11), 661–677. DOI: 10.1038/s41580-020-00288-9 PMID: 32973356

Chapter 3

Precision Genome Engineering Technologies: A Comparative Review

T. Naveen Kumar
http://orcid.org/0009-0008-4232-494X
University of Bonn, Germany

S. Eswari
http://orcid.org/0009-0009-1187-9253
Vels Institute of Science, Technology, and Advanced Studies, India

A. Sai Ramesh
http://orcid.org/0000-0002-5592-875X
Vels Institute of Science, Technology, and Advanced Studies, India

ABSTRACT

Gene-editing tools including zinc finger nucleases (ZFNs), transcription activator-like effector nucleases (TALENs), and CRISPR/Cas systems have transformed plant biotechnology. Initially developed to improve crop adaptability, stress tolerance, and productivity, ZFNs enabled the first precise genome modifications in plants, while TALENs offered improved targeting efficiency and specificity. CRISPR/Cas, derived from bacterial adaptive immunity, has become the leading genome-editing platform due to its simplicity and effectiveness. Advances such as Cas9, Cas12, Cas13, base editing, and prime editing have expanded editing precision. Comparative studies highlight the strengths and limitations of each platform. Applications in rice (Oryza), wheat (Triticum), and soybean (Glycine) have enhanced crop improvement

DOI: 10.4018/979-8-2600-3575-7.ch003

and disease resistance. Future innovations in repair pathway engineering, delivery systems, and next-generation editing technologies are expected to accelerate sustainable agriculture and reshape global farming practices.

1. INTRODUCTION

Modifying plant genomes at the nucleotide level constitutes a basic change of paradigm, not an extension of traditional breeding methods. Modern molecular biology allows us to manipulate genes with unprecedented ease and specificity. We can now make changes that could never have been made using conventional genetic manipulation techniques. Thus, this review is a compilation of more than a decade of uninterrupted progress in plant genomics due to the enormous growth of research in the field. Interestingly, the number of publications related to CRISPR applications in plants has increased two-fold from 2019 to 2023 (Mao et al., 2024). Against this background of considerable literature, this book chapter aims to trace critical methodological advances and identify persistent research gaps within the discipline.

The subsequent sections explore four interconnected themes in plant genome editing: the transition toward enhanced molecular precision, the emergence of RNA-guided systems, ongoing practical limitations, and recent post-2023 innovations beginning to reshape the field. Attention is also directed toward overlooked areas in the current literature and the broader biotechnological implications of these omissions. A foundational debate persists regarding whether CRISPR constitutes an entirely novel frontier in plant breeding or merely a refinement of preexisting techniques, a point of ongoing contention among plant biologists (Puchta, 2005). Resolving this discourse is critical, as it directly influences regulatory frameworks, research funding, and public perception. Therefore, this review carefully evaluates both perspectives. Drawing upon foundational research from the early 2000s through contemporary reports, this synthesis integrates insights from molecular biology, crop science, regulatory affairs, and delivery technologies without functioning as a systematic species catalog or a meta-analysis of editing efficiencies (Anzalone et al., 2020; Chen et al., 2019). This manuscript aims to connect established insights to clarify the contribution of the experimental chapters that follow.

The structural organization follows a chronological sequence in its initial sections before transitioning to a thematic framework. This shift intentionally reflects the historical progression of the field; the trajectory of plant genome editing is defined not only by the development of superior molecular tools but equally by an evolving understanding of precision and biological constraints. Establishing this historical context is necessary prior to examining specific methodological details. Furthermore, given the rapid pace of advancement in this discipline, ongoing discoveries may continuously redefine current paradigms. Thus, this review should be interpreted as a substantive contemporary synthesis of a highly dynamic scientific landscape.

2. HISTORICAL FOUNDATIONS OF TARGETED SITE GENE EDITING

At its core, plant genetic improvement involves the deliberate manipulation of inherited traits. Historically, before the mechanics of heredity were fully understood, early breeding efforts relied on prolonged selection cycles and limited natural genetic variation. The introduction of mutagens, such as gamma radiation, X-rays, and ethyl methanesulfonate (EMS), significantly expanded the mutation spectrum, culminating in the release of hundreds of novel crop varieties by the 1990s. As noted by Yaşar et al. (2026), while these historical mutagenic approaches expanded genetic diversity, they were fundamentally limited by their reliance on unpredictable, genome-wide alterations and protracted backcrossing phases. However, these methods induced random, genome-wide mutations, requiring extensive screening procedures to link specific traits to their genetic loci. Between the 1970s and 1990s, the advent of recombinant DNA technologies and *Agrobacterium tumefaciens*-mediated transformation provided new mechanisms for specific genomic modification (Gelvin, 2003). These advancements led to the successful commercialization of transformative transgenic crops, including Bt-expressing varieties, herbicide-tolerant soybeans, and virus-resistant papayas. Moreover, this era also established the foundational delivery mechanics for modern biotechnology; as noted by Li Y. et al. (2025), *Agrobacterium* has since evolved from its historical role of depositing permanent transgenic gene cassettes into acting as a optimized, transient delivery mode for current genome editing machinery.

The progression toward site-specific genomic editing was initially catalyzed by research into meganucleases. As detailed in recent historical retrospectives of site-directed nucleases (Brock et al, 2025), this era marked a transition from random mutagenesis to protein-guided cleavage, and established a foundational strategy of hijacking the cell's endogenous double-strand break repair pathways. These enzymes utilize large recognition sites of approximately 20 base pairs, offering unmatched site-specificity compared to standard restriction endonucleases (Chevalier & Stoddard, 2001). While engineered meganucleases theoretically allowed for directed cleavage at chosen genomic loci, practical applications were severely hindered by the structural constraints of the protein scaffold, which made the alteration of target specificity without the loss of cleavage efficiency highly challenging. A 2025 review on precision genetics by Joo, J. H. et al. (2025) emphasizes the evolutionary bottleneck, noting that the rigid reprogramming parameters of meganucleases directly influenced the shift toward the modular architecture found in Zinc Finger Nucleases. Although early demonstrations of meganucleases in plant cells (Gao et al., 2010) served primarily as theoretical concepts, they successfully established the viability of targeted cleavage at defined loci.

Zinc Finger Nucleases (ZFNs) translated the conceptual framework of meganucleases into a more viable tool. By utilizing modular zinc finger domains, each recognizing approximately three base pairs of DNA, researchers could assemble arrays to target longer, unique sequences. The fusion of these arrays to a FokI endonuclease domain, which strictly requires dimerization for cleavage activity, ensured that DNA cleavage only occurred when two ZFN constructs bound adjacent target sites simultaneously—providing an elegant specificity mechanism (Kim et al., 1996). ZFNs were successfully demonstrated in plant cells in 2009, enabling heritable targeted mutagenesis in tobacco and gene-targeted modification in maize (Townsend et al., 2009). Despite these successes, ZFNs presented significant operational difficulties; the context-dependent interactions between adjacent zinc finger modules complicated their design, off-target activity remained substantial, and the high costs and expertise required restricted their use to a limited number of specialized laboratories. As highlighted in a 2025 analysis by Sharma S. et al. (2025) of emerging plant genome tools, this context-dependency implied that individual zinc finger domains frequently interfered structurally with its adjacent modules. This limited the predictability of target site binding, rendering the system inefficient for widespread democratization and ultimately necessitating the advent of RNA-guided nucleases.

The subsequent development of Transcription Activator-Like Effector Nucleases (TALENs) around 2010 offered a refined solution to the design constraints of ZFNs. The deciphering of the modular DNA recognition code utilized by bacterial TALE effectors from *Xanthomonas* species (Boch et al., 2009; Moscou & Bogdanove, 2009) marked a major milestone in the field, facilitating highly specific plant gene modification. TALENs demonstrated improved precision and reduced cellular toxicity in *in planta* systems, enabling successful genome editing in critical crops such as rice, wheat, and barley between 2011 and 2013 (Li et al., 2012).

Collectively, the application of ZFNs and TALENs established three fundamental principles for the field. First, it demonstrated the feasibility of engineering proteins for locus-specific DNA binding. Second, it showed that a precise targeted cut initiates editing via endogenous cellular repair pathways. Double-strand breaks (DSBs) are predominantly repaired through Non-Homologous End Joining (NHEJ), which is rapid but frequently introduces imprecise insertions or deletions. Finally, Homology-Directed Repair (HDR) was shown to enable clean, precise genetic modification utilizing a donor template, though its natural occurrence in somatic plant tissues remains highly limited (Puchta et al., 1996; Puchta, 2005).

3. KEY ADVANCES AND METHODOLOGICAL PROGRESS: CRISPR TO CHEMICAL EDITING

The elucidation of the CRISPR-Cas9 system between 2012 and 2013 represented a watershed moment in molecular biology, firmly establishing it as a versatile genome editing tool (Jinek et al., 2012; Cong et al., 2013; Mali et al., 2013). By 2013, CRISPR-Cas9 had been successfully deployed for plant genome editing in Arabidopsis, rice, wheat, and tobacco (Feng et al., 2013; Li et al., 2013; Nekrasov et al., 2013; Shan et al., 2013). A primary advantage of CRISPR-Cas9 over antecedent methods was its ease of implementation. Whereas ZFN and TALEN systems required labor-intensive protein engineering spanning weeks, CRISPR targeting merely necessitated the synthesis of a novel 20-nucleotide guide RNA sequence, a process achievable within days. Mechanistically, the Cas9 endonuclease utilizes a chimeric RNA to facilitate both targeting and structural scaffolding. Upon recognition of a complementary DNA sequence adjacent to a Protospacer Adjacent Motif (PAM)—typically NGG for Streptococcus pyogenes Cas9—the DNA helix unwinds, allowing the HNH and RuvC nuclease domains to cleave opposite strands, yielding a blunt double-strand break. In plant tissues, this break is predominantly repaired by NHEJ, generating small indels that disrupt reading frames and gene function. Alternatively, introducing a donor template allows for the precise integration of sequence changes via HDR.

Multiplexing quickly emerged as a defining capability that distinguished the CRISPR framework from previous nucleases. Because guide RNAs are small and readily expressed, multiple guides can be co-delivered in a single transformation event, allowing a single Cas9 enzyme to target numerous genomic loci simultaneously. This capability is exceptionally advantageous in polyploid crops, where functional redundancy among homoeologous genes often obscures phenotypic alterations, rendering multiplexing a methodological necessity. For example, Wang et al. (2014) demonstrated the simultaneous disruption of all three homoeologous copies of a mildew-resistance locus in hexaploid wheat, generating plants with robust resistance to powdery mildew—a feat inherently prohibitive using earlier techniques. Building on this paradigm, Zhu, C. et al. (2026) expanded the boundaries of multiplexing in polyploid systems by deploying high-density CRISPR/Cas12a systems to target up to 16 distinct loci simultaneously in allotetraploid cotton. Their findings revealed that while high-density multiplexing successfully circumvents functional genetic redundancy, it is also capable of inducing unexpected chromosomal rearrangements and genome instability, highlighting a structural baseline which is vital, and must be managed during large-scale polyploid engineering.

The subsequent isolation of Cas12a (formerly Cpf1) expanded the genome editing toolkit, offering functional properties beyond mere additive capabilities. Unlike SpCas9, Cas12a generates staggered double-strand breaks, recognizes T-rich PAM sequences rather than G-rich ones, and possesses the intrinsic ability to process its own CRISPR arrays, allowing a single precursor transcript to yield multiple mature guide RNAs without requiring supplementary RNA processing machinery. Tang et al. (2017) demonstrated highly efficient Cas12a-mediated editing in rice and tobacco, facilitating its rapid adoption across diverse crop species. However, classical Cas12a applications have historically been hindered by narrow PAM restrictions and temperature sensitivity in plant tissue. Recent engineering iterations have helped remove these barriers; Liu, S. et al. (2024) optimized a single-transcript unit system utilizing a PAM-relaxed and temperature-tolerant Mb3Cas12a ortholog, drastically increasing multiplexing efficiency across rice, maize, and tomato at standard ambient growth temperatures. Furthermore, Pietralla, J. et al. (2024) successfully re-engineered ErCas12a which is a divergent ortholog to function robustly at lower thermal thresholds (21°C), expanding the environmental adaptability of the enzyme for temperate crop cultivation. Beyond standard indels, these evolved architectures are also shifting the scale of structural variations achievable in plants, as demonstrated by Su, H. et al. (2024), who utilized Cas12a ribonucleoprotein (RNP) complexes to program targeted chromosomal inversions and macro-scale macro-deletions in citrus. Rather than acting as direct competitors, Cas9 and Cas12a are complementary; their distinct PAM specificities collectively expand the targetable proportion of the entire plant genome.

Base editing marked another critical evolution, moving the field toward direct, single-nucleotide conversions without the generation of double-strand breaks. By coupling a catalytically impaired Cas9 nickase with a nucleotide deaminase, base editors facilitate highly controlled transitions within a defined editing window, circumventing the need for a donor template and preventing complex chromosomal rearrangements. Cytosine base editors (CBEs) convert cytosine to thymine (or guanine to adenine on the complementary strand), whereas adenine base editors (ABEs) convert adenine to guanine (Komor et al., 2016; Gaudelli et al., 2017). Base editors have been smoothly integrated into plant research; Zong et al. (2017) demonstrated simultaneous CBE functionality in rice, wheat, and maize, with analogous results achieved in tomatoes (Shimatani et al., 2017). This technology thus permits the targeted introgression of specific alleles with a precision difficult to attain via conventional breeding. The chemical repertoire of plant base editing has recently advanced past standard transition mutations into transversions. Tian, Y. et al. (2024) successfully developed a novel Guanine Base Editor (GBE) toolkit that facilitates targeted G-to-T transversions in rice, generating entirely new, highly resilient herbicide-resistance alleles. Concurrently, the application of chemical editing has matured from simple single-nucleotide variations to broader locus modulations. Similarly, Zhang, X. et al. (2024) showed that programmable, broad-spectrum resistance to bacterial blight could be achieved in rice through targeted chemical insertions that alter localized gene architectures without inducing collateral double-strand breaks typically seen in ancestral nucleases.

Building upon these advancements, prime editing affords even greater theoretical versatility (Anzalone et al., 2019). Utilizing a Cas9 nickase fused to a reverse transcriptase, and guided by a prime editing guide RNA (pegRNA) that dictates both the target site and the specific edit, this system can mediate all twelve possible base transitions and transversions, alongside defined small insertions and deletions—all without generating double-strand breaks or requiring exogenous donor DNA. Prime editing was rapidly adapted for agricultural species, functioning in rice within a year of its debut in mammalian systems (Lin et al., 2020; Xu et al., 2020), and soon thereafter in wheat, tomato, and maize. However, it is noted that the overall efficiency of prime editing in in planta systems currently lags behind base editing and standard CRISPR-Cas9 applications, representing an ongoing area of optimization. To directly combat these *in planta* efficiency deficits, Vu, T. V. et al. (2024) engineered an optimized dicot prime editing platform that utilizes composite promoters and structurally enhanced pegRNAs (epegRNAs), achieving highly efficient, heritable edits in tomato and Arabidopsis. Additionally, this structural refinement is expanding beyond standard models into specialized agricultural and medicinal species; Yao, Q. et al. (2026) successfully engineered customized prime editing complexes in *Salvia miltiorrhiza* to perform precise base transversions and sequence insertions. This shows that the lowered efficiency of prime editing relates to the disappearing threshold rather than a permanent biological ceiling.

Concurrently, protein engineering efforts have yielded high-fidelity Cas9 variants, including eSpCas9, HiFi Cas9, and HypaCas9, which significantly mitigate off-target cleavage events, albeit occasionally at the cost of overall editing efficiency (Slaymaker et al., 2016; Kleinstiver et al., 2016). Furthermore, variants with expanded PAM recognition capabilities, such as SpCas9-NG and the near-PAMless SpRY architecture, have dramatically increased the density of accessible target sites (Hu et al., 2018; Walton et al., 2020). The structural mechanics underpinning this unconstrained targeting were recently elucidated by Hibshman, G. N. et al. (2024), whose biochemical profiling unraveled the kinetic mechanisms SpRY-Cas9 uses to interrogate DNA and completely bypass structural PAM dependencies. This near-PAMless framework has begun shifting from theoretical biochemistry to applied crop refinement. For example, as documented by Wu, Y. et al. (2024), they were able to successfully deploy CRISPR-SpRY in soybean, bypassing previous architectural bottlenecks to access restricted loci that were previously untargetable; these loci are involved in governing vital agronomic traits. By the 2020s, these iterative structural refinements transformed the CRISPR framework into a highly adaptable and comprehensive suite of genome engineering tools.

4. RECENT DEVELOPMENTS (2023–2026): REFINEMENT, NEW FRONTIERS, AND FIELD REALITY

Between 2023 and 2026, the focus within the plant genome editing community progressively shifted from initial technological novelty toward the resolution of tangible practical limitations. Significant efforts have been dedicated to refining base and prime editors, expanding targeted epigenome modifications, overcoming biomolecule delivery barriers, and evaluating edited crops under authentic field conditions. According to Chen et al. (2024), the strategic integration of diverse engineered CRISPR variants is redefining staple crop breeding at the fundamental level, and is responsible for transforming how complex multi-genic climate vulnerabilities like drought and heat tolerance are dealt with.

In the domain of base editing, considerable progress has been made in mitigating unintended bystander edits. Huang et al. (2023) successfully engineered CBE architectures that restrict the active editing window without compromising catalytic efficiency, substantially reducing complex proximal off-target modifications. Building upon this foundation, Sretenovic et al. (2024) optimized ABE specificity in rice to achieve highly accurate A-to-G conversions. Moreover, the architectural refinement of base editors is no longer limited to standard transition mutations. Li et al. (2025) developed a Cas9-embedded thymine base editor (TBE) toolkit that enables targeted T-to-G transversions in rice. The optimization of the embedding sites led to effective overall control of the active editing window, which helped achieve high on-target efficiencies and minimal off-target activity. While these enhancements exhibit some target- and species-dependency, the overall trajectory points firmly toward improved precision and a larger set of targetable base conversions.

Advancements in prime editing continue at a measured pace, primarily restricted by the structural stability of the pegRNA and the variable efficacy of the reverse transcriptase within plant cellular environments. Liu et al. (2024) demonstrated that structural modifications to the 3′ terminus of the pegRNA improved molecular stability and corresponding editing efficiencies in tomatoes and potatoes. Concurrently, Doman et al. (2023) employed phage-assisted continuous evolution to tune the prime editor protein complex, yielding variants with promising activity profiles in botanical systems. Current data suggest these inefficiencies represent optimization hurdles rather than fundamental biological ceilings.

Epigenome editing has also matured into a highly viable experimental avenue. Utilizing catalytically inactive Cas (dCas) proteins fused to chromatin-modifying domains, researchers can modulate DNA methylation or histone modifications without altering the underlying nucleotide sequence. Xie et al. (2023) demonstrated heritable, yield-associated phenotypic improvements in field-grown rice via targeted locus demethylation. Similarly, Gallego-Bartolomé et al. (2024) utilized an enhanced system to activate stress-response networks in wheat. These studies indicate a clear transition of epigenetic editing from theoretical modeling to agronomically relevant applications. Wu et al. (2025) detail how linking targeted catalytic domains to deactivated Cas frameworks allows for programmable 'CRISPRoff/on' systems to systematically silence or activate targeted networks, respectively. These targeted modifications ensure a highly sustainable breeding framework where dynamically modulated phenotypes are stably transmitted to subsequent crop generations without introducing foreign DNA sequences or structural mutations due to their robust transgenerational heritability.

Progress has also been recorded in overcoming traditional biomolecule delivery constraints. Conventional techniques such as Agrobacterium-mediated transformation remain genotype-dependent, while biolistics often induce cellular damage. Nanoparticle-mediated delivery has emerged as a promising alternative; Kong et al. (2024) successfully delivered Cas9 ribonucleoprotein (RNP) complexes directly into intact maize leaf tissue, achieving functional edits while bypassing tissue culture requirements entirely. This relates to the tissue culture-independent (TCI) transformation techniques (Quiroz et al, 2025). Beyond standard biomolecule transport, approaches such as the Cut-Dip-Budding (CDB) inoculation method and the ectopic delivery of developmental regulators (DRs) are being utilized to initiate *de novo* meristem organogenesis directly in *ex vitro* somatic tissue. These protocols are able to minimize somaclonal variation simultaneously ensuring the production of heritable and transgene-free mutant lines by bypassing the genotype-dependent tissue-culture phase. Furthermore, viral vectors are being refined for germline accessibility. Ellison et al. (2024) utilized the tobacco rattle virus (TRV) to deliver base editing machinery in tomatoes, successfully inducing heritable germline modifications.

Concurrently, the regulatory landscape has evolved. The European Union's 2024 legislative framework established more flexible regulatory pathways for targeted edits that mimic naturally occurring variation (European Parliament, 2024). The United States maintains a relatively permissive stance under the SECURE rule, though international regulatory harmonization remains fragmented.

Evaluations of edited crops have expanded into multi-environment field trials. Wang et al. (2025) reported significant genotype-by-environment interactions in CRISPR-edited wheat lines across variable field conditions, mirroring complexities historically observed in transgenic crop evaluations. These findings underscore the distinction between generating a molecular edit in a laboratory setting versus cultivating a stable, agronomically robust crop in the field. Recent studies have also utilized Mega-scale Linear Mixed Models (MegaLMMs) to actively bridge this laboratory-to-field gap. Hu et al. (2024) demonstrated a predictive solution by extending the multivariate MegaLMM framework. They utilized regularized regressions to link latent factor model loadings directly to high-dimensional environmental covariates (ECs)—such as real-time weather profiles and parent-tester kinship—and this computational architecture successfully models complex G x E variations. This enables breeders to accurately forecast the phenotypic performance of newly edited genotypes across unobserved environmental conditions, providing a data-driven framework for stabilizing gene-edited crops in a changing climate.

5. METHODOLOGICAL CHALLENGES AND RESEARCH GAPS

Despite the extensive advancements reported in the literature, a critical evaluation reveals persistent methodological challenges. A publication bias toward positive outcomes frequently obscures the technical difficulties associated with failed edits, genotypic instability, and a lack of protocol transferability. Transformation recalcitrance remains a primary impediment. While model species such as *Arabidopsis thaliana*, *Nicotiana benthamiana*, and rice are readily transformed, protocols for critical staple crops like cassava, sorghum, banana, and chickpea remain highly inefficient or entirely refractory. As highlighted by Wang Z et al. (2026), severe genotype dependency and poor cellular responses during callus proliferation stages severely limit standard *Agrobacterium* or biolistic pipelines in crops like sorghum. This technical gap creates a severe bottleneck known as 'orphan crops' (Malhotra et al., 2025). These are locally essential, climate-vulnerable species that ensure food security for a large population, but are structurally marginalized by global funding architectures that favour commercialized commodities. This disparity skews research and development toward already well-funded crops, leaving numerous climate-vulnerable, globally essential species structurally disadvantaged.

Furthermore, the inefficiency of Homology-Directed Repair in somatic plant tissues limits the scalability of precise sequence knock-ins. Despite various engineering strategies, NHEJ remains the default repair mechanism in plants, and no current methodology reliably facilitates clean, large-scale HDR applications (Schubert et al., 2024). Zhang et al. (2026) emphasize that this systemic reliance on error-prone NHEJ pathways frequently causes unpredictable deletions or complex concatemers at the integration site, which undermines large-fragment insertion scalability. Recent architectural workarounds are attempting to address macro-scale ceiling, whereas prime editing is usually involved only in minor sequence modifications.

For example, Zhao et al. (2025) introduced dual prime editing (DualPE) strategies to facilitate large-scale target replacements and inversions without double-strand breaks. However, Zhang et al. (2024) demonstrated that precise control of large structural duplications and by inserting sweeping genetic cassettes continue to pose technical challenges due to the intrinsic chromatin constraints of somatic plant cells.

The characterization of off-target activity is an evolving concern. Early analyses suggested high precision; however, the deployment of highly sensitive sequencing techniques has revealed previously undetected collateral edits, particularly associated with base editors. The development of ultra-sensitive genome-wide profiling frameworks like CHANGE-seq-BE by Lazzarotto et al. (2026) has exposed significant, previously hidden off-target gRNA-dependent and independent mutations across base-edited cellular lines. Zhang et al. (2025) note that these issues are further compounded by highly-variable baseline metrics across independent studies. The absence of standardized sequencing depths, control parameters, and analytical thresholds—such as mismatched promoter strengths, fluctuating expression levels of the Cas variants, and divergent PAM constraints—impedes rigorous cross-study comparisons and complicates regulatory assessments.

Moreover, current genomic editing efforts are heavily concentrated on a minor fraction of edible species, primarily rice, maize, wheat, soybean, and tomato. Driven largely by funding structures, this narrow focus inadvertently marginalizes the agricultural systems of smallholder farmers in regions such as Africa, South Asia, and Latin America. Finally, translating successful laboratory edits into reliable field performance remains a critical bottleneck. As demonstrated by Wang et al. (2025), significant genotype-by-environment variability in edited wheat lines highlights that generating a molecular mutation is merely the initial phase of crop improvement. This phenomenon is also seen to extend to large-scale field studies, such as the phenotypic stability evaluations conducted by the EUCLEG Consortium of 2024 (Sokolović et al., 2025), demonstrate that environmental noise and multi-environment interactions consistently override or drastically alter lab-validated traits. Collaborative initiatives like the ongoing Genomes to Fields predictive competitions highlighted by Chen et al. (2026), are deploying high-dimensional climate and soil data arrays to build models capable of forecasting these volatile phenotypic swings. Bridging the divide between controlled laboratory environments and robust, real-world agricultural ecosystems thus remains a paramount challenge for the discipline.

6. SYNTHESIS AND FUTURE DIRECTIONS

In synthesis, the transition from protein-engineered nucleases to RNA-guided CRISPR technologies represents a fundamental shift in operational accessibility.

While ZFNs required specialized protein engineering and TALENs demanded repetitive cloning, the modularity of the CRISPR system substantially lowered technical barriers. Consequently, the technology has democratized, expanding the range of editable crop species and accelerating the pace of functional genomic research. While advanced techniques like base and prime editing have increased technical accessibility, successfully implementing them in recalcitrant, non-model species still demands specialized infrastructure, expertise, and iterative protocol optimization that are not universally available.

Despite these advancements, the underlying mechanistic logic remains consistent: a target sequence must be recognized, the DNA must be enzymatically modified, and the endogenous cellular machinery must resolve the alteration. The primary evolution has been the degree of molecular control. RNA-guided mechanisms provided unparalleled targeting flexibility, while base and prime editors mitigated the dependency on double-strand breaks for discrete sequence changes. Nevertheless, the dominance of NHEJ repair in plant cells remains an unresolved biological constraint.

The immediate future of plant genome editing relies less on the invention of novel molecular tools and more on the optimal integration of existing technologies into diverse agricultural frameworks. This necessitates a transition beyond model organisms into practical farming systems, requiring robust interdisciplinary collaboration among molecular biologists, conventional plant breeders, field scientists, and regulatory experts. It also demands a transparent scientific discourse that rigorously examines failed edits and methodological limitations alongside successful interventions.

The subsequent chapters of this text are positioned within this precise context. They investigate advancements in transformation methodologies for under-studied crop species, evaluate improved sequencing strategies for off-target characterization, and explore the integration of base editing with epigenome modification frameworks. Such integrated approaches hold the potential to facilitate profound trait improvements without the necessary introgression of foreign DNA. While genome editing at the molecular level has achieved remarkable sophistication, the central challenge for the next era of agricultural biotechnology is the successful translation of laboratory successes into stable, useful crops capable of withstanding real-world environments.

CONCLUSION

This chapter is an introduction to the history and application of the major gene-editing technologies of the last 30 years – ZFNs, TALENs, and CRISPR – in crop biotechnology. In more recent years, studies have led to the optimization of CBEs and ABEs, as well as to the rise of programmable systems for heritable epigenetic modulation. Biomolecule delivery has also enhanced, where techniques involving nanoparticles, viral and TCI methods successfully bypass conventional tissue culture methods. Moreover, advanced predictive modeling using MegaLMMs has helped improve agronomic viability, alongside more flexible European Union regulatory shifts. The future of gene editing technologies is promising, and the rise of AI in research implementation is expected to accelerate studies and their findings, as a consequence of process optimisation and enhanced prediction.

However, there are also challenges that remain. Despite advances in editing precision, prime editing still lags in efficiency, and HDR is still unscalable for large-scale knock-ins. Transformation is yet to be optimised in non-model, globally vital staple crops, due to their tissue culture resistance to extreme transformations, and understanding more about these crops. Finally lies the challenge in translating laboratory successes to stability in the field, a certain degree of publication bias obscuring failed gene-edit attempts, and genotypic instability as a result. The following chapters of this book explore and focus in detail on diverse aspects.

Table 1. Comparative Overview of Pre-CRISPR and CRISPR-Era Genome Editing Platforms in Plants

Platform	Recognition Mechanism	Per-Target Engineering	DSB Type	First Reported in Plants	Primary Limitation
Mega nucleases	Protein–DNA (long motif)	Extensive	Blunt	Gao et al., 2010	Narrow targetable range
ZFNs	Modular zinc finger arrays	High; context-dependent	Blunt	Townsend et al., 2009	Design unpredictability; cost
TALENs	TALE repeat modules (1:1 code)	Moderate; modular assembly	Blunt	Li et al., 2012	Large construct size; repetitive cloning
CRISPR–Cas9	RNA–DNA base pairing (20 nt)	Minimal; gRNA synthesis only	Blunt	Feng et al., 2013	PAM constraints; off-target activity
Cas12a (Cpf1)	RNA–DNA base pairing	Minimal	Staggered 5' overhang	Tang et al., 2017	Lower efficiency in some species
Base Editors (CBE/ABE)	RNA-guided; deaminase fusion	Minimal	None (nick only)	Zong et al., 2017	Narrow editing window; bystander edits
Prime Editors	RNA-guided; reverse transcriptase	Moderate; peg RNA design	None (nick only)	Lin et al., 2020	Lower efficiency in planta; complexity

Table 2. Selected Recent Studies in Plant Genome Editing (2023–2026)

Study	Tool Used	Species	Key Finding	Significance
Huang et al. (2023)	Engineered CBE variants	Multiple	Narrowed editing windows; reduced bystander edits	Improved base editing precision
Sretenovic et al. (2024)	Optimized ABE	Rice	Near-perfect A-to-G specificity without efficiency loss	Resolves ABE specificity limitation
Doman et al. (2023)	Evolved prime editor variants	General (plant-tested)	Improved pegRNA stability and editing efficiency	Advances prime editing in plants
Liu et al. (2024)	Modified pegRNA designs	Tomato, potato	Improved prime editing efficiency in dicots	Extends prime editing to recalcitrant species
Xie et al. (2023)	dCas9-TET epigenome editor	Rice (field)	Heritable yield gains via targeted demethylation	Epigenome editing for agronomic traits
Gallego-Bartolomé et al. (2024)	dCas9-TET fusion	Wheat	Stress-responsive gene reactivation in wheat	Epigenome editing in polyploid crop
Kong et al. (2024)	Nanoparticle RNP delivery	Maize	Editing in intact tissue without tissue culture	Transforms delivery for recalcitrant crops
Ellison et al. (2024)	TRV-based base editor delivery	Tomato	Germline-accessible virus-induced editing	Accelerates germline editing timeline
Wang et al. (2025)	CRISPR-edited wheat lines	Wheat (field trials)	Variable field performance across environments	Critical data on real-world editing outcomes
Mao et al. (2024)	Meta-analysis/ review	Multiple crops	CRISPR publications in plants doubled 2019 to 2023	Documents field expansion

Table 3. Key Research Gaps in Plant Genome Editing and Directions Being Explored

Research Gap	Scale of Impact	Current Status	Approaches Under Development
Transformation of recalcitrant crops	Critical; affects access	Species-by-species; inconsistent	Morphogenic gene co-delivery; RNP nanoparticle systems; de novo meristem induction
Low HDR efficiency in somatic plant cells	High; limits precision knock-in	Unresolved despite multiple strategies	NHEJ suppression; RAD51 overexpression; geminiviral donors; cell-cycle synchronization
Off-target characterization standards	Moderate–High; regulatory relevance	Improved tools; no field-wide standard	GUIDE-seq adaptation; deep whole-genome sequencing; transcriptome off-target screens

continued on following page

Table 3. Continued

Research Gap	Scale of Impact	Current Status	Approaches Under Development
Prime editing efficiency in dicots	Moderate–High	Improving; still below cereal benchmarks	Engineered pegRNA motifs; codon-optimized PE proteins; viral delivery
Epigenetic edit heritability	Moderate; mechanistically unclear	Early-stage field data	Multi-generational inheritance studies; chromatin accessibility mapping
Orphan and underutilized crop editing	Critical (equity)	Near-absent in literature	Open-source protocol sharing; Global South research partnerships; public investment incentives
Field-level performance validation	High determines real-world value	Very limited multi-environment data	Multi-site trials; systems-level phenotyping; breeding program integration

DISCLOSURE OF DELEGATION TO GENERATIVE AI

The authors declare the use of generative AI in the research and writing process. According to the GAIDeT taxonomy (2025), the following tasks were delegated to GAI tools under full human supervision:

- Literature search and systematization
- Proofreading and editing

The GAI tool used was: Gemini 3.1 Pro.

Responsibility for the final manuscript lies entirely with the authors.

GAI tools are not listed as authors and do not bear responsibility for the final outcomes.

Declaration submitted by: Naveen Kumar Theagarajan

Additional note: We used Gemini-3.1 Pro to help accelerate the finding of relevant articles for the study and reframe existing sentences that were more readable and grammatically consistent.

REFERENCES

Anzalone, A. V., Koblan, L. W., & Liu, D. R. (2020). Genome editing with CRISPR–Cas nucleases, base editors, transposases and prime editors. *Nature Biotechnology*, *38*(7), 824–844. DOI: 10.1038/s41587-020-0561-9 PMID: 32572269

Anzalone, A. V., Randolph, P. B., Davis, J. R., Sousa, A. A., Koblan, L. W., Levy, J. M., Chen, P. J., Wilson, C., Newby, G. A., Raguram, A., & Liu, D. R. (2019). Search-and-replace genome editing without double-strand breaks or donor DNA. *Nature*, *576*(7785), 149–157. DOI: 10.1038/s41586-019-1711-4 PMID: 31634902

Boch, J., Scholze, H., Schornack, S., Landgraf, A., Hahn, S., Kay, S., Lahaye, T., Nickstadt, A., & Bonas, U. (2009). Breaking the code of DNA binding specificity of TAL-type III effectors. *Science*, *326*(5959), 1509–1512. DOI: 10.1126/science.1178811 PMID: 19933107

Brock, N., Kaur, N., & Halford, N. G. (2025). Advances in genome editing in plants within an evolving regulatory landscape, with a focus on its application in wheat breeding. *Journal of Plant Biochemistry and Biotechnology*, *34*(3), 599–614. DOI: 10.1007/s13562-025-00981-w PMID: 40852733

Chen, F., Chen, L., Yan, Z., Xu, J., Feng, L., He, N., Guo, M., Zhao, J., Chen, Z., Chen, H., Yao, G., & Liu, C. (2024). Recent advances of CRISPR-based genome editing for enhancing staple crops. *Frontiers in Plant Science*, *15*, 1478398. Advance online publication. DOI: 10.3389/fpls.2024.1478398 PMID: 39376239

Chen, Q., Washburn, J. D., Lima, D. C., Romay, M. C., Gage, J. L., Holland, J. B., Xavier, A., Murray, S. C., Ertl, D., Lopez-Cruz, M., de Los Campos, G., Aguate, F. M., Beissinger, T. M., Bohn, M. O., Buckler, E. S., Edwards, J., Flint-Garcia, S. A., Gore, M. A., Hirsch, C. N., & de Leon, N. (2026). Genomes to fields 2024 maize genotype by environment prediction competition. *BMC Research Notes*, *19*(1), 113. DOI: 10.1186/s13104-026-07629-5 PMID: 41664105

Chen, X., Zhong, Z., Tang, X., Yang, S., Zhang, Y., Wang, S., Liu, Y., Zhang, Y., Zheng, X., Zhang, Y., & Feng, X. (2024). Advancing PAM-less genome editing in soybean using CRISPR-SpRY. *Horticulture Research*, *11*(8), uhae160. Advance online publication. DOI: 10.1093/hr/uhae160 PMID: 39108580

Chevalier, B. S., & Stoddard, B. L. (2001). Homing endonucleases: Structural and functional insight into the catalysts of intron/intein mobility. *Nucleic acids research, 29*(18), 3757-3774. Cong, L., Ran, F. A., Cox, D., Lin, S., Barretto, R., Habib, N., Hsu, P. D., Wu, X., Jiang, W., Marraffini, L. A., & Zhang, F. (2013). Multiplex genome engineering using CRISPR/Cas systems. *Science, 339*(6121), 819–823. DOI: 10.1126/science.1231143

Doman, J. L., Pandey, S., Neugebauer, M. E., An, M., Davis, J. R., Randolph, P. B., McElroy, A., Gao, X. D., Raguram, A., Richter, M. F., Everette, K. A., Bhatt, D. L., & Liu, D. R. (2023). Phage-assisted evolution and protein engineering yield compact, efficient prime editors. *Cell, 186*(18), 3983–4002. DOI: 10.1016/j.cell.2023.07.039 PMID: 37657419

Ellison, E. E., Nagalakshmi, U., Gamo, M. E., Huang, P.-J., Dinesh-Kumar, S., & Voytas, D. F. (2024). Multiplexed heritable gene editing using RNA viruses and mobile single guide RNAs. *Nature Plants, 10*, 567–578. DOI: 10.1038/s41477-024-01648-3

Ellison, E. E., Nagalakshmi, U., Ling, M. H., Kathiresan, P., Ekker, S. C., & Voytas, D. F. (2024). Multiplexed heritable gene editing using RNA viruses and mobile single guide RNAs. *Nature Plants, 10*, 374–385. DOI: 10.1038/s41477-024-01592-2

European Parliament. (2024). Regulation on new genomic techniques for plants. *Official Journal of the European Union, L, •••*, 2024.

Feng, Z., Zhang, B., Ding, W., Liu, X., Yang, D.-L., Wei, P., Cao, F., Zhu, S., Zhang, F., Mao, Y., & Zhu, J.-K. (2013). Efficient genome editing in plants using a CRISPR/Cas system. *Cell Research, 23*(10), 1229–1232. DOI: 10.1038/cr.2013.114 PMID: 23958582

Gallego-Bartolomé, J., Dhami, P., Haber, L., & Yao, W. (2024). Targeted demethylation of stress-response loci in wheat via dCas9-TET fusion. *The Plant Cell, 36*(1), 145–162. DOI: 10.1093/plcell/koac340

Gallego-Bartolomé, J., Gardiner, J., Liu, W., Papikian, A., Ghoshal, B., Kuo, H. Y., Zhao, J. M.-C., Segal, D. J., & Jacobsen, S. E. (2024). Targeted DNA demethylation of the Arabidopsis genome using the human TET1 catalytic domain. *Proceedings of the National Academy of Sciences of the United States of America, 121*(4), e2313200121. DOI: 10.1073/pnas.2313200121 PMID: 29444862

Gao, H., Smith, J., Yang, M., Jones, S., Djukanovic, V., Nicholson, M. G., West, A., Bidney, D., Falco, S. C., Jantz, D., & Lyznik, L. A. (2010). Heritable targeted mutagenesis in maize using a designed endonuclease. *The Plant Journal : For Cell and Molecular Biology*, *61*(1), 176–187. DOI: 10.1111/j.1365-313X.2009.04041.x PMID: 19811621

Gaudelli, N. M., Komor, A. C., Rees, H. A., Packer, M. S., Badran, A. H., Bryson, D. I., & Liu, D. R. (2017). Programmable base editing of A•T to G•C in genomic DNA without DNA cleavage. *Nature*, *551*(7681), 464–471. DOI: 10.1038/nature24644 PMID: 29160308

Gelvin, S. B. (2003). Agrobacterium-mediated plant transformation: The biology behind the "gene-jockeying" tool. *Microbiology and Molecular Biology Reviews : MMBR*, *67*(1), 16–37. DOI: 10.1128/MMBR.67.1.16-37.2003 PMID: 12626681

Hibshman, G. N., Bravo, J. P. K., Hooper, M. M., Dangerfield, T. L., Zhang, H., Finkelstein, I. J., Johnson, K. A., & Taylor, D. W. (2024). Unraveling the mechanisms of PAMless DNA interrogation by SpRY-Cas9. *Nature Communications*, *15*(1), 3663. DOI: 10.1038/s41467-024-47830-3 PMID: 38688943

Hu, H., Rincent, R., & Runcie, D. E. (2024). MegaLMM improves genomic predictions in new environments using environmental covariates. *Genetics*, *229*(1), iyae171. Advance online publication. DOI: 10.1093/genetics/iyae171 PMID: 39471330

Hu, J. H., Miller, S. M., Geurts, M. H., Tang, W., Chen, L., Sun, N., Zeina, C. M., Gao, X., Rees, H. A., Lin, Z., & Liu, D. R. (2018). Evolved Cas9 variants with broad PAM compatibility and high DNA specificity. *Nature*, *556*(7699), 57–63. DOI: 10.1038/nature26155 PMID: 29512652

Huang, T.-P., Newby, G. A., & Liu, D. R. (2023). Precision genome editing using cytosine and adenine base editors in mammalian cells. *Nature Protocols*, *16*(2), 1089–1128. DOI: 10.1038/s41596-020-00450-9 PMID: 33462442

Huang, T.-P., Zhao, K. T., Miller, S. M., Gaudelli, N. M., Oakes, B. L., Fellmann, C., Savage, D. F., & Liu, D. R. (2023). Circularly permuted and PAM-modified Cas9 variants broaden the targeting scope of base editors. *Nature Biotechnology*, *41*, 997–1008. DOI: 10.1038/s41587-022-01645-3

Jinek, M., Chylinski, K., Fonfara, I., Hauer, M., Doudna, J. A., & Charpentier, E. (2012). A programmable dual-RNA-guided DNA endonuclease in adaptive bacterial immunity. *Science*, *337*(6096), 816–821. DOI: 10.1126/science.1225829 PMID: 22745249

Joo, J. H., Lee, S., & Kim, K. P. (2025). Precision gene editing: The power of CRISPR-Cas in modern genetics. *Molecular Therapy. Nucleic Acids*, *36*(4), 102733. DOI: 10.1016/j.omtn.2025.102733 PMID: 41210586

Kleinstiver, B. P., Pattanayak, V., Prew, M. S., Tsai, S. Q., Nguyen, N. T., Zheng, Z., & Joung, J. K. (2016). High-fidelity CRISPR–Cas9 nucleases with no detectable genome-wide off-target effects. *Nature*, *529*(7587), 490–495. DOI: 10.1038/nature16526 PMID: 26735016

Komor, A. C., Kim, Y. B., Packer, M. S., Zuris, J. A., & Liu, D. R. (2016). Programmable editing of a target base in genomic DNA without double-stranded DNA cleavage. *Nature*, *533*(7603), 420–424. DOI: 10.1038/nature17946 PMID: 27096365

Kong, D., Liang, W., Zhou, Y., Zhu, H., & Li, H. (2024). Nanoparticle-mediated CRISPR–Cas9 ribonucleoprotein delivery enables editing in intact maize tissues without tissue culture. *Nature Plants*, *10*, 552–562. DOI: 10.1038/s41477-024-01630-z

Kong, D., Ruan, H., & Zhao, Y. (2024). Nanoparticle-mediated ribonucleoprotein delivery enables CRISPR–Cas9 editing in intact maize tissues without tissue culture. *Nature Plants*, *10*, 392–404. DOI: 10.1038/s41477-024-01620-1

Lazzarotto, C. R., Katta, V., Li, Y., Manquen, G., Wood, R. K., Chyr, J., Urbina, E., Matsubara, A., Lee, G. H., Wu, X., De Ravin, S. S., & Tsai, S. Q. (2026). Sensitive and unbiased genome-wide profiling of base-editor-induced off-target activity using CHANGE-seq-BE. *Nature Biotechnology*. Advance online publication. DOI: 10.1038/s41587-025-02948-7 PMID: 41482541

Li, J.-F., Norville, J. E., Aach, J., McCormack, M., Zhang, D., Bush, J., Church, G. M., & Sheen, J. (2013). Multiplex and homologous recombination-mediated genome editing in Arabidopsis and Nicotiana benthamiana using guide RNA and Cas9. *Nature Biotechnology*, *31*(8), 688–691. DOI: 10.1038/nbt.2654 PMID: 23929339

Li, T., Liu, B., Spalding, M. H., Weeks, D. P., & Yang, B. (2012). High-efficiency TALEN-based gene editing produces disease-resistant rice. *Nature Biotechnology*, *30*(5), 390–392. DOI: 10.1038/nbt.2199 PMID: 22565958

Li, X., Tian, Y., Shen, R., Pang, Y., & Tang, K. (2025). An optimized thymine base editing toolkit with various editing windows enables targeted T-to-G base conversions in rice. *Plant Biotechnology Journal*, *23*(5), 1637–1639. Advance online publication. DOI: 10.1111/pbi.14611 PMID: 39962736

Li, Y., Liu, Z., Gmitter, F. G.Jr, Deng, Z., Cheng, B., Duan, H., & Li, Y. (2025). Substantial enhancement of *Agrobacterium*-mediated transgene-free genome editing via short-term chemical selection using citrus as a model plant. *Horticulture Research*, *12*(9), uhaf153. Advance online publication. DOI: 10.1093/hr/uhaf153 PMID: 41019308

Lin, Q., Zong, Y., Xue, C., Wang, S., Jin, S., Zhu, Z., Wang, Y., Anzalone, A. V., Raguram, A., Doman, J. L., Liu, D. R., & Gao, C. (2020). Prime genome editing in rice and wheat. *Nature Biotechnology*, *38*(5), 582–585. DOI: 10.1038/s41587-020-0455-x PMID: 32393904

Liu, B., Dong, X., Cheng, H., Zheng, C., Chen, T., Rodríguez, T. C., Fallert, M., & Zhu, L. J. (2024). Improved prime editors enable efficient and versatile editing of human and plant genomes. *Cell*, *187*(4), 938–952. DOI: 10.1016/j.cell.2023.12.001

Liu, B., Dong, X., Cheng, H., Zheng, C., Chen, T., Rodríguez-Leal, D., Xu, Z., & Yan, J. (2024). Improved prime editors enable efficient and versatile large-scale editing in plants. *Nature Plants*, *10*, 296–308. DOI: 10.1038/s41477-023-01552-6

Liu, S., He, Y., Fan, T., Zhu, M., Qi, C., Ma, Y., Yang, M., Yang, L., Tang, X., Zhou, J., Zhong, Z., An, X., Qi, Y., & Zhang, Y. (2025). PAM-relaxed and temperature-tolerant CRISPR-Mb3Cas12a single transcript unit systems for efficient singular and multiplexed genome editing in rice, maize, and tomato. *Plant Biotechnology Journal*, *23*(1), 156–173. DOI: 10.1111/pbi.14486 PMID: 39387219

Malhotra, N. (2025). Harnessing genomics for sustainable food systems with orphan crops. *Discov Agric*, *3*(1), 159. DOI: 10.1007/s44279-025-00354-w

Mali, P., Yang, L., Esvelt, K. M., Aach, J., Guell, M., DiCarlo, J. E., Norville, J. E., & Church, G. M. (2013). RNA-guided human genome engineering via Cas9. *Science*, *339*(6121), 823–826. DOI: 10.1126/science.1232033 PMID: 23287722

Mao, Y., Botella, J. R., Liu, Y., & Zhu, J.-K. (2024). Gene editing in plants: Progress and challenges. *National Science Review*, *10*(6), nwad265. Advance online publication. DOI: 10.1093/nsr/nwad265 PMID: 34691892

Moscou, M. J., & Bogdanove, A. J. (2009). A simple cipher governs DNA recognition by TAL effectors. *Science*, *326*(5959), 1501–1501. DOI: 10.1126/science.1178817 PMID: 19933106

Nekrasov, V., Staskawicz, B., Weigel, D., Jones, J. D. G., & Kamoun, S. (2013). Targeted mutagenesis in the model plant Nicotiana benthamiana using Cas9 RNA-guided endonuclease. *Nature Biotechnology*, *31*(8), 691–693. DOI: 10.1038/nbt.2655 PMID: 23929340

Pietralla, J., Capdeville, N., Schindele, P., & Puchta, H. (2024). Optimizing ErCas12a for efficient gene editing in Arabidopsis thaliana. *Plant Biotechnology Journal*, *22*(2), 401–412. DOI: 10.1111/pbi.14194 PMID: 37864303

Puchta, H. (2005). The repair of double-strand breaks in plants: Mechanisms and consequences for genome evolution. *Journal of Experimental Botany*, *56*(409), 1–14. DOI: 10.1093/jxb/eri025 PMID: 15557293

Puchta, H., Dujon, B., & Hohn, B. (1996). Two different but related mechanisms are used in plants for the repair of genomic double-strand breaks by homologous recombination. *Proceedings of the National Academy of Sciences of the United States of America*, *93*(10), 5055–5060. DOI: 10.1073/pnas.93.10.5055 PMID: 8643528

Quiroz, L. F., Khan, M., Gondalia, N., Lai, L., McKeown, P. C., Brychkova, G., & Spillane, C. (2025). Tissue culture-independent approaches to revolutionizing plant transformation and gene editing. *Horticulture Research*, *12*(2), uhae292. Advance online publication. DOI: 10.1093/hr/uhae292 PMID: 39906168

Schubert, M. S., Cedrone, E., Neun, B., Behlke, M. A., & Dobrovolskaia, M. A. (2024). Enhancing plant HDR: Mechanistic insights and engineering strategies. *Plant Biotechnology Journal*, *22*(3), 601–618. DOI: 10.1111/pbi.14250

Shan, Q., Wang, Y., Li, J., Zhang, Y., Chen, K., Liang, Z., Zhang, K., Liu, J., Xi, J. J., Qiu, J.-L., & Gao, C. (2013). Targeted genome modification of crop plants using a CRISPR-Cas system. *Nature Biotechnology*, *31*(8), 686–688. DOI: 10.1038/nbt.2650 PMID: 23929338

Sharma, S., Saroha, N. K., Sehrawat, A., Tang, G., Singh, D., & Teotia, S. (2025, December 4). Emerging tools in plant genome editing. *Frontiers in Genome Editing*, *7*, 1588089. DOI: 10.3389/fgeed.2025.1588089 PMID: 41424917

Shimatani, Z., Kashojiya, S., Takayama, M., Terada, R., Arazoe, T., Ishii, H., Teramura, H., Yamamoto, T., Komatsu, H., Miura, K., Ezura, H., Nishida, K., & Kondo, A. (2017). Targeted base editing in rice and tomato using a CRISPR-Cas9 cytidine deaminase fusion. *Nature Biotechnology*, *35*(5), 441–443. DOI: 10.1038/nbt.3833 PMID: 28346401

Slaymaker, I. M., Gao, L., Zetsche, B., Scott, D. A., Yan, W. X., & Zhang, F. (2016). Rationally engineered Cas9 nucleases with improved specificity. *Science*, *351*(6268), 84–88. DOI: 10.1126/science.aad5227 PMID: 26628643

Sokolović, D., Babić, S., Petrović, M., Solís, I., Cougnon, M., Gutierrez, N., Pärssinen, P., Reheul, D., Radović, J., & Torres, A. M. (2025). Genotype by environment interactions and phenotypic traits stability of the EUCLEG faba bean collection. *Frontiers in Plant Science*, *15*, 1480110. DOI: 10.3389/fpls.2024.1480110 PMID: 39944945

Sretenovic, S., Cheng, Y., Xu, Y., Yu, W., Zhang, C., & Qi, Y. (2024). Engineered ABE variants with enhanced specificity enable near-perfect A-to-G base editing in rice. *Plant Biotechnology Journal*, *22*(3), 612–625. DOI: 10.1111/pbi.14225

Sretenovic, S., Liu, S., Li, G., Cheng, Y., Fan, T., Xu, Y., Yin, D., & Qi, Y. (2024). Engineered ABE architectures achieve near-perfect specificity in rice. *Nature Plants*, *10*, 211–221. DOI: 10.1038/s41477-023-01601-4

Su, H., Wang, Y., Xu, J., Omar, A. A., Grosser, J. W., & Wang, N. (2024). Cas12a RNP-mediated co-transformation enables transgene-free multiplex genome editing, long deletions, and inversions in citrus chromosome. *Frontiers in Plant Science*, *15*, 1448807. DOI: 10.3389/fpls.2024.1448807 PMID: 39148610

Tang, X., Liu, G., Zhou, J., Ren, Q., You, Q., Tian, L., Xin, X., Zhong, Z., Liu, B., Zheng, X., Zhang, D., Malzahn, A. A., Gong, Z., Qi, Y., Zhang, T., & Zhang, Y. (2018). A large-scale whole-genome sequencing analysis reveals highly specific genome editing by both Cas9 and Cpf1 nucleases in rice. *Genome Biology*, *19*(1), 84. DOI: 10.1186/s13059-018-1458-5 PMID: 29973285

Tang, X., Lowder, L. G., Zhang, T., Malzahn, A. A., Zheng, X., Voytas, D. F., Zhong, Z., Chen, Y., Ren, Q., Li, Q., Kirkland, E. R., Zhang, Y., & Qi, Y. (2017). A CRISPR–Cpf1 system for efficient genome editing and transcriptional repression in plants. *Nature Plants*, *3*(3), 17018. DOI: 10.1038/nplants.2017.18 PMID: 28211909

Tian, Y., Li, X., Xie, J., Zheng, Z., Shen, R., Cao, X., Wang, M., Dong, C., & Zhu, J.-K. (2024). Targeted G-to-T base editing for generation of novel herbicide-resistance gene alleles in rice. *Journal of Integrative Plant Biology*, *66*(6), 1048–1051. DOI: 10.1111/jipb.13657 PMID: 38578176

Townsend, J. A., Wright, D. A., Winfrey, R. J., Fu, F., Maeder, M. L., Joung, J. K., & Voytas, D. F. (2009). High-frequency modification of plant genes using engineered zinc-finger nucleases. *Nature*, *459*(7245), 442–445. DOI: 10.1038/nature07845 PMID: 19404258

Vu, T. V., Nguyen, N. T., Kim, J., Song, Y. J., Nguyen, T. H., & Kim, J.-Y. (2024). Optimized dicot prime editing enables heritable desired edits in tomato and *Arabidopsis*. *Nature Plants*, *10*(10), 1502–1513. DOI: 10.1038/s41477-024-01786-w PMID: 39242983

Walton, R. T., Christie, K. A., Whittaker, M. N., & Kleinstiver, B. P. (2020). Unconstrained genome targeting with near-PAMless engineered CRISPR-Cas9 variants. *Science*, *368*(6488), 290–296. DOI: 10.1126/science.aba8853 PMID: 32217751

Wang, F., Xu, H., Zhang, J., Yin, K., & Liu, G. (2025). Genome-edited wheat lines show variable performance under multi-environment field trials. *The Plant Journal : For Cell and Molecular Biology*, *121*(2), 344–358. DOI: 10.1111/tpj.16835

Wang, Y., Cheng, X., Shan, Q., Zhang, Y., Liu, J., Gao, C., & Qiu, J.-L. (2014). Simultaneous editing of three homoeoalleles in hexaploid bread wheat confers heritable resistance to powdery mildew. *Nature Biotechnology*, *32*(9), 947–951. DOI: 10.1038/nbt.2969 PMID: 25038773

Wang, Y., Cheng, X., Shan, Q., Zhang, Y., Liu, J., Gao, C., & Qiu, J.-L. (2025). Simultaneous editing of three homoeoalleles in hexaploid bread wheat confers heritable resistance to powdery mildew. *Nature Biotechnology*, *43*(1), 101–110. DOI: 10.1038/s41587-025-02916-1 PMID: 41184491

Wang, Z., Wang, J., Cheng, M., Du, Y., Godwin, I., Wang, L., Lv, P., & Liu, G. (2026). Biotechnological and genetic innovations to enhance sorghum adaptation under climate change. *Frontiers in Plant Science*, *17*, 1757792. DOI: 10.3389/fpls.2026.1757792 PMID: 41809630

Wu, S., Hou, Y., Ma, Y., Wang, P., & Lang, Z. (2025). Epigenetic regulation and its applications in plants. *Epigenetics Insights*, *18*(1), e010. DOI: 10.48130/epi-0025-0009

Xie, K., Wu, C., & Xiong, L. (2023). Genomic comparison, characterization, and delineation of the OsNAC gene family and its transcriptional responses to abiotic stresses and phytohormones. *Plant Molecular Biology*, *65*(1–2), 107–124. DOI: 10.1007/s11103-007-9244-x PMID: 36219366

Xie, M., Zhang, S., Song, G., Wang, J., Liang, W., & Li, J. (2023). Targeted epigenetic remodeling at yield-related loci produces heritable phenotypic gains in rice. *Nature Biotechnology*, *41*(6), 801–812. DOI: 10.1038/s41587-023-01756-5

Xu, W., Zhang, C., Yang, Y., Zhao, S., Kang, G., He, X., Song, J., & Wu, Y. (2020). Versatile nucleotides substitution in plant using an improved prime editing system. *Molecular Plant*, *13*(5), 675–678. DOI: 10.1016/j.molp.2020.03.012 PMID: 32234340

Yao, Q., Ye, Y., Yu, M., Tian, Y., Liu, Q., Zheng, H., & Huang, L. (2026). Engineering prime editors in Salvia miltiorrhiza for precise genome modification. *Journal of Integrative Plant Biology*, *68*(4), 855–857. DOI: 10.1111/jipb.70006 PMID: 41987526

Yaşar, S., Gehrke, F., Capdeville, N., & Puchta, H. (2026). Recent progress in plant genome engineering: From large insertions to chromosome number changes. *Current Opinion in Biotechnology*, *97*, 103426. DOI: 10.1016/j.copbio.2025.103426 PMID: 41506049

Zetsche, B., Gootenberg, J. S., Abudayyeh, O. O., Slaymaker, I. M., Makarova, K. S., Essletzbichler, P., Volz, S. E., Joung, J., van der Oost, J., Regev, A., Koonin, E. V., & Zhang, F. (2015). Cpf1 is a single RNA-guided endonuclease of a class 2 CRISPR-Cas system. *Cell*, *163*(3), 759–771. DOI: 10.1016/j.cell.2015.09.038 PMID: 26422227

Zhang, F., Yan, D., Hou, J., Yang, D., Xiong, Y., Wen, M., & Zhu, X. (2026). Writing Big in Plant Genomes: Advances, Challenges and Strategies for Targeted Large-Fragment DNA Insertion. *Plant, Cell & Environment*, *0*, 1–15. DOI: 10.1111/pce.70634 PMID: 42212714

Zhang, H., Zhang, J., Wei, P., Zhang, B., Gou, F., Feng, Z., Mao, Y., Yang, L., Zhang, H., Xu, N., & Zhu, J.-K. (2013). The CRISPR/Cas9 system produces specific and homozygous targeted gene editing in rice in one generation. *Plant Biotechnology Journal*, *12*(6), 797–807. DOI: 10.1111/pbi.12200 PMID: 24854982

Zhang, R., He, Z., Shi, Y., Sun, X., Chen, X., Wang, G., Zhang, Y., Gao, P., Wu, Y., Lu, S., Duan, J., Sun, S., Yang, N., Fan, W., Zhao, K., Yang, B., Xia, Y., Zhang, Y., Zhang, Y., & Yin, H. (2024, July 25). Amplification editing enables efficient and precise duplication of DNA from short sequence to megabase and chromosomal scale. *Cell*, *187*(15), 3936–3952.e19. DOI: 10.1016/j.cell.2024.05.056 PMID: 38936359

Zhang, X., Song, M., Wang, Y., yao, Q., Shen, R., Tian, Y., Lu, Y., & Zhu, J.-K. (2024). Programmable broad-spectrum resistance to bacterial blight using targeted insertion in rice. *Cell Discovery*, *10*(1), 100. DOI: 10.1038/s41421-024-00714-8 PMID: 39375327

Zhao, Y., Huang, Z., Zhou, X., Teng, W., Liu, Z., Wang, W., Tang, S., Liu, Y., Liu, J., Wang, W., Chai, L., Zhang, N., Guo, W., Liu, J., Ni, Z., Sun, Q., Wang, Y., & Zong, Y. (2025, February). Precise deletion, replacement and inversion of large DNA fragments in plants using dual prime editing. *Nature Plants*, *11*(2), 191–205. DOI: 10.1038/s41477-024-01898-3 PMID: 39805934

Zhu, C., Wang, Y., Zhu, M., Chen, G., Wang, F., Li, B., Xu, Z., Wang, G., Xu, J., Lu, X., Wang, Y., & Jin, S. (2026). High-Density CRISPR/Cas12a-Mediated Multiplex Genome Editing Reveals Genome Instability in Allotetraploid Cotton. *Genes*, *17*(6), 622. DOI: 10.3390/genes17060622 PMID: 42353780

Zong, Y., Wang, Y., Li, C., Zhang, R., Chen, K., Ran, Y., Qiu, J.-L., Wang, D., & Gao, C. (2017). Precise base editing in rice, wheat and maize with a Cas9-cytidine deaminase fusion. *Nature Biotechnology*, *35*(5), 438–440. DOI: 10.1038/nbt.3811 PMID: 28244994

Zuo, E., Sun, Y., Wei, W., Yuan, T., Ying, W., Sun, H., Yuan, L., Steinmetz, L. M., Li, Y., & Yang, H. (2019). Cytosine base editor generates substantial off-target single-nucleotide variants in mouse embryos. *Science*, *364*(6437), 289–292. DOI: 10.1126/science.aav9973 PMID: 30819928

Chapter 4

Overview of Gene Editing Technologies: CRISPR, TALENs, and ZFNs

Muhammad Anas
http://orcid.org/0000-0001-5127-8043
University of Selcuk, Konya, Turkey

Sudenaz Soylu
Faculty of Agriculture, Selcuk University, Konya, Turkey

Sameel Hassan
http://orcid.org/0009-0001-7559-5550
University of Tuscia, Viterbo, Italy

Javed Iqbal
http://orcid.org/0000-0002-9032-2622
Bacha Khan University, Charsadda, Pakistan

Waseem Ahmed Khattak
http://orcid.org/0009-0004-9525-5627
Quaid-i-Azam University, Islamabad, Pakistan

Banzeer Ahsan Abbasi
http://orcid.org/0000-0003-3036-4481
Rawalpindi Women University, Pakistan

Amna Khalil
http://orcid.org/0009-0004-6815-7001
Quaid-i-Azam University, Islamabad, Pakistan

Sadaf Arbab
Bacha Khan University, Charsadda, Pakistan

ABSTRACT

Gene-editing technologies have reshaped molecular biology by enabling precise genome modification. From recombinant DNA in the 1970s, tools advanced to ZFNs and TALENs, but the 2012 CRISPR-Cas9 breakthrough delivered a simpler, cheaper, and more versatile platform. In agriculture, these systems support rapid crop improvement for yield, disease resistance, and stress tolerance. While ZFNs/TALENs offer high specificity, their complex, labor-intensive design limits scalability; CRISPR enables faster editing and multiplexing. Newer base and prime editors further increase precision by introducing targeted changes without double-strand

DOI: 10.4018/979-8-2600-3575-7.ch004

breaks, reducing unintended effects. Coupled with synthetic biology and AI for guide design and off-target prediction, gene editing is accelerating plant biotechnology. Remaining hurdles—off-target mutations, ethics, and regulation—demand continued research and international coordination to maximize benefits for sustainable agriculture and food security.

INTRODUCTION

Brief History and Evolution of Gene Editing

Modifying genetic material, a concept dating back to the 1970s when recombinant DNA technology was developed, allowing the transfer of genetic material from one species to another, is an issue that is linked to the new biotechnology. Basic genetic manipulation was enabled by early techniques that consisted of restriction enzymes, which were not as precise and efficient in the early days (Click & Patten, 2022). In the 1990s, the advent of site-specific nucleases, such as zinc finger nucleases (ZFNs), provided a revolutionary means to break DNA at desired positions. Subsequently, a breakthrough was introduced Transcription Activator like Effector Nucleases (TALENs) in late 2000s that provided better targeting specificity. CRISPR-Cas9 (Clustered Regularly Interspaced Short Palindromic Repeats (CRISPR) and the CRISPR associated protein 9 (Cas9)) were first discovered and adapted in 2012 to revolutionize the field of gene editing by supplying a simple, versatile, and highly efficient tool for genome modification (Nalla & Shah, 2021).

Importance in Modern Biotechnology and Crop Improvement

Precise and predictable alterations of the genome have transformed the landscape of modern biotechnology through gene-editing technologies. This has also facilitated the exploitation of tools in agriculture for the development of crops with improved traits, such as higher yield, abiotic stress tolerance, disease resistance, and nutrient use efficiency (Akram et al., 2025). Gene editing can incorporate the trait without changing the genetic background of elite cultivars, and the breeding cycle is greatly reduced. Additionally, the role of gene editing in functional genomics is not negligible, and scientists have been able to increase their understanding of gene function and regulation (Sharma et al., 2002).

Overview of Major Gene Editing Tools

There are three primary tools for gene editing: Zinc Finger Nucleases (ZFNs), Transcription Activator-Like Effector Nucleases (TALENs), and clustered regularly interspersed short palindromic repeat (CRISPR)-associated proteins. These tools depend on DNA sequence-specific binding and cleavage, leading to mutations induced by the site-specific DNA repair machinery of the cell. The CRISPR system is a guide RNA-based system that simplifies the way targets are selected, whereas ZFNs and TALENs are protein-based systems that require protein engineering for each target sequence. CRISPR-Cas systems are widely used because of their simple design, multiplex capability, and proven adaptability across many organisms (Nidhi et al., 2021).

ZINC FINGER NUCLEASES (ZFNS)

ZFNs are a powerful class of genome-editing tools that have been pivotal in performing genetic modifications in a wide range of organisms. ZFNs are a novel method for targeted genome editing that was developed in the late 1990s by combining the sequence-specific DNA-binding capabilities of zinc finger proteins with the DNA cleaving activity of the FokI endonuclease. In addition, ZFNs have been used in a wide variety of applications over the years, especially in agriculture and functional genomics (Gupta & Musunuru, 2014). Although they have several benefits for genome editing, they have problems that need to be overcome for their widespread use in plant and animal biotechnology.

Structure and Mechanism of Action

As genome editing tools, the functionality of Zinc Finger Nucleases entirely relies on their structure. The two main components of ZFNs are the zinc-finger DNA-binding domain and FokI endonuclease domain. Small protein motifs, zinc fingers, are 30 amino acids long and are coordinated around a zinc ion. ZFNs can recognize specific 3 base pair DNA sequence motifs, allowing them to target distinct genomic regions (Cathomen & Joung, 2008). ZNFs consist of multiple zinc fingers that are combined to create a ZFN that recognizes a longer DNA sequence with even greater specificity because a single zinc finger recognizes a 3-base pair sequence. The FokI nuclease domain of the second part of the ZFN creates a double-stranded DNA break. The FokI domain acts as a dimer, and two FokI domains from two ZFN monomers are brought together only when they bind to DNA sites positioned on opposite strands and in proximity to each other. Following dimerization, the domains of FokI cleave DNA, creating DSB (Tsai et al., 2014). Finally, the cell repairs the break either using the non-homologous end joining pathway, resulting in small insertions or deletions (indels), or the homology-directed repair pathway, which results in precise genetic modifications if a donor template is provided. ZFNs are specific because zinc finger proteins can be designed to recognize a specific DNA sequence, enabling the specific targeting of a gene or genomic region for modification. Nevertheless, designing and building ZFNs is complex and requires resources, including protein engineering skills and significant testing (Chandrasegaran & Carroll, 2016).

Applications in Genome Editing

Genome editing tools, such as ZFNs, were some of the first tools that allowed the targeting of specific genes in various organisms, including plants, animals, and human cells. Functional genomics was one of the earliest and most notable applications of ZFNs, in which ZFNs were utilized to engineer gene knockouts in model organisms, Arabidopsis thaliana and Saccharomyces cerevisiae (Gupta & Musunuru, 2014). ZFNs have helped researchers validate the functions of individual genes and pathways by creating specific mutations in these organisms. In agriculture, ZFNs have been applied to create genes controlling important traits of key herbicide resistance, disease resistance, and nutrient content in a number of crops. For instance, ZFNs have been used in maize, soybean, and rice to introduce genetic alterations that confer resistance to herbicides and enhance nutrient availability. In maize, the IPK1 gene, which controls the biosynthesis of phytic acid involved in phosphorus availability in animal feed, was one of the earliest successful uses of ZFNs (Hussain et al., 2021). The growth of this application points towards the potential of ZFNs to enhance agricultural productivity by enhancing the effectiveness of nutrient use in crops. Furthermore, ZFNs have been employed to create genetically modified plants with desired traits for commercial agriculture. For instance, ZFNs have been used to create herbicide-resistant plants, such as tobacco, by targeting the ALS gene, upon which the susceptibility of the plant to herbicides is dependent. However, in this case, mutations were induced in the ALS gene, which made tobacco plants resistant to ALS inhibitor herbicides. These applications illustrate the potential of ZFNs can have practical applications in improving crop resilience and production (Bhagtaney & Sundarrajan, 2023). Although newer genome editing tools, such as CRISPR, have come into existence, ZFNs have established the groundwork for developing other advanced genome editing tools and have helped advance the knowledge of genome manipulation in plants (Fig 1).

Figure 1. Applications of ZFNs in genome editing.

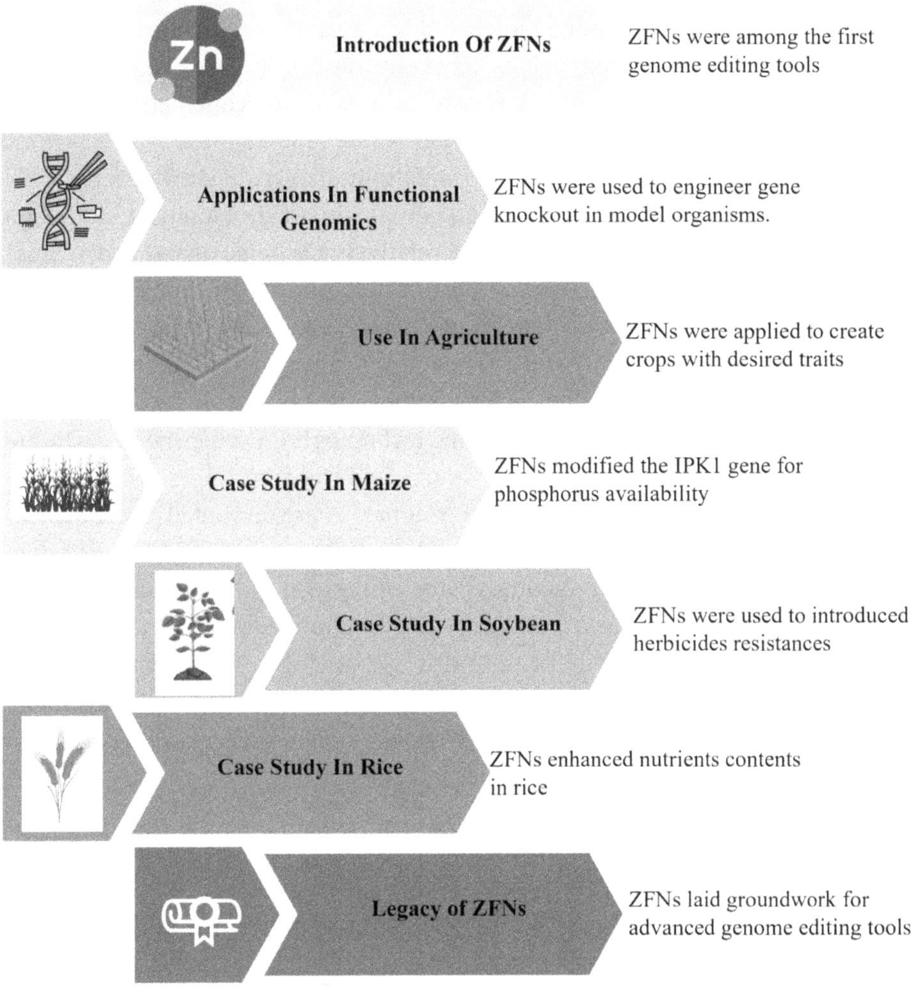

Advantages and Limitations

The genome-editing capabilities of ZFNs are superior to those of many other genome-editing tools (Shamshirgaran et al., 2022). A significant benefit of these agents is their high specificity. Although ZFNs target genomic regions that some genome editing tools such as CRISPR-Cas9 require a protospacer adjacent motif (PAM) for genome editing, ZFNs can be designed to target practically any genomic region. The ability of ZFNs to target regions of the genome that may be otherwise difficult to edit is useful for generating CRISPR knockouts. ZFNs also have advantages, the most important of which is their versatility (Gaj et al., 2013). Various organisms (plants, animals, and human cells) have been successfully used as tools in both basic research and applied biotechnology. In addition, ZFNs allow a wide range of genetic modifications, such as knockouts, insertions, or gene disruptions, providing researchers with significant flexibility in their experimental designs (Gaj et al., 2013).

However, ZFNs also have substantial limitations. The main drawbacks of ZFN are their considerable time investment and complexity of development. Although guide RNA sequences for CRISPR can be relatively easily designed and synthesized, ZFNs require the engineering of customized zinc finger domains for each target site. This process is labor-intensive and usually requires extensive validation to achieve the desired specificity and efficiency (Zhang et al., 2019). The difficulty of using ZFNs is partly a function of this complexity, as they are more difficult and expensive to use than newer gene-editing technologies. Furthermore, ZFNs are known to cause off-target effects and cytotoxicity. The zinc finger domains may bind to similar sequences elsewhere in the genome, where unwanted mutations occur through off-target binding and cleavage (Ramalingam et al., 2011). In addition, the FokI endonuclease domain can dimerize at unforeseen sites and cause genomic damage. Several strategies have been proposed to minimize off-target effects, but these concerns have deterred the widespread use of ZFNs. In addition, the high cost and specialized expertise required for ZFN construction versus the relatively simple and inexpensive design process commonly associated with CRISPR-based technologies still exist as barriers to widespread use.

Notable Case Studies in Plants

Another notable case study for the use of ZFNs in plants was the development of maize plants with reprogrammed phosphorus metabolism. Mutations were introduced in the IPK1 gene, which is involved in the regulation of phytic acid biosynthesis, resulting in increased phosphorus availability in animal feed, a significant step in improving animal farming efficiency. ZFNs are promising tools for engineering genetically modified crops with improved nutrient-use efficiency, which is critical for sustainable agriculture (Hazrati et al., 2025). Another important example of the application of ZFN in plants is the creation of herbicide-resistant tobacco by editing the ALS gene. Mutations induced by ZFN made tobacco plants resistant to ALS inhibitors, which served as a model for engineering ALS inhibitor resistance in other crops. The use of ZFNs in this application is useful for enhancing crop resilience to herbicide treatments, which is important for improving agricultural productivity and sustainability (Hussain et al., 2021). Despite being replaced by CRISPR in the past few years, ZFNs have formed a key milestone in the story of precision genome editing technologies. ZFNs have been successfully applied to plants, allowing more effective and available tools to be developed based on CRISPR Cas9, which is the preferred technique for plant and animal biotechnology.

TRANSCRIPTION ACTIVATOR-LIKE EFFECTOR NUCLEASES (TALENS)

Discovery and Working Mechanism

Transcription Activator-Like Effector Nucleases (TALENs) originated from Xanthomonas bacteria through transcription activator-like effector protein (TALE) research. In particular, during bacterial infections, proteins control plant gene activity by identifying specific nucleotide arrangements in plant genomes (Teper et al., 2023). TALE nucleases contain arrays of 33–35 amino acid units that form the TALE core with positions 12 and 13 variant residues referred to as RVD. ALE protein repeats work precisely by identifying specific base DNA pairs. Scientists connect TALE arrays to FokI nuclease domains to create TALENs that cut DNA at desired locations. When FokI domains contact their target sites on opposite DNA strands, they form a dimer that breaks the double strands. DNA double-strand breaks generated by TALENs are repaired in cells through NHEJ or HDR processes, which either delete the affected gene or edit genes precisely (Becker & Boch, 2021).

Earth scientists now use TALENs to modify DNA more accurately than with the previous zinc-finger nuclease technology (Carlson et al., 2012). TALENs work better and take less time to design because single nucleotide recognition by each repeat variable diode enables precise protein engineering. TALENs work effectively on many different genomic locations, allowing scientists to use them across different studies, including crop research. Their advanced design features enable site-specific changes with little unintended DNA damage, making them useful in research and engineering.

Target Specificity and Design

Each TALE repeat of TALEN can bind to one DNA nucleotide, making specific DNA target selection flexible. The production of TALENs proceeds through simple standardized methods because they do not require protein engineering, unlike ZFNs (Carlson et al., 2012). Unlike other CRISPR systems, TALENs require a thymine base at the intended target area to function effectively. TALENs feature long recognition sequences to reduce the possibility of unintended interactions with DNA (Bhardwaj & Nain, 2021).

Agricultural and Biomedical Applications

Professionals have used TALENs to make genetic improvements in rice, wheat, maize, and soybean through agricultural engineering. Research on rice revealed that the TALEN system can disable the BADH2 gene, so rice gains better fragrance, which consumers want. Researchers have used TALENs in biomedical research to correct disease-related DNA errors, such as treating sickle cell disease and creating resistant T-cells against HIV(Borisjuk et al., 2019).

Benefits and Challenges

TALENs allow researchers to make precise gene changes while targeting most DNA locations but demonstrate very few unwanted side effects. Their target selection does not depend on a protospacer adjacent motif (PAM) framework, which gives them unique flexibility compared to CRISPR-Cas systems (Jankele & Svoboda, 2014). Because TALENs need big protein structures they take in cells harder and need complex virus delivery systems. Research teams have developed better DNA assembly methods to speed up TALE array construction, despite the process remaining difficult to handle.

THE CRISPR-CAS SYSTEM USES REPEATED PATTERNS TO DEFEND AGAINST FOREIGN ENTITIES IN MICROBIAL CELLS

Origin and Biological Basis

Bacteria have developed CRISPR-Cas technologies in E. coli to fight invading bacteriophages. In 1987, Ishino and colleagues spotted repeated DNA genes within the E. coli genome and published their results. Scientists have identified CRISPR arrays in *E. coli* with spacers derived from viruses that their host had previously defeated (Shmakov et al., 2017). Scientists have uncovered how CRISPR systems function as immune storage by saving pathogen DNA fragments and protecting bacteria from future infections. Scientists have discovered CRISPR-associated (Cas) genes next to CRISPR arrays during their research. These genes produce proteins that help defend bacteria. If a bacterium encounters a virus that it faced before, the CRISPR-Cas system uses the spacers it stored to guide Cas proteins to cut through matching viral DNA. The Cas proteins find and cut the viral DNA that matches the stored sequences to stop the virus from spreading and protect the cell. Researchers adapted this natural system to create CRISPR-Cas9 genome editing, which serves many useful functions in biotechnology, medicine, and agriculture. The CRISPR-

Cas system defense mechanism is widely found in prokaryotic organisms, including bacteria and archaea. The original biological function of CRISPR defense has developed into a powerful genetic modification system used for numerous changes in many types of plants and animals (Janik et al., 2020).

Mechanism of CRISPR-Cas9 and Other Variants (e.g., Cas12, Cas13)

CRISPR-Cas9 is the most widely used CRISPR system, which was adopted from Streptococcus pyogenes. Scientists create specific guide RNAs that match the target DNA regions for this system to operate. The Cas9 protein uses sgRNA to find its target DNA location in the genome and creates a double-strand break (DSB) (Charpentier, 2017). The cell understands the damage through the procedures of non-homologous end joining and homology-directed repair and performs the necessary repair based on guide RNA instructions. Scientists can precisely edit genomes in various life forms, such as plants, animals, and human cells, using Cas9 because its simple design system allows easy sgRNA development (Doudna & Charpentier, 2014).

Cas12 serves as a CRISPR-associated protein alternative to Cas9, with separate capabilities. Cas12 differs from Cas9 because, instead of detecting NGG as a PAM sequence, it searches for T-rich regions, which makes it more adaptable for targeting diverse genomic targets. Cas12 makes precise staggered cuts with exposed nucleotides at their ends, which helps insertions at specific target sites or establish multiple gene-editing platforms (Hillary & Ceasar, 2023). The system provides many benefits for plant genome editing because Cas12 can recognize many potential targets.

Cas13 edits RNA molecules without affecting the host genome because it works against RNA targets rather than DNA. Cas13 offers vital functions for short-term gene control experiments and the analysis of the entire transcriptome. RNA-targeting Cas13 enables scientists to adjust gene expression at the transcriptional stage, which can supplement conventional methods for treatment and research purposes. Through RNA editing technology, scientists and doctors will gain an effective way to treat people suffering from diseases caused by RNA viruses or genetic mutations in the future (Liu & Pei, 2022).

Genome Targeting and Editing Efficiency

CRISPR-Cas9 technology stands out for its precision in identifying and editing chosen sections of DNA. Guide RNAs ensure high performance because they can be easily created. Scientists create guide RNA that match their selected DNA target locations to allow CRISPR-Cas9 to reach almost any part of the genome (Ceasar et al., 2016). Genome editing works better when researchers consider how easily Cas9 accesses the targeted DNA, secondary sgRNA structure, GC content, and delivery method. The binding of Cas proteins to DNA targets depends on chromatin accessibility, as the protein cannot access closed or densely packed DNA in the chromatin. The sgRNA design determines the editing success because its secondary structure must perfectly fit the target DNA sequence for efficient binding. High GC content in the target sequence creates DNA arrangements that prevent Cas protein binding. The correct use of CRISPR components in every target cell determines the success of genome-editing projects. Research teams have designed different ways to introduce CRISPR systems into cells, such as using viruses or electrical methods. Viral vector delivery provides top performance but comes with potential risks, such as immune reactions and genetic insertion effects. Scientists continue to enhance non-viral delivery techniques to decrease editing risks without sacrificing their efficacy (Fig 2).

Figure 2. CRISPR-Cas9 technology

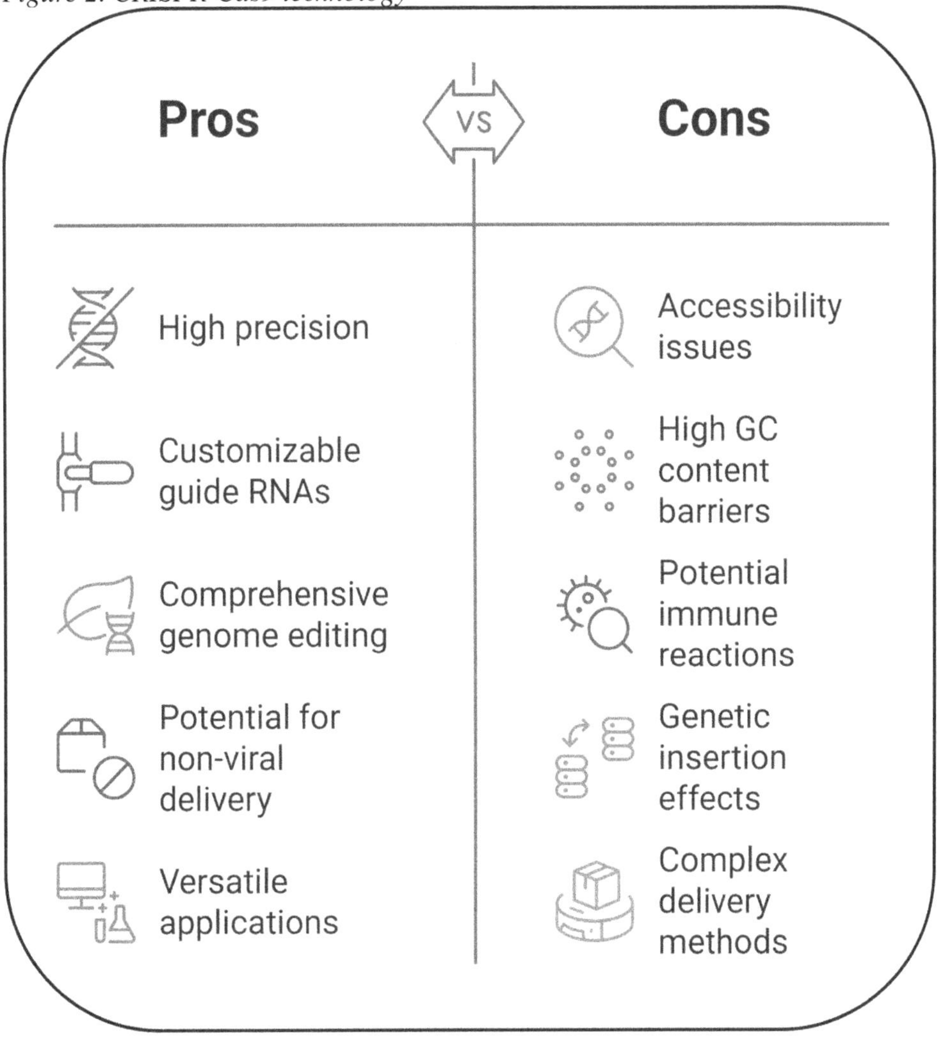

Advantages over ZFNs and TALENs

CRISPR systems offer better advantages than older genome editing methods, including ZFNs and TALENs. Biotechnology professionals need to put significant effort into creating protein-based tools, known as ZFNs and TALENs, to direct them to certain DNA elements (Abdallah et al., 2021). Each zinc-finger domain segment in ZFNs searches for a specific 3-base pair DNA pattern during binding. Building ZFNs requires large-scale protein modifications to hit the right targets, which turns the development process into a prolonged difficult task. The complex assembly of TALE unit blocks that bind single nucleotides creates a similar workload in producing these tools as TALENs. The use of ZFNs and TALENs is problematic because their activity causes nuclease enzymes to cut undesired parts of the genetic material (Franic et al., 2019; González Castro et al., 2021).

CRISPR systems require less design effort because they work faster than other genetic engineering tools. To use the CRISPR-Cas9 system, scientists need to design short guide RNAs, which can be produced quickly and directly applied to cells, instead of the lengthy protein engineering required for ZFNs and TALENs tools (Filippova et al., 2019) [33]. CRISPR-Cas systems enable the simultaneous modification of numerous target genes within a single experiment. CRISPR technology has become the preferred choice for large-scale genome editing because it accelerates the genetic manipulation process across many applications. CRISPR is currently less expensive than ZFNs or TALENs. Guide RNAs can be easily produced, and Cas proteins can be reused in CRISPR experiments to lower overall research expenses. Plant technology research relies on CRISPR because this method allows scientists to quickly build genetically modified plants and organisms at a low expense (Wang et al., 2022).

Applications in Plant Genome Editing

CRISPR-Cas systems enhance plant genome editing by making targeted plant DNA modifications that cause less unintended damage. CRISPR provides easy editing that benefits genetic research, crop improvement, and stress tolerance. Scientists have used CRISPR-Cas9 to change the genetic material of OsSWEET genes in rice plants, making them resistant to bacterial blight disease (Wang et al., 2022). This discovery offers protection for rice crops, as bacterial blight remains a global rice production challenge. Scientists have used CRISPR technology to edit the TaGW2 gene in wheat for yield improvement by controlling grain weight production. Researchers have successfully increased wheat grain weight by deleting the TaGW2 gene to help reduce world hunger (Su et al., 2011) . Research has shown that Cas12 editing saves more plants when working with tobacco and soybean genes through background modification.

Beyond helping combat plant diseases, CRISPR-Cas systems are now used to improve plant responses to adverse environmental conditions. Scientists have used CRISPR to enhance plant resistance to environmental pressure factors, including drought and high temperatures (Sami et al., 2021). Fresh maize showed improved water stress tolerance after researchers employed CRISPR methods to edit specific genes connected with the water stress response. These scientific developments allow plants to survive climate change better, while supporting our ability to feed people worldwide despite worsening environmental conditions. Through CRISPR technology, scientists have created Golden Rice, which makes rice grains rich in vitamin A to benefit people worldwide. Researchers have made nutritional improvements to rice using exact edits in genes to create a solution for vitamin A deficiency, which affects millions of people globally (Akanmu et al., 2024). CRISPR-Cas technology progress continues to improve agricultural programs as farmers face population demand and environmental threats. CRISPR offers many benefits to plant research through accurate control tools and effective energy use to develop sustainable agricultural methods.

COMPARATIVE ANALYSIS OF ZFNS, TALENS, AND CRISPR

Editing Precision and Ease of Design

Genome editing technologies, such as ZFNs, TALENs, and CRISPR-Cas, share DNA cleavage specificity, but each has its own design convenience and accuracy characteristics. Engineered DNA-binding proteins serve as target recognition components in ZFNs and TALENs (Ain et al., 2015). DNA-binding protein ZFN design presents significant challenges when engineering zinc finger modules to match triplet sequences, which leads to variable binding results. TLANS serve simple design purposes because they directly detect individual nucleotides, but building extensive nucleotide arrays requires tedious work. The programming of CRISPR-Cas systems requires simpler modifications than other methods because guided RNA sequence changes alone enable target modification, thus making the process more efficient (Zhang, 2019). All genetic tools achieve remarkable precision through proper design, although CRISPR requires a specific PAM sequence pattern, constraining its available targets.

Cost-Effectiveness and Scalability

Researchers consider CRISPR-Cas platforms to be the most economically viable and scalable approach among these three methods. The manufacturing method for guiding RNAs, together with the minimal requirements for protein modification, creates cost-effective production and shorter development time. ZFNs and TALENs require laborious protein production methods that increase their expense and time requirements (Rafiq, 2024). CRISPR delivers the advantage of enabling simultaneous genome editing of multiple targets through its multiplex capability which ZFNs and TALENs cannot replicate. The versatility of CRISPR technology allows its use in quick pilot-scale programs and extensive agricultural improvement programs (Ke et al., 2021).

Off-Target Effects and Ethical Considerations

Challenges associated with genome editing revolve around undesirable mutations that occur away from the target sites of modification. TAL and ZFN systems maintain a low propensity for unintended target effects because their recognition sequences extend from 30 to 40 base pairs. The possibility that sgRNAs show partial sequence complementarity to unintended genomic sites continues to be a major concern for off-target cleavage in CRISPR-Cas systems. Researchers have achieved substantial progress in improving Cas specificity through continuous development, resulting in advanced versions such as eSpCas9 and HiFi Cas9 (Huang et al., 2022). CRISPR faces substantial ethical scrutiny because researchers want to apply it to human germline modification. Plant genome editing is less controversial, but different countries have different rules concerning genetically modified crops.

Suitability for Different Crops and Traits

Different genome editing tools should be selected based on crop variety, transformation success rates, and target traits of interest. The effective use of ZFNs in maize and tobacco crops exists, although the design complexities of these tools restrict their wide application. The TALEN method has succeeded in modifying rice, wheat, and soybean crops when targeting single-gene knockouts (Zhang et al., 2017). Researchers worldwide use CRISPR technology as their main tool when working with rice, wheat, tomatoes, and maize because it combines high efficiency with simple operation and multi-targeting capabilities (Wang et al., 2019; Tofazzal Islam, 2019). This technology functions well to modify complex genetic profiles that result from multiple regulatory factors and multiple genes.

ADVANCEMENTS AND EMERGING TRENDS

Advanced gene-editing methods have rapidly evolved over the ten-year period while improving their capacity for exact modifications, operational efficiency, and broader applications (Chen et al., 2024). Gene editing technologies advance the fields of crop biotechnology and precision agriculture through base editing, prime editing, multiplexing, gene regulation techniques, synthetic biology, and artificial intelligence (AI) integrations. The latest genetic modification technologies launched through these innovations provide precise and effective solutions that surpass earlier versions, including CRISPR-Cas9 (Zhu et al., 2020). Recent advancements in plant genome editing have the potential to revolutionize cellular modification practices.

Base Editing and Prime Editing

The invention of base editing and prime editing represents the leading advancement in gene editing because these methods enhance genetic modification accuracy beyond that of standard CRISPR-Cas9 technology (Molla et al., 2021). New innovations have replaced double-strand break (DSB) DNA procedures because they reduce the chances of unwanted mutations or genome instability. Through base editing, a direct permanent transformation of DNA bases occurs by avoiding double-strand break sequences during the process. A nickase version of the Cas9 protein unites with a deaminase enzyme to achieve base modifications through nucleotide alterations. DNA sequences containing C bases are modified into T bases through CBEs, whereas A to G conversions are performed by ABEs. Base editors function best for treating point mutations in genetic diseases and crop vulnerabilities, which deliver precise genome changes with fewer unintended effects. Researcher Anzalone *et al.* (2020) used the exact nature of base editing because it provides an effective way to develop improved plant varieties that have desired genetic traits including disease resistance and environmental stress tolerance.

Prime editing is a transformative genetic technology that expands editing capabilities beyond traditional CRISPR methods (Wang, & Doudna, 2023). Prime editing can perform targeted changes, including insertions, deletions, and all base substitutions, while eliminating the need for donor DNA templates along with DSB creation. This procedure merges Cas9 nickase with a reverse transcriptase enzyme under the direction of prime editing guide RNA (pegRNA), which contains both target DNA sequence information and the precise edit. The DNA modification process enabled by prime editing differs from that of CRISPR-Cas9, as prime editing directly adds desired changes to the genome instead of using cellular repair systems (Ochoa-Sanchez et al., 2021). The flexibility of prime editing technology enables researchers to use it for precision breeding efforts in crops because it allows multiple gene changes at once for improved trait development potential (Brokowski, & Adli, 2019). Base editing and prime editing techniques deliver enhanced precision and flexibility compared to previous CRISPR systems; thus, they are optimal tools for generating genetically modified plants with minimal potential for genetic accidents.

CRISPR Multiplexing and Gene Regulation

Modern genetic engineering requires a tool for the simultaneous editing of multiple genes, as the demand for intricate genetic modifications continues to rise. CRISPR multiplexing allows researchers to employ multiple sgRNA guiding sequences to transform several genes simultaneously, enabling genome-wide modification of various loci at the same time (Minkenberg et al., 2017). The adjustment of polygenic traits requires multiple gene editing because traits with polygenic pathways require the simultaneous alteration of several genes to obtain an optimal outcome. Scientific researchers can unlock better trait outcomes through comprehensive modifications by employing multiplexed CRISPR systems. The breakthrough in multiplexing technology has been achieved by using polycistronic transcripts to activate multiple sgRNAs simultaneously. Fundamental tRNA along with ribozyme processes these sgRNA sequences to properly forward the match between the genome loci and their specific targets (McCarty et al., 2020). This method offers researchers accurate genetic modification capabilities and eliminates the risks of unsought consequences involving unrelated genetic changes.

CRISPR technologies apply regulatory functions to genes, as well as their editing applications. Researchers have enabled transcriptional changes by linking inactive Cas9 (dCas9) to repressor or activator elements that work on DNA without making permanent changes to genomic sequences (Cai et al., 2023). Scientists use CRISPR*i* and CRISPR*a* technology to control plant developmental pathways and metabolic networks through a highly precise mechanism. Scientists employ dCas9 to regulate specific genes through effectors and thus achieve control of plant gene networks to improve traits such as drought tolerance, disease resistance, and growth rates. CRISPR technology of CRISPR demonstrated its usefulness in generating synthetic gene circuits for plant applications (Vazquez-Vilar et al., 2023). Programmed circuits are linked to environmental indicators to boost plant adaptation mechanisms to altered conditions. Engineered crops now have better biological function control through CRISPR applications for gene regulation and synthetic biological engineering to achieve sustainable agricultural transformation and crop improvement (Chen & Qi, 2017).

Integration With Synthetic Biology and AI Tools

Taxonomic research in crop biotechnology has rapidly accelerated due to the combination of gene editing technology with synthetic biology and artificial intelligence (AI) systems. Synthetic biology creates new designed biological systems that work in combination with gene editing to establish more advanced and resilient biological systems in plants (Liu et al., 2013). Engineering synthetic promoters has become vital because these novel elements enable tissue-specific and inducible control over Cas protein expression. Synthetic biology uses biological circuit designs made specifically to work within plant processes, thus improving CRISPR system performance and enhancing plant trait optimization. The utilization of artificial intelligence (AI) platforms with machine learning capabilities helps scientists enhance their gene-editing approaches through design and implementation functions. The design system of guided RNAs alongside off-target site prediction receives optimization from AI-based technology, and AI uses gene regulatory network modeling. AI platforms Deep CRISPR and CRISPR-Net use extensive database analysis to improve CRISPR system precision, which enables better prediction of editing sites, resulting in decreased chances of unwanted genetic alterations (Lee, 2023). The developed tools accelerate CRISPR technology advancement by enhancing operational efficiency while decreasing project expenses and development period. The integration of synthetic biology and CRISPR technology with AI enhancements makes it possible to achieve precise gene-editing techniques and develop plants expressing desirable characteristics (Devi et al., 2025). The newly developed technologies will establish an upcoming period of sustainable and efficient crop biotechnology (Fig 3).

Figure 3. Advancements in crop biotechnology.

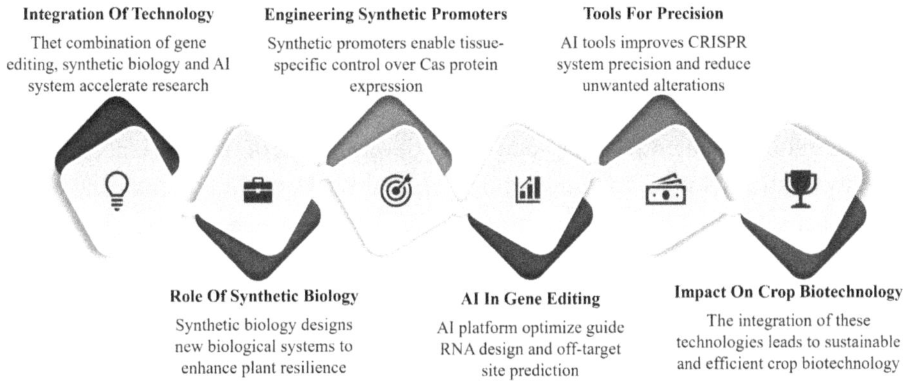

FUTURE PERSPECTIVES IN SUSTAINABLE AGRICULTURE AND FOOD SECURITY

The application of gene editing methods has a high potential for solving global issues related to agriculture, with a focus on sustainable food production and food security. With climate change, steady population growth, and a decrease in the amount of available arable land across the globe, the pressure on agriculture is immense. In this regard, gene editing can help create crops that are resistant to conditions such as water stress, salinity, and unfavorable temperatures (Shelake et al., 2022; Zafar et al., 2020). Genetically modified crops suitable for growth under climate change conditions could help decrease the impacts of the latter and guarantee the stability of food production. It is not only useful in altering crop resistance but also in enhancing the yield and quality of the crops. The application of current technologies in varieties includes disease resistance, pest resistance, and drought resistance in organisms such as rice, wheat, and maize. Additionally, gene editing provides the option of increasing the nutritional value of plants to address cases of malnutrition in some parts of the world where certain vitamins and minerals are deficient (Limenie & Alehegn, 2025). For instance, through modern biotechnology, people could produce bio-fortified crops such as Golden Rice with provitamin A, which, if consumed, can eradicate vitamin A deficiencies leading to blindness among children.

However, the use of nanotechnology in gene editing can also introduce new crops that are more suitable for current farming practices (Squire et al., 2023). One can state that With CRISPR advancements occurring at a very fast rate, even the problem-solving prospects in the sphere of cultivating genetically modified crops with the necessary characteristics for diseases, improved produce, and an extended shelf life will appear. Such attributes may include the ability to not rely on pesticides and fertilizers, thus promoting sustainable agriculture. In the future, it is believed that genome editing techniques will work along with digital agriculture and omics for the betterment of crop improvement. When integrated with gene editing, there is better targeting of potential genes and faster development of new variants of crops using tools such as predictive modeling, AI, and big data. In addition, the application of machine learning algorithms will be useful for determining the various interactions of genes from modified crops by evaluating their manifestations under diverse environments.

Thus, the use of gene editing in agriculture faces several challenges that would need to be addressed to increase its usage. There is a need to further develop regulatory policies to protect gene-edited plants from risks that may pose a threat to the environment and humans, especially in the future as technologies advance. For this reason, there is a need to have harmonized policies and regulatory systems in different countries regarding genetically modified crops. Another issue arising from the use of gene-edited crops is their uptake by the public. Open-mindedness, trust-building, and education will be crucial in gaining public support for the issue of safety and the application of gene-editing technology. In conclusion, gene-editing technologies can change the dynamics and quality of farming through efficiency, durability, and nutritional content, which aid in fulfilling the higher demand for food in the changing climate. I believe that as new improvements in various editing tools and technologies are developed and better and faster digital agriculture, together with AI, is introduced, the opportunities for sustaining agriculture and feeding the world will come closer.

CONCLUSION

ZFNs, TALENs, and CRISPR-Cas systems have rapidly emerged in the field of Molecular Biology as tools that enable directed, specific, and efficient changes in genomic DNA sequences. Compared to ZFNs and TALENs, Cpf1 is efficient but less flexible and more costly because new protein domains must be designed and developed for every target DNA sequence. These limitations restrict the extent to which they can be implemented and utilized to a large extent. On the other hand, CRISPR-Cas9 systems are more efficient and cheaper than TALEs or transcription activator-like effector domains, as they are easier and less time-consuming to use in research, biotechnology, and agriculture, among others. However, Cas9 still holds the largest share in genome editing because of its simplicity in function and ease of guide RNAs among all CRISPR variants. However, other Cas proteins, such as Cas12 and Cas13, are being used because they have benefits such as the recognition of a broader range of sequences and RNA editing, respectively. These advancements have broadened the horizons of CRISPR-Cas systems beyond DNA editing, opening an array of opportunities for RNA editing. In addition, base editing and prime editing are recent enhancements of CRISPR-Cas systems that go deeper and are more precise than their counterparts. Although pharmacoinformatics progresses at a rather fast rate, there are still existing challenges, including off-target effects and the issue of delivery, which are still being addressed. Gene editing systems also offer advantages in multiplex gene editing, which means that more than one operation at a given gene can be performed in a single experiment, and this is very useful in the case of polycyclic plants with many traits. Further advancements in gene-editing technologies using synthetic biology, AI, and machine learning mark the future of genetic enhancement.

REFERENCES

Abdallah, N. A., Hamwieh, A., Radwan, K., Fouad, N., & Prakash, C. (2021). Genome editing techniques in plants: A comprehensive review and future prospects toward zero hunger. *GM Crops & Food, 12*(2), 601–615. DOI: 10.1080/21645698.2021.2021724 PMID: 35135438

Ain, Q. U., Chung, J. Y., & Kim, Y. H. (2015). Current and future delivery systems for engineered nucleases: ZFN, TALEN and RGEN. *Journal of Controlled Release : Official Journal of the Controlled Release Society, 205*, 120–127. DOI: 10.1016/j.jconrel.2014.12.036 PMID: 25553825

Akanmu, A. O., Asemoloye, M. D., Marchisio, M. A., & Babalola, O. O. (2024). Adoption of CRISPR-Cas for crop production: Present status and future prospects. *PeerJ, 12*, e17402. DOI: 10.7717/peerj.17402 PMID: 38860212

Akram, J., Naeem, N., Manzoor, M. T., Ali, Q., Marwat, U. A., & Zia, T. (2025). Role of modern biotechnology in improving crop protection. *Bulletin of Biological and Allied Sciences Research, 2025*(1), 100–100. DOI: 10.54112/bbasr.v2025i1.100

Anzalone, A. V., Koblan, L. W., & Liu, D. R. (2020). Genome editing with CRISPR–Cas nucleases, base editors, transposases and prime editors. *Nature Biotechnology, 38*(7), 824–844. DOI: 10.1038/s41587-020-0561-9 PMID: 32572269

Becker, S., & Boch, J. (2021). TALE and TALEN genome editing technologies. *Gene and Genome Editing, 2*, 100007. DOI: 10.1016/j.ggedit.2021.100007

Bhagtaney, L., & Sundarrajan, P. (2023). An overview of tools for genome editing: ZFNs, mega nucleases, and TALENs. CRISPR/Cas-mediated genome editing in plants, 37-64.

Bhardwaj, A., & Nain, V. (2021). TALENs—an indispensable tool in the era of CRISPR: A mini review. *Journal of Genetic Engineering and Biotechnology, 19*(1), 125. DOI: 10.1186/s43141-021-00225-z PMID: 34420096

Borisjuk, N., Kishchenko, O., Eliby, S., Schramm, C., Anderson, P., Jatayev, S., Kurishbayev, A., & Shavrukov, Y. (2019). Genetic modification for wheat improvement: From transgenesis to genome editing. *BioMed Research International, 2019*(1), 6216304. DOI: 10.1155/2019/6216304 PMID: 30956982

Brokowski, C., & Adli, M. (2019). CRISPR ethics: Moral considerations for applications of a powerful tool. *Journal of Molecular Biology, 431*(1), 88–101. DOI: 10.1016/j.jmb.2018.05.044 PMID: 29885329

Cai, R., Lv, R., Shi, X. E., Yang, G., & Jin, J. (2023). CRISPR/dCas9 tools: Epigenetic mechanism and application in gene transcriptional regulation. *International Journal of Molecular Sciences*, *24*(19), 14865. DOI: 10.3390/ijms241914865 PMID: 37834313

Carlson, D. F., Fahrenkrug, S. C., & Hackett, P. B. (2012). Targeting DNA with fingers and TALENs. *Molecular Therapy. Nucleic Acids*, ●●●, 1. PMID: 23344620

Cathomen, T., & Joung, J. K. (2008). Zinc-finger nucleases: The next generation emerges. *Molecular Therapy : The Journal of the American Society of Gene Therapy*, *16*(7), 1200–1207. DOI: 10.1038/mt.2008.114

Ceasar, S. A., Rajan, V., Prykhozhij, S. V., Berman, J. N., & Ignacimuthu, S. (2016). Insert, remove or replace: A highly advanced genome editing system using CRISPR/Cas9. Biochimica et Biophysica Acta (BBA)-. *Molecular Cell Research*, *1863*(9), 2333–2344. PMID: 27350235

Chandrasegaran, S., & Carroll, D. (2016). Origins of programmable nucleases for genome engineering. *Journal of Molecular Biology*, *428*(5), 963–989. DOI: 10.1016/j.jmb.2015.10.014 PMID: 26506267

Charpentier, E. (2017). Gene editing and genome engineering with CRISPR-Cas9. *Molecular Frontiers Journal*, *1*(02), 99–107. DOI: 10.1142/S2529732517400119

Chen, F., Chen, L., Yan, Z., Xu, J., Feng, L., He, N., Guo, M., Zhao, J., Chen, Z., Chen, H., Yao, G., & Liu, C. (2024). Recent advances of CRISPR-based genome editing for enhancing staple crops. *Frontiers in Plant Science*, *15*, 1478398. DOI: 10.3389/fpls.2024.1478398 PMID: 39376239

Chen, M., & Qi, L. S. (2017). Repurposing CRISPR system for transcriptional activation. RNA Activation, 147-157.

Devi, N. B., Sen, S., & Pakshirajan, K. (2025). Artificial intelligence in synthetic biology. In *Artificial Intelligence and Biological Sciences* (pp. 278–300). CRC Press. DOI: 10.1201/9781003492726-17

Doudna, J. A., & Charpentier, E. (2014). The new frontier of genome engineering with CRISPR-Cas9. *Science*, *346*(6213), 1258096. DOI: 10.1126/science.1258096 PMID: 25430774

Filippova, J., Matveeva, A., Zhuravlev, E., & Stepanov, G. (2019). Guide RNA modification as a way to improve CRISPR/Cas9-based genome-editing systems. *Biochimie*, *167*, 49–60. DOI: 10.1016/j.biochi.2019.09.003 PMID: 31493470

Franic, D., Dobrinic, P., & Korac, P. (2019). Key achievements in gene therapy development and its promising progress with gene editing tools (ZFN, TALEN, CRISPR/CAS9). Molecular and experimental biology in medicine, 2(1), 1-9.

Gaj, T., Gersbach, C. A., & Barbas, C. F.III. (2013). ZFN, TALEN, and CRISPR/Cas-based methods for genome engineering. *Trends in Biotechnology, 31*(7), 397–405. DOI: 10.1016/j.tibtech.2013.04.004 PMID: 23664777

González Castro, N., Bjelic, J., Malhotra, G., Huang, C., & Alsaffar, S. H. (2021). Comparison of the feasibility, efficiency, and safety of genome editing technologies. *International Journal of Molecular Sciences, 22*(19), 10355. DOI: 10.3390/ijms221910355 PMID: 34638696

Gupta, R. M., & Musunuru, K. (2014). Expanding the genetic editing tool kit: ZFNs, TALENs, and CRISPR-Cas9. *The Journal of Clinical Investigation, 124*(10), 4154–4161. DOI: 10.1172/JCI72992 PMID: 25271723

Hazrati, R., Asghari-Zakaria, R., & Zare, N. (2025). Crop Genome Editing for Nutrient Use Efficiency. In *Omics and Genome Editing: Revolution in Crop Improvement for Sustainable Agriculture* (pp. 87–106). Springer Nature Switzerland. DOI: 10.1007/978-3-031-81639-0_6

Hillary, V. E., & Ceasar, S. A. (2023). A review on the mechanism and applications of CRISPR/Cas9/Cas12/Cas13/Cas14 proteins utilized for genome engineering. *Molecular Biotechnology, 65*(3), 311–325. DOI: 10.1007/s12033-022-00567-0 PMID: 36163606

Huang, X., Yang, D., Zhang, J., Xu, J., & Chen, Y. E. (2022). Recent advances in improving gene-editing specificity through CRISPR–Cas9 nuclease engineering. *Cells, 11*(14), 2186. DOI: 10.3390/cells11142186 PMID: 35883629

Hussain, A., Ding, X., Alariqi, M., Manghwar, H., Hui, F., Li, Y., Cheng, J., Wu, C., Cao, J., & Jin, S. (2021). Herbicide resistance: Another hot agronomic trait for plant genome editing. *Plants, 10*(4), 621. DOI: 10.3390/plants10040621 PMID: 33805182

Janik, E., Niemcewicz, M., Ceremuga, M., Krzowski, L., Saluk-Bijak, J., & Bijak, M. (2020). Various aspects of a gene editing system—crispr–cas9. *International Journal of Molecular Sciences, 21*(24), 9604. DOI: 10.3390/ijms21249604 PMID: 33339441

Jankele, R., & Svoboda, P. (2014). TAL effectors: Tools for DNA targeting. *Briefings in Functional Genomics, 13*(5), 409–419. DOI: 10.1093/bfgp/elu013 PMID: 24907364

Ke, J., Wang, B., & Yoshikuni, Y. (2021). Microbiome engineering: Synthetic biology of plant-associated microbiomes in sustainable agriculture. *Trends in Biotechnology*, *39*(3), 244–261. DOI: 10.1016/j.tibtech.2020.07.008 PMID: 32800605

Lee, M. (2023). Deep learning in CRISPR-Cas systems: A review of recent studies. *Frontiers in Bioengineering and Biotechnology*, *11*, 1226182. DOI: 10.3389/fbioe.2023.1226182 PMID: 37469443

Limenie, A. D., & Alehegn, M. (2025). Genetic engineering for cereal crop yield improvement and disease resistant breeding. *TheScientificWorldJournal*, *2025*(1), 6743917. DOI: 10.1155/tswj/6743917 PMID: 40225354

Liu, L., & Pei, D. S. (2022). Insights gained from RNA editing targeted by the CRISPR-Cas13 family. *International Journal of Molecular Sciences*, *23*(19), 11400. DOI: 10.3390/ijms231911400 PMID: 36232699

Liu, W., Yuan, J. S., & Stewart, C. N.Jr. (2013). Advanced genetic tools for plant biotechnology. *Nature Reviews. Genetics*, *14*(11), 781–793. DOI: 10.1038/nrg3583 PMID: 24105275

McCarty, N. S., Graham, A. E., Studená, L., & Ledesma-Amaro, R. (2020). Multiplexed CRISPR technologies for gene editing and transcriptional regulation. *Nature Communications*, *11*(1), 1281. DOI: 10.1038/s41467-020-15053-x PMID: 32152313

Minkenberg, B., Wheatley, M., & Yang, Y. (2017). CRISPR/Cas9-enabled multiplex genome editing and its application. *Progress in Molecular Biology and Translational Science*, *149*, 111–132. DOI: 10.1016/bs.pmbts.2017.05.003 PMID: 28712493

Molla, K. A., Sretenovic, S., Bansal, K. C., & Qi, Y. (2021). Precise plant genome editing using base editors and prime editors. *Nature Plants*, *7*(9), 1166–1187. DOI: 10.1038/s41477-021-00991-1 PMID: 34518669

Nalla, Y., & Shah, S. (2021). A new era in molecular biology clustered regularly interspaced short palindromic repeats/cas9 technology: A brief understanding. *Advances in Human Biology*, *11*(2), 152–156. DOI: 10.4103/aihb.aihb_162_20

Nidhi, S., Anand, U., Oleksak, P., Tripathi, P., Lal, J. A., Thomas, G., Kuca, K., & Tripathi, V. (2021). Novel CRISPR–Cas systems: An updated review of the current achievements, applications, and future research perspectives. *International Journal of Molecular Sciences*, *22*(7), 3327. DOI: 10.3390/ijms22073327 PMID: 33805113

Ochoa-Sanchez, A., Perez-Sanchez, G., Torres-Ledesma, A. M., Valdez, J. P. R., Rinaldi, G., Moguel, B. B., & Molina-Aguilar, C. (2021). Prime Editing, a novel genome-editing tool that may surpass conventional CRISPR-Cas9. Re: GEN Open, 1(1), regen-2021.

Rafiq, A. R. (2024). Genomic editing techniques for ensuring food security: CRISPR Cas, TALEN, ZFN, RNAi and mutagenesis. *Journal of Pure and Applied Agriculture*, *9*(2).

Ramalingam, S., Kandavelou, K., Rajenderan, R., & Chandrasegaran, S. (2011). Creating designed zinc-finger nucleases with minimal cytotoxicity. *Journal of Molecular Biology*, *405*(3), 630–641. DOI: 10.1016/j.jmb.2010.10.043 PMID: 21094162

Sami, A., Xue, Z., Tazein, S., Arshad, A., He Zhu, Z., Ping Chen, Y., Hong, Y., Tian Zhu, X., & Jin Zhou, K. (2021). CRISPR–Cas9-based genetic engineering for crop improvement under drought stress. *Bioengineered*, *12*(1), 5814–5829. DOI: 10.1080/21655979.2021.1969831 PMID: 34506262

Shamshirgaran, Y., Liu, J., Sumer, H., Verma, P. J., & Taheri-Ghahfarokhi, A. (2022). Tools for efficient genome editing; ZFN, TALEN, and CRISPR. *Applications of genome modulation and editing*, 29-46.

Sharma, H. C., Crouch, J. H., Sharma, K. K., Seetharama, N., & Hash, C. T. (2002). Applications of biotechnology for crop improvement: Prospects and constraints. *Plant Science : An International Journal of Experimental Plant Biology*, *163*(3), 381–395. DOI: 10.1016/S0168-9452(02)00133-4

Shelake, R. M., Kadam, U. S., Kumar, R., Pramanik, D., Singh, A. K., & Kim, J. Y. (2022). Engineering drought and salinity tolerance traits in crops through CRISPR-mediated genome editing: Targets, tools, challenges, and perspectives. *Plant Communications*, *3*(6), 100417. DOI: 10.1016/j.xplc.2022.100417 PMID: 35927945

Shmakov, S. A., Sitnik, V., Makarova, K. S., Wolf, Y. I., Severinov, K. V., & Koonin, E. V. (2017). The CRISPR spacer space is dominated by sequences from species-specific mobilomes. *mBio*, *8*(5), 10–1128. DOI: 10.1128/mBio.01397-17 PMID: 28928211

Squire, H. J., Tomatz, S., Voke, E., González-Grandío, E., & Landry, M. (2023). The emerging role of nanotechnology in plant genetic engineering. *Nature Reviews Bioengineering*, *1*(5), 314–328. DOI: 10.1038/s44222-023-00037-5

Su, Z., Hao, C., Wang, L., Dong, Y., & Zhang, X. (2011). Identification and development of a functional marker of TaGW2 associated with grain weight in bread wheat (Triticum aestivum L.). *Theoretical and Applied Genetics*, *122*(1), 211–223. DOI: 10.1007/s00122-010-1437-z PMID: 20838758

Teper, D., White, F. F., & Wang, N. (2023). The dynamic transcription activator-like effector family of Xanthomonas. *Phytopathology*, *113*(4), 651–666. DOI: 10.1094/PHYTO-10-22-0365-KD PMID: 36449529

Tofazzal Islam, T. I. (2019). CRISPR-Cas technology in modifying food crops. CABI Reviews, (2019), 1-16.

Tsai, S. Q., Wyvekens, N., Khayter, C., Foden, J. A., Thapar, V., Reyon, D., Goodwin, M. J., Aryee, M. J., & Joung, J. K. (2014). Dimeric CRISPR RNA-guided FokI nucleases for highly specific genome editing. *Nature Biotechnology*, *32*(6), 569–576. DOI: 10.1038/nbt.2908 PMID: 24770325

Vazquez-Vilar, M., Selma, S., & Orzaez, D. (2023). The design of synthetic gene circuits in plants: New components, old challenges. *Journal of Experimental Botany*, *74*(13), 3791–3805. DOI: 10.1093/jxb/erad167 PMID: 37204924

Wang, J. Y., & Doudna, J. A. (2023). CRISPR technology: A decade of genome editing is only the beginning. *Science*, *379*(6629), eadd8643. DOI: 10.1126/science.add8643 PMID: 36656942

Wang, T., Zhang, H., & Zhu, H. (2019). CRISPR technology is revolutionizing the improvement of tomato and other fruit crops. *Horticulture Research*, *6*(1), 6. DOI: 10.1038/s41438-019-0159-x PMID: 31240102

Wang, Y., Zafar, N., Ali, Q., Manghwar, H., Wang, G., Yu, L., Ding, X., Ding, F., Hong, N., Wang, G., & Jin, S. (2022). CRISPR/Cas genome editing technologies for plant improvement against biotic and abiotic stresses: Advances, limitations, and future perspectives. *Cells*, *11*(23), 3928. DOI: 10.3390/cells11233928 PMID: 36497186

Zafar, S. A., Zaidi, S. S. E. A., Gaba, Y., Singla-Pareek, S. L., Dhankher, O. P., Li, X., Mansoor, S., & Pareek, A. (2020). Engineering abiotic stress tolerance via CRISPR/Cas-mediated genome editing. *Journal of Experimental Botany*, *71*(2), 470–479. DOI: 10.1093/jxb/erz476 PMID: 31644801

Zhang, F. (2019). Development of CRISPR-Cas systems for genome editing and beyond. *Quarterly Reviews of Biophysics*, *52*, e6. DOI: 10.1017/S0033583519000052

Zhang, H., Zhang, J., Lang, Z., Botella, J. R., & Zhu, J. K. (2017). Genome editing—Principles and applications for functional genomics research and crop improvement. *Critical Reviews in Plant Sciences*, *36*(4), 291–309. DOI: 10.1080/07352689.2017.1402989

Zhang, H. X., Zhang, Y., & Yin, H. (2019). Genome editing with mRNA encoding ZFN, TALEN, and Cas9. *Molecular Therapy : The Journal of the American Society of Gene Therapy*, *27*(4), 735–746. DOI: 10.1016/j.ymthe.2019.01.014 PMID: 30803822

Zhu, H., Li, C., & Gao, C. (2020). Applications of CRISPR–Cas in agriculture and plant biotechnology. *Nature Reviews. Molecular Cell Biology*, *21*(11), 661–677. DOI: 10.1038/s41580-020-00288-9 PMID: 32973356

Chapter 5

Integrating Gene Editing With Traditional Breeding Methods

Sumreen Dawood
University of Mianwali, Pakistan

Syeda Anber Zahra
http://orcid.org/0009-0009-2933-9001
Shanghai Zhongqiao Vocational and Technical University, Shanghai, China

Salman Majeed
University of Mianwali, Pakistan

Xu Shujun
Shanghai Zhongqiao Vocational and Technical University, Shanghai, China

ABSTRACT

Crop development underpins agricultural progress and food security. Traditional breeding—selection, hybridization, and mutation—has delivered durable, high-yield varieties but is slow and limited by available genetic diversity. Gene-editing tools such as CRISPR-Cas9, TALENs, and ZFNs now enable precise genetic changes to improve yield, nutrition, disease resistance, and stress tolerance. This chapter reviews gene editing in modern agriculture, highlighting biofortification, extended shelf life, and enhanced resistance in major crops, as well as applications for climate-resilient crops and livestock. It also discusses barriers to adoption, including public perception, ethical concerns, and regulatory complexity. Case studies on climate-resilient

DOI: 10.4018/979-8-2600-3575-7.ch005

rice, heat-tolerant cattle, and pest-resistant maize illustrate integrated breeding strategies. Future gains will depend on advances such as pangenomics and genomic selection, alongside clearer regulation and improved public engagement.

INTRODUCTION

Crop improvement is a key invention that has contributed significantly to the advancement of agriculture and, consequently, human civilization. Crop improvement is an ongoing process of changing the genetically based characteristics of crops to better meet human needs. It must be defined from the perspective of humans growing crops for food and fibers (Kalaitzandonakes et al., 2023). Applied genetics has been introduced to agriculture through plant breeding, which has made it possible to produce hybrid types with increased agricultural yields. However, exploiting existing gene pools to create new varieties is a challenge for crop breeders (Nerkar et al., 2022). Breeders have introduced heritable mutations into plant genomes in various ways to produce new plant types. According to Mao et al. (2019), several mutagens have enabled researchers to generate vast genetic diversity during the last century.

New genes introduced via genetic engineering help expand the existing selection of genetic traits in plants. The long-standing objective of plant breeders has been to accurately alter crop genomes at specific locations in the genome, and genome editing technologies have become extremely effective tools for this purpose. This research investigates the operational methods of genome editing tools together with an assessment of CRISPR/Cas9 as the most frequently used genome editing system for enhancing crop characteristics through resistance to diseases and tolerance to herbicides and stress tolerance and improvements in yield attributes, along with reduced anti-nutrient content and extended shelf life, as well as insights into regulatory standards for genome-edited agricultural products (Nerkar et al., 2022). Figure 1 illustrates various strategies for crop genetic modification, including classical breeding, mutagenesis, RNA interference (RNAi), transgenesis and gene editing. Classical breeding relies on hybridization and selection, whereas mutagenesis introduces random genetic changes using radiation or chemicals. RNAi enables post-transcriptional gene silencing without altering the DNA sequence, whereas transgenesis involves the insertion of foreign DNA into plant genomes.

Figure 1. Strategies for crop genetic modification, including classical breeding, mutagenesis, RNA interference (RNAi), transgenesis, and gene editing, in terms of mechanisms, outcomes, and regulatory aspects.

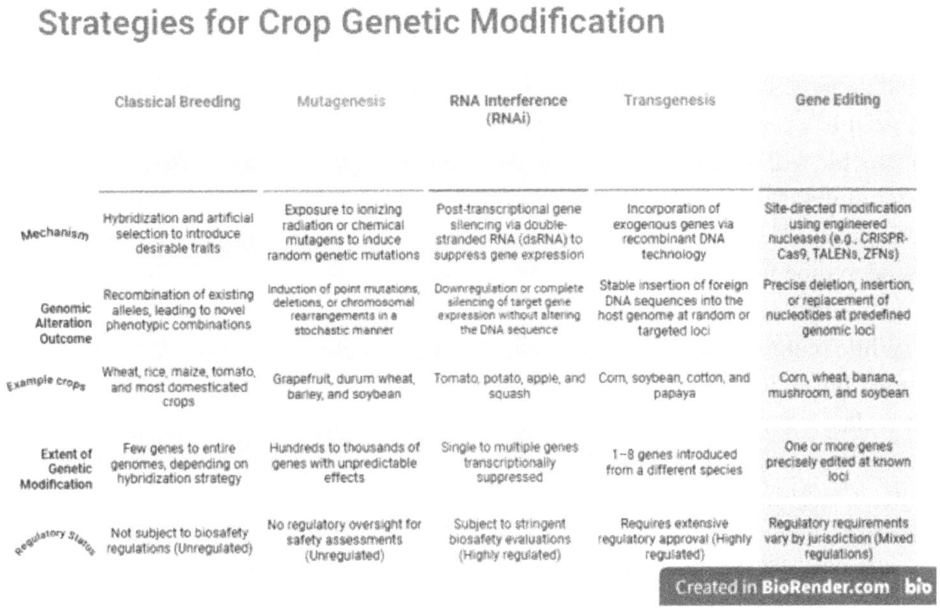

Strategies for Crop Genetic Modification

	Classical Breeding	Mutagenesis	RNA Interference (RNAi)	Transgenesis	Gene Editing
Mechanism	Hybridization and artificial selection to introduce desirable traits	Exposure to ionizing radiation or chemical mutagens to induce random genetic mutations	Post-transcriptional gene silencing via double-stranded RNA (dsRNA) to suppress gene expression	Incorporation of exogenous genes via recombinant DNA technology	Site-directed modification using engineered nucleases (e.g., CRISPR-Cas9, TALENs, ZFNs)
Genomic Alteration Outcome	Recombination of existing alleles, leading to novel phenotypic combinations	Induction of point mutations, deletions, or chromosomal rearrangements in a stochastic manner	Downregulation or complete silencing of target gene expression without altering the DNA sequence	Stable insertion of foreign DNA sequences into the host genome at random or targeted loci	Precise deletion, insertion, or replacement of nucleotides at predefined genomic loci
Example crops	Wheat, rice, maize, tomato, and most domesticated crops	Grapefruit, durum wheat, barley, and soybean	Tomato, potato, apple, and squash	Corn, soybean, cotton, and papaya	Corn, wheat, banana, mushroom, and soybean
Extent of Genetic Modification	Few genes to entire genomes, depending on hybridization strategy	Hundreds to thousands of genes with unpredictable effects	Single to multiple genes transcriptionally suppressed	1–8 genes introduced from a different species	One or more genes precisely edited at known loci
Regulatory Status	Not subject to biosafety regulations (Unregulated)	No regulatory oversight for safety assessments (Unregulated)	Subject to stringent biosafety evaluations (Highly regulated)	Requires extensive regulatory approval (Highly regulated)	Regulatory requirements vary by jurisdiction (Mixed regulations)

Plant breeding systems have faced difficulties in the twenty-first-century breeding processes used for developing climate-resistant crops. Climate change exacerbates the problems facing the world's food systems. Climate change has already affected agricultural outputs. enhanced temperature-moisture coupling intensifies the effects of climate change in semi-arid areas and increases the frequency of extreme events that threaten production stability. Rising temperatures and their fluctuations decrease agricultural yields worldwide (Xiong et al., 2007). The main challenges at the moment are to reduce pesticide use and agricultural practices and activities that are detrimental to both humans and the environment, find innovative ways to combat crop diseases and pests, and update agricultural management models for the information era. Modern agricultural output can be made sustainable and dependable even in the face of calamities such as pandemics or recurring problems such as climate change. A detailed examination of the problems faced by modern agriculture, including population growth, COVID-19, seed resource destruction, lack of water, dietary changes, air pollution, global warming, and changes in food prices, demonstrates that it is the shared duty of all nations to make the greatest investments possible to address these issues. Furthermore, crop growing seasons continue to be influenced by droughts and floods, which restrict water supplies, promote the growth of weeds, pests, and fungi, and decrease crop yields (Devlet, 2021).

The main obstacle lies in using gene-editing techniques on species that do not have transformation methods because it prevents the modification of non-crop species with nutritional value and essential crops' wild relatives and orphan crops. Generating gene-edited progeny from non-transformable species becomes feasible when genome-editing components are delivered to either shoot meristem cells or the germline (Hua et al., 2019). Population projections show that more than 9 billion people will inhabit the planet by 2050, posing significant challenges to modern agriculture that demands crops with better quality and higher yields while utilizing fewer inputs. The widespread application of conventional breeding methods requires extended time for the evolution from early screening phases until the initial crosses reach commercial use (Zhang et al., 2018).

While mutation breeding, molecular breeding, and in vitro culture are examples of non-conventional breeding, crossbreeding and distant hybridization are considered more classic forms of conventional breeding. Unconventional breeding strategies, such as molecular breeding and gene editing, are relatively quick and profitable. In contrast, traditional plant breeding techniques include crossing/hybridization, selection, and domestication (Sun et al., 2024). In contrast, genome-editing techniques provide the capacity to rapidly and precisely change genomes, enabling the direct introduction of enhancements into elite varieties (Hua et al., 2019). Genome editing tools can alter the genome structure with precision and accuracy at specific points. All of these techniques have been used to generate plants that can adapt to biotic and abiotic stresses and have high agricultural yields. Conventional plant breeding has significantly improved the global food supply and aided the development of modern civilization. Whole-genome sequencing, together with transcriptome sequencing, has made it easier to perform the fundamental tools of genetic and lineage mapping. These methods have made it easier to identify possible quantitative trait loci (QTLs) that are important for agriculture. The use of molecular breeding tactics increasingly depends on genomics (Mohanta et al., 2017). Many genome-editing techniques have been developed to address plant-related obstacles in food supply for future generations. The future production efficiency of essential foods, including rice, wheat, soybeans, and maize (*Zea mays*), should increase by 38 to 67 percent.

The genomes of bryophytes and angiosperms have been effectively modified using TAL effector nucleases (TALENs). The features we examined regarding genome-editing technologies demonstrate their advantages, including their operational effectiveness and multitarget gene-editing capabilities, as well as security aspects and ease of use, thereby enabling crop-breeding advancement (Ahmar et al., 2020a).

Research on agricultural genes has become possible through the implementation of CRISPR/Cas technology; nevertheless, observing changes can be challenging if two genes are identical. To overcome this, scientists may use a technique called CAFRI-Rice, which speeds up rice research and could also be useful for other plants, to compare genes and choose which ones to look into (Ahmar et al., 2020b). Plant breeding using genome editing could sustainably increase the food supply by 2050. The use of tools for genome editing is increasing with greater knowledge and fewer limitations. With the current agricultural techniques, there would never have been a population and food supply balance. Global warming restricts plant growth and food production. We are working to reduce these impacts to support agriculture. Improved agricultural practices and abiotic stress response systems have enabled the development of more tolerant crops. Despite the time required, both conventional and molecular breeding methods have successfully produced new crop varieties. Breeders are interested in genome editing using CRISPR/Cas9 technology. Enticing plant species for a secure supply (Zhang et al., 2018). The industrial implementation of single-atom catalysts requires guaranteed stability. Li et al. presented a perspective work that examined the theoretical aspects of single-atom catalyst stability challenges (Zhang et al., 2018).

OBJECTIVES

Crop Improvement

Gene editing has been linked to increased productivity through growth and reproduction processes. Rebuilding crops that can withstand abiotic challenges, such as salinity, drought, and high temperatures. By altering genes related to immunity or susceptibility, it can resist bacterial, fungal, and viral infections. Nutritional enhancement involves altering genes to increase the number of vitamins, minerals, or vital fatty acids in food.

Modification Without Transgenics

By causing mutations without adding foreign DNA, it can create non-transgenic plants, allaying concerns about public acceptance and regulations. By changing the genes that regulate flowering, plants can adapt better to local climates and growing seasons.

Climate Adaptation

Crop engineering to establish in a changing environment is known as climate adaptation (CA). It can create high-yield, hardy crops to satisfy the needs of the expanding global population. With the ongoing advancement of CRISPR-Cas technology, more accurate and multiplexed editing is now possible, leading to even more creative applications in agriculture and plant research.

Enhancement of Shelf Life

One of the most important characteristics of ornamentals, vegetables, and fleshy fruits is their long shelf life, which also has a better effect on market potential. One of the major challenges in breeding and genome engineering initiatives is maintaining and extending shelf life. Advanced gene integration technology is used to control the expression of genes linked to shelf life to preserve the flavor, aroma, and qualitative attributes of crops.

Biofortification

To address serious global issues such as malnutrition, especially in underdeveloped nations, CRISPR/Cas technology has been used in biofortification initiatives, such as raising the iron and zinc levels in wheat and the pro-vitamin A content in rice. CRISPR/Cas is used in biofortification to improve photosynthetic efficiency, nutrient uptake, and lodging resistance.

APPLICATIONS OF INTEGRATED APPROACHES

Enhancing Disease Resistance in Crops

Improving disease resistance is important for growing crops, as plant illnesses lead to significant economic losses worldwide. For a long time, most resilient crop varieties have been developed using conventional breeding methods. However, these methods can take a lot of time, require many resources, and depend on much genetic variety there is within a crop species. Conventional breeding techniques, along with genetic engineering tools such as CRISPR-Cas, can be used to develop pathogen-resistant crops to overcome these challenges.

Mechanisms of Disease Resistance

The modifications obtained through gene editing enable crops to develop resistance to diseases.

Adding Resistance Genes

Genetic engineering technology has been extensively used to introduce resistance genes to boost crops' immunity against pathogens from unfarmed crop varieties. The Xa21 gene, which was transplanted from wild rice into cultivated rice, protects against bacterial blight (*Xanthomonas oryzae*).

Disabling Susceptibility Genes

This technology prevents infections by inhibiting the S gene. Culture stocks demonstrate the operation of the MLO gene in wheat and barley plants, whereby disabling MLO results in fungal disease resistance to powdery mildew.

Boosting Immune Pathways

The plant receives higher activation of immunological signaling genes, enhancing its capability to identify and respond to infections.

Applications Across Major Crops

Table 1 summarizes the gene transfer methods and enzymes used in various breeding programs.

Table 1. Overview of gene transfer method in various crops

Sr. No.	Plant name	Gene transfer method	Used enzymes during gene transfer.
1	Golden rice	Particle bombardment, Agrobacterium-mediated transformation, and marker-assisted backcrossing (MABC).	Restrictions enzymes.
2	Bt cotton	Agrobacterium-mediated gene transfer. This method uses the bacterium Agrobacterium to transfer DNA into cotton plant cells.	Do not use enzymes.
3	Flavsavr tomato	Agrobacterium-mediated transformation in which the transfer-DNA (T-DNA).	Neomycin phosphotransferase II (NPTII)
4	BtBrinjal	Agrobacterium-mediated gene transfer.	Transferase enzymes.
5	Potato	CRISPR technology	Nucleases known as transcription activator-like effectors (TALENs).

Wheat

MLO genetic factor in wheat has been modified through CRISPR-Cas intervention to yield variants that fight off powdery mildew. Wheat growers benefit from this technique because it eliminates the requirement for fungicides, thus preserving environmental health while retaining steady harvest numbers (Numan et al., 2021).

Rice

Gene editing procedures modify the bacterial infection points of SWEET genes within the promoter regions to enhance rice plants against bacterial blight. According to Pedrozo et al., (2025), scientists have enhanced blast disease resistance through the genetic modification of essential resistance genes.

Tomatoes

Scientists have modified tomatoes to enhance their resistance to bacterial spot diseases by targeting genes that show susceptibility, such as SlJAZ2. Scientists have made changes to the Ty-1 and Ty-3 regions to enhance plant resistance against the tomato yellow margin curl virus (Chen et al., 2019).

Maize

CRISPR-Cas works by targeting viral proteins to protect maize against maize lethal necrosis (MLN). Scientists continue to explore gene modifications for better northern leaf blight disease mitigation (Zhang et al., 2018).

BENEFITS OF GENE MANIPULATION FOR DISEASE RESISTANCE

Precision

The modern method of gene editing achieves targeted gene selection with minimal risk of introducing unintended changes.

Speed

Gene editing expedites the process of breeding disease-resistant crops, which can take years using traditional methods.

Sustainability

Gene-edited crops promote more ecologically friendly farming by lowering the demand for chemical pesticides and fungicides.

Integration with Conventional Breeding

The combination of several resistance genes is made possible by combining gene editing and conventional breeding. This practice increases the ability of crops to tolerate various threats. To combat the evolving agricultural disease and pest situations, both institutions and governments need to collaborate under a unified strategy (Numan et al., 2021; Pedrozo et al., 2025).

OVERVIEW OF TRADITIONAL BREEDING METHODS

Traditional breeding methods are used to improve the varieties of plants and animals. The primary objective of traditional breeding methods is to achieve the highest possible yield. Different traditional breeding methods are used for this purpose. Some limitations of this study are as follows:

Selection And Breeding

The choice of parents in cultivar breeding methods is crucial. Selection and breeding are the most common traditional breeding methods. In this method, animals or plants with desired qualities, such as size, color, or resistance to disease, are selected and then bred to obtain benefits. Selection and breeding can be performed on wide crosses or distantly related organisms to obtain novel cultivars.

Cross-Pollination

In this method, pollen from the anther of one flower of a plant is transferred to the stigma of the flower of a different plant. Pollination can be achieved by using chemicals, treatment of pollen mixtures, in vitro pollination, etc. Cross-pollination results in a variety of traits in organisms, thus increasing genetic diversity. Cross-pollination may occur through a variety of agencies, such as insects, wind, birds, and water.

Hybridization

This technique is widely used to produce hybrid cultivars by crossing inbred lines with each other. It is used for crops such as maize, sunflower, and sugar beet, and for vegetables such as carrot, onion, tomato, and cabbage. New offspring arise from the cross of two individuals of the same or different species. The new generation has the desired characteristics of both parents. For example, the crossing of a high-yielding rice variety with one that is resistant to any disease.

Mutation Breeding

Mutation breeding is a method that uses physical or chemical means to produce or induce genetic variations in animals or plants to produce new varieties. This mutation results in evolution. Many techniques are important for manipulating chromosomes and genomes. This includes haploidization, polyploidization, translocation breeding, mutagenesis cell fusion, and partial genome transfer.

Mass Selection

As the name indicates, in this method of traditional breeding, a large group with specific traits is selected. Among these individuals, the best-performing individual is selected. Subsequently, the offspring of this individual were used for further selection. This method is used to obtain better quality crops.

Clonal Selection

This is a method of asexual reproduction in which genetic uniformity is maintained. Cutting and grafting are used for clonal selection. Trees, crops, and other plants are used for clonal selections.

KEY DIFFERENCES IN PLANT BREEDING TECHNIQUES

Traditional plant breeding methods have been the backbone of agriculture, allowing the creation of superior crop varieties with desirable features. Major techniques for traditional plant breeding include selection, hybridization, inbreeding, and backcrossing. Plant breeders can use these techniques to recognize, select, and propagate plants with superior qualities, combine complementary traits to create hybrids, inbreed desirable traits to fix them, and introduce specific traits through backcrossing. Traditional breeding relies heavily on phenotype-based selection, where observable physical traits are assessed to produce wise selections. Community-based methods, known as mass selection and pure line selection, allow for the formation of genetically homogeneous lines or the multiplication of seeds from numerous individuals with desirable features. Furthermore, field testing and evaluation are very important parameters for determining how plants respond in a variety of environmental situations as well as in the face of potential pests and diseases.

Differences Between Traditional Breeding and Modern Breeding

Traditional Breeding Methods

Traditional breeding methods rely on the natural genetic variation within crop populations and involve the selection of plants with desirable traits to breed the next generation.

Selection

The process of selecting plants with desirable traits or characteristics for reproduction. This could be as simple as a farmer saving seeds from the healthiest plants for the next season's planting.

Hybridization

This involves crossing two different plant varieties to produce offspring that may inherit the desirable traits of both parents. This method has been used to combine disease resistance from one plant with a high yield from another plant.

Landraces

They are locally adapted plant varieties that have developed over time through traditional farming. They often exhibit natural resistance to local pests and diseases.

Modern Breeding techniques

Plant breeders have used powerful tools to enhance the disease resistance of crops. Modern techniques allow for the precise manipulation of a plant's genetic material, enabling the introduction of disease-resistant traits from a broader range of sources, including different species.

Marker-Assisted Selection (MAS)

This technique uses molecular markers to identify plants carrying genes for specific traits such as disease resistance. MAS can significantly accelerate the breeding process by allowing breeders to select plants based on their genetic potential, rather than waiting for the trait to be expressed.

Genetic Engineering (GE)

Genetic engineering involves the direct insertion or modification of genes within the plant genome. This can introduce disease-resistant traits from unrelated species, offering possibilities that are beyond the reach of traditional breeding methods. It is also known as genetic modification.

Genome Editing

Gene editing introduces foreign DNA, and genome editing modifies plant genes, potentially sidestepping some of the regulatory and public acceptance issues associated with GMOs.

GENE EDITING

Biotechnological techniques, such as gene editing, enable researchers to precisely alter an organism's DNA (Joo et al., 2025). The discipline of gene editing is not new; in fact, studies in this area date back to the 1970s. In traditional genetics, gene-modifying operations are performed by choosing specific genomic locations (Mao et al., 2019). In molecular genetics, researchers examine the phenotypic impacts of particular modified gene sequences to understand gene functions. The process of altering and designing genomic DNA sequences is known as genome engineering. It focuses on single-gene addition and deletion (Boti et al., 2023). Figure 2 illustrates the mechanisms of genome editing tools, including ZFNs, TALENs, and CRISPR-Cas9, which enable site-specific DNA modifications. ZFNs and TALENs utilize engineered DNA-binding proteins fused to FokI nuclease to introduce double-strand breaks at target sites, requiring dimerization for cleavage.

Figure 2. The mechanisms of genome editing tools, including Zinc Finger Nucleases (ZFNs), Transcription Activator-Like Effector Nucleases (TALENs), and CRISPR-Cas9, involve DNA recognition and cleavage.

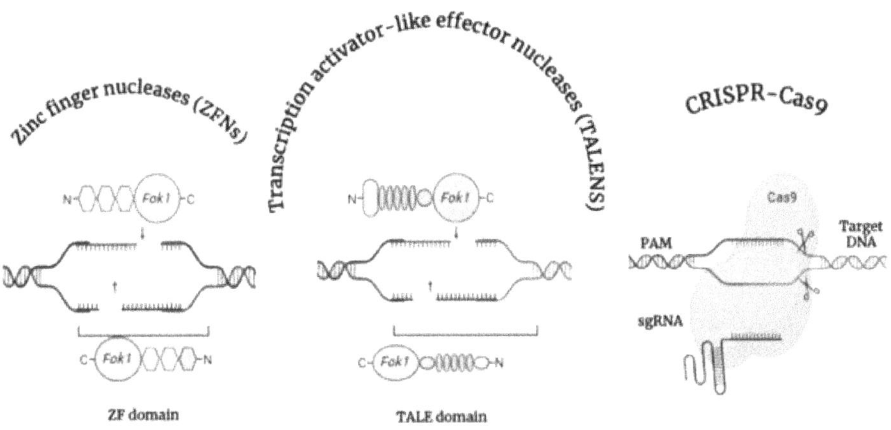

CRISPER-Cas9 Techniques

The gene-editing technique CRISPR-Cas9 has several applications and has gained popularity in biological research because of its affordability and benefits (Figure 3). The Cas9 factor, which functions similarly to molecular scissors, translates the RNA (gRNA), which points the Cas9 enzyme in a direction of a particular spot on the DNA, making up this system, which is inspired by a natural defense mechanism in bacteria (Yao et al., 2023). Because the gRNA sequence complements that of the target DNA, it can precisely cut the genome (Kumar et al., 2024). Scientists can use this technique to induce particular alterations, such as gene disruption, to correct genetic abnormalities once the Cas9 enzyme slices the DNA and activates the repair systems (Kumar et al., 2024).

Various fields have been changed by this technology, but biomedical research is particularly important, as it has made it possible to create infection models, study gene function, and diagnose new treatments and approaches to curing genetic diseases (Song et al., 2026). Genes associated with cancer have been recognized for immunotherapy and novel therapies using CRISPR Cas9 (Song et al., 2026).

Figure 3. Applications of CRISPR-Cas9 in genomic engineering, highlighting its role in medicine, biology, and biotechnology, including genetic disorder correction, pharmaceutical development, animal models, biofuel production, and crop improvement.

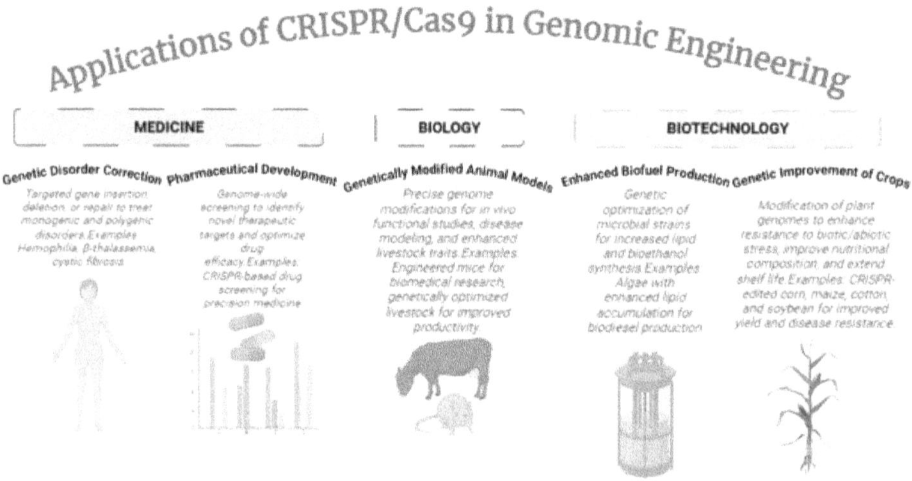

TALENs

Gene editing is also possible through engineered proteins and transcription activators, such as effector nucleases (TALENs), which can accurately modified accurately (Joung et al., 2013). Two fundamental components of TALENs are the restriction enzyme FokI from the non-specific DNA-cleaving nuclease domain and transcription activator-like effectors (TALEs), which are DNA-binding domains (Joung et al., 2013).

Fok I nuclease domain helps generate double-strand breaks by binding to opposing DNA strands (Zhang et al., 2021; Moiani, et al., 2024). Specific DNA with 14-20 nucleotides targeted by TALENs with specially designed DNA binding domains for a range of genome editing applications, including insertions, gene knockout, and knockin (Zhang et al., 2021). TALENs have been successfully applied to a variety of organisms, including animal, human, and plant cells (Joung et al., 2013). Sickle cell mutations have recently been rectified in human cells using TALENs (Moiani, et al., 2024).

ZFNs

Zinc finger nucleases are artificial restriction enzymes used to alter or manipulate the target, a specific region of DNA (Jaganathan et al., 2018). It has primary functional domains that are DNA requisite domain, such proteins identified as (ZFP) zinc finger protein and Folk I restriction enzyme-based cleavage domain. Each ZFP recognizes a sequence of three base pairs is recognized by each zinc finger protein (ZFP) (Holmes-Davis et al., 2005). Three to six finger pairs bind to 18–36 nucleotides in the genome. Two ZFNs are attached to the opposing DNA strands at an exacting space and cut by the Fok nuclease domain (Holmes-Davis et al., 2005).

Double-strand breaks (DSBs) are caused by ZFNs for accurate innate DNA repair at particular genome locations (Zagelbaum et al., 2023). Following DSB, precision gene editing can be accomplished with ZFNs using homology-concentrating repair or non-homologous ends (Bonawitz et al., 2019). ZFNs can become more efficient and selective by increasing the number of fingers, although off-target effects must still be considered. ZFNs have been used in clinical trials to fight HIV by interfacing with the CCR5 gene in T-cells. HR-mediated gene changes can be accomplished by ZFN nickases (ZFNickases); furthermore, mutagenic NHEJ repairs can be decreased due to single-strand breaks (Ain et al., 2015).

Meganucleases

Meganucleases are molecular DNA scissors that are helpful tools for genome editing because they can detect long DNA sequences using a type of highly specific endonuclease (Vaschetto et al., 2021). Meganucleases are members of the LAGL-IDADG family of homing endonucleases that cause DSB and trigger the repair process through homologous recombination (HR) (Nerkar et al., 2022). The most frequently employed meganucleases in genome engineering are I-SceI and I-CreI (Taylor et al., 2012).

Due to their high selectivity, I-SceImeganuclease reduces toxicity and promotes homologous recombination for successive cutting of both DNA strands. Meganucleases help create a particular DSB through the HR repair pathway. New opportunities for gene editing have been made possible by meganucleases, which can alter the genome of microbes, animals, and plants (Vaschetto et al., 2021). By meganuclease gene insertion, gene correction, DNA sequence removal, and gene inactivation have been manipulated (Taylor et al., 2012).

ETHICAL CONSIDERATIONS OF TRADITIONAL METHODS AND GENE EDITING

Disease and Disability

Potential harm to those who currently live with illness or loss is one of the most commonly raised issues regarding the effects of technologies such as HHGE and PGT. There exists criticism of people living with that condition or disability expressed in the future.

Governance

The Human Fertilization & Embryology Act must be updated by Parliament for HHGE to be legal in a nation such as the UK. Some governments will need equivalent methods, whereas others will have different requirements. Political and public support for HHGE will likely depend on this technology being supervised.

Deontology

Deontology is a moral theory founded on actions based on rules that some actions are right or wrong.

Rights

Due to rights, there are many ethical difficulties surrounding methods by which biotechnology could prevent the widespread killing of day-old chicks or decrease animal diseases, or it could cause discomfort by leading to some body parts growing more than others (meat cattle races that cannot give birth naturally; too heavy chicken, breaking their legs). Parallel analyses are available for plant breeding. From an organic/biodynamic perspective, Lammerts van Bueren and Struik arrived at the following rights of crop plants.

Public Acceptability

Gene-editing techniques are complicated, and people face difficulties in using this technology. To make this technology successful, there should be great awareness about it, and the public should be satisfied. Perceptions were usually favorable, particularly regarding medical conditions and deadly illnesses, but they were also influenced by a sense of risk. The acceptability of somatic therapies was higher than that of germline therapies. Although there are several ways to quantify opinions on the latest techniques, there are limitations.

Moral Responsibility of Breeder

Breeders have the moral responsibility to inform customers about the breeding methods used to create new varieties when requested. It is up to the customers to decide when it is appropriate to transfer that information into the value chain and ultimately reach consumers.

APPLICATION OF INTEGRATED APPROACHES IN DEVELOPING CLIMATE-RESILIENT AGRICULTURE SYSTEM IN TRADITIONAL BREEDING SYSTEM

Breeding harvest plants with increased productivity and tolerance to harsh settings is essential for world nutrition in light of current and anticipated climate change, which includes maximum heat and unpredictable weather extremes worldwide in large areas of the earth. Climate change has complex effects on crop production, affecting yields both directly and indirectly. There will need to be a major rise in crop-plant output (such as food, feed, fibers, and bioenergy) to cater to the constantly expanding human population. The most promising approaches to guarantee global food security involve identifying the physiological and genetic processes that underlie crop resistance to less-than-ideal soil states and creating harvests with an improved productivity ratio.

Traditional Breeding Systems and Climate Challenges

Slow Progress

To create new varieties, traditional breeding techniques require drawn-out cycles of crossing, selection, and testing, which may take years. This makes it challenging to adapt to the difficulties of a changing climate. Ahmar et al. (2020a and 2020b).

Limited Genetic Diversity

The ability to introduce novel features required for climate resilience may be limited if current crop types are relied upon for their genetic foundations (Atlin et al., 2017).

Phenotypic Selection Limitations

Conventional breeding frequently depends on outwardly observable features (phenotypes), which do not always correctly represent the underlying genetic pathways for stress tolerance, particularly for subterranean traits, such as root architecture (Ndour et al., 2017).

Environmental Variability

Finding climate-resilient cultivars can be difficult because climate conditions can change dramatically from year to year, making testing and selection across multiple environments problematic (H. M. Paniza, 2024).

Integrated Approaches in Breeding for Climate Resilience

Marker-Assisted Selection (MAS)

Transgenics have revolutionized current crop enhancement, such as the remarkable launch of bio-fortified rice to fight hypovitaminosis A and genetically modified insect-resistant corn. The MAS complemented the vast plant hybridization programs worldwide. With a focus on climate resilience, the next phase of crop enhancement will likely have advantages in utilizing large-scale data and greater hereditary variety. Using the most recent developments, we analyze the potential for crop advances in the future.

Expanding the Gene Pool by Utilizing CWRs

The genetic diversity of contemporary crops has been reduced by domestication and breeding bottlenecks. In contrast, because they have not been subjected to anthropogenic selection, CWRs retain noticeably greater genetic variety than their cultivated progeny. Furthermore, they possess a diverse set of adaptive features, such as tolerance to numerous living and nonliving stresses. CWRs are valuable sources that can be used to enhance the resistance of high-yielding varieties.

Genome Diversity and Unexplored Genetic Variation

Genetic modification has accelerated agricultural enhancement initiatives by using unique mutations and manufactured changes in the gene pool with greater accuracy. The manipulation of genetic diversity is at the heart of crop improvement. Significant advancements in genome sequencing have shown that a species' variety is not fully represented by the genetic information gathered from a single reference genome. As a result, pangenomes have been built for several crop species. Super-pangenomes, or genus-wide pangenomes that include wild germplasm repertoires, have been proposed recently. Single nucleotide polymorphisms (SNPs), mutations, and genes with structural variants (SVs) that regulate significant population features are among the genetic factors highlighted by crop pangenomics. To increase confidence in SVs as the causative feature linked to powdery mildew resistance, researchers have investigated the correlation between mapping genetic markers and disease-resistant genes in thale cress (Weigel and Barragan, 2021).

Genetic Selection for Next-Generation Agriculture

Genomics-assisted breeding has consistently led the way in terms of genetically enhancing harvest. However, the rate of improvement must keep up with the pressures of demographic expansion and environmental changes. Genomic selection (GS) is an innovation that significantly increases genetic gain by quickly identifying superior lines using both phenotypic and genome-wide marker information. GS has been extensively studied in many economically significant crops. It has proven particularly helpful in enhancing traits with intermediate phenotypes that are challenging to breed, like disease resistance, and non-living stress tolerance are crucial for upcoming crop improvement initiatives (Singh et al., 2025).

CHALLENGES IN INTEGRATION

Obtaining Efficient GE Transformants

The most significant obstacle is the acquisition of effective GE transformants. The ability of GE plants to regenerate is a key factor in determining their potential, although their genetic transformability is often genotype-dependent. While several developmental regulators (including WUSCHEL, BABY BOOM, and GROWTH-REGULATING FACTOR) encourage the regeneration of shoots in certain agricultural plants, it is unknown whether they do the same for other species (Wang et al., 2025).

Challenging for Very Complex Polyploid Genomes

With extremely complex polyploid genomes, such as those of sugarcane, cotton, wheat, and potatoes, it is extremely difficult and time-consuming to introduce desirable traits into popular crop variants using traditional breeding techniques. In polyploid plants, it may be challenging to introduce several characteristics and alter metabolic pathways using traditional breeding techniques (Arora et al.,2017).

Increased Susceptibility to Abiotic Stress and Other Pathogenic Agents

To promote disease resistance using genome editing, plant susceptibility (S) and/or resistance (R) genes must be identified and functionally annotated. This requires access to whole-genome sequencing and is a time-consuming process. However, little or no knowledge is available regarding the molecular functions of S/R genes in most non-model plant species. Furthermore, many S genes are essential for the continued survival of the host because they have dual roles in plant physiology and pathogen susceptibility. Inactivating genes usually leads to disease resistance in addition to pleiotropic effects, such as reduced plant development, morphological defects, and greater susceptibility to environmental stresses and/or other pathogenic agents. (Cardi et al.,2023).

Off-Target Gene Mutation With CRISPR-Cas9

One major issue with CRISPR-Cas9 is the potential for editing incorrect regions of the genome, which could lead to undesirable genetic changes in the target organism. The characteristics and development of plants may be affected by these mistakes. Although there are methods to help identify mistakes and upgraded Cas variants, such as Cas9-HF1 and eSpCas9, to reduce them, coping with the complexity of plant genomes remains challenging (Nerkar et al.,2022; Mao et al.,2019).

Deleting Non-Coding Genes is Difficult

Compared with protein-coding genes, ncRNA genes are more difficult to suppress or remove using the CRISPR/Cas9 method. Knockdown/knockout of small non-coding genes is difficult because of the restricted design space for CRISPR-Cas9 to target non-coding genes without interfering with neighboring genes. This is particularly troublesome when it comes to silencing miRNAs because many of them are found within the introns of host genes that code for proteins (Basak et al.,2015).

REGULATORY AND ETHICAL CONCERNS

Regulatory Challenges

Diverse Regulatory Frameworks

Various countries enforce separate legislative systems for genetically modified crops. The United States and Canada maintain less strict regulations on gene-edited crops, while the European Union applies GMO-style regulations to them (Idris et al., 2023). Divergent laws create trade barriers that restrict the international commerce of modified plants derived from genetic editing (Lyzenga et al., 2021).

Classification Issues

Regulatory systems face major challenges in defining whether gene-edited crops should be categorized as GMOs or non-GMOs. The classification system determines the regulatory requirements for safety evaluations and labeling protocols for producers (Lyzenga et al., 2021). A universal classification system is necessary to develop unified policies because no such system currently exists (Vaschetto et al., 2021).

Intellectual Property Rights

The questioning of ownership rights emerges when gene-editing technologies and their resulting crops receive patent protection. Conditions within intellectual property rules restrict both farmers and researchers from essential scientific exchange activities, impeding conventional breeding processes (Idris et al., 2023; Lyzenga et al., 2021).

Ethical Concerns

Public Perception and Acceptance

The public perception of gene editing depends on language components, including precision and editing, and their use in this field. The technology suffers from an obscure nomenclature, which enables it to conceal potential threats and ethical complications, thus preventing people from engaging in knowledgeable discussions about it (Idris et al., 2023). According to Lyzenga et al. (2021), society needs genuine and honest dialog to solve existing problems.

Consumer Rights

The ethical examination of food product information must respect the rights of consumers as part of its process. The demand for the clear identification of gene-edited products exists in markets that prefer organic and non-GMO foods (Idris et al., 2023; Lyzenga et al., 2021).

Impact on Organic Farming

The arrival of gene editing poses a risk to the authenticity of organic farming. The blending of biotechnological treatments with genetic modification techniques arising from gene editing raises concerns about regulatory conflicts in the organic industry (Idris et al., 2023).

PUBLIC ACCEPTANCE AND MARKET CHALLENGES

Public Acceptance

Differences in Acceptance

The general public shows better acceptance of gene-edited foods than genetically modified organisms, although it varies depending on the specific location and population characteristics. Public interpretation of food technology relies heavily on two main aspects: knowledge of the technologies and their benefits (Strobbe et al., 2023). People tend to adopt gene-edited products once they understand the benefits of these products for their health and sustainability (Spök et al., 2022).

Human Health Impacts

All foods produced using new technological methods require extensive testing before being declared suitable for human consumption. Science must be used to determine the safety of food products by integrating exposure evaluations to confirm risk levels with hazard identification and characterization. Regulations in many jurisdictions require safety testing of food made from genetically modified crops. Science-based frameworks for regulating food safety should prioritize ensuring that food is healthy, secure, and nourishing rather than being convoluted by politics or opinion, which breeds polarized opinions, misunderstandings, and mistrust (Caradus, 2023).

Knowledge About New Technology

Customers frequently lack the knowledge necessary to determine whether genetically modified items are dangerous or to evaluate the risks of a particular technology against its possible advantages. It is interesting to note that when labeling is mandated, a specific customer advantage is necessary for product adoption because a positive association has been shown between consumer sentiments about GMO-free foods and purchase behavior (Caradus, 2023).

Market Challenges

Regulatory Obstacles

Different nations have different laws governing gene-edited crops. Strict laws, similar to those governing genetically modified organisms (GMOs), prevent gene-edited items from entering some markets. In contrast, nations such as the United States, Canada, and Australia have implemented more relaxed laws, facilitating the rapid marketing of genetically modified crops (Spök et al., 2022; Strobbe et al., 2023).

Customer Preferences

The success of gene-edited foods in the market depends on many consumers who prefer traditional breeding approaches to contemporary techniques. Studies have shown that consumers primarily choose foods derived from conventional plant farming over genetically altered alternatives or edited versions when they consider new products to be less authentic (Strobbe et al., 2023). This choice prevents the product from achieving satisfactory market penetration.

Price Sensitivity

The prices set by retailers influence customers' responses to gene-edited foods. Research findings have established that customers tend to purchase products when they cost less than their non-gene-edited counterparts (Spök et al., 2022). Price sensitivity within the market threatens producers' investments in gene-editing technologies.

Market Confusion Challenge

CRISPR-Cas-produced plants and naturally changed plants are indistinguishable because they possess identical traits. The public finds it difficult to differentiate between distinct mutations, such as those caused by cisgenesis, CRISPR, or chemical/physical approaches, because genetic analysis is required for both plants, and consumers may become confused. The public finds it difficult to differentiate between CRISPR-driven mutations and spontaneous or chemical/physically induced mutagenic plants because both require genetic analysis (Jiang, 2020).

International Trade Challenge

The commerce of gene-edited plants between the European Union and other nations is adversely affected by strict GMO laws in the agricultural sector. The free market is restricted, and customer choices are limited by rigorous GMO laws. (Jiang, 2020).

Need for Public Dialogue

Experts stress the significance of continuing public discourse to address concerns and misunderstandings regarding gene editing, notwithstanding the optimistic outlook. The historical background of the introduction of GMOs, which was met with criticism because of poor communication, emphasizes the necessity of an open dialogue regarding the advantages and safety of gene-editing technologies (McFadden et al., 2024; Kato-Nitta et al., 2019).

Case Study

IRRI is creating rice cultivars that are resilient to climate change-related circumstances, which are expected to become increasingly frequent and severe. This covers soil issues such as salinity and iron toxicity, as well as water shortage, floods, heat, and cold. (Costa-Font et al., 2008)

Examples of Heat-Resilient Wheat

PBW 803: A wheat cultivar that can withstand heat and yields typical amounts

Dbw 187: A wheat cultivar that can withstand high temperatures and yields typical amounts

DBW 222: A wheat cultivar that can withstand heat and gives typical amounts

Wheat varieties that can tolerate temperatures up to 4°C above the average are known as heat- and drought-tolerant varieties (Bhandari et al. 2024).

Transgenic Cotton

i) Luminayan 28: A BT transgenic cotton cultivar that can withstand drought and is impervious to insect infestations ii)Luminayan 29, a transgenic cotton cultivar that can withstand waterlogging and insect infestations

iii) Luminayan 30, a transgenic hybrid cotton cultivar that can withstand waterlogging and insect infestations

CONCLUSION AND FUTURE DIRECTIONS

Plant biotechnology has undergone a revolution because of CRISPR-Cas technology, which makes genome editing accurate, effective, and flexible in plants. The first directed evolution platform based on CRISPR/Cas has been established for plants, aiming to modify the plant genome without serious side effects. This technology is crucial for improving plants in the context of environmental issues and climate change. The CRISPR/dcas9 activation mechanism is still in its early stages, but research is ongoing in plant species, such as rice and *Arabidopsis*. New gene-editing methods, such as Cas12a T-containing PAM with short RNA guide sequences, are being used to enhance complex proteins. Moreover, Plant-Usable Light-Switch Elements (PULSE) have been created to address compatibility issues with plant growth requirements. Single-cell sequencing technology in plants is enhancing our understanding of transcriptional regulation and creating new opportunities for CRISPR-activation systems. CRISPR, initially discovered in bacteria and archaea, has revolutionized life sciences and plant improvement. This simplifies basic research and accelerates plant germplasm growth, addressing ecological, agricultural, and environmental issues. However, delivering CRISPR libraries to plants requires considerable effort. CRISPR-Cas technology can be used to develop insect-resistant plants, thereby reducing production costs and increasing farm profitability. Genome-editing methods use in vitro and cellular processes to precisely alter genes; however, their use for crop development should be limited to breeding goals because of existing variability. Legal concerns and biosafety issues remain regarding genetically modified insect pests. CRISPR-Cas gene editing is simple, effective, and adaptable, making it useful for precise gene editing in medicinal plants. However, it has drawbacks, such as limitations with the protospacer-adjacent motif (PAM) sequence, increased off-targeting, and a larger Cas9 gene. DNA repair mechanisms in plants to predict and control the editing outcomes. CRISPR technology can be used to create new crops, enhancing research in plant biology and food security. However, challenges such as improved gene-targeting efficiency, identification of essential plant characteristics, and effective delivery of the CRISPR construct have been addressed by CRISPR/Cas9 variations and Cas9 orthologs. Its efficiency, high target specificity, and flexibility allow for precise genetic modification, leading to new farming techniques and genetic differences in germplasm development. High-throughput phenotyping, which utilizes advanced sensors and cameras, helps address this issue by providing precise data for breeding goals. CRISPR-related plant biotechnologies have been created or enhanced, including morphogenetic modification, transgene-free plant genome editing, multiplexing, and functional genomics. Agriculture, which is crucial for the survival of the Earth, produces essential goods such as milk, meat, eggs, wood, clothing, honey, silk, waxes, and medicines. With CRISPR technology transforming plant genetics, the future of molecular markers is expected to be more accurate, efficient, and cost-effective, providing deeper insights into desirable traits. This tool enables the full utilization of genomic and systems biology data to discover novel traits in numerous plant species. Moreover, the use of CRISPR guide molecular libraries for gene and trait screening in human health could be adapted to improve crop yield, pest resistance and disease tolerance. A major challenge for CRISPR in agriculture is the effective delivery of the CRISPR-Cas complex to plant cells, which requires the development of advanced delivery systems to enable high-efficiency genome editing. With CRISPR/Cas-based techniques enabling multiplex editing and de novo domestication, plant breeders now have extraordinary opportunities to create genetic diversity and develop elite crops, thereby overcoming the limitations of traditional breeding methods. While advancements in genome editing (GE) continue to progress in plant science, challenges remain, particularly in achieving high-efficiency edits in polyploid plants, where multiple gene copies complicate homozygous plant production. To overcome this, large-scale screening is required after GE experiments. High-throughput phenotyping is essential for assessing edits, and increasing recombination frequencies is crucial for making GE more applicable to crop plants. Pooled CRISPR libraries for model plants would aid in functional genomics and crop improvement, especially in coding and non-coding regions. CRISPR/Cas has advanced genome editing, enabling the creation of designer crops; however, challenges such as improving recombination frequency and reducing off-target effects remain. As CRISPR/Cas evolves, it is essential to update regulations and inform the public about genome-edited plants. This technology, which offers alternatives to strict GMO laws, has already been successful with products such as high-oleic soybean oil. The development of smaller or improved Cas9 proteins is crucial for advancing genome editing, particularly for applications in agricultural genome editing with CRISPR-Cas9, which raises concerns about off-target mutations and unwanted genetic changes. Efforts are needed to refine gRNA design, protein engineering, and delivery methods to minimize these risks. Orthogonal gene targeting has not yet been explored in plant systems and requires further research to determine its cross-species adaptability. Improved bioinformatic tools for gRNA design can reduce off-target effects and enhance editing efficiency; however, more CRISPR-Cas datasets are needed to refine these methods.

REFERENCES

Ahmar, S., Gill, R. A., Jung, K. H., Faheem, A., Qasim, M. U., Mubeen, M., & Zhou, W. (2020b). Conventional and Molecular Techniques from Simple Breeding to Speed Breeding in Crop Plants: Recent Advances and Future Outlook. *International Journal of Molecular Sciences*, *21*(7), 2590. DOI: 10.3390/ijms21072590 PMID: 32276445

Ahmar, S., Saeed, S., Khan, M. H. U., Ullah Khan, S., Mora-Poblete, F., Kamran, M., & Jung, K. H. (2020a). A Revolution Toward Gene-Editing Technology and Its Application to Crop Improvement. *International Journal of Molecular Sciences*, *21*(16), 5665. DOI: 10.3390/ijms21165665 PMID: 32784649

Ain, Q. U., Chung, J. Y., & Kim, Y. H. (2015). Current and future delivery systems for engineered nucleases: ZFN, TALEN and RGEN. *Journal of Controlled Release : Official Journal of the Controlled Release Society*, *205*, 120–127. DOI: 10.1016/j.jconrel.2014.12.036 PMID: 25553825

Arora, L., & Narula, A. (2017). Gene Editing and Crop Improvement Using CRISPR-Cas9 System. *Frontiers in Plant Science*, *8*, 1932. DOI: 10.3389/fpls.2017.01932 PMID: 29167680

Barragan, A. C., & Weigel, D. (2021). Plant NLR diversity: The known unknowns of pan-NLRomes. *The Plant Cell*, *33*(4), 814–831. DOI: 10.1093/plcell/koaa002 PMID: 33793812

Basak, J., & Nithin, C. (2015). Targeting Non-Coding RNAs in Plants with the CRISPR-Cas Technology Is a Challenge Yet Worth Accepting. *Frontiers in Plant Science*, *6*, 1001. DOI: 10.3389/fpls.2015.01001 PMID: 26635829

Bhandari, R., Paudel, H., Nyaupane, S., & Poudel, M. R. (2024). Climate resilient breeding for high yields and stable wheat (Triticum aestivum L.) lines under irrigated and abiotic stress environments. *Plant Stress*, *11*, 100352. https://doi.org/https://doi.org/10.1016/j.stress.2024.100352. DOI: 10.1016/j.stress.2024.100352

Bonawitz, N. D., Ainley, W. M., Itaya, A., Chennareddy, S. R., Cicak, T., Effinger, K., Jiang, K., Mall, T. K., Marri, P. R., Samuel, J. P., Sardesai, N., Simpson, M., Folkerts, O., Sarria, R., Webb, S. R., Gonzalez, D. O., Simmonds, D. H., & Pareddy, D. R. (2019). Zinc finger nuclease-mediated targeting of multiple transgenes to an endogenous soybean genomic locus via non-homologous end joining. *Plant Biotechnology Journal*, *17*(4), 750–761. DOI: 10.1111/pbi.13012 PMID: 30220095

Boti, M. A., Athanasopoulou, K., Adamopoulos, P. G., Sideris, D. C., & Scorilas, A. (2023). Recent Advances in Genome-Engineering Strategies. *Genes*, *14*(1), 129. DOI: 10.3390/genes14010129 PMID: 36672870

Caradus, J. R. 2023. Processes for Regulating Genetically Modified and Gene-Edited Plants. *GM Crops & Food* 14 (1): 1–41. DOI: 10.1080/21645698.2023.2252947

Cardi, T., Murovec, J., Bakhsh, A., Boniecka, J., Bruegmann, T., Bull, S. E., & Van Laere, K. (2023). CRISPR/Cas-Mediated Plant Genome Editing: Outstanding Challenges a Decade after Implementation. *Trends in Plant Science*, *28*(10), 1144–1165. DOI: 10.1016/j.tplants.2023.05.012 PMID: 37331842

Chen, K., Wang, Y., Zhang, R., Zhang, H., & Gao, C. (2019). CRISPR/Cas Genome Editing and Precision Plant Breeding in Agriculture. *Annual Review of Plant Biology*, *70*(1), 667–697. DOI: 10.1146/annurev-arplant-050718-100049 PMID: 30835493

Costa-Font, M., Gil, J. M., & Traill, W. B. (2008). Consumer acceptance, valuation of and attitudes towards genetically modified food: Review and implications for food policy. *Food Policy*, *33*(2), 99–111. https://doi.org/https://doi.org/10.1016/j.foodpol.2007.07.002. DOI: 10.1016/j.foodpol.2007.07.002

Devlet, A. (2021). Modern Agriculture and Challenges. *Frontiers in Life Sciences and Related Technologies*, *2*(1), 21–29. DOI: 10.51753/flsrt.856349

Food and Agriculture Organization (FAO). (2018). *Climate-Smart Agriculture: Policies, Practices, and Financing for Food Security*. FAO.

Holmes-Davis, R., Li, G., Jamieson, A. C., Rebar, E. J., Liu, Q., Kong, Y., Case, C. C., & Gregory, P. D. (2005). Gene regulation in planta by plant-derived engineered zinc finger protein transcription factors. *Plant Molecular Biology*, *57*(3), 411–423. DOI: 10.1007/s11103-004-7820-x PMID: 15830130

Hua, K., Zhang, J., Botella, J. R., Ma, C., Kong, F., Liu, B., & Zhu, J. K. (2019). Perspectives on the Application of Genome-Editing Technologies in Crop Breeding. *Molecular Plant*, *12*(8), 1047–1059. DOI: 10.1016/j.molp.2019.06.009 PMID: 31260812

Idris, S. H., Mat Jalaluddin, N. S., & Chang, L. W. (2023). Ethical and Legal Implications of Gene Editing in Plant Breeding: A Systematic Literature Review. *Journal of Zhejiang University. Science. B*, *24*(12), 1093–1105. DOI: 10.1631/jzus.B2200601 PMID: 38057267

International Rice Research Institute (IRRI). (2020). *Stress-Tolerant Rice for Africa and South Asia (STRASA): Final Project Report*. IRRI.

Jaganathan, D., Ramasamy, K., Sellamuthu, G., Jayabalan, S., & Venkataraman, G. (2018). CRISPR for crop improvement: An update review. *Frontiers in Plant Science*, *9*, 985. DOI: 10.3389/fpls.2018.00985 PMID: 30065734

Jiang, L. 2020. Commercialization of the Gene-Edited Crop and Morality: Challenges from the Liberal Patent Law and the Strict GMO Law in the EU. *New Genetics and Society* 39 (2): 191–218. https://doi.org/DOI: 10.1080/14636778.2019.1686968

Joo, J. H., Lee, S., & Kim, K. P. (2025). Precision gene editing: The power of CRISPR-Cas in modern genetics. *Molecular Therapy. Nucleic Acids*, *36*(4), 102733. DOI: 10.1016/j.omtn.2025.102733 PMID: 41210586

Joung, J. K., & Sander, J. D. (2013). TALENs: A widely applicable technology for targeted genome editing. *Nature Reviews. Molecular Cell Biology*, *14*(1), 49–55. DOI: 10.1038/nrm3486 PMID: 23169466

Kalaitzandonakes, N., Willig, C., & Zahringer, K. (2023). The Economics and Policy of Genome Editing in Crop Improvement. *The Plant Genome*, *16*(2), e20248. DOI: 10.1002/tpg2.20248 PMID: 36321718

Kato-Nitta, N., Maeda, T., Inagaki, Y., & Tachikawa, M. 2019. Expert and Public Perceptions of Gene-Edited Crops: Attitude Changes in Relation to Scientific Knowledge. *Palgrave Communications* 5: 1. https://doi.org/DOI: 10.1057/s41599-019-0328-4

Kumar, R., Shanker, R., Singh, P., Yadav, M. K., Chaudhary, V., & Kumar, M. (2024). Genome editing towards pests and disease management in agricultural crops: Recent developments, challenges and future prospects. *Physiological and Molecular Plant Pathology*, *134*, 102402. DOI: 10.1016/j.pmpp.2024.102402

Li, Z., Liu, Z. B., Xing, A., Moon, B. P., Koellhoffer, J. P., Huang, L., & Cigan, A. M. (2015). Cas9-Guide RNA Directed Genome Editing in Soybean. *Plant Physiology*, *169*(2), 960–970. DOI: 10.1104/pp.15.00783 PMID: 26294043

Lyzenga, W. J., Pozniak, C. J., & Kagale, S. (2021). Advanced Domestication: Harnessing the Precision of Gene Editing in Crop Breeding. *Plant Biotechnology Journal*, *19*(4), 660–670. DOI: 10.1111/pbi.13576 PMID: 33657682

Mao, Y., Botella, J. R., Liu, Y., & Zhu, J. K. (2019). Gene Editing in Plants: Progress and Challenges. *National Science Review*, *6*(3), 421–437. DOI: 10.1093/nsr/nwz005 PMID: 34691892

McFadden, P., McColgan, M., MacLochlainn, J., Davies, H., Kirby, K., Currie, D., Schroder, H., Nicholl, P., McGrory, S., Naylor, R., Kirby, K., Mullineux, J., & Mallett, J. (2024). *Safer and Effective Staffing Research and Policy Development Older People's and Children's Social Work in Northern Ireland: Report 1- The Starting Point: Baseline Analysis*. Advance online publication.

Mohanta, T. K., Bashir, T., & Hashem, A., Abd_Allah, E. F., and Bae, H. (. (2017). Genome Editing Tools in Plants. *Genes*, *8*(12), 399. DOI: 10.3390/genes8120399 PMID: 29257124

Moiani, A., Letort, G., Lizot, S., Chalumeau, A., Foray, C., Felix, T., Le Clerre, D., Temburni-Blake, S., Hong, P., Leduc, S., Pinard, N., Marechal, A., Seclen, E., Boyne, A., Mayer, L., Hong, R., Pulicani, S., Galetto, R., Gouble, A., & Valton, J. (2024). Non-viral DNA delivery and TALEN editing correct the sickle cell mutation in hematopoietic stem cells. *Nature Communications*, *15*(1), 4965. DOI: 10.1038/s41467-024-49353-3 PMID: 38862518

Nadeem, M. A., Nawaz, M. A., Shahid, M. Q., Doğan, Y., Comertpay, G., Yıldız, M., & Baloch, F. S. (2018). DNA Molecular Markers in Plant Breeding: Current Status and Recent Advancements in Genomic Selection and Genome Editing. *Biotechnology, Biotechnological Equipment*, *32*(2), 261–285. DOI: 10.1080/13102818.2017.1400401

Ndour, A., Vadez, V., Pradal, C., & Lucas, M. (2017). Virtual plants need water too: Functional-structural root system models in the context of drought tolerance breeding. *Frontiers in Plant Science*, *8*, 1577. DOI: 10.3389/fpls.2017.01577 PMID: 29018456

Nerkar, G., Devarumath, S., Purankar, M., Kumar, A., Valarmathi, R., Devarumath, R., & Appunu, C. (2022). Advances in Crop Breeding through Precision Genome Editing. *Frontiers in Genetics*, *13*, 880195. DOI: 10.3389/fgene.2022.880195 PMID: 35910205

Numan, M., Khan, A. L., Asaf, S., Salehin, M., Beyene, G., Tadele, Z., & Ligaba-Osena, A. (2021). From Traditional Breeding to Genome Editing for Boosting Productivity of the Ancient Grain Tef (*Eragrostis tef*). *Plants*, *10*(4), 628. DOI: 10.3390/plants10040628 PMID: 33806233

Paniza, H. M. (2024). Challenges in plant breeding under climate change: a review. *Plant Quarantine Challenges under Climate Change Anxiety*, 533-556.

Pedrozo, R., Osakina, A., Huang, Y., Nicolli, C. P., Wang, L., & Jia, Y. (2025). Status on genetic resistance to rice blast disease in the post-genomic era. *Plants*, *14*(5), 807. DOI: 10.3390/plants14050807 PMID: 40094775

Robertson, G., Burger, J., & Campa, M. (2022). CRISPR/Cas-Based Tools for the Targeted Control of Plant Viruses. *Molecular Plant Pathology*, *23*(11), 1701–1718. DOI: 10.1111/mpp.13252 PMID: 35920132

Singh, G., Raigar, O. P., Kaur, S., Bishnoi, R., Mondal, K., Abreha, K. B., Nayak, A. K., Athar, T., Sharma, V., Aziz, D., Gudi, S., Saini, P., Kumar, A., Bhardwaj, R., & Riar, A. (2025). Advances in genomics-assisted breeding strategies for enhancing nutrient uptake and use efficiency in cereals: A pathway toward sustainable agriculture. *Plant Stress*, *18*, 101002. https://doi.org/https://doi.org/10.1016/j.stress.2025.101002. DOI: 10.1016/j.stress.2025.101002

Song, X., Dang, C., Wang, F., Yao, H., Song, Q., & Ye, G. (2026). CRISPR/Cas gene editing for plant resistance to pests, diseases, and herbicide: Application, risk assessment, and safety evaluation. *GM Crops & Food*, *17*(1), 2641893. DOI: 10.1080/21645698.2026.2641893

Spök, A., ... 2022. Towards Social Acceptability of Genome-Edited Plants in Industrial Crop Production. *Nature Sustainability* 4 (5): 391–98. https://doi.org/ DOI: 10.3389/fgeed.2022.899331

Strobbe, S., Wesana, J., Van Der Straeten, D., & De Steur, H. (2023). Public acceptance and stakeholder views of gene edited foods : a global overview. https://doi.org/DOI: 10.1016/j.tibtech.2022.12.011

Sun, L., Lai, M., Ghouri, F., Nawaz, M. A., Ali, F., Baloch, F. S., Nadeem, M. A., Aasim, M., & Shahid, M. Q. (2024). Modern plant breeding techniques in crop improvement and genetic diversity: From molecular markers and gene editing to artificial intelligence—A critical review. *Plants*, *13*(19), 2676. DOI: 10.3390/plants13192676 PMID: 39409546

Taylor, G. K., Petrucci, L. H., Lambert, A. R., Baxter, S. K., Jarjour, J., & Stoddard, B. L. (2012). LAHEDES: The LAGLIDADG homing endonuclease database and engineering server. *Nucleic Acids Research*, *40*(W1), W110–W116. DOI: 10.1093/nar/gks365 PMID: 22570419

U.S. Department of Agriculture. 2023. Biotechnology and USDA's Role in Regulating Genetically Engineered Organisms. https://www.usda.gov/topics/biotechnology

Vaschetto, R., Pecere, A., Perkins, G. D., Mistry, D., Cammarota, G., Longhini, F., Ferrer, M., Pletsch-Assunção, R., Carron, M., Moretto, F., Qiu, H., Della Corte, F., Barone-Adesi, F., & Navalesi, P. (2021). Effects of early extubation followed by noninvasive ventilation versus standard extubation on the duration of invasive mechanical ventilation in hypoxemic non-hypercapnic patients: A systematic review and individual patient data meta-analysis of randomized controlled trials. *Critical Care : The Official Journal of the Critical Care Forum*, *25*(1), 189. DOI: 10.1186/s13054-021-03595-5 PMID: 34074314

Wang, P., Si, H., Li, C., Xu, Z., Guo, H., Jin, S., & Cheng, H. (2025). Plant genetic transformation: Achievements, current status and future prospects. *Plant Biotechnology Journal*, *23*(6), 2034–2058. DOI: 10.1111/pbi.70028 PMID: 40052992

Xiong, W., Lin, E., Ju, H., & Xu, Y. (2007). Climate change and critical thresholds in China's food security. *Climatic Change*, *81*(2), 205–221. DOI: 10.1007/s10584-006-9123-5

Yao, D., Zhou, J., Zhang, A., Wang, J., Liu, Y., Wang, L., Pi, W., Li, Z., Yue, W., Cai, J., Liu, H., Hao, W., & Qu, X. (2023). Advances in CRISPR/Cas9-based research related to soybean [Glycine max (Linn.) Merr] molecular breeding. *Frontiers in Plant Science*, *14*, 1247707. DOI: 10.3389/fpls.2023.1247707 PMID: 37711287

Zagelbaum, J., & Gautier, J. (2023). Double-strand break repair and mis-repair in 3D. *DNA Repair*, *121*, 103430. DOI: 10.1016/j.dnarep.2022.103430 PMID: 36436496

Zhang, H., Qin, C., An, C., Zheng, X., Wen, S., Chen, W., Liu, X., Lv, Z., Yang, P., Xu, W., Gao, W., & Wu, Y. (2021). Application of the CRISPR/Cas9-based gene editing technique in basic research, diagnosis, and therapy of cancer. *Molecular Cancer*, *20*(1), 1–22. DOI: 10.1186/s12943-021-01431-6 PMID: 34598686

Zhang, Y., Massel, K., Godwin, I. D., & Gao, C. (2018). Applications and potential of genome editing in crop improvement. *Genome Biology*, *19*(1), 210. DOI: 10.1186/s13059-018-1586-y PMID: 30501614

Chapter 6
The Role of Gene Editing in Sustainable Agriculture

Nilotpal Das
http://orcid.org/0009-0005-0083-4093
Uttaranchal University, India

Atin Kumar
http://orcid.org/0000-0002-1653-2146
Uttaranchal University, India

Rohit Kumar
http://orcid.org/0009-0003-4713-2625
GLA University, India

Sharad Sachan
Lovely Professional University, India

ABSTRACT

Gene editing is a groundbreaking technology in sustainable agriculture, whereby plant genomes can be strictly edited to produce quality, high-yielding, nutritious, and stress-tolerant crops. The CRISPR/Cas and other technologies enhance resistance to pests, diseases, drought, salinity, and heat, as well as resource-use efficiency and decrease the use of chemical inputs. Gene editing, unlike traditional genetic modification, enables a targeted modification without the use of foreign DNA, which simplifies the approval of regulatory and societal acceptance. Although it is promising, issues of biosafety, ethics, regulation, and accessibility should be overcome to ensure rampant adoption.

DOI: 10.4018/979-8-2600-3575-7.ch006

1. INTRODUCTION

To a large extent, gene editing has become one of the relevant technologies in the pursuit of sustainable agriculture, and it has provided a degree of speed and accuracy in crop and livestock enhancement that has not been seen before. The current method of gene editing technologies, most notably CRISPR–Cas systems, allows targeting the modification of a particular sequence of DNA, based on previous generations of so-called designer nucleases, and now allows easy and low-cost editing of almost any genome (Rustgi et al., 2023; Nascimento et al., 2023). This technological development has seen the growth of microbes and model organisms into large crops like rice, wheat, maize, soybean, and horticultural crops, where editing is currently applied to develop new traits, including disease resistance and abiotic stress tolerance, in addition to the quality and shelf life (Rustgi et al., 2023; Nascimento et al., 2023). Meanwhile, the world's agriculture has experienced increasing sustainability challenges: population explosion, climate change, land and water scarcity, loss of biodiversity, requirements to decrease greenhouse gas emissions and the use of agrochemicals, without reducing or halting yields (Ndudzo et al., 2024; Boros et al., 2024). Sustainable farming, then, involves the need to combine genetic, environmentally efficient, and socio-economically advantageous aspects, i.e., increasing the yield and resilience per hectare, maintaining the environmental health, and ensuring food and nutrition security throughout and beyond the current generations (Boros et al., 2024). There has been a significant role of traditional breeding in driving productivity and resilience, but expandable breeding cycles, intricate inheritance of essential characteristics, genotype-environment interactions, and the inability to quickly pyramid several stress tolerance and quality characteristics with the unwanted drag limit its ability to achieve high yields (Nascimento et al., 2023; Boros et al., 2024). Although effective, transgenic strategies have some other obstacles: high cost of development, regulatory barriers, and ongoing public and market rejections of transplants with foreign DNA, and environmental effects and gene flow (Bhattacharjee et al., 2023; Nascimento et al., 2023). In contrast, gene editing has the capacity to incorporate precise and small modifications, which can be indistinguishable to natural variation, and so-called transgene free or DNA free editing solutions (such as through ribonucleoprotein delivery) is being designed specifically to avoid permanent foreign DNA insertion and reduce off target effects to ease regulatory and social acceptance issues (Bhattacharjee et al., 2023; Nascimento et al., 2023). In practice, gene editing is already being implemented to produce crops that are more resistant to drought, salinity, temperature extremes, etc., and are more nutritionally advantageous and less prone to post harvest losses, which in combination can reduce inputs and associated environmental impacts per unit of food output (Nascimento et al., 2023; Ndudzo et al., 2024). Hypothetically, it makes gene editing a revolutionary agent of sustainable food systems: it has the potential to hasten breeding pipelines, aid in climate smart and resource efficient food production, and help in overcoming food insecurity, particularly to smallholder farmers in climatically vulnerable areas, as long as sufficient attention is paid to governance, ethical, and socio-economic challenges (Ndudzo et al., 2024; Idris et al., 2023). In this respect, gene editing is not intended to substitute more comprehensive agroecological and policy changes but is a potent supplementary lever in sustainable agriculture, which makes it possible to move faster and more precisely towards genetic adaptation of crops and livestock to the multidimensional issues of productivity, resilience, environmental protection, and equitable access to food (Nascimento et al., 2023; Ndudzo et al., 2024).

2. OVERVIEW OF GENE EDITING TECHNOLOGIES

The principles behind gene editing technologies lie in the site-specific genome modification, where the engineered nucleases are targeted at specific sequences of DNA to cause either a double-strand break (DSBs) or a base mutation, which is then repaired with the help of the endogenous cell DNA repair pathways, resulting in a specific insertion, deletion, or replacement. Early systems like Zinc Finger Nucleases (ZFNs) and Transcription Activator-Like Effector Nucleases (TALENs) use customizable DNA recognition protein domains - zinc-finger motifs that identify triplet nucleotides in ZFNs, and TALE repeats that identify single nucleotides in TALENs - fused to the FokI nuclease; two of these monomers bind flanking sequences to form a DSB at a specific locus in the genome (Salavatipour et al., 2024; Rahimmanesh et al., 2022) (Figure 1). These systems showed that genomes might be edited in a highly specific manner, although each new target requires complex and time-consuming protein engineering, and that they have poor targeting density, and are associated with off-target cleavage and cytotoxicity, especially when using ZFNs (Salavatipour et al., 2024; Rahimmanesh et al., 2022). The discovery of CRISPR/Cas systems, which displaces the protein based recognition with an RNA guided system, has changed the face of genome editing, with a single guide RNA (sgRNA) affinity binding to a genomic target of a size of about 20 bp near a protospacer adjacent motif (PAM), which recruits the Cas nuclease (which is typically Cas9) to cause a DSB or, in engineered versions, a single strand nick, base conversion, or prime edit (Binnie et al., 2021; Rahim CRISPR technologies have swiftly expanded to cover Cas12a and additional nucleases with different PAM needs and cutting sites and base and prime editors that permit specific nucleotide alterations devoid of an DSB (Binnie et al., 2021). Relative to the previous genetic modification approaches of random mutagenesis or transgenic insertion, recent gene editing technology has a number of benefits: sequence level accuracy at predefined locus; one can create small and precise mutations or allelic substitutions that can be indistinguishable to natural variation; it is much easier and quicker to retarget than the classical GMOs, just by changing the guide RNA sequence instead of repeating the whole DNA binding protein; and it can be done in a way that is transgene free or DNA free so that no foreign DNA is present in the end product, Although all of the key tools, such as ZFNs, TALENs, and CRISPR/Cas, are based on the identical idea of programmable nuclease mediated breaks and subsequent host repair, CRISPR/Cas systems are characterized by high versatility, scalability, and reduced cost, becoming the dominating platform in modern genome engineering in medicine, agriculture, and biotechnology (Nascimento et al., 2023; Rahimmanesh et al., 2022).

Figure 1. Gene Editing Technologies

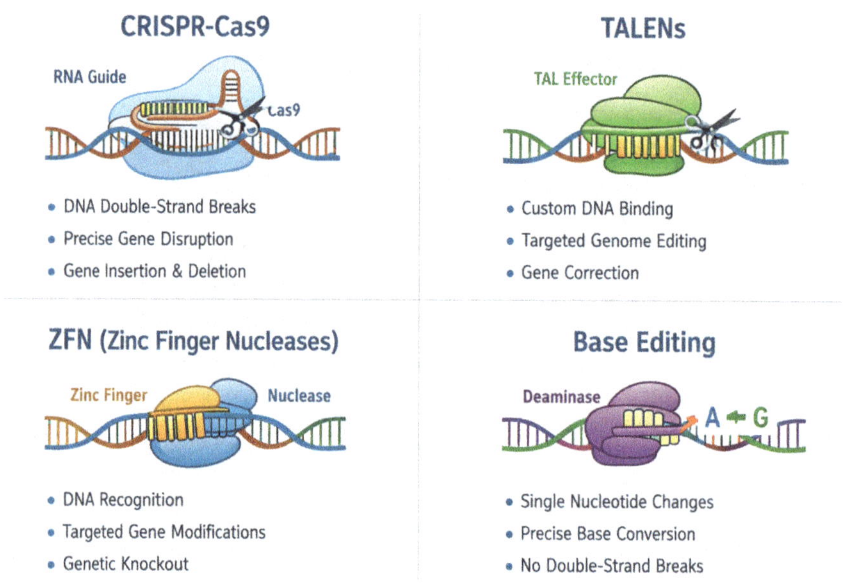

3. GENE EDITING FOR CROP YIELD IMPROVEMENT

The advancement of gene editing technology, especially CRISPR/Cas systems and new prime editing systems, is changing the way crops will be improved, as it is now possible to efficiently and frequently without transgenes modify key genes to increase crop productivity, reduce resource utilization and stress adaptation in major cereals and horticultural crops (Ukhatova et al., 2023; Long et al., 2023) (Table 1).

3.1 Manipulation of Yield-Related Genes

One of the main possibilities to reach greater yield is to directly edit genes that control grain size, grain number, thousand grain weight, and assimilate partitioning. There are at least 287 target genes edited in 28 crop species that are systematically catalogued, with a substantial portion of them focused on the yield and productivity characteristics (Ukhatova et al., 2023). CRISPR/Cas9 knockout of negative regulators (GW2, GW5, TGW6) in rice and wheat has increased grain weight and yield per panicle by 3068 per cent in some triple mutant lines, and TaGW2 knockouts in wheat have increased thousand grain weight and dimensions of grain by over 25 per cent (Long et al., 2023). Major applications are based on loss-of-function mutations produced via non-homologous end joining, but base editing and prime editing are becoming more popular to create desirable alleles without large insertions or foreign DNA, which can be difficult to regulate (Ukhatova et al., 2023).

3.2 Enhancement of Photosynthetic Efficiency

In addition to isolated yield components, gene editing is also being incorporated in the synthetic biology approaches to enhance the efficiency of photosynthesis and downstream carbon utilization. Light harvesting, photorespiration, Rubisco activities, and chloroplast formation are altered with synthetic circuit design and CRISPR-based editing, and so are regulatory networks that control photosynthesis in relation to nutrient uptake and stress responses (Zhang et al., 2024).

3.3 Reduction of Yield Losses Under Suboptimal Conditions

A significant portion of the world's yield losses is due to abiotic stressors: drought, salinity, and extremes of temperature, as well as diseases and pests. The systematic reviews demonstrate that CRISPR/Cas is widely used to modify transcription factors, hormone signaling proteins, ion transporters, osmotic adjustment genes, and immune receptors to increase tolerance to both abiotic and biotic stresses, reducing the gap between the potential and actual yield (Nascimento et al., 2023; Mascarenhas et al., 2024). In the case of abiotic stresses, numerous edits enhance photosynthesis, membrane stability, and reproductive success amid water scarcity or high salinity, whereas in the case of biotic stresses, editing the genes of susceptibility to stress or increasing defense signaling leads to long-term resistance to fungi, bacteria, viruses and nematodes (Nascimento et al., 2023; Mascarenhas et al., 2024). The use of ribonucleoprotein delivery or transient expression, DNA-free and transgene-free methods, also justifies the usage of such traits in elite cultivars and alleviates regulatory and consumer anxieties (Bhattacharjee et al., 2023).

Table 1. Major gene-editing levers for boosting crop yield.

Focus area	Example targets/strategies	Expected impact on yield	Citations
Yield-related genes	Knockout of GW2, GW5, TGW6; novel yield loci in 28 crops	Larger grains, higher grain number, faster breeding	(Ukhatova et al., 2023; Long et al., 2023)
Photosynthetic efficiency	Editing photosynthetic and regulatory genes; AI-guided circuit design	More biomass per light and nutrient input	(Zhang et al., 2024)
Architecture & flowering	Height, tillering, leaf angle, root traits, FT-like genes	Dense planting, better resource capture, optimized crop cycles	(Ukhatova et al., 2023; Ebrahimi & Hashemi, 2024)
Stress tolerance	TFs, ion transporters, immune receptors, susceptibility genes	Stable yields under drought, salinity, disease, and pests	(Nascimento et al., 2023; Mascarenhas et al., 2024)

4. ENHANCING RESISTANCE TO BIOTIC STRESSES

4.1 Gene Editing for Disease Resistance (Fungal, Bacterial, Viral)

The type of gene editing based on CRISPR/Cas is very common to improve the resistance of plants to a wide spectrum of pathogens by knocking out susceptibility (S) genes, boosting immune component signaling, or targeting the genome of the pathogen (Mascarenhas et al., 2024). The systematic review of 296 CRISPR studies in the interaction of plants and pathogens/pests indicates the use in major crop species, including rice, wheat, and tomato, among others, with edits providing resistance to key genera of fungi, bacteria, and viruses, and demonstrating intricate signaling pathways underlying the biotic stress responses (Mascarenhas et al., 2024) (Figure 2). Cas9 is the most widely used nuclease, with Cas12a and Cas13 becoming more widespread, with the latter having the best application in targeting RNA viruses in planta, thereby disrupting viral genomes or interfering with viral replication (Mascarenhas et al., 2024).

Figure 2. Gene editing for disease resistance

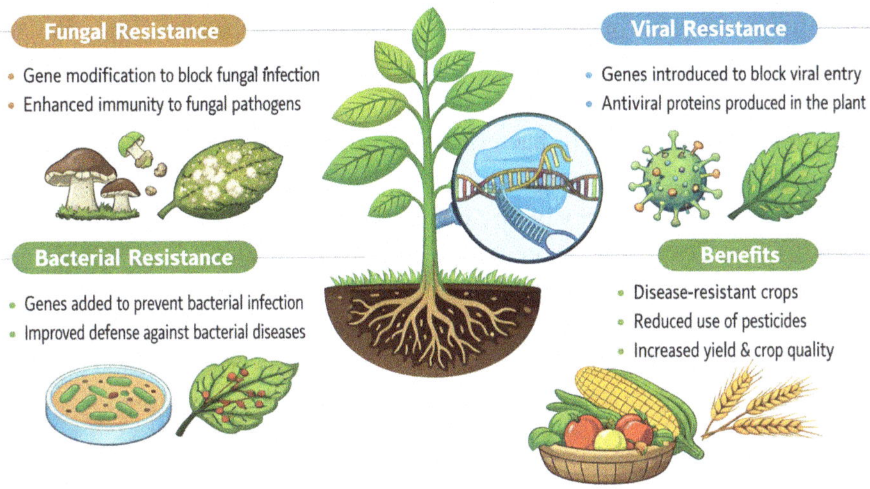

4.2 Development of Insect-Resistant Crop Varieties

In the case of insect resistance gene editing supplements the classical transgenic approaches like Bt crops. CRISPR is used to manipulate genes in plants (modulates insect herbivory, e.g., defense signal, secondary metabolites, cell wall traits), and to optimise or layer insecticidal traits in a wider and more permanent range of control (Mascarenhas et al., 2024; Zhang et al., 2024). Synthetic biology solutions incorporate edited defense signaling and engineered microbial endophytes to provide plants with immunity to insect pests, and are directed towards multiple resistance design solutions where multiple biotic stress factors, such as insects, are addressed in one solution (Zhang et al., 2024). Simultaneously, insect-resistant crops have been engineered by genome engineering through classical transgenesis (e.g., Bt cotton and Bt maize), and this, too, is showing evidence of concept yet provides a template on which gene editing can reduce major lepidopteran pests with more targeted and transgene-free edits (Yadav et al., 2025; Zhang et al., 2024).

4.3 Case Studies of Edited Disease-Resistant Crops

Over 10 years of experience in use, CRISPR/Cas editing has produced lines of plants resistant to diseases in rice, tomato, wheat, barley, and other crops, and commonly by knockout of the S genes, or modifications to immunity regulators (Mascarenhas et al., 2024; Ndudzo et al., 2024). The comprehensive overview of the biotic stress tolerance findings indicates many instances of single or multiplex edits that provide resistance to bacterial blight, rice blast, powdery mildew, viral diseases, and pest insects, often without significant yield losses (Mascarenhas et al., 2024). There is a wider survey of 287 edited target genes in 28 cultivated species, and again, a significant proportion of these targets are focused on the amplification of disease resistance, which confirms that disease resistance traits are now one of the most prevalent gene-editing breeding targets (Ukhatova et al., 2023). Combined, these papers indicate that CRISPR-based improvement of resistance to fungi, bacteria, viruses, and insects is quickly transitioning from proof-of-concept to the center of the product, integrated, pesticide-free crop protection.

5. IMPROVING TOLERANCE TO ABIOTIC STRESSES

Enhanced crop resistance to abiotic stresses is a key to maintaining yields in the face of climate change since drought, salinity, temperature extremes, and flooding can all cause average crop productivity to decline by half in susceptible systems (Nascimento et al., 2023). The field is undergoing increased application of gene editing and broader synthetic biology methods to increase drought resistance and water use efficiency through altering regulators of abscisic acid (ABA) signaling, stomatal behavior, root architecture, and osmotic adjustment. A systematic review of CRISPR/Cas use in the abiotic stresses documents widespread editing of transcription factors and stress-responsive genes controlling antioxidant defense, osmolyte accumulation, and hydraulic conductance to allow plants to sustain photosynthesis and growth during water deficit but to be more water efficient (Nascimento et al., 2023; Wahab et al., 2022). Likewise, the resilience to salinity and heat stress is also being engineered through the engineering of ion transporters, compatible solute pathways, and heat shock/ membrane stability modules. Engineered or edited versions of important gene modules like class II trehalose 6 phosphate synthase (e.g. osTPS8) and the methyltransferase osTRM13 increase salt tolerance in rice by modulating osmotic balance and ion homeostasis, and regulation of PLASTID GALACTOGLYCEROLIPID DEGRADATION genes (PGD1/PGD2) and dehydration responsive regulator DREB2A stabilizes grain production and heat shock behaviours under high temperatures (Nawaz and Satterfield, Cold and flooding tolerance are also on the list since they are being targeted through the regulators of freezing response, energy metabolism and the presence of hypoxia signalling. An example of how these networks can tune core cold response mechanisms to extend freezing tolerance of cells or sustain growth under low temperature is overexpression or editing of C repeat binding factors such as CBF4 and LeCBF1 to confer enhanced freezing tolerance in Arabidopsis and tomato (Nawaz et al., 2022). Simultaneously, systematic CRISPR experiments have shown that numerous tolerance phenotypes are quantitative and that numerous signalling networks are complex and interconnected, implying that multiplex editing of multiple genes is usually required to obtain a strong response of cold, drought, salinity, or flood tolerance in field-like conditions (Nascimento et al., 2023). Together, these genetic approaches combine to constitute a key backbone of climatically resilient agriculture: by improving tolerance to various abiotic stress, gene edited crops enable farmers, in particular smallholders in climate prone areas, to cope with the increasingly unpredictable rainfall, soil erosion and temperature shocks without high yield costs, thus preventing downward shifts in the food production curve and resulting in increased reliance on high-externuity inputs (Nascimento et al., 2023; Ndudzo et al., 2024; Nawaz & Satterfield, 2022) (Table 2).

Table 2. Key Abiotic Stress Targets and Climate Role

Stress focus	Main genetic/physiological targets	Contribution to climate-resilient agriculture	Citations
Drought & water-use efficiency	ABA signaling, roots, antioxidant, and osmolyte pathways	Stable yields under water scarcity, better water productivity	(Nascimento et al., 2023; Wahab et al., 2022)
Salinity & heat	Ion transporters, osmotic balance genes, DREB2A, PGD1/2	Productivity on saline, hot soils expands usable land	(Nascimento et al., 2023; Nawaz & Satterfield, 2022)
Cold & flooding	CBF transcription factors, hypoxia, and energy pathways	Protection from late frosts, waterlogging losses	(Nascimento et al., 2023; Nawaz & Satterfield, 2022)
Multi-stress climate resilience	Multiplex CRISPR edits in complex networks	Crops adapted to variable, extreme climates with less risk	(Nascimento et al., 2023; Ndudzo et al., 2024; Nawaz & Satterfield, 2022)

6. GENE EDITING AND RESOURCE USE EFFICIENCY

6.1 Improved Nutrient Uptake and Utilization (N, P, K)

The reprogramming of nutrient uptake and internal utilization pathways is being performed with the aid of gene editing and more general synthetic biology technologies to enable crops to become more efficient in terms of the nitrogen (N), phosphorus (P), and potassium (K) intake and use. Strategies to re-engineer genetic circuits that regulate nutrient transporters, carbon- nitrogen metabolism, and hormonal signaling are described as the approaches to transforming root absorption capacity and remobilization of nutrients and maintaining or maximizing the yield with the same or reduced supply of fertilizers (Ofori et al., 2022). Multiscale studies of nitrogen use efficiency have demonstrated efficacy of maximizing N uptake efficiency (root acquisition) and N use efficiency (conversion into biomass and grain) fundamentally relies on root morphology and activity of certain N transport systems, which are high-priority targets of precision editing (Yang et al., 2025a). Root-directed systematic reviews show that conserved genes that mediate auxin transport, root hair development, and inorganic cation uptake across the plasma membrane are promising targets in editing to enhance the ability of roots to scavenge N, P, and K in heterogeneous soils (Justamante et al., 2025).

6.2 Enhancement of Root System Architecture

The root system architecture plays a central role in determining efficiency in resource use since it dictates the amount of soil to be explored, as well as the contact with the available nutrients and water. Analyses of RSA in abiotic stress conditions reveal that the response of plastic changes in primary roots, lateral roots, and root hairs to drought and nutrient deficiency is regulated by a relatively limited set of genes, including those that regulate cell division, auxin polar transport, and root hair development (Justamante et al., 2025). Modifying these genes can provide the means to develop more complex or branched root systems that enhance contact with soil and, therefore, the efficiency of nutrient uptake through both increased instantaneous uptake rates and maximum rooting volume (Yang et al., 2025b; Justamante et al., 2025). With the assistance of synthetic biology and AI-directed design, it is now possible to modify traits of roots to promote deeper rooting to take in nitrate and water, or to promote dense topsoil foraging to acquire immobile phosphorus using multi-omics data as the basis of gene editing (Ofori et al., 2022).

6.3 Contribution to Sustainable Soil and Water Management

Enhanced nutrient uptake and enhanced RSA are direct ways of achieving sustainable soil and water management through decreased requirement of fertilizers, minimization of nutrient loss, and productive growing of marginal soils. According to reviews of RSA, more efficient root systems can simplify the requirement of fertilizers and irrigation, reduce nutrient leaching to the aquifers, assist in the conservation of finite phosphate rock reserves, and facilitate crop production in unfavorable environments (Justamante et al., 2025). Increased efficiency of N use saves the proportion of the applied N that is leached or released as a greenhouse gas, which eases the burden of water pollution and nitrous oxide emissions that occur as a result of intensive fertilization (Yang et al., 2025). Gene-edited crops that enhance nutrient intake and tolerance to stress are a central feature of sustainable and climate-smart agricultural systems in synthetic-biology visions, as the high yields are offset by reduced environmental impacts in the water and soil system (Ofori et al., 2022).

7. QUALITY IMPROVEMENT AND NUTRITIONAL ENHANCEMENT

7.1 Biofortification of Essential Micronutrients

Biofortification aims to increase the abundance of staple crops of essential vitamins and minerals (e.g., FE, Zn, Se, provitamin A, folate) directly into edible tissues, and it is a sustainable, low-cost approach to eradicate hidden hunger especially in settings that are not well-equipped to access a wide range of diets, supplements, and industrial fortification (Avnee et al., 2023; Dhaliwal et al., 2022). Through genetic and agronomic biofortification, there have been crops of rice, wheat, maize, pulses, and other staples, which contain significantly increased levels of iron, zinc, provitamin A, and B vitamins, in many cases without yield loss, thus enhancing micronutrient density in the diet of a population (Scharff et al., 2021; Dhaliwal et al., 2022).

7.2 Improvement of Protein, Oil, and Starch Quality

Bettering of macronutrients has been aimed at the quantity and performance of proteins, oils, and starches in grains and oilseeds. Breeding and transgenic techniques are considered to enhance the content of major amino acids (e.g., lysine, tryptophan), fatty-acid ratios (greater polyunsaturated and omega-3 fatty acids), and starch structure to increase the content of resistant starch or amylose as a dietary fiber and glycemic properties (Scharff et al., 2021). Cereal quality reviews have shown progress in the molecular regulation of endosperm filling, starch granule synthesis, and protein body formation, which allows the rational design of super crops, which are high-yielding with improved nutritional and processing characteristics (Yang et al., 2025).

7.3 Reduction of Anti-Nutritional Factors

Anti-nutritional compounds like phytate, tannins, lectins, enzyme inhibitors, saponins, and some alkaloids bind nutrients or suppress digestive enzymes, decreasing protein digestibility and mineral bioavailability, and therefore leading to micronutrient malnutrition (Samtiya et al., 2020). Improvement of nutritional quality is hence a priority where the reduction of anti-nutrient contents is done without decreasing agronomic performance either by breeding, molecular tools, or optimised post-harvest processing (e.g., soaking, fermentation, germination, debranning, autoclaving), which can reduce phytate, trypsin inhibitors, and other antinutrients by a significant margin and subsequently lead to increased actual uptake of iron, zinc, and proteins by plant foods (Scharff et al., 2021; Samtiya et al., 2020).

The whole issue of biofertilized crops being handled and processed post-harvest is a strong determinant of the proportionality of the added nutrient content to the final products consumed by consumers. Destruction of large proportions of minerals concentrated in the bran and outer layers with milling or polishing of cereals, thus quality improvement programs focus on genotypes with high levels of micronutrients in edible fractions, and processing methods that do not destroy these nutrients (Scharff et al., 2021; Dhaliwal et al., 2022) (Table 3). Likewise, anti-nutrient reduction in post-harvest processing enhances the bioavailability of minerals and functional quality of plant foods, which are complementary to biofortification and breeding of high-quality macronutrients, producing stable and nutrient-dense products with satisfactory sensory and storage properties (Scharff et al., 2021; Samtiya et al., 2020).

Table 3. Shelf-life enhancement and post-harvest quality

Aspect	Key nutritional/quality focus	Example strategies	Citations
Retention of micronutrients	Minimize losses of Fe, Zn, vitamins during milling/ cooking	Selecting genotypes with higher micronutrients in endosperm, optimizing polishing and cooking	(Scharff et al., 2021; Dhaliwal et al., 2022)
Stability of biofortified traits	Maintain nutrient levels through storage and processing	Breeding for favorable mineral localization and compound stability	(Scharff et al., 2021; Dhaliwal et al., 2022)

8. ENVIRONMENTAL SUSTAINABILITY AND ECOSYSTEM SERVICES

8.1 Reduction of Agrochemical Use and Environmental Pollution

The shift towards agroecological and regenerative systems, which is primarily driven by social and environmental concerns, is the key to reducing the use of agrochemicals and the pollution that follows. Agroecology focuses on diversified crops, cover crops, agroforestry, and integrated pest management, which, respectively, reduce the use of synthetic fertilizer and pesticides and do not reduce productivity (Vikas et al., 2024; Duru et al., 2015). Similarly, regenerative and organic farming methods do not or severely reduce synthetic inputs, replenish soil organic matter, and enhance soil structure, thereby reducing leaching and surface runoff of nutrients to surface and groundwater (U. et al., 2025). Total involvement of agroforestry systems in nutrient interception and recycling is especially good: tree roots can reduce soil nitrogen and phosphorus residues by 20-100%, and pesticide runoff is significantly reduced (LaCanne and Lundgren, 2018). They are directly related to the documented global issues of pesticide and fertilizer contamination of the air, soil, and water that represent a threat to the environment and human health (Pavlidis and Tsihrintzis, 2017).

8.2 Lower Greenhouse Gas Emissions From Agriculture

Regenerative agricultural and conservation agricultural practices, which focus on sustainable land management, can play a major role in mitigating greenhouse gas (GHG) emissions and increasing carbon sinks. Minimized soil disturbance, permanent soil cover, and diversified rotations reduce CO_2 and N_2O emissions, which enhances soil organic matter and prevents erosion better than conventional tillage (U. et al., 2025; Sokal and Kachel, 2025). Cover crops and agroforestry, organic amendments, and integrated livestock regenerative systems can promote the formation of soil carbon and more effective recycling of nutrients, which is associated with climate change mitigation (U. et al., 2025; Duru et al., 2015). Wider scans of agricultural GHGs emphasize that methods like lesser tillage, efficient management of nitrogen, diversified rotations, and precise use of inputs can collaboratively reduce the emissions and increase the soil carbon capacity, and that agriculture can be a source of mitigation and a carbon repository (Sokal & Kachel, 2025).

8.3 Support for Biodiversity-Friendly Farming Systems

Biodiversity is explicitly used to provide ecosystem services through agroecology and agricultural diversification. A variety of systems, such as intercropping, agroforestry, mixed cropping, and elements of habitat, can increase above and below ground biodiversity, enhance pollination, natural pest management, nutrient recycling, and water management, without reducing yields (Duru et al., 2015; Kamyab et al., 2023) (Figure 3).

Figure 3. Support for biodiversity-friendly farming systems

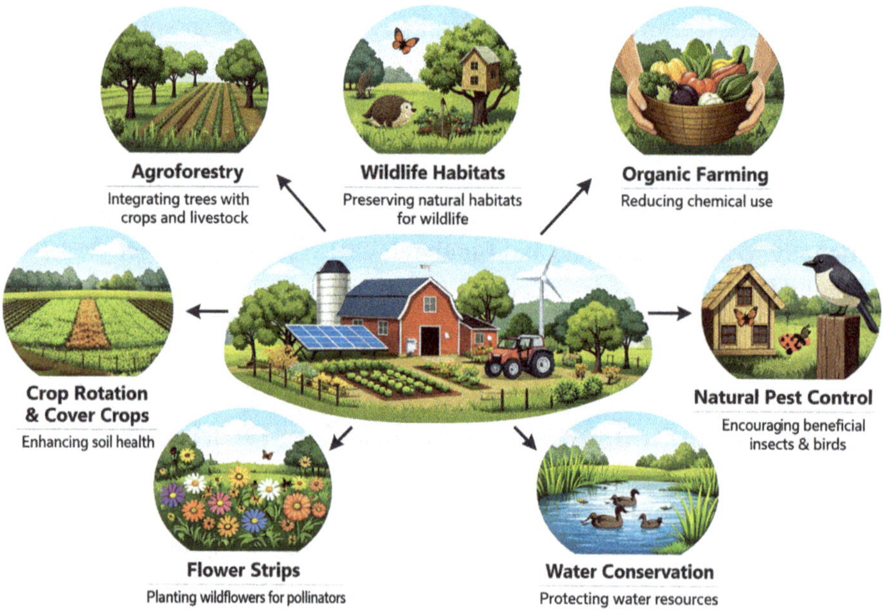

Agroecological landscapes comprising semi-natural habitats (hedges, grass strips, woodlots) with a cropped field promote the presence of natural enemies, pollinators, and other useful organisms by replacing the ecologically controlled environment with the use of chemicals (Zhang et al., 2025; Duru et al., 2015). The examples of different regions demonstrate that such practices help to preserve native species and functional groups and keep or enhance farm productivity and resilience (Vikas et al., 2024; Zhang et al., 2025).

8.4 Role in Regenerative and Conservation Agriculture

Regenerative agriculture takes soil conservation as the access point to regenerate several ecosystem services, provisioning, regulating, and supporting, and it aims to achieve social and economic sustainability (U. et al., 2025; Tamburini et al., 2020). Some of the core practices encompass conservation tillage, cover crops, crop rotations, composting, livestock integration, and agroforestry, among many other practices that are aimed at replenishing healthy soils, improving water infiltration, increasing nutrient cycling, and replenishing biodiversity (U. et al., 2025). These systems are consistent with agroecological concepts of reducing inputs, recycling, soil health, and biodiversity, and reflect increased levels of transition in which farms are restructured around ecological processes instead of simple input substitution (Duru et al., 2015; Tamburini et al., 2020). When regeneratively and conservatively practiced, regenerative and conservation agricultural systems develop resilient agroecosystems with carbon sequestration, water purification, soil development, pest control, and habitat capabilities that form the core of environmentally sustainable and ecosystem service dense agricultural landscapes (U. et al., 2025; Duru et al., 2015; Tamburini et al., 2020).

9. SOCIO-ECONOMIC IMPACTS OF GENE-EDITED CROPS

9.1 Benefits for Smallholder and Marginal Farmers

Compared to transgenic GM crops, gene-edited crops can be developed faster and at lower R&D cost and thus are economically viable to target traits that matter to small and low-value markets and so-called orphan crops that are central to smallholder and marginal farmers in low- and middle-income countries (Kalaitzandonakes et al., 2022; Pixley et al., 2022). Due to the relatively low price and technical accessibility of tools such as CRISPR, a public research institute and/or a CGIAR center can create farmer-targeted non-proprietary varieties with specific characteristics, including drought tolerance, disease resistance, and improved nutritional quality, which may contribute to reducing losses and risk of production in smallholder situations (Ndudzo et al., 2024; Pixley et al., 2022). Initial experience on GM crops indicates that an improvement in yield and a decrease in input costs can substantially increase farm incomes and prevent more countries from becoming poor; the same or even larger pro-poor benefits are anticipated in gene-edited crops with traits designed to suit smallholder limitations (Aziz et al., 2022; Pixley et al., 2022).

9.2 Economic Viability and Cost-Effectiveness

Genome editing is more productive in terms of breeding pipelines and increases the number of economically viable traits and crops due to several factors: an innovation point of view requires less time to develop traits, capital, labor, and regulatory costs are reduced in comparison to conventional GM (Kalaitzandonakes et al., 2022; Bullock et al., 2021). According to economic modeling, given a specific trait value, gene editing would assume up to approximately 96% less planted area than genetic modification to create a break-even on investment of R&D at the private level, or it is much cheaper and more appropriate to the small market (Bullock et al., 2021). Macro-level cross-country patent-based analysis is a connection between CRISPR diffusion and the motivations of substantial positive outcomes on the economic effectiveness of agriculture, which implies that extensive use of gene-editing technology can enhance the productivity and profitability of agricultural sectors as a whole (Zhu et al., 2025).

9.3 Impact on Food Security and Rural Livelihoods

There is the deployment of gene-edited crops to enhance climate change resistance to drought, heat, flooding, and pests, as well as improve yield, nutritional value, and shelf life, which directly contribute to food availability, stability, and quality of the diet (Ndudzo et al., 2024; Chen et al., 2024a; Ahmad, 2023). The analyses conducted by experts and policy makers worldwide determine that genome editing has the potential to deliver better varieties to smallholders faster, diversify crop systems (such as forgotten and unsocked species), and help to achieve Sustainable Development Goals of ending hunger, enhancing food security, and eliminating poverty, most especially when used as a common good in Africa and other areas (Tripathi et al., 2022; Aziz et al., 2022; Pixley et al., 2022). Gene-edited crops can enhance the living conditions of rural people by reducing losses, stabilizing production, reducing the cost of inputs, and enhancing the quality of products available on the market, but the resulting benefits will rely on access to seeds, complementary services, and accessible markets (Aziz et al., 2022; Pixley et al., 2022).

9.4 Public Acceptance and Farmer Adoption

The regulation, market signals, as well as societal perceptions are strong conditioners of socio-economic benefits. According to expert surveys, unclear or limiting regulatory frameworks along with the lack of popular and retailer acceptance are the most significant impediments to commercialization and integration of the value chain of genome-edited crops despite the evident agronomic and economic advantages

(Maaß et al., 2019; Lassoued et al., 2021). Meanwhile, according to the experts and consumer research, the attitude towards gene-edited products is much more positive than that regarding conventional GM, particularly in cases where the edits do not imply the inclusion of foreign DNA and are positioned in the context of evidently positive consumer or societal benefits, which are safer oils, less waste, or being resistant to climatic changes (Lassoued et al., 2021; Lassoued et al., 2019). Where technologies have a proven ability to increase income and decrease risk, following trusted communication, availability of risk-benefit information, and consistent and science-based regulatory frameworks, which grant gene-edited crops a social license and market access, the adoption is likely to be far more widespread (Maaß et al., 2019; Lassoued et al., 2021; Pixley et al., 2022).

10. REGULATORY FRAMEWORKS AND BIOSAFETY CONSIDERATIONS

10.1 Global Regulatory Perspectives on Gene-Edited Crops

The regulation of gene-edited crops is fractured across the world, as some areas take them as normal plants and others as GMOs. Several nations have now excluded some products of site-directed nuclease 1 (SDN1) that may happen during natural breeding by full GMO regulations (e.g., several American and Asian states), whilst others, most notably the EU and New Zealand, regard all GMO-edited plants as GMOs, initiating intense regulatory procedures (Ahmad et al., 2021a; Gupta et al., 2021). A survey of over 30 nations demonstrates a range between product-based schemes targeting the ultimate characteristic (e.g., USA, Argentina) and process-based programs that control any application of modern biotechnology, with the Cartagena Protocol on Biosafety serving as a guide to this (Ahmad et al., 2021b; Turnbull et al., 2021). Recent studies explain that there is a tendency of limited convergence, which may be the imposition of simplified GMO-like regulations over some edits by many states or leave them unregulated but demand some notification or confirmation, which complicates harmonization and trade (Tachikawa & Matsuo, 2023; Sprink et al., 2022) (Table 4).

Table 4. Key Claims & Evidence

Claim	Evidence Strength	Reasoning	Papers
Many countries deregulate some transgene-free gene-edited crops	Evidence strength: Strong (8/10)	Comparative legal reviews show exemptions for SDN1/2 in several American, Asian, and African systems, while others retain full GMO status (Ahmad et al., 2021a; Turnbull et al., 2021; Sprink et al., 2022)	(Ahmad et al., 2021b; Turnbull et al., 2021; Sprink et al., 2022)
GMO and gene-edited crops increasingly share risk-assessment logic	Evidence strength: Moderate (7/10)	Both are assessed via case-by-case, science-based ERA focusing on phenotype and exposure, though triggers differ (Ahmad et al., 2021a; Agapito-Tenfen et al., 2018; Duensing et al., 2018)	(Ahmad et al., 2021b; Agapito-Tenfen et al., 2018; Duensing et al., 2018)

10.2 Comparison Between GMO and Gene-Edited Crop Regulations

Most traditional GM crops have been regulated by certain biosafety laws, which were constructed around transgenesis, with pre-market environmental and food safety evaluation, traceability, and labeling (Ahmad et al., 2021a; Turnbull et al., 2021). Genes edited crop system blurs this line: whether it is a small deletion or point mutation, there may be no difference between gene editing and regular mutagenesis, and thus the products should be regulated by focusing on their novel traits and not by using the editing tool (Gupta et al., 2021; Duensing et al., 2018). Under product based systems (e.g. USA, Argentina, some Latin American countries), most SDN1/2 edits that do not introduce foreign DNA are not subject to GMO, whereas SDN3 or those that imitate transgenesis are treated as GMO; whereas the EU has a strictly process based definition, SDN1000/1003 countries of the same type fall under the same directives (2001/18/EC, 1829/2003) as GM plants (Ahmad et al., 2021b; Gupta et al., 2021; Ahmad et al., 2021). This deviation gives rise to non-synchronous approvals, trade tensions, and dissimilar motivations to innovation.

10.3 Biosafety Assessment and Risk Management

Gene-edited plants risk governance typically expands on GMO models, through case-by-case environmental risk assessment (ERA) and stepwise testing to assess the impact of gene-edited plants on human, animal, and ecosystem health (Ahmad et al., 2021a; Agapito-Tenfen et al., 2018). The important questions of biosafety revolve around on and off target, genetic changes, the possible effects of genetic change on non-target organisms, the gene flow, the phenomenon of weediness or invasiveness, and many scholars state that with simple edits the risk profile is more of a phenomenon than the editing instrument (Agapito-Tenfen et al., 2018; Duensing et al., 2018). Among them are traceability, as most edits are analytically indistinguishable from the conventional mutations, and a lack of transparency surrounding the commercial pipelines, which can make them harder to monitor and post-market (Agapito-Tenfen et al., 2018; Choriyeva, 2024). It has been noted that biosafety research requires improvement; the creation of knowledge gaps must be addressed with science-based, proportionate ERA, and it should be international to avoid the increasing risk -benefit gap between precaution and innovation (Agapito-Tenfen et al., 2018; Duensing et al., 2018).

11. CHALLENGES AND LIMITATIONS

The development of gene-edited crops is subject to various and intersecting issues and constraints at all four levels of technical, regulatory, social, and equity. Technically, despite the precision and specificity of targeted modifications, as of 2025, plant genome editing remains plagued by both efficient delivery of reagents to appropriate explants and regeneration of edited plants, as well as production of truly transgene-free lines in most species, particularly those of vegetative propagation (Lokya et al., 2025; Goralogia et al., 2021). Even with improved guide design and better Cas variants, off target mutations are still a matter of concern to risk communication: next generation sequencing has shown that off targets normally introduce only a small number of variants to the already extensive load of natural and induced mutations in breeding populations, and better care of guide design can reduce them, although developers continue to recommend that off targets should be well characterized to maintain both scientific and popular confidence (Lokya et al., 2025; Graham et al., 2020). These technical problems are enhanced by regulatory uncertainty. There is a stark divide in countries that regulate or not regulate gene edited crops; some product-based systems (notably the EU) regulate all plants genome edited as GMOs, whilst others do not (including some of those that do regulate GMOs) creating a patchwork of global regulation which complicates compliance, traceability and trade (Lokya et al., 2025; Turnbull et al., 2021). In recent studies, it has been stressed that divergent and changing rules raise the cost of approval, slow down commercialization, cause non-tariff trade barriers, particularly to small and medium-sized developers, and influence R&D decisions toward low-risk markets (Ríos et al., 2025; Turnbull et al., 2021). There is another level of restriction, which is the opinion of the public. The history of GMOs has planted the seeds of distrust, and initial research indicates that a significant number of citizens have little power to differentiate between gene editing and older forms of genetic modification; hence, negative perceptions can be transferred between GM and gene-edited crops, with encouragement by vocal critics and fake news (Ishii & Araki, 2016; Folta, 2025). Polls in other countries indicate that acceptance is heavily contingent on perception of benefits, trust in regulators and scientists, and easily available and clear balanced information, but most individuals report inadequate or incomprehensible information on genome editing, creating gaps that misinformation and fear-based stories can fill (Ishii and Araki, 2016; Bearth et al., 2024). Lastly, the ability to access genome editing technology in third-world countries is limited by capacity, infrastructure, and governance. Although the reduced costs of R&D and the possibility of making edits that are too pure and clean in theory democratize biotechnology, the risk of repeating earlier GMO related scandals, tight or uncertain regulation and reliance on outsourced expertise makes people fear that poorer people and breeders in the Global South would once again be marginalized or delayed in obtaining access to the benefits (Lassoued et al., 2021; Venkataraman et al., 2025).

12. FUTURE PROSPECTS OF GENE EDITING IN SUSTAINABLE AGRICULTURE

The future of gene editing in sustainable agriculture can be determined by the rapid development of precision tools, integration of data, and targeting of crops with the aim of facilitating food security and environmental ambitions. More precise and simultaneous (at multiple loci) edits are possible with new generations of CRISPR-based systems, including base and prime editors, engineered Cas variants, and multiplex platforms, making it possible to engineering polygenic traits of complex traits, e.g. yield, nutrient use efficiency and climate resilience (Chen et al., 2019; Li et al., 2024; Abdelrahman et al., 2021). Multiplex genome editing can be used to introduce mutations into whole gene families or pathways, such as to enhance grain quality, stress resistance, or de novo domestication of wild relatives in a single breeding, which was fundamentally not possible using previous tools (Chen et al., 2019; Abdelrahman et al., 2021). Simultaneously, genome editing is becoming more and more consistent with genomics, phenomics, and artificial intelligence to develop data-rich pipelines that associate the variation of the sequence with the performance of the field. Identification of thousands of loci affecting yield, disease resistance, and adaptation through large scale sequencing and multi omics, coupled with high throughput phenotyping, is accompanied by prioritization of targets by AI and machine learning, designing of more specific and efficient guide RNAs, modeling on and off target prediction of genotype phenotype relationships in changing climatic conditions (Chavhan et al., 2025; Bradbury et al., 2025; Li et al., 2025; Chen et al., 2024b). Such convergence enables the movement towards systems-level crop design using precision farming technology, e.g., sensors and decision support tools to manage deployed edited varieties in climate-smart systems (Chavhan et al., 2025; Bradbury et al., 2025; Li et al., 2025). One of the largest frontiers is in the area of gene editing in the underutilized and orphan products, most of which are foods in food-insecure countries, but have not been genetically developed. The recent decreases in the cost of sequencing, open-source genomic tools and flexible CRISPR workflows can now be used to cause targeted mutagenesis and rapid domestication of crops like millets, sorghum and other legumes, which have resulted in new allelic variation and enhanced traits such as yield stability, nutritional quality and stress tolerance (Yaqoob et al., 2023; Rajpal et al., 2023; Venezia and Krainer, 2021). The process of editing has been highlighted in reviews as having the ability to transform semi-domesticated or wild species into agronomically competitive crops in a short period of time compared to conventional breeding and maintain their natural benefits in resilience and dietary diversity (Yaqoob et al., 2023; Venezia and Krainer, 2021). Taken together, these trends make genome editing one of the fundamental technologies towards the improvement of several Sustainable Development Goals, in particular, SDG2 (Zero Hunger) and SDG13 (Climate Action). Gene editing may help achieve food and nutritional security with potentially reduced input requirements and environmental impacts by enabling better yield on small areas of land, greater resistance to biotic and abiotic stressors, and biofortification and diversification through orphan crops (Ahmadikhah et al., 2025; Saxena et al., 2025; Hamdan et al., 2022; Fiaz et al., 2021).

13. CONCLUSION

Gene editing is becoming an efficient and accurate instrument in the development of sustainable agriculture through the enhancement of crop productivity, hardiness, and efficiency of resource utilization. With precise adjustment of particular genes, it allows building climate-resistant, nutrient-effective, and high-quality crops, as well as decreasing the use of chemical additives. This strategy is a direct contributor to environmental sustainability, economic viability, and world food security. Nevertheless, effective implementation of gene editing in agricultural systems must be well balanced with scientific innovation, bio-safety, and trust of the people. Clear regulatory systems, efficient risk management, and science communication should be applied to tackle this ethical issue and promote social acceptance. In the future, gene editing will become one of the pillars of future agriculture. Together with genomics, digital agriculture, and precision farming, it has the transformative potential to address the needs of climate change, population increase, and scarce natural resources. The prudent application of gene editing technologies will be central to the creation of resilient, sustainable, and inclusive food systems for future generations.

REFERENCES

Abdelrahman, M., Wei, Z., Rohila, J. S., & Zhao, K. (2021). Multiplex Genome-Editing technologies for revolutionizing plant biology and crop improvement. *Frontiers in Plant Science, 12,* 721203. DOI: 10.3389/fpls.2021.721203 PMID: 34691102

Agapito-Tenfen, S. Z., Okoli, A. S., Bernstein, M. J., Wikmark, O.-G., & Myhr, A. I. (2018). Revisiting risk governance of GM plants: The need to consider new and emerging Gene-Editing techniques. *Frontiers in Plant Science, 9,* 1874. DOI: 10.3389/fpls.2018.01874 PMID: 30622546

Ahmad, A., Ghouri, M. Z., Munawar, N., Ismail, M., Ashraf, S., & Aftab, S. O. (2021a). Regulatory, Ethical, and Social Aspects of CRISPR Crops. In *CRISPR Crops* (pp. 261–287). DOI: 10.1007/978-981-15-7142-8_9

Ahmad, A., Munawar, N., Khan, Z., Qusmani, A. T., Khan, S. H., Jamil, A., Ashraf, S., Ghouri, M. Z., Aslam, S., Mubarik, M. S., Munir, A., Sultan, Q., Abd-Elsalam, K. A., & Qari, S. H. (2021b). An Outlook on Global Regulatory Landscape for Genome-Edited Crops. *International Journal of Molecular Sciences, 22*(21), 11753. DOI: 10.3390/ijms222111753

Ahmad, M. (2023). Plant breeding advancements with "CRISPR-Cas" genome editing technologies will assist future food security. *Frontiers in Plant Science, 14,* 1133036. DOI: 10.3389/fpls.2023.1133036 PMID: 36993865

Ahmadikhah, A., Zarabizadeh, H., Nayeri, S., & Abbasi, M. S. (2025). Advancements in genome editing tools for genetic studies and crop improvement. *Frontiers in Plant Science, 15,* 1370675. DOI: 10.3389/fpls.2024.1370675 PMID: 39963359

Avnee, S., Sood, S., Chaudhary, D. R., Jhorar, P., & Rana, R. S. (2023). Biofortification: An approach to eradicate micronutrient deficiency. *Frontiers in Nutrition, 10,* 1233070. DOI: 10.3389/fnut.2023.1233070 PMID: 37789898

Aziz, M. A., Brini, F., Rouached, H., & Masmoudi, K. (2022). Genetically engineered crops for sustainably enhanced food production systems. *Frontiers in Plant Science, 13,* 1027828. DOI: 10.3389/fpls.2022.1027828 PMID: 36426158

Bearth, A., Drummond Otten, C., & Segrè Cohen, A. (2024). Consumers' perceptions and acceptance of genome editing in agriculture: Insights from the United States of America and Switzerland. *Food Research International, 178,* 113982. DOI: 10.1016/j.foodres.2024.113982 PMID: 38309884

Bhattacharjee, S., Bhowmick, R., Kant, L., & Paul, K. (2023). Strategic transgene-free approaches of CRISPR-based genome editing in plants. *Molecular Genetics and Genomics*, *298*(3), 507–520. DOI: 10.1007/s00438-023-01998-3 PMID: 36840794

Binnie, A., Fernandes, E., Almeida-Lousada, H., De Mello, R. A., & Castelo-Branco, P. (2021). CRISPR-based strategies in infectious disease diagnosis and therapy. *Infection*, *49*(3), 377–385. DOI: 10.1007/s15010-020-01554-w PMID: 33393066

Boros, A., Szólik, E., Desalegn, G., & Tőzsér, D. (2024). A Systematic review of opportunities and limitations of innovative practices in sustainable agriculture. *Agronomy (Basel)*, *15*(1), 76. DOI: 10.3390/agronomy15010076

Bradbury, A., Clapp, O., Biacsi, A.-S., Kuo, P., Gaju, O., Hayta, S., Zhu, J.-K., & Lambing, C. (2025). Integrating genome editing with omics, artificial intelligence, and advanced farming technologies to increase crop productivity. *Plant Communications*, *6*(7), 101386. DOI: 10.1016/j.xplc.2025.101386 PMID: 40443034

Bullock, D. W., Wilson, W. W., & Neadeau, J. (2021). Gene editing versus genetic modification in the research and development of new crop traits: An Economic comparison. *American Journal of Agricultural Economics*, *103*(5), 1700–1719. DOI: 10.1111/ajae.12201

Chavhan, R. L., Jaybhaye, S. G., Hinge, V. R., Deshmukh, A. S., Shaikh, U. S., Jadhav, P. K., Kadam, U. S., & Hong, J. C. (2025). Emerging applications of gene editing technologies for the development of climate-resilient crops. *Frontiers in Genome Editing*, *7*, 1524767. DOI: 10.3389/fgeed.2025.1524767 PMID: 40129518

Chen, F., Chen, L., Yan, Z., Xu, J., Feng, L., He, N., Guo, M., Zhao, J., Chen, Z., Chen, H., Yao, G., & Liu, C. (2024a). Recent advances of CRISPR-based genome editing for enhancing staple crops. *Frontiers in Plant Science*, *15*, 1478398. DOI: 10.3389/fpls.2024.1478398 PMID: 39376239

Chen, K., Wang, Y., Zhang, R., Zhang, H., & Gao, C. (2019). CRISPR/CAS genome editing and precision plant breeding in agriculture. *Annual Review of Plant Biology*, *70*(1), 667–697. DOI: 10.1146/annurev-arplant-050718-100049 PMID: 30835493

Chen, L., Liu, G., & Zhang, T. (2024b). Integrating machine learning and genome editing for crop improvement. *aBIOTECH*, *5*(2), 262–277. DOI: 10.1007/s42994-023-00133-5 PMID: 38974863

Choriyeva, Z. (2024). Genetically modified ecosystems: Innovative approaches in agriculture and their environmental impact. *BIO Web of Conferences, 149*, 01009. DOI: 10.1051/bioconf/202414901009

Dhaliwal, S. S., Sharma, V., Shukla, A. K., Verma, V., Kaur, M., Shivay, Y. S., Nisar, S., Gaber, A., Brestic, M., Barek, V., Skalicky, M., Ondrisik, P., & Hossain, A. (2022). Biofortification—A frontier novel approach to enrich micronutrients in field crops to encounter the nutritional security. *Molecules (Basel, Switzerland)*, *27*(4), 1340. DOI: 10.3390/molecules27041340 PMID: 35209127

Duensing, N., Sprink, T., Parrott, W. A., Fedorova, M., Lema, M. A., Wolt, J. D., & Bartsch, D. (2018). Novel features and considerations for ERA and regulation of crops produced by genome editing. *Frontiers in Bioengineering and Biotechnology*, *6*, 79. DOI: 10.3389/fbioe.2018.00079 PMID: 29967764

Duru, M., Therond, O., Martin, G., Martin-Clouaire, R., Magne, M.-A., Justes, E., Journet, E.-P., Aubertot, J.-N., Savary, S., Bergez, J.-E., & Sarthou, J. P. (2015). How to implement biodiversity-based agriculture to enhance ecosystem services: A review. *Agronomy for Sustainable Development*, *35*(4), 1259–1281. DOI: 10.1007/s13593-015-0306-1

Ebrahimi, V., & Hashemi, A. (2024). CRISPR-based gene editing in plants: Focus on reagents and their delivery tools. *BioImpacts*, *15*, 30019. DOI: 10.34172/bi.30019 PMID: 39963563

Fiaz, S., Ahmar, S., Saeed, S., Riaz, A., Mora-Poblete, F., & Jung, K.-H. (2021). Evolution and application of genome editing techniques for achieving food and nutritional security. *International Journal of Molecular Sciences*, *22*(11), 5585. DOI: 10.3390/ijms22115585 PMID: 34070430

Folta, K. M. (2025). Acceptance of crop biotechnology requires a change in communication strategy. *Plant Physiology*, *198*(3), kiaf167. Advance online publication. DOI: 10.1093/plphys/kiaf167 PMID: 40314159

Goralogia, G. S., Redick, T. P., & Strauss, S. H. (2021). Gene editing in tree and clonal crops: Progress and challenges. *In Vitro Cellular & Developmental Biology. Plant : Journal of the Tissue Culture Association*, *57*(4), 683–699. DOI: 10.1007/s11627-021-10197-x

Graham, N., Patil, G. B., Bubeck, D. M., Dobert, R. C., Glenn, K. C., Gutsche, A. T., Kumar, S., Lindbo, J. A., Maas, L., May, G. D., Vega-Sanchez, M. E., Stupar, R. M., & Morrell, P. L. (2020). Plant genome editing and the relevance of Off-Target changes. *Plant Physiology*, *183*(4), 1453–1471. DOI: 10.1104/pp.19.01194 PMID: 32457089

Gupta, S., Kumar, A., Patel, R., & Kumar, V. (2021). Genetically modified crop regulations: Scope and opportunity using the CRISPR-Cas9 genome editing approach. *Molecular Biology Reports*, *48*(5), 4851–4863. DOI: 10.1007/s11033-021-06477-9 PMID: 34114124

Hamdan, M. F., Karlson, C. K. S., Teoh, E. Y., Lau, S.-E., & Tan, B. C. (2022). Genome editing for sustainable crop improvement and mitigation of biotic and abiotic stresses. *Plants*, *11*(19), 2625. DOI: 10.3390/plants11192625 PMID: 36235491

Idris, S. H., Jalaluddin, N. S. M., & Chang, L. W. (2023). Ethical and legal implications of gene editing in plant breeding: A systematic literature review. *Journal of Zhejiang University. Science. B*, *24*(12), 1093–1105. DOI: 10.1631/jzus.B2200601 PMID: 38057267

Ishii, T., & Araki, M. (2016). Consumer acceptance of food crops developed by genome editing. *Plant Cell Reports*, *35*(7), 1507–1518. DOI: 10.1007/s00299-016-1974-2 PMID: 27038939

Justamante, M. S., Larriba, E., Luque, A., Nicolás-Albujer, M., & Pérez-Pérez, J. M. (2025). A systematic review to identify target genes that modulate root system architecture in response to abiotic stress. *Scientific Reports*, *15*(1), 13219. DOI: 10.1038/s41598-025-97266-y PMID: 40246916

Kalaitzandonakes, N., Willig, C., & Zahringer, K. (2022). The economics and policy of genome editing in crop improvement. *The Plant Genome*, *16*(2), e20248. DOI: 10.1002/tpg2.20248 PMID: 36321718

Kamyab, H., SaberiKamarposhti, M., Hashim, H., & Yusuf, M. (2023). Carbon dynamics in agricultural greenhouse gas emissions and removals: A comprehensive review. *Carbon Letters*, *34*(1), 265–289. DOI: 10.1007/s42823-023-00647-4

LaCanne, C. E., & Lundgren, J. G. (2018). Regenerative agriculture: Merging farming and natural resource conservation profitably. *PeerJ*, *6*, e4428. DOI: 10.7717/peerj.4428 PMID: 29503771

Lassoued, R., Macall, D. M., Hesseln, H., Phillips, P. W. B., & Smyth, S. J. (2019). Benefits of genome-edited crops: Expert opinion. *Transgenic Research*, *28*(2), 247–256. DOI: 10.1007/s11248-019-00118-5 PMID: 30830581

Lassoued, R., Phillips, P. W. B., Macall, D. M., Hesseln, H., & Smyth, S. J. (2021). Expert opinions on the regulation of plant genome editing. *Plant Biotechnology Journal*, *19*(6), 1104–1109. DOI: 10.1111/pbi.13597 PMID: 33834596

Li, B., Sun, C., Li, J., & Gao, C. (2024). Targeted genome-modification tools and their advanced applications in crop breeding. *Nature Reviews. Genetics*, *25*(9), 603–622. DOI: 10.1038/s41576-024-00720-2 PMID: 38658741

Li, G., An, L., Yang, W., Yang, L., Wei, T., Shi, J., Wang, J., Doonan, J. H., Xie, K., Fernie, A. R., Lagudah, E. S., Wing, R. A., & Gao, C. (2025). Integrated biotechnological and AI innovations for crop improvement. *Nature*, *643*(8073), 925–937. DOI: 10.1038/s41586-025-09122-8 PMID: 40702261

Lokya, V., Singh, S., Chaudhary, R., Jangra, A., & Tiwari, S. (2025). Emerging trends in transgene-free crop development: Insights into genome editing and its regulatory overview. *Plant Molecular Biology*, *115*(4), 84. DOI: 10.1007/s11103-025-01600-x PMID: 40632351

Long, Y., Wang, C., Liu, C., Li, H., Pu, A., Dong, Z., Wei, X., & Wan, X. (2023). Molecular mechanisms controlling grain size and weight and their biotechnological breeding applications in maize and other cereal crops. *Journal of Advanced Research*, *62*, 27–46. DOI: 10.1016/j.jare.2023.09.016 PMID: 37739122

Maaß, O., Consmüller, N., & Kehlenbeck, H. (2019). Socioeconomic impact of genome editing on agricultural value chains: The case of Fungal-Resistant and Coeliac-Safe wheat. *Sustainability (Basel)*, *11*(22), 6421. DOI: 10.3390/su11226421

Mascarenhas, M. S., Nascimento, F. S., Rocha, A. J., Ferreira, M. S., Oliveira, W. D. S., Morais Lino, L. S., Mendes, T. A. O., Ferreira, C. F., Santos-Serejo, J. A., & Amorim, E. P. (2024). Use of CRISPR Technology in Gene Editing for Tolerance to Biotic Factors in Plants: A Systematic Review. *Current Issues in Molecular Biology*, *46*(10), 11086–11123. DOI: 10.3390/cimb46100659 PMID: 39451539

Nascimento, F. D. S., De Jesus Rocha, A., Da Silva Soares, J. M., Mascarenhas, M. S., Ferreira, M. D. S., Lino, L. S. M., De Souza Ramos, A. P., Diniz, L. E. C., De Oliveira Mendes, T. A., Ferreira, C. F., Santos-Serejo, J. A. D., & Amorim, E. P. (2023). Gene Editing for Plant Resistance to Abiotic Factors: A Systematic review. *Plants*, *12*(2), 305. DOI: 10.3390/plants12020305 PMID: 36679018

Nawaz, S., & Satterfield, T. (2022). Climate solution or corporate co-optation? US and Canadian publics' views on agricultural gene editing. *PLoS One*, *17*(3), e0265635. DOI: 10.1371/journal.pone.0265635 PMID: 35313327

Ndudzo, A., Makuvise, A. S., Moyo, S., & Bobo, E. D. (2024). CRISPR-Cas9 genome editing in crop breeding for climate change resilience: Implications for smallholder farmers in Africa. *Journal of Agriculture and Food Research*, *16*, 101132. DOI: 10.1016/j.jafr.2024.101132

Ofori, K. F., Antoniello, S., English, M. M., & Aryee, A. N. A. (2022). Improving nutrition through biofortification–A systematic review. *Frontiers in Nutrition*, *9*, 1043655. DOI: 10.3389/fnut.2022.1043655 PMID: 36570169

Pavlidis, G., & Tsihrintzis, V. A. (2017). Environmental Benefits and Control of Pollution to Surface Water and Groundwater by Agroforestry Systems: A Review. *Water Resources Management*, *32*(1), 1–29. DOI: 10.1007/s11269-017-1805-4

Pixley, K. V., Falck-Zepeda, J. B., Paarlberg, R. L., Phillips, P. W. B., Slamet-Loedin, I. H., Dhugga, K. S., Campos, H., & Gutterson, N. (2022). Genome-edited crops for improved food security of smallholder farmers. *Nature Genetics*, *54*(4), 364–367. DOI: 10.1038/s41588-022-01046-7 PMID: 35393597

Rahimmanesh, I., Boshtam, M., Kouhpayeh, S., Khanahmad, H., Dabiri, A., Ahangarzadeh, S., Esmaeili, Y., Bidram, E., Vaseghi, G., Javanmard, S. H., Shariati, L., Zarrabi, A., & Varma, R. S. (2022). Gene Editing-Based Technologies for beta-hemoglobinopathies treatment. *Biology (Basel)*, *11*(6), 862. DOI: 10.3390/biology11060862 PMID: 35741383

Rajpal, V. R., Sehgal, D., Valluru, R., & Singh, S. (2023). Editorial: Current advances in genomics and gene editing tools for crop improvement in a changing climate scenario. *Frontiers in Genetics*, *14*, 1214679. DOI: 10.3389/fgene.2023.1214679 PMID: 37377598

Ríos, D. F., Quintana, S. A., Paniagua, P. G., Arrúa, A. A., Brozón, G. R., Hicar, M. S. B., Alegría, A. C., & Goberna, M. F. (2025). Regulatory challenges and global trade implications of genome editing in agriculture. *Frontiers in Bioengineering and Biotechnology*, *13*, 1609110. DOI: 10.3389/fbioe.2025.1609110 PMID: 40613005

Rustgi, S., Zhang, H., & Oz, T. M. (2023). Editorial: Overcoming genome editing challenges in plants: new tools and nanotechnologies. *Frontiers in Genome Editing*, *5*, 1230424. DOI: 10.3389/fgeed.2023.1230424 PMID: 37324649

Salavatipour, M. S., Poursalehi, Z., Rouzbahani, N. H., Mohammadyar, S., & Vasei, M. (2024). CRISPR-Cas9 in basic and translational aspects of cancer therapy. *BioImpacts*, *14*(6), 30087. DOI: 10.34172/bi.2024.30087 PMID: 39493894

Samtiya, M., Aluko, R. E., & Dhewa, T. (2020). Plant food anti-nutritional factors and their reduction strategies: An overview. *Food Production, Processing and Nutrition*, *2*(1), 6. Advance online publication. DOI: 10.1186/s43014-020-0020-5

Saxena, R., Kumar, M., Tomar, R. S., Yadav, N., Kaur, N., Puri, P., Yadav, A., Negi, R., & Yadav, A. N. (2025). Genome editing: A new age technology for sustainable agriculture and developing abiotic stress resilience in crops. *Journal of Applied Biology and Biotechnology*. Advance online publication. DOI: 10.7324/JABB.2025.232158

Scharff, L. B., Saltenis, V. L. R., Jensen, P. E., Baekelandt, A., Burgess, A. J., Burow, M., Ceriotti, A., Cohan, J., Geu-Flores, F., Halkier, B. A., Haslam, R. P., Inzé, D., Lankhorst, R. K., Murchie, E. H., Napier, J. A., Nacry, P., Parry, M. J., Santino, A., Scarano, A., & Pribil, M. (2021). Prospects to improve the nutritional quality of crops. *Food and Energy Security*, *11*(1), e327. Advance online publication. DOI: 10.1002/fes3.327

Sokal, K., & Kachel, M. (2025). Impact of Agriculture on Greenhouse Gas Emissions—A Review. *Energies*, *18*(9), 2272. DOI: 10.3390/en18092272

Sprink, T., Wilhelm, R., & Hartung, F. (2022). Genome editing around the globe: An update on policies and perceptions. *Plant Physiology*, *190*(3), 1579–1587. DOI: 10.1093/plphys/kiac359 PMID: 35976141

Tachikawa, M., & Matsuo, M. (2023). Divergence and convergence in international regulatory policies regarding genome-edited food: How to find a middle ground. *Frontiers in Plant Science*, *14*, 1105426. DOI: 10.3389/fpls.2023.1105426 PMID: 36794228

Tamburini, G., Bommarco, R., Wanger, T. C., Kremen, C., Van Der Heijden, M. G. A., Liebman, M., & Hallin, S. (2020). Agricultural diversification promotes multiple ecosystem services without compromising yield. *Science Advances*, *6*(45), eaba1715. Advance online publication. DOI: 10.1126/sciadv.aba1715 PMID: 33148637

Tripathi, L., Dhugga, K. S., Ntui, V. O., Runo, S., Syombua, E. D., Muiruri, S., Wen, Z., & Tripathi, J. N. (2022). Genome editing for sustainable agriculture in Africa. *Frontiers in Genome Editing*, *4*, 876697. DOI: 10.3389/fgeed.2022.876697 PMID: 35647578

Turnbull, C., Lillemo, M., & Hvoslef-Eide, T. K. (2021). Global regulation of genetically Modified crops Amid the gene edited crop boom – A review. *Frontiers in Plant Science*, *12*, 630396. DOI: 10.3389/fpls.2021.630396 PMID: 33719302

Ukhatova, Y. V., Erastenkova, M. V., Korshikova, E. S., Krylova, E. A., Mikhailova, A. S., Semilet, T. V., Tikhonova, N. G., Shvachko, N. A., & Khlestkina, E. K. (2023). Improvement of crops using the CRISPR/CAS system: New target genes. *Molecular Biology*, *57*(3), 375–397. DOI: 10.1134/S0026893323030135 PMID: 37326044

Venezia, M., & Krainer, K. M. C. (2021). Current advancements and limitations of gene editing in orphan crops. *Frontiers in Plant Science*, *12*, 742932. DOI: 10.3389/fpls.2021.742932 PMID: 34630494

Venkataraman, S., Zaruk, D., & Hefferon, K. (2025). Recent developments, challenges and opportunities in genome editing for crop science from a societal perspective. *Frontiers in Genome Editing*, *7*, 1568072. DOI: 10.3389/fgeed.2025.1568072 PMID: 40528898

Vikas, and Ranjan, R. (2024). Agroecological approaches to sustainable development. *Frontiers in Sustainable Food Systems, 8*. DOI: 10.3389/fsufs.2024.1405409

Wahab, A., Abdi, G., Saleem, M. H., Ali, B., Ullah, S., Shah, W., Mumtaz, S., Yasin, G., Muresan, C. C., & Marc, R. A. (2022). Plants' Physio-Biochemical and Phyto-Hormonal Responses to Alleviate the Adverse Effects of Drought Stress: A Comprehensive Review. *Plants*, *11*(13), 1620. DOI: 10.3390/plants11131620 PMID: 35807572

Yadav, H., Roberts, P. A., & Lopez-Arredondo, D. (2025). Combating Root-Knot Nematodes (Meloidogyne spp.): From Molecular Mechanisms to Resistant Crops. *Plants*, *14*(9), 1321. DOI: 10.3390/plants14091321 PMID: 40364350

Yang, C., Huang, L., Wang, B.-C., Zhong, Y., Ma, X., Zhang, C., Sun, Q., Wu, Y., Yao, Y., & Liu, Q. (2025a). Enhancing quality traits in staple crops: Current advances and future perspectives. *Journal of Genetics and Genomics*. *Journal of Genetics and Genomics*, *52*(12), 1438–1459. DOI: 10.1016/j.jgg.2025.05.001 PMID: 40348082

Yang, Z., Yan, H., Liu, H., Yang, L., Mi, G., & Wang, P. (2025b). Enhancing crop nitrogen efficiency: The role of mixed nitrate and ammonium supply in plant growth and development. *Biology (Basel)*, *14*(5), 546. DOI: 10.3390/biology14050546 PMID: 40427735

Yaqoob, H., Tariq, A., Bhat, B. A., Bhat, K. A., Nehvi, I. B., Raza, A., Djalovic, I., Prasad, P. V., & Mir, R. A. (2023). Integrating genomics and genome editing for orphan crop improvement: A bridge between orphan crops and modern agriculture system. *GM Crops & Food*, *14*(1), 1–20. DOI: 10.1080/21645698.2022.2146952 PMID: 36606637

Zhang, D., Xu, F., Wang, F., Le, L., & Pu, L. (2024). Synthetic biology and artificial intelligence in crop improvement. *Plant Communications*, *6*(2), 101220. DOI: 10.1016/j.xplc.2024.101220 PMID: 39668563

Zhang, Y., Zhang, H., Tao, T., Chen, J., Li, P., Wang, Y., Liu, P., Zhu, Y., Routledge, M. N., Yang, C., & Zhang, C. (2025). Investigation and insights into the technical strategies for reducing agrochemical inputs in rice farming to enhance agroecosystem resilience and food safety, A case study in China. *Journal of Environmental Management*, *392*, 126619. DOI: 10.1016/j.jenvman.2025.126619 PMID: 40695181

Zhu, H., Wu, X., Zheng, R., & Zhu, Y. (2025). The impact of gene editing technology on agricultural economic efficiency Biomechanical Applications: An empirical analysis based on international CRISPR patent data. *Molecular & Cellular Biomechanics*, *22*(3), 1031. DOI: 10.62617/mcb1031

Chapter 7
Regulatory Framework for Gene-Edited Plants

Khansa Ayub
http://orcid.org/0009-0007-2169-2033
Department of Plant Sciences, Quaid-i-Azam University, Islamabad, Pakistan

Hafiza Rana
http://orcid.org/0009-0009-3164-6908
Department of Plant Sciences, Quaid-i-Azam University, Islamabad, Pakistan

Zakir Ullah
Department of Plant Sciences, Quaid-i-Azam University, Islamabad, Pakistan

Muhammad Abid
Bacha Khan University, Charsadda, Pakistan

Zahid Ullah
Center for Plant Sciences and Biodiversity, University of Swat, Charbagh, Pakistan

Ahmad Ali
Center for Plant Sciences and Biodiversity, University of Swat, Charbagh, Pakistan

ABSTRACT

Gene-editing tools such as CRISPR/Cas9 are reshaping agriculture by enabling precise, efficient, and relatively low-cost changes to plant genomes, accelerating development of higher-yielding, disease-resistant, and climate-adapted crops to support global food security. To manage risks and enable deployment, major regions have adopted distinct regulatory approaches. This chapter compares frameworks in the European Union, United States, China, and India, highlighting differences between product-based systems (focused on the final trait and risk profile) and process-based systems (focused on how genetic changes are introduced). The resulting regulatory fragmentation can complicate innovation, trade, and public acceptance, yet remains central to balancing biosafety, ethics, and public trust as gene editing expands.

DOI: 10.4018/979-8-2600-3575-7.ch007

INTRODUCTION

The field of plant breeding has been greatly improved by the promising influence of gene editing technology, which has produced impressive advancements in a short duration (Kumar *et al.,* 2020). CRISPR/Cas9 has revolutionized genome editing and has had a significant impact on industries such as biotechnology, agriculture, and medicine (Aljabali *et al.*, 2024).

The scope of plant genetic improvement, including modifications for higher yields, stress tolerance, and insect resistance, has been expanded by these developments. Gene editing speed and accuracy have addressed urgent challenges such as soil erosion, climate change, and the nutritional needs of a growing global population (Atia *et al.,* 2024). Despite its transformative potential, problems such as immunological responses and off-target effects remain significant. (Wen and Zhang, 2022; Guo *et al.,* 2023).

Overview of Gene Editing Technologies (e.g., CRISPR/Cas9)

The bacterial immunological processes that gave rise to CRISPR/Cas9 allow precise gene variations using guide RNAs, which route the Cas9 enzyme to the DNA sequences. This technology enables insertions, deletions, or modifications with unparalleled accuracy and efficiency (Singh, 2024). Advancements such as prime and base editing can be made without causing double-strand breaks. CRISPR technology has been applied to a few crops and industrial fungi, in addition to food crops (Wilson & Harrison, 2021). Refinements can be made without breaking double strands, with achievements such as base and prime editing (Yuan *et al.*, 2024).

Importance of Gene-Edited Plants in Agriculture and Food Security

Enhancing Crop Productivity

Gene editing makes it feasible to accurately alter fundamental genetic characteristics, produce higher-yield crops, and reduce reliance on chemical inputs. Examples of successful uses include the development of disease-resistant wheat, drought-tolerant maize, and improved tomatoes. With techniques such as CRISPR/Cas9, these improvements can now be produced more quickly than with traditional breeding techniques (Karavolias *et al.,* 2021).

Climate Change Resilience

According to an estimate, the world population will reach 10 billion by the end of 2050. In this scenario for food security, food crop production should be increased by 60% (Bhatta *et al.*, 2021). However, crop yield is being hampered by abiotic stresses, which leads to a 50% annual yield, leading to 14-19$ million losses (Bhat *et al.*, 2021; Kaur & Pati, 2019). Developing stress-resistant crops that endure stress guarantees steady agricultural production and protects food supply lines. Examples include rice cultivars that can withstand flooding and maize engineered to use water more efficiently (Karavolias *et al.,* 2021).

Improving Nutritional Content

Gene editing resolves nutritional inadequacies, which is crucial in areas where hunger is prevalent. Improvements include increasing the amount of iron and zinc in a variety of staple crops and increasing the amount of vitamin A in rice (Golden Rice 2.0) (Zhang & Zhu, 2023).

Reducing Dependency on Agrochemicals

Gene editing minimizes the environmental impact of chemical use by producing plants that are resistant to pests and diseases, thereby reducing the need for pesticides and herbicides. CRISPR, for instance, has been used to create non-browning mushrooms, extending their shelf life and lowering food waste (Zhu 2022).

Speeding up Crop Development

New traits can be introduced using traditional breeding methods over more than ten years. However, the time required to produce the desired traits in a few crop generations can be significantly reduced by gene editing (Tan *et al.,* 2020). Adapting to swift changes in agricultural demands, such as climatic stressors and emerging pest threats, requires a quick turnover.

Contributions to Sustainable Agriculture

Gene editing is expected to produce more resilient and useful crops, supporting larger sustainability efforts. The latest analysis indicates that with the help of genome-edited crops, several Sustainable Development Goals (SDGs), such as promoting health, preventing hunger, well-being, and combating climate change, can be accomplished (Hüdig *et al.,* 2022). By focusing on the SWEET gene family, a rice line modified with the CRISPR/Cas9 gene-editing technique showed broad-spectrum resistance to bacterial blight to a greater extent, lowering yield loss under disease stress, as shown in (Figure 1) (Oliva *et al.,* 2019).

Genome editing technologies, such as CRISPR/Cas9, provide an accurate and rapid method for creating climate-resilient cultivars, which are being extensively investigated to increase heat stress tolerance in important cereal crops, including rice, wheat, and maize (Pandey *et al.*, 2024).By focusing on specific genes involved in ripening and flavor, gene editing can be used to improve fruit quality characteristics, such as shelf life, taste, and texture.

To increase the shelf life of strawberries and tomatoes, ethylene-related genes have been knocked out using CRISPR/Cas9 (Gao *et al.*, 2021). Studies on public opinion report conflicting perspectives on gene-edited foods, with acceptance being lower in Europe and Japan and higher in the US. This is often related to perceived label transparency (Kato-Nitta *et al.,* 2023;) and.

Figure 1. Importance of Gene editing technologies in Agriculture

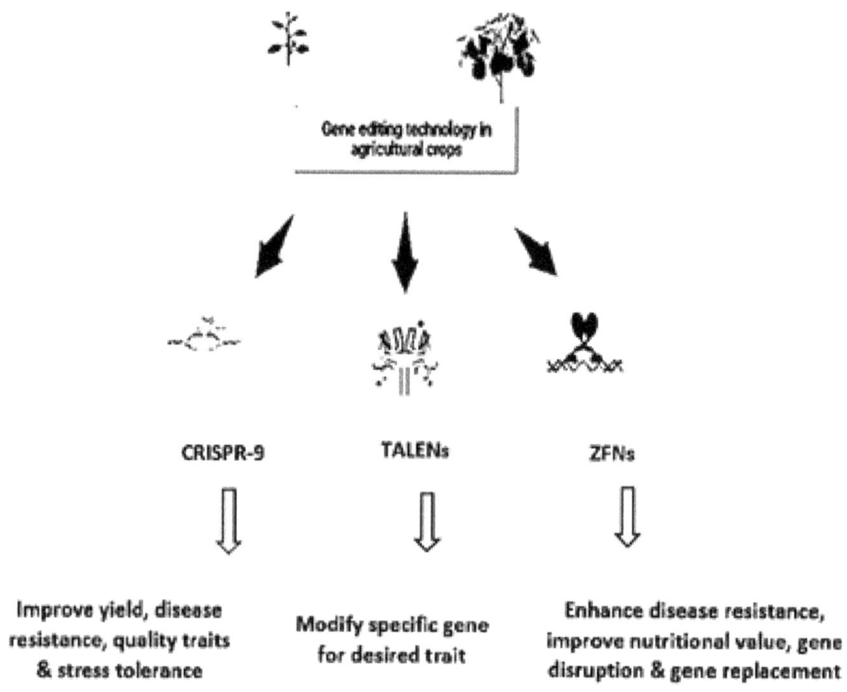

Overview of Regulatory Challenges and Frameworks

Gene editing can fulfill the vision of the Sustainable Development Goals (SDGs) of the United Nations and bring in a new Green Revolution (Bain *et al.*, 2019) within the framework of SDG 13: Climate Action and "SDG 2: Zero Hunger". New Plant Breeding Techniques (NPBTs) may be developed using gene-editing techniques (Idris *et al.*, 2023). Gene-editing techniques in plant breeding have sparked debates about the ethics and law of whether or not gene-edited crops should be classified as genetically modified organisms (GMOs) and the necessity of establishing a regulatory framework for this approach were the main points of disagreement in the early gene-editing discussion (Whelan and Lema, 2015). Many nations have clarified their laws governing the release of gene-edited crops, leading to differences in national regulatory strategies. Some countries, such as the European Union and New Zealand, have strict laws, while others have laws that are tolerant of gene editing, such as Canada and the United States (Menz *et al.*, 2020). CRISPR/Cas9 has transformed plant breeding, offering unmatched effectiveness and accuracy. However, the adoption of these technologies has been delayed by substantial regulatory hurdles. Globally, governments are attempting to maintain a balance between biosafety, ethical issues, public acceptance, and innovation in their regulatory frameworks for gene-edited crops (Mbaya *et al.*, 2022).

Difficulty in Detection and Traceability

Different novel genomic techniques (NGTs) have been used to evaluate practical approaches for finding and classifying plants that are important to the market. This emphasizes the challenges of discriminating between plants whose genomes have been modified and those with naturally occurring mutations, emphasizing the requirement of information-sharing mechanisms and detection techniques to guarantee traceability and regulatory compliance (Guertler *et al.*, 2023).

Public Trust and Transparency

Consumers may become more doubtful when regulatory requirements seem excessive or out of proportion to the perceived risk because they may consider this a sign of obscured dangers (Herman *et al.*, 2021). Japan has established a regulatory framework that places a high value on transparency to maintain consumer trust in goods that have undergone gene editing. Developers are required to submit a notice form and participate in a pre-submission consultation with the Ministry of Labor and Welfare (MHLW) and the Ministry of Health for products created using genome editing technology. Following this process, the MHLW posts summaries of the authorized items on its website, guaranteeing that the public has immediate access to information regarding the launch of gene-edited foods in the marketplace. This strategy seeks to maintain a balance between innovation and public approval by providing customers with timely and easily available information (Matsuo & Tachikawa, 2022). People are skeptical of gene-edited crops, as they were of earlier GMO products. Surveys show that popular acceptability varies greatly among countries. Americans are generally more tolerant, whereas Germans and Japanese are more cautious (Shew *et al.*, 2018). Public involvement and label transparency are regarded as essential tactics for increasing confidence in and acceptance of genetically modified goods (Hill *et al.,* 2022).

Ethical and Societal Concerns

Beyond technical risk assessments, there is growing interest in ethical problems related to environmental impact, biodiversity, and equal access to technology. For example, The African continent faces two particular issues: resource constraints and fragmented regulatory readiness (Rabuma *et al.,* 2024). Global regulatory frameworks must be developed to accommodate these, ensuring inclusive, socio-ethical dimensions and sustainable innovation.

Defining Gene Edited Crops

The major regulatory obstacle is the classification of gene-edited plants. In contrast to conventional genetically modified organisms (GMOs) that integrate foreign DNA, transgenic material is not used in many gene-edited plants. However, regulatory bodies often find it difficult to determine whether these goods should be considered under the current GMO regulations.

Despite scientific controversy, the European Court of Justice of the European Union decided that gene-edited crops should be regulated equivalently to genetically modified organisms (GMOs), which greatly impeded innovation (Puchta, 2024). By regulatory regimes In the USA and Japan, gene-edited crops with and without foreign DNA are distinguished, and the latter is frequently exempted from full GMO regulation (Turnbull *et al.*, 2021). Public communication, intellectual property protection, and international trade are all obstructed by this classification disagreement.

Regulatory Complexity and Delays

This study examines the economic effects of regulatory restrictions on genetically modified crops. It draws attention to the fact that longer approval processes, brought about by stricter regulations, have left developers and investors in the dark.

According to the authors, these kinds of delays could endanger future public sector investments in agricultural R&D, especially for smaller organizations that do not have the means to handle complicated regulatory environments (Smyth *et al.*, 2016). The lack of defined global criteria, lengthy examinations, and uncertainty surrounding risk assessments continue to impede global commercialization efforts (Strobbe *et al.*, 2023).

Intellectual Property Rights

IP conflicts have taken on new dimensions because of gene editing. Patent rights for altered characteristics and CRISPR technologies raise concerns regarding accessibility for small-scale farmers and public-sector breeders (Lukasiewicz *et al.*, 2024a). One major global regulatory concern is ensuring equitable access to gene-editing technologies without restricting innovation by focusing on opportunities and challenges (Figure 2).

Figure 2. Key regulatory challenges of GE plants & their opportunities

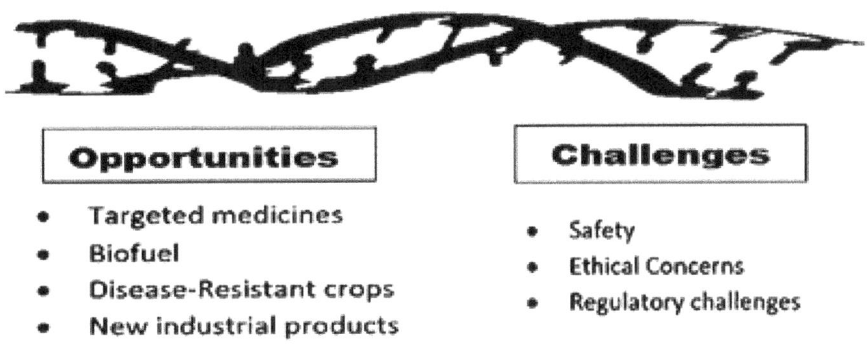

GLOBAL COMPARISONS OF REGULATORY FRAMEWORKS

United States

As long as no foreign DNA is added, many gene-edited plants are free from stringent pre-market inspections under the USDA's new secure regulation. The emphasis is shifting from process-based evaluation to product-based evaluation (Hoffman,

2021). Although innovation is accelerated by this approach, it raises concerns about reduced public transparency and potential market disruptions.

European Union

Similar to first-generation GMOs, genome-edited organisms must pass environmental and food and feed risk assessments, according to this academic article that examines the EU's present regulatory framework. The authors contend that to address the unique unintended consequences connected to genome editing techniques, the current risk assessment guidelines must be revised (Kawall *et al.*, 2020).

China

The "Guidelines for Safety Evaluation of Gene-Edited Plants for Agricultural Use (Trial)" were released by the Ministry of Agriculture and Rural Affairs (MARA) on January 24, 2022, and they provide specifications and application processes for gene-edited plants. In contrast to traditional GMOs, the recommendations cover gene-edited plants that do not introduce external genes, suggesting a faster regulatory process (Liang *et al.*, 2022).

India

India's new guidelines (2022) exempt gene-edited plants generated using SDN-1 and SDN-2 procedures (without foreign DNA) from strict GMO regulations. This continuous approach seeks to promote swift local innovation while protecting biosafety (Prasad & Chimata, 2023).

Latin America

Countries such as Argentina, Chile, and Brazil have adopted case-by-case strategies to control gene-edited plants, prioritizing product quality over editing techniques. This adaptability has facilitated a faster market entry for CRISPR-edited crops (Hernández-Soto & Gatica-Arias, 2024).

REGULATORY FRAMEWORKS AROUND THE WORLD (INTERNATIONAL AGREEMENTS AND GUIDELINES)

As gene-editing technologies such as CRISPR/Cas9 become more widely used, discussions over proper regulatory control are increasing. In addition to offering rational, empirically based approaches for evaluating, regulating, and permitting the use of gene-edited plants, international agreements and standards seek to maintain a balance between innovation, safety, and trade issues (Lassoued *et al.*, 2021).

Cartagena Protocol on Biosafety

The Cartagena Protocol, adopted in 2000 under the Convention on Biological Diversity (CBD), governs the transboundary movement of genetically modified organisms (GMOs). Despite being older than current gene-editing technologies, certain gene-edited plants may fall under its broad categories (Lassoued *et al.*, 2021). Some argue that regulating simple gene changes in the same way as genetically modified organisms (GMOs) could unduly impede innovation, and they favor protocol revisions that would classify gene-edited organisms based on their risk profiles (Brookes & Smyth, 2024).

Codex Alimentarius Guidelines

The Codex General Principles of Food Hygiene (GPFH), which established the minimal requirements for food production, processing, and distribution, was originally published in 1969 as a result of cooperation among specialists from 168 countries. The Codex Alimentarius has general, commodity-specific, and regional committee sections. Experts from practically every country participated in the committees, which helped to set science-based guidelines and promote risk communication (Hasani *et al.*, 2024)

Emerging Proposals for International Coordination

Anticipatory Detection and Information Sharing Framework

One of the major regulatory challenges is the difficulty in detecting genome edits that do not involve foreign DNA. To address this, experts have proposed a voluntary information-sharing database and the establishment of an anticipatory system where developers submit genome editing information, aiding regulatory enforcement without stifling innovation (Eckerstorfer *et al.*, 2021). Trade disruptions and support compliance could be minimized with labeling schemes using this anticipatory detection model.

Calls for Harmonized Global Standards

Many scholars advocate for international guidelines that balance safety and scientific innovation. Proposals suggest risk-proportionate regulations that focus on plants with novel traits posing environmental or health risks and lighter oversight for smaller genome edits (Hernández-Soto & Gatica-Arias, 2024).Furthermore, countries such as Chile have been highlighted for their science-based, flexible frameworks that could serve as models for widespread global adoption (Sánchez, 2024).

Challenges in Developing Global Guidelines

Countries differ in their perception of the risks associated with gene editing. For example, Europe tends toward a precautionary principle, while the U.S. and Latin American nations favor a product-based approach (Yang *et al.,* 2023). Genome-edited plants that mimic natural mutations challenge traceability and the implementation of labeling and import/export controls (Kalaitzandonakes *et al.*, 2023). African nations confront challenges in using genome editing technology, highlighting the necessity of strong regulatory frameworks. This highlights how the adoption and efficient regulation of new technologies are hindered by skepticism, uneven adherence to international agreements, knowledge gaps among stakeholders, and uneven compliance with international agreements (Akinbo *et al.*, 2025).

Regulatory Frameworks in Key Countries

This study examines the different laws that nations have applied to agricultural plants with altered genomes. This highlights the fact that whereas some countries use process-based evaluations that highlight the techniques used in development, others use product-based assessments that focus on the qualities of the finished product (Entine *et al.*, 2021).

United States

U.S. In 2020, the Department of Agriculture (USDA) implemented a SECURE rule that exempts gene-edited crops from regulatory assessment in cases where the genetic alterations could be natural or the result of conventional breeding. This product-based strategy accelerates invention and commercialization by emphasizing the characteristics rather than the means of achieving them (Jenkins *et al.*, 2021). Food safety and environmental risks are supervised by the Food and Drug Administration (FDA) and Environmental Protection Agency (EPA), respectively.

European Union

In contrast, the European Court of Justice ruled in 2018 that the European Union must control gene-edited plants using the same strict guidelines as GMOs. This process-based approach concentrates on the method employed, despite the presence of foreign DNA in the finished product (Voigt, 2023). The introduction of a tiered risk-based system was presented in 2023, but a complete legislative change has not yet been enforced.

Latin America

Latin American nations with adaptable, case-by-case regulatory regimes include Colombia, Argentina, Brazil, and Chile. For example, Argentina has developed a paradigm that exempts specific gene-edited plants from GMO laws if no foreign DNA is present (Zarate *et al.,* 2023). The implementation of a science-driven framework that encourages domestic research and commercialization has made Colombia a pioneer in agricultural biotechnology in the region (Sánchez, 2024). Certain nations, such as Peru, have a moratorium on genetically modified crops, including plants that have had their genes altered to illustrate regional regulatory variations (Hernández-Soto & Gatica-Arias, 2024).

Japan

Japan has enforced a modest, product-based regulatory system that distinguishes between different genome-editing methods. Goods produced using SDN-1 techniques, which do not require the insertion of foreign DNA, are not subject to strict GMO labelling guidelines and regulations. Rather, before going on sale, developers must inform the Ministry of Health, Labour, and Welfare (MHLW) to ensure transparency and maintain customer confidence (Tsuda *et al.*, 2019).

Southeast Asia

Various gene-edited crops, such as non-browning bananas, have been deregulated in nations such as the Philippines under product-based frameworks. However, Southeast Asia's legislative frameworks vary, and a lack of uniformity might make trading challenging (Jalaluddin *et al.,* 2024). Efforts are ongoing to align regional standards to facilitate cross-border trade in agricultural biotechnology (Table 1).

Table 1. Comparative summary of regulatory framework approaches in key countries

Region/Country	Regulatory Approach	Highlights
USA	Product-based	Focus on final traits; SECURE rule accelerates approvals
EU	Process-based	All gene edits treated as GMOs; possible future tiered system
Argentina, Brazil, Chile	Product-based (case-by-case)	Innovation-friendly frameworks; fast market access
Japan	Product-based (notification system)	Moderate approach; excludes non-transgenic edits from strict rules
Philippines/Southeast Asia	Emerging mixed models	Movement toward deregulation, but still fragmented

Gene Editing Technologies and Their Applications

The constraints of traditional breeding methods have been successfully overcome by genome editing, a new technology that uses sequence-specific nucleases (SSNs) to introduce targeted changes in crops with great efficiency and precision. Zinc finger nucleases (ZFNs), transcriptional activator-like effector nucleases (TALENs), and clustered regularly interspaced short palindromic repeats (CRISPR)-associated endonuclease Cas9 (CRISPR/Cas9) are examples of artificially engineered SSNs that have demonstrated exceptional efficacy in targeted mutagenesis in a variety of crops and model plants (Mishra *et al.*,2018).[REMOVED HYPERLINK FIELD]

CRISPR (Clustered regularly Interspaced Short Palindromic Repeats)

CRISPR/Cas9 genome editing has been performed in rice, wheat, maize, barley, and other significant cereal crops. It depicts the accuracy and effectiveness of the technology in producing the desired attributes, which makes it easier for cereal crop development to adopt it widely (Ansari *et al.*, 2020). Recent developments, such as base editing and prime editing, have significantly improved the accuracy of CRISPR. Base editors introduce point changes without causing double-strand breaks, whereas prime editors allow targeted insertions, deletions, and sequence replacements, as shown in (Figure 3) (Li *et al.,* 2021).

Figure 3. Gene editing through CRISPR/Cas9

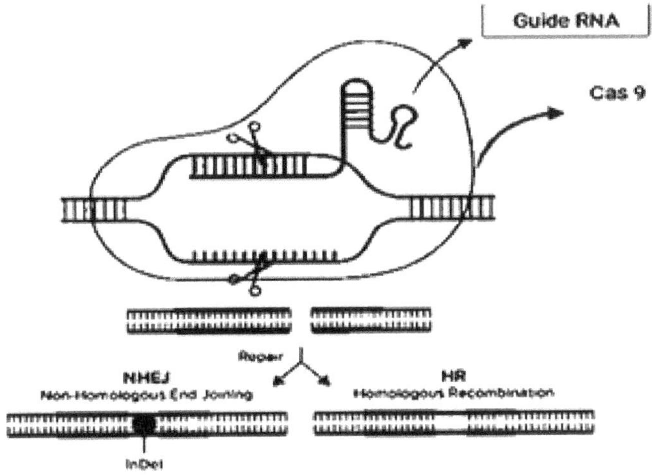

Zinc Finger Nucleases (ZFNs) and Transcription Activator-Like Effector Nuclease (TALENs)

TALENs and ZFNs can modify genomes with high specificity and efficiency. It also diverted attention to factors affecting the selection of various gene-editing technologies, such as delivery systems and intellectual property rights (Figure 4) (Carroll, 2021).

Figure 4. Gene editing through Zinc Finger Nucleases (ZFNs) and TALENs

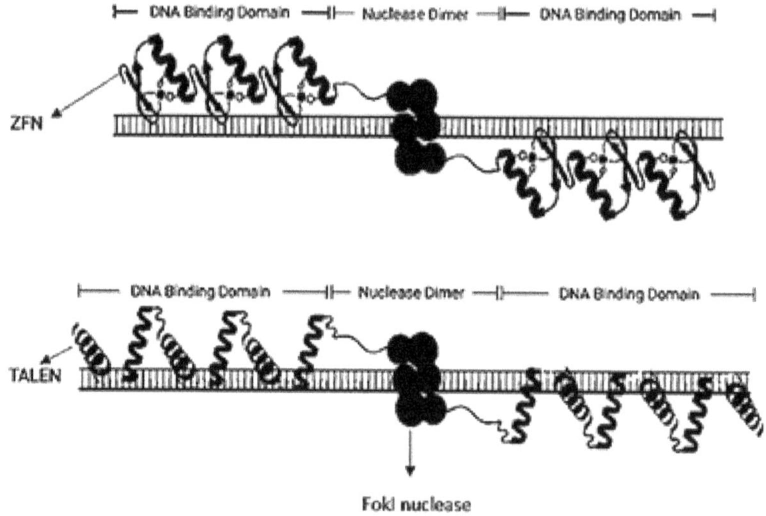

Virus-Induced Genome Editing (VIGE)

For species that are challenging to modify using conventional Agrobacterium or biolistic techniques, the use of viral vectors to introduce CRISPR components into plants is becoming increasingly popular (Figure 5). VIGE makes it possible to produce transgene-free edited plants and increases their editing ability (Zhang *et al.,* 2022).

Figure 5. Viral-induced genome editing

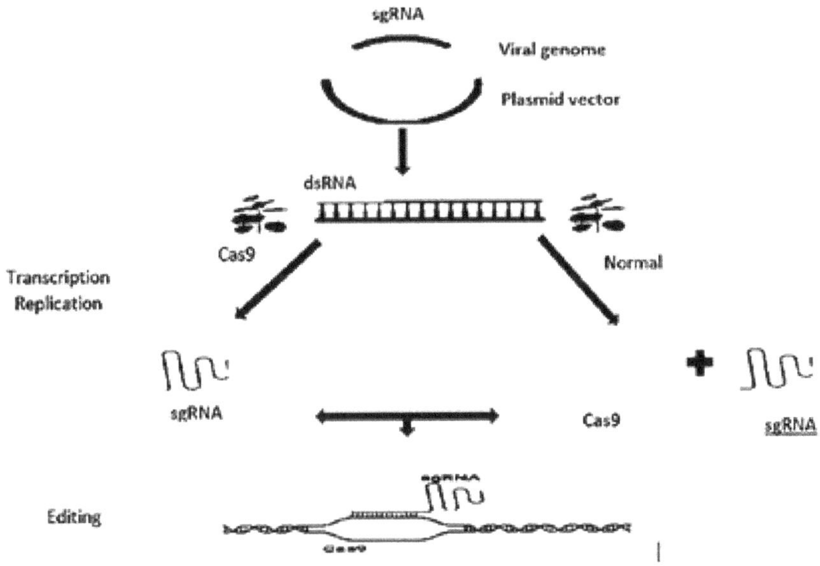

APPLICATIONS IN PLANT RESEARCH AND CROP IMPROVEMENT

Enhancing Abiotic Stress Tolerance

Plants with increased resistance to environmental challenges, such as drought, salinity, and severe temperatures, have been created using gene-editing techniques. For example, wheat and rice, which have shown improved resilience to drought and saline soils, have been CRISPR-edited (Puchta *et al.,* 2022).

Modifying Flowering Time and Seed Dormancy

To maximize crop adaptation to different conditions, precise control over flowering time and seed dormancy is necessary. These characteristics of key cereals have been successfully modulated by CRISPR/Cas9, which increases yield and flexibility across regions (Kishchenko *et al.,* 2020).

Increasing Nutritional Value

In 2021, the first consumer-available CRISPR-edited food was introduced in Japan when a Tokyo-based business, Sanatech Seed, started selling tomatoes with high GABA levels. According to the company, GABA can encourage relaxation and help decrease blood pressure. Sanatech, after expanding distribution in Japan, aims to bring their tomatoes to the US and complete regulatory paperwork to introduce them in the Philippines (Matsuo & Tachikawa, 2022). Research is ongoing to increase the micronutrient content, such as iron, zinc, and vitamin A, in staple crops.

De Novo Domestication

The current study demonstrates the use of CRISPR-Cas9 genome editing on the ground cherry orphan crop, *Physalis pruinosa.* Researchers have quickly improved plant architecture, flower production, and fruit size by focusing on the orthologs of tomato domestication genes, which has made it easier to domesticate this wild species (Lemmon *et al.*, 2018).

Technical Challenges and Future Perspectives

However, it remains difficult to efficiently deliver editing reagents to various plant tissues. Innovations such as VIGE and nanotechnology-based delivery provide promising solutions (Ebrahimi & Hashemi, 2024). The complete removal of off-target mutations remains a scientific endeavor, even though new generations of CRISPR variants have reduced their occurrence (Li *et al.,* 2021). Enhancing crop characteristics, including pest resistance, disease tolerance, drought resilience, and nutritional value, is the goal of horticulture's adoption of genetic modification techniques. However, there are several complicated advantages, hazards, and public perceptions associated with the use of GM horticultural crops that need to be considered (Vishnu, 2024).

Risk Assessment and Safety Evaluation Frameworks

Most regulatory bodies take a case-by-case approach to risk assessment for gene-edited plants, judging them based on product attributes rather than the production process. The European Food Safety Authority (EFSA) advises that, similar to evaluations for conventional GMOs, environmental and food safety evaluations of genome-edited plants should concentrate on characteristics, unexpected consequences, and off-targ*et al*terations (Eckerstorfer *et al.,* 2023). Non-transgenic genome-edited products from regulatory oversight are exempted in a few countries, such as the United States, while in the European Union (EU), novel products equivalent to genetically modified (GM) traits are regulated. In Canada, the novel food-based regulatory regime is unique as it is based on the principle of examining product attributes that affect food safety instead of focusing on breeding technology (Lassoued *et al*., 2024).

Broader Ethical and Societal Risk Considerations

In addition to technological hazards, several frameworks, such as Norway's Gene Technology Act, require evaluations of societal, ethical, and sustainability effects prior to the approval of gene-edited species (Myskja & Myhr, 2020). This broader view recognizes the disruptive potential of gene-editing technologies, not only biologically but also socioeconomically.

Specific Risk Assessment Innovations

Novel bioinformatics frameworks are being developed to evaluate off-target consequences more accurately in products that have undergone genome editing. For example, to enhance regulatory safety assessments, a human health risk assessment approach was presented that used bioinformatics to forecast off-target effects from RNA-based biopesticides (Devisetty *et al.,* 2024).

Labeling Requirements (Divergent Global Policies)

1. Labeling is generally not required if foreign DNA is absent in the United States.
2. According to present laws, the European Union mandates that foods that have undergone genetic modification must be labeled.
3. If the change does not introduce exogenous genes, Japan requires public disclosure and notification but does not require GMO-style labeling (Kondo & Taguchi, 2022).

Challenges in Traceability Requirements

Conventional GMO traceability and detection techniques are complicated by the fact that gene-edited crops frequently lack detectable foreign DNA sequences. It is nearly impossible to differentiate gene-edited products from conventionally produced crops using laboratory examinations alone without clear disclosure or specialist databases. Experts have advised the creation of a global framework for anticipatory detection and identification, which includes voluntarily informing regulatory bodies about genome-editing incidents. By doing this, trade disruptions would be avoided, and labeling methods would be enforced (Ribarits *et al.,* 2021). Furthermore, although it increases complexity and expense, the use of distinct molecular markers during editing procedures has been suggested as a technical method to support traceability (Eckerstorfer *et al.,* 2021).

CASE STUDIES AND EXAMPLES

United States: A Regulatory Structure Based on Products

In the United States, regulatory decisions regarding gene-edited crops are based on the characteristics of the product, not the process. The SECURE rule, which was put into effect by the USDA's Animal and Plant Health Inspection Service (APHIS) in 2020, exempts many gene-edited crops from regulation if they do not contain foreign DNA (Kuzma & Grieger, 2020). By eliminating pre-market field testing for qualified crops, this rule expedites approvals. For instance, CRISPR-created high-oleic acid soybean cultivars were deregulated as they did not have any foreign transgenes and did not pose any new plant pest threats (Hartman *et al.*, 2020). The FDA (food safety), EPA (pesticides), and USDA (plant health) are all given roles in the U.S. regulatory system (Figure 6), although gene-edited crops that do not fit certain requirements, such as novel protein expression, may pass some of these evaluations (Wolt *et al.*, 2016).

Figure 6. Regulatory review process by regulatory authorities

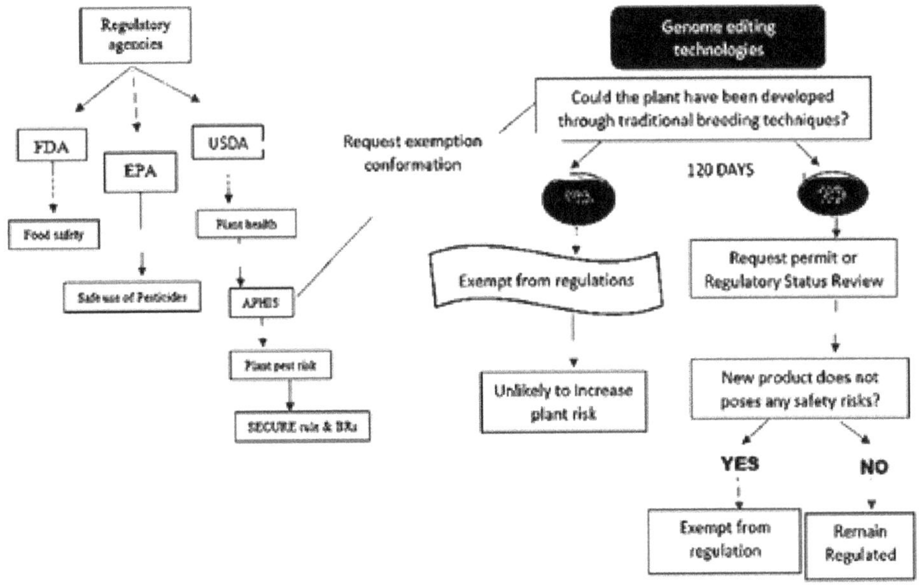

Argentina: Developing a Model for Case-by-Case Assessment

Drought-tolerant soybeans and wheat that have been altered without transgenic components, for example, have advanced quickly through Argentina's approval process. Due to reduced regulatory constraints, public sector developers and entrepreneurs have participated more (Entine *et al.,* 2021).

European Union: Legal Complexity and Process-Based Regulation

In contrast, the other hand, follows a process-based regulatory framework. Even in cases where no foreign DNA is added, gene-altered organisms are subject to current GMO laws, according to a 2018 ruling by the Court of Justice of the European Union (CJEU). To distinguish novel genomic techniques (NGTs) from genetically modified organisms (GMOs), the EU has suggested new legislation in 2023 (San-Epifanio *et al.,* 2023). There will likely be two regulatory streams: one for transgenic changes and another for gene-edited plants without transgenes (mutagenesis, cisgenesis, etc.). However, gene-altered crops must undergo the same drawn-out and expensive regulatory procedures as conventional GMOs until new laws are approved (Ludlow *et al.*, 2021).

Philippines: The First Gene-Edited Banana to Be Deregulated

The Philippines has become a leader in biotechnology in Southeast Asia. Under a regulatory framework that distinguishes gene-edited creatures from genetically modified organisms (GMOs) when no foreign DNA is added, the nation authorized its first gene-edited crop in 2023, a non-browning banana (Jalaluddin *et al.,* 2024). Clear criteria for new breeding techniques (NBTs) were added to the Department of Agriculture's usage of the existing biotech review processes already in place. Other Southeast Asian nations investigating regulatory modifications

Canada: Modification of Regulations without Structural Change

Regardless of the breeding technique, plants with novel traits (PNTs) are assessed using Canada's regulatory framework. Even in the absence of foreign DNA, gene-edited crops may be subject to regulatory scrutiny if they produce unexpected characteristics (Smyth, 2019). Canada has previously approved gene-edited canola. However, effective stakeholders have demanded more precise regulations on what qualifies as “novelty to increase developers' predictability and openness (Ludlow *et al.*, 2021).

CRISPR-Edited Soybean Commercialization in the U.S.

A notable advancement in agricultural biotechnology has been the creation of CRISPR-edited soybeans with increased oil content. Researchers have used genome-editing tools to target fatty acid production pathways, specifically altering genes such as GmFATA1 and GmFATA2 to increase oleic acid levels without introducing foreign DNA (Liao *et al.,* 2024). The U.S. Department of Agriculture (USDA) decided not to regulate them as GMOs because these CRISPR-edited soybeans lack transgenic material, which accelerated their commercialization (Zheng *et al.,* 2024). The oleic acid levels in soybean cultivars reached 87.6%, significantly enhancing oil stability and health advantages compared to traditional soybean oils (Li *et al.,* 2023). In addition to improving oil quality, CRISPR-edited soybeans preserve similar plant vigor and seed output, which has attracted producers and the food industry (Yao *et al.,* 2023).

Argentina's Drought-Resistant Gene-Edited Corn

Argentina has led the way in the regulation of gene-edited crops independently of conventional GMOs. The approval of drought-resistant corn created by CRISPR-based changes targeting water-use efficiency genes is one of the notable successes (Kaul *et al.,* 2024). Argentina's success with gene-edited drought-tolerant corn has highlighted the advantages of product-based regulatory frameworks for encouraging agricultural innovation and guaranteeing food security in areas vulnerable to climate change (Kar *et al.,* 2022). Gene editing made it possible to change stress response pathways without introducing extraneous genetic material, and the modified corn was exempted from Argentina's GMO regulations. Evaluations of CRISPR-edited corn line performance in the field showed improved yield stability during drought without sacrificing grain quality (Sony *et al.,* 2023).

PROBLEMS AND DEBATES: PUBLIC PERCEPTIONS OF GLYPHOSATE-RESISTANT CROPS

Public concerns are still shaped by past problems involving glyphosate-resistant crops, even if CRISPR-edited plants promise high accuracy and possibly fewer regulatory requirements. Transgenic glyphosate-resistant (GR) crops have increased dependence on glyphosate herbicides, which in turn has sparked the evolution of herbicide-resistant weeds (Baek *et al.,* 2021). Therefore, communication efforts must clearly convey the hazards and benefits of gene editing while distinguishing it from conventional GMOs. If this is not done, gene-altered crops may face the same kind of market resistance as glyphosate-resistant cultivars (Zhang *et al.,* 2023). Despite the widespread use of glyphosate, weeds, including *Amaranthus palmeri* and *Conyza spp.,* have been able to flourish because of mechanisms such as EPSPS gene amplification and target-site alterations (Koo *et al.,* 2021). Although regulatory agencies such as the European Food Safety Authority (EFSA) and the Environmental Protection Agency (EPA) assert that glyphosate is safe when used in accordance with recommendations, public outrage has also resulted from concerns about the chemical's environmental persistence and possible health effects (Pan *et al.,* 2021). Multiple resistance traits that endanger sustainable agriculture have developed as a result of gene flow from GR crops into wild relatives, further complicating weed management techniques (Kalsing *et al.,* 2024). Field research in South America has demonstrated that the emergence of difficult-to-control species, such as *Digitaria insularis,* is exacerbated by an overreliance on glyphosate-tolerant systems without herbicide rotation techniques. Concerns regarding environmental toxicity have also emerged, particularly regarding the impact of glyphosate residues on aquatic ecosystems, non-target plant species, and soil microbiota (Hertel *et al.,* 2021). Pollen-mediated gene transfer from glyphosate-resistant to susceptible populations of common water hemp was empirically demonstrated in this study. The results showed that herbicide resistance traits might spread more easily because of gene flow, which exacerbates the problem of resistant weed populations (Sarangi *et al.*, 2017). This paper highlights the rise of glyphosate-resistant weed species, including *Conyza bonariensis, Conyza canadensis,* and *Digitaria insularis, i*n Brazil, and examines the agricultural effects of glyphosate-resistant soybean cultivation in South America. According to scientists, the widespread use of glyphosate has increased the prevalence of these resistant weeds, making weed control more difficult and perhaps resulting in yield losses (Cerdeira *et al*., 2011). Although CRISPR-edited plants have the potential to lower ecological dangers when managed appropriately, these past difficulties still influence how the general public views all biotechnology crops. However, critics argue that these technical solutions might only postpone, rather than address, the underlying problems. In the absence of varied weed control techniques, resistance is likely to continue developing (Beckie & Martin, 2021). More complexity is introduced when several features (such as insect resistance and herbicide tolerance) are combined in a single variety. Research indicates that although stacked crops can increase resilience, they may also change how plants respond to stress and metabolism, which could impact the general health and nutritional profiles of the plants (Zanatta *et al.,* 2020).

FUTURE PROSPECTS, DIFFICULTIES, AND REGULATORY IMPLICATIONS OF GENE EDITING TECHNOLOGIES

CRISPR-Cas9, base editing, and prime editing are gene-editing technologies that have drastically changed the future of crop development and food production in the field of agriculture. Future developments that allow accurate nucleotide changes without causing double-strand breaks, such as CRISPR and next-generation tools such as base editors and prime editors, reduce the chances of off-target alterations (Waghmode *et al.,* 2024). These developments have the great capacity to speed up the generation of features, especially for complex polygenic traits in climate-robust crops, and might provide a long-term way to increase agricultural output without causing environmental damage (Rogo *et al.,* 2024). Gene editing techniques in field crops, such as fruits, nuts, and vegetables, have been enhanced by the biotechnology revolution because of longer breeding cycles and complex genomes is equally promising (Bate *et al.,* 2021).

Base Editing: Accuracy without Breaking Double-Strands

Base editing enables the direct and irreversible conversion of one DNA nucleotide to another without causing double-strand breaks. While adenine base editors (ABEs) change A•T to G•C, cytosine base editors (CBEs) change C•G to T•A. Glycosylate base editors and dual editors that can target both cytosine and adenine bases are more recent developments (Yu *et al.,* 2021). Base editing has been utilized in agriculture to increase the nutritional value of crops and introduce herbicide resistance. As these modifications are similar to naturally occurring mutations, it is questionable whether they tend to be regulated as genetically modified organisms (Waghmode *et al.*, 2024). When no foreign DNA is added, regulators in nations such as the U.S. and Argentina have typically adopted a more lenient approach, excluding such goods from GMO laws (Prado *et al.,* 2024).

Prime Editing: Versatile, Search-and-Replace Genetic Engineering

All types of point mutations, tiny insertions and deletions, and combinations of them can be inserted through prime editing by combining a reverse transcriptase and a catalytically impaired Cas9 nickase, which introduces edits using a primary editing guide RNA (gRNA) (Liu *et al.,* 2022). Prime editing provides specificity and versatility in plant breeding by enhancing disease resistance and fine-tuning metabolic pathways for nutritional benefits (Yu *et al.,* 2023). Despite its potential and regulatory classification, prime editing is still in its infancy. Certain regulatory frameworks might manage various jurisdictions differently because they focus more on the sophistication of the editing tool than the final phenotype (Hao *et al.,* 2021).

CRISPR Delivery Techniques: Increasing Usability and Access

While traditional approaches, such as *Agrobacterium*-mediated transformation and biolistics, are still widely used, newer approaches use ribonucleoprotein (RNP) complexes and nanomaterials, which reduce the risk of transgene introduction (Chen *et al.,* 2024). The consistent recognition of these techniques remains a global challenge (Prado *et al.*, 2024). To accomplish transgene-free genome editing in plants, including the use of transient expression systems and ribonucleoproteins (RNPs). It emphasizes that genome editing can produce mutations that are indistinguishable from natural variants, which is why some nations exempt such products from GMO regulations; however, it also points out that the absence of international agreement on these presents difficulties for global harmonization (Gu *et al.*, 2021).

Process vs. Product-Based Models and Their Regulatory Consequences

The difficulties in regulating genome editing emphasize the distinctions between regulatory frameworks based on processes and those based on products. The United States adopts a product-based approach, emphasizing the qualities of the finished product, whereas the European Union (EU) employs a process-based approach, regulating organisms according to the methods used to generate them (Sprink *et al.*, 2016).

Effects Off-Target and Ethical Monitoring

Expert editors reduce the possibility of off-target consequences, but do not completely eliminate them. As more advanced technologies become available, regulatory scrutiny is likely to increase. To preserve public confidence, open safety reviews, thorough risk assessments, and public involvement are essential (Deneault, 2024).

HARMONIZATION OF THE REGULATORY FRAMEWORK ON AN INTERNATIONAL LEVEL

Initiating regulatory alignment in the clinical and agricultural biotechnology sectors, for example, has been made possible by the International Pharmaceutical Regulators Programme (IPRP) and International Conference on Harmonization (ICH). These organizations encourage the adoption of standardized data requirements, training, and technical exchanges (Arcidiacono, 2023). Brazil, Colombia, and other Latin American nations' regulatory strategies concerning genome-edited plants. These nations have implemented case-by-case consultation procedures to decide whether a product derived from novel breeding methods qualifies as a genetically modified organism (GMO), with the main criterion being the lack of foreign DNA. This regulatory convergence facilitates the creation of accurate, scientifically grounded standards, fostering both domestic innovation and global trade (Sánchez, 2024).

The major goal of this collaboration is to pave the way for the trade of gene-edited crops and avoid obstacles (Jones *et al.*, 2022). By enabling regulatory frameworks to counter scientific discoveries without completely revising legislative frameworks, these tools foster innovation while preserving the environment and public health (Lema, 2021). International coordinating initiatives must consider ethical values and sociopolitical contexts in addition to harmonizing technical requirements (Hernández-Soto & Gatica-Arias, 2024).

PUBLIC INVOLVEMENT AND GENE EDITED PLANT ACCEPTANCE

Government regulations' effects on genome editing and attitudes towards food have changed. This highlights that, despite their ignorance of the process and understanding of the technique, consumers have a positive attitude towards and favor genome-edited foods over genetically modified ones (Atimango *et al.*, 2024). This study addresses the general view of how awareness of genome editing technologies, political ideology, and religiosity influence overall attitudes towards science. However, it revealed that conservative political people are less worried, while people with more positive views towards science and greater scientific literacy are more tolerant of gene editing (Calabrese *et al.*, 2021).

According to cross-national studies, the regional context is also significant. According to surveys conducted in the US, Germany, and Japan, there are varying levels of confidence in gene-edited crops. Respondents in the US showed the most recognition and the least disagreement between gene-edited and conventional crops, while Japanese and German respondents showed more careful views, considering it unnatural (Yamaguchi *et al.*, 2024). Research supports participatory government, two-way communication, and the co-creation of values. Public discussion promotes more validity and awareness of gene-editing techniques (Milne *et al.*, 2023). The need for context-specific commitment is demonstrated by case studies from Argentina, India, and Burkina Faso (Barbosa *et al.*, 2021). The acceptance of gene-edited crops by consumers is influenced by the purported benefits and transparency of the technology. People accept gene-edited crops when they are aware of their specific benefits and environmental sustainability. This emphasizes that critical labelling and clear communication are important to gain public trust (Bearth *et al.*, 2024).

CONCLUSION

Although gene-edited plants provide substantial benefits for food security and agriculture, global regulatory frameworks are still fragmented, with the main problems being public trust, precise definitions, and safety evaluations. Facilitating the responsible implementation of these technologies and resolving disputes continue to demand public trust and international cooperation. Future regulations must encourage innovation while maintaining global consistency and transparency, and in the agricultural sector, concerted efforts are essential to fully utilize gene editing.

REFERENCES

Akinbo, O., Nkhabindze, B., Amedu, J., Ebegba, R., Asagbra, A., Ratemo, B. O., & Dada, S. A.. (2025). Africa and zero hunger agenda: Genome editing policy landscape, challenges and opportunities. *Frontiers in Bioengineering and Biotechnology*, *13*, 1526851. DOI: 10.3389/fbioe.2025.1526851 PMID: 40144392

Aljabali, A. A. A., El-Tanani, M., & Tambuwala, M. M. (2024). Aljabali, Alaa AA, Mohamed El-Tanani, and Murtaza M. Tambuwala. "Principles of CRISPR-Cas9 technology: Advancements in genome editing and emerging trends in drug delivery.". *Journal of Drug Delivery Science and Technology*, *92*, 105338. DOI: 10.1016/j.jddst.2024.105338

Ansari, W. A., Chandanshive, S. U., Bhatt, V., Nadaf, A. B., Vats, S., Katara, J. L., Sonah, H., & Deshmukh, R. (2020). Genome editing in cereals: Approaches, applications and challenges. *International Journal of Molecular Sciences*, *21*(11), 4040. DOI: 10.3390/ijms21114040 PMID: 32516948

Arcidiacono, J. (2023). International Harmonization for Cell and Gene Therapy Products. In *Regulatory Aspects of Gene Therapy and Cell Therapy Products: A Global Perspective* (pp. 235–240). Springer International Publishing., DOI: 10.1007/978-3-031-34567-8_14

Atia, M., Jiang, W., Sedeek, K., Butt, H., & Mahfouz, M. (2024). Crop bioengineering via gene editing: Reshaping the future of agriculture. *Plant Cell Reports*, *43*(4), 98. DOI: 10.1007/s00299-024-03183-1 PMID: 38494539

Atimango, A. O., Wesana, J., Kalule, S. W., Verbeke, W., & De Steur, H. (2024). Genome editing in food and agriculture: From regulations to consumer perspectives. *Current Opinion in Biotechnology*, *87*, 103127. DOI: 10.1016/j.copbio.2024.103127 PMID: 38564970

Baek, Yousoon, Lucas K. Bobadilla, Darci A. Giacomini, Jacob S. Montgomery, Brent P. Murphy, and Patrick J. Tranel. "Evolution of glyphosate-resistant weeds." *Reviews of Environmental Contamination and Toxicology Volume 255: Glyphosate* (2021): 93-128. https://doi.org/DOI: 10.1007/398_2020_55

Bain, C., Lindberg, S., & Selfa, T. (2019). Emerging sociotechnical imaginaries for gene edited crops for foods in the United States: Implications for governance. *Agriculture and Human Values*, *37*(2), 265–279. DOI: 10.1007/s10460-019-09980-9

Barbosa, S., Toe, L. P., Thizy, D., Vaz, M., & Carter, L. (2021). Engagement and social acceptance in genome editing for human benefit: Reflections on research and practice in a global context. *Wellcome Open Research*, *5*, 244. DOI: 10.12688/wellcomeopenres.16260.2 PMID: 34095505

Bate, N. J., Dardick, C. D., de Maagd, R. A., & Williams, R. W. (2021). Opportunities and challenges applying gene editing to specialty crops. *In Vitro Cellular & Developmental Biology. Plant : Journal of the Tissue Culture Association*, *57*(4), 709–719. DOI: 10.1007/s11627-021-10208-x

Bearth, A., Otten, C. D., & Cohen, A. S. (2024). Consumers' perceptions and acceptance of genome editing in agriculture: Insights from the United States of America and Switzerland. *Food Research International*, *178*, 113982. DOI: 10.1016/j.foodres.2024.113982 PMID: 38309884

Beckie, H. J., & Martin, S. L. (2021). Monitoring herbicide resistance gene flow in weed populations. In *Gene flow: monitoring, modeling and mitigation* (pp. 86–102). CABI., DOI: 10.1079/9781789247480.0006

Bhat, M. A., Mir, R. A., Kumar, V., Shah, A. A., Zargar, S. M., Rahman, S., & Jan, A. T. (2021). Mechanistic insights of CRISPR/Cas-mediated genome editing towards enhancing abiotic stress tolerance in plants. *Physiologia Plantarum*, *172*(2), 1255–1268. DOI: 10.1111/ppl.13359 PMID: 33576013

Bhatta, M., Sandro, P., Smith, M. R., Delaney, O., Voss-Fels, K. P., Gutierrez, L., & Hickey, L. T. (2021). Need for speed: Manipulating plant growth to accelerate breeding cycles. *Current Opinion in Plant Biology*, *60*, 101986. DOI: 10.1016/j.pbi.2020.101986 PMID: 33418268

Brookes, G., & Smyth, S. J. (2024). Risk-appropriate regulations for gene-editing technologies. *GM Crops & Food*, *15*(1), 1–14. DOI: 10.1080/21645698.2023.2293510 PMID: 38215017

Calabrese, C., Featherstone, J. D., Robbins, M., & Barnett, G. A. (2021). Examining the relationship between gene editing knowledge, value predispositions, and general science attitudes among US farmers, scientists, policymakers, and the general public. *JCOM, Journal of Science Communication*, *20*(2), A02. DOI: 10.22323/2.20020202

Carroll, D. (2021). A short, idiosyncratic history of genome editing. *Gene and Genome Editing*, *1*, 100002. DOI: 10.1016/j.ggedit.2021.100002

Cerdeira, A. L., & Dionsio, L. P. (2011). Gazziero, Stephen O. Duke, and Marcus B. Matallo. "Agricultural impacts of glyphosate-resistant soybean cultivation in South America.". *Journal of Agricultural and Food Chemistry*, *59*(11), 5799–5807. DOI: 10.1021/jf102652y PMID: 20839871

Chen, F., Chen, L., Yan, Z., Xu, J., Feng, L., He, N., Guo, M., Zhao, J., Chen, Z., Chen, H., Yao, G., & Liu, C. (2024). Recent advances of CRISPR-based genome editing for enhancing staple crops. *Frontiers in Plant Science*, *15*, 1478398. DOI: 10.3389/fpls.2024.1478398 PMID: 39376239

Deneault, E. (2024). Recent therapeutic gene editing applications to genetic disorders. *Current Issues in Molecular Biology*, *46*(5), 4147–4185. DOI: 10.3390/cimb46050255 PMID: 38785523

Devisetty, U. K., De Neef, E., & Eric, R. L. (2024). Gordon, Valeria Velasquez-Zapata, Kenneth Narva, Laurent Mézin, Peter Mc Cahon, Kenneth W. Witwer, and Krishnakumar Sridharan. "A bioinformatics framework for human health risk assessment of externally applied dsRNA-based biopesticides.". *Computational Toxicology (Amsterdam, Netherlands)*, *100340*. Advance online publication. DOI: 10.1016/j.comtox.2024.100340

Ebrahimi, V., & Hashemi, A. (2024). CRISPR-based gene editing in plants: Focus on reagents and their delivery tools. *BioImpacts*, *15*, 30019. DOI: 10.34172/bi.30019 PMID: 39963563

Eckerstorfer, M. F., Dolezel, M., Engelhard, M., Giovannelli, V., Grabowski, M., Heissenberger, A., Lener, M., Reichenbecher, W., Simon, S., Staiano, G., Wüst Saucy, A. G., Zünd, J., & Lüthi, C. (2023). Recommendations for the assessment of potential environmental effects of genome-editing applications in plants in the EU. *Plants*, *12*(9), 1764. DOI: 10.3390/plants12091764 PMID: 37176822

Eckerstorfer, M. F., Grabowski, M., Lener, M., Engelhard, M., Simon, S., Dolezel, M., Heissenberger, A., & Lüthi, C. (2021). Biosafety of genome editing applications in plant breeding: Considerations for a focused case-specific risk assessment in the EU. *Biotech*, *10*(3), 10. DOI: 10.3390/biotech10030010 PMID: 35822764

Entine, J., Maria Sueli, S. F., Groenewald, J.-H., Kershen, D. L., Lema, M., McHugh-en, A., & Lima Nepomuceno, A.. (2021). Regulatory approaches for genome edited agricultural plants in select countries and jurisdictions around the world. *Transgenic Research*, *30*(4), 551–584. DOI: 10.1007/s11248-021-00257-8 PMID: 33970411

Gao, Y., Chen, S., Halene, S., & Tebaldi, T. (2021). Transcriptome-wide quantification of double-stranded RNAs in live mouse tissues by dsRIP-Seq. *STAR Protocols*, *2*(1), 100366. DOI: 10.1016/j.xpro.2021.100366 PMID: 33778776

Gu, X., Liu, L., & Zhang, H. (2021). Transgene-free genome editing in plants. *Frontiers in Genome Editing*, *3*, 805317. DOI: 10.3389/fgeed.2021.805317 PMID: 34927134

Guertler, P., Pallarz, S., Belter, A., Eckermann, K. N., & Grohmann, L. (2023). Detection of commercialized plant products derived from new genomic techniques (NGT)-Practical examples and current perspectives. *Food Control*, *152*, 109869. DOI: 10.1016/j.foodcont.2023.109869

Guo, C., Ma, X., Gao, F., & Guo, Y. (2023). Off-target effects in CRISPR/Cas9 gene editing. *Frontiers in Bioengineering and Biotechnology*, *11*, 1143157. DOI: 10.3389/fbioe.2023.1143157 PMID: 36970624

Hao, L., Pu, X., & Song, J. (2021). Introduction of mutations in plants with prime editing. *Methods : A Companion to Methods in Enzymology*, *194*, 83–93. DOI: 10.1016/j.ymeth.2021.03.014 PMID: 33774158

Hartman, S., Horner, W., Jackson, C., Kovak, E., & Velan, V. (2020). Streamlining USDA Regulation of Gene Editing to Benefit US Agriculture. *Journal of Science Policy & Governance*, *17*(01). Advance online publication. DOI: 10.38126/JSPG170108

Hasani, M., Camacho-Martinez, S. V., Zai, B., Warriner, L. J., Boesch, C., Kopko, C., Marafiotti, F., & Warriner, K. (2024). Development of an online food safety toolbox based on the Codex Alimentarius General Principles of Food Hygiene: Engaging users through mapping, chunking, and learning-by-asking. *Food Quality and Safety*, *8*, fyae030. Advance online publication. DOI: 10.1093/fqsafe/fyae030

Herman, R. A., Storer, N. P., Anderson, J. A., Amijee, F., Cnudde, F., & Raybould, A. (2021). Transparency in risk-disproportionate regulation of modern crop-breeding techniques. *GM Crops & Food*, *12*(1), 376–381. DOI: 10.1080/21645698.2021.1934353 PMID: 34107854

Hernández-Soto, A., & Gatica-Arias, A. (2024). Genome editing in Latin America: Research achievements and regulatory evolution. [PCTOC]. *Plant Cell, Tissue and Organ Culture*, *159*(3), 55. DOI: 10.1007/s11240-024-02904-4

Hertel, R., Gibhardt, J., Martienssen, M., Kuhn, R., & Commichau, F. M. (2021). Molecular mechanisms underlying glyphosate resistance in bacteria. *Environmental Microbiology*, *23*(6), 2891–2905. DOI: 10.1111/1462-2920.15534 PMID: 33876549

Hill, N., Meyers, C., Li, N., Doerfert, D. L., & Mendu, V. (2022). Persuasive effects of metaphors regarding gene-editing in agriculture. *Journal of Applied Communications*, *106*(1), 1. https://newprairiepress.org/jac/vol106/iss1/1/. DOI: 10.4148/1051-0834.2416

Hoffman, N. E. (2021). Revisions to USDA biotechnology regulations: The SECURE rule. *Proceedings of the National Academy of Sciences of the United States of America, 118*(22), e2004841118. DOI: 10.1073/pnas.2004841118 PMID: 34050018

Hüdig, M., Laibach, N., & Hein, A.-C. (2022). Genome editing in crop plant research—Alignment of expectations and current developments. *Plants, 11*(2), 212. DOI: 10.3390/plants11020212 PMID: 35050100

Idris, S. H., Nurzatil, S. M. J., & Lee, W. C. (2023). Ethical and legal implications of gene editing in plant breeding: A systematic literature review. *Journal of Zhejiang University. Science. B, 24*(12), 1093–1105. DOI: 10.1631/jzus.B2200601 PMID: 38057267

Jalaluddin, M., Sharleeza, N., Abdulah, A.-H. A. F., & Othman, R. Y. (2024). Regulatory landscape and public perception for gene-edited bananas in the Southeast Asian region. *Transgenic Research, 33*(3), 89–97. DOI: 10.1007/s11248-024-00379-9 PMID: 38600337

Jenkins, D., Dobert, R., Atanassova, A., & Pavely, C. (2021). Impacts of the regulatory environment for gene editing on delivering beneficial products. *In Vitro Cellular & Developmental Biology. Plant : Journal of the Tissue Culture Association, 57*(4), 609–626. DOI: 10.1007/s11627-021-10201-4 PMID: 34429575

Jones, M. G., Fosu-Nyarko, J., Iqbal, S., Adeel, M., Romero-Aldemita, R., Arujanan, M., ... & Khoo, K. (2022). Enabling trade in gene-edited produce in Asia and Australasia: The developing regulatory landscape and future perspectives. Plants, 11(19), 2538.

Kalaitzandonakes, N., Willig, C., & Zahringer, K. (2023). The economics and policy of genome editing in crop improvement. *The Plant Genome, 16*(2), e20248. DOI: 10.1002/tpg2.20248 PMID: 36321718

Kalsing, A., Velini, E. D., Merotto, A.Jr, & Carbonari, C. A. (2024). The population genomics of Conyza spp. in soybean macroregions suggest the spread of herbicide resistance through intraspecific and interspecific gene flow. *Scientific Reports, 14*(1), 19536. DOI: 10.1038/s41598-024-70153-8 PMID: 39174662

Kar, S. R., Choudhury, S., & Chakraborty, A. (2022). CRISPR/Cas9 for soybean improvement: A review. *Asia-Pacific Journal of Molecular Biology and Biotechnology, 30*, 40–56. DOI: 10.35118/apjmbb.2022.030.3.05

Karavolias, N. G., Horner, W., Abugu, M. N., & Evanega, S. N. (2021). Application of gene editing for climate change in agriculture. *Frontiers in Sustainable Food Systems, 5*, 685801. DOI: 10.3389/fsufs.2021.685801

Kato-Nitta, N., Tachikawa, M., Inagaki, Y., & Maeda, T. (2023). Public perceptions of risks and benefits of gene-edited food crops: An international comparative study between the US, Japan, and Germany. *Science, Technology & Human Values*, *48*(6), 1360–1392. DOI: 10.1177/01622439221123830

Kaul, T., Thangaraj, A., Jain, R., Bharti, J., Kaul, R., Verma, R., Sony, S. K., Abdel Motelb, K. F., Yadav, P., & Agrawal, P. K. (2024). CRISPR/Cas9-mediated homology donor repair base editing system to confer herbicide resistance in maize (Zea mays L.). *Plant Physiology and Biochemistry : PPB*, *207*, 108374. DOI: 10.1016/j.plaphy.2024.108374 PMID: 38310724

Kaur, N., & Pati, P. K. (2019). Harnessing the potential of brassinosteroids in abiotic stress tolerance in plants. In Brassinosteroids: plant growth and development (pp. 407-423). Singapore: Springer Singapore.

Kawall, K., Cotter, J., & Then, C. (2020). Broadening the GMO risk assessment in the EU for genome editing technologies in agriculture. *Environmental Sciences Europe*, *32*(1), 106. DOI: 10.1186/s12302-020-00361-2

Kishchenko, O., Zhou, Y., Jatayev, S., Shavrukov, Y., & Borisjuk, N. (2020). Gene editing applications to modulate crop flowering time and seed dormancy. *aBIOTECH*, *1*(4), 233–245. DOI: 10.1007/s42994-020-00032-z PMID: 36304127

Kondo, K., & Taguchi, C. (2022). Japanese regulatory framework and approach for genome-edited foods based on latest scientific findings. *Food Safety (Tokyo, Japan)*, *10*(4), 113–128. DOI: 10.14252/foodsafetyfscj.D-21-00016 PMID: 36619008

Koo, D.-H., Sathishraj, R., Friebe, B., & Gill, B. S. (2021). Deciphering the mechanism of glyphosate resistance in Amaranthus palmeri by cytogenomics. *Cytogenetic and Genome Research*, *161*(12), 578–584. DOI: 10.1159/000521409 PMID: 35021177

Kumar, A., Kumar, R., Singh, N., & Mansoori, A. (2020). "Regulatory framework and policy decisions for genome-edited crops." *CRISPR/Cas Genome Editing: Strategies And Potential For. Crop Improvement*, •••, 193–201. DOI: 10.1007/978-3-030-42022-2_9

Kuzma, J., & Grieger, K. (2020). Community-led governance for gene-edited crops. *Science*, *370*(6519), 916–918. DOI: 10.1126/science.abd1512 PMID: 33214269

Lassoued, R., & Peter, W. B. (2021). Phillips, Diego Maximiliano Macall, Hayley Hesseln, and Stuart J. Smyth. "Expert opinions on the regulation of plant genome editing.". *Plant Biotechnology Journal*, *19*(6), 1104–1109. DOI: 10.1111/pbi.13597 PMID: 33834596

Lassoued, R., Smyth, S. J., & Phillips, P. W. (2024). Regulatory outlook of CRISPR-edited plants in Canada. In Global regulatory outlook for CRISPRized plants (pp. 231-250). Academic Press.

Lema, M. (2021). Regulatory assessment of off-target changes and spurious DNA insertions in gene-edited organisms for agri-food use. *Journal of Regional Science*, *9*(1), 1–15. DOI: 10.21423/JRS-V09I1LEMA

Lemmon, Z. H., Reem, N. T., Dalrymple, J., Soyk, S., Swartwood, K. E., Rodriguez-Leal, D., Van Eck, J., & Lippman, Z. B. (2018). Rapid improvement of domestication traits in an orphan crop by genome editing. *Nature Plants*, *4*(10), 766–770. DOI: 10.1038/s41477-018-0259-x PMID: 30287957

Li, H., Zhou, R., Liu, P., Yang, M., Xin, D., Liu, C., Zhang, Z., Wu, X., Chen, Q., & Zhao, Y. (2023). Design of high-monounsaturated fatty acid soybean seed oil using GmPDCTs knockout via a CRISPR-Cas9 system. *Plant Biotechnology Journal*, *21*(7), 1317–1319. DOI: 10.1111/pbi.14060 PMID: 37084283

Li, Y., Li, W., & Li, J. (2021). The CRISPR/Cas9 revolution continues: From base editing to prime editing in plant science. *Journal of Genetics and Genomics*, *48*(8), 661–670. DOI: 10.1016/j.jgg.2021.05.001 PMID: 34362681

Liang, J., Yang, X., Jiao, Y., Wang, D., Zhao, Q., Sun, Y., Li, Y., & Wu, K. (2022). The evolution of China's regulation of agricultural biotechnology. *aBIOTECH*, *3*(4), 237–249. DOI: 10.1007/s42994-022-00086-1 PMID: 36533267

Liao, W., Guo, R., Qian, K., Shi, W., Whelan, J., & Shou, H. (2024). The acyl–acyl carrier protein thioesterases GmFATA1 and GmFATA2 are essential for fatty acid accumulation and growth in soybean. *The Plant Journal : For Cell and Molecular Biology*, *118*(3), 823–838. DOI: 10.1111/tpj.16638 PMID: 38224529

Liu, Yao, Xian-Hui Zhou, Shu-Hong Huang, and Xiao-Long Wang. "Prime editing: a search and replace tool with versatile base changes." *Yi Chuan= Hereditas* 44, no. 11 (2022): 993-1008. https://doi.org/.DOI: 10.16288/j.yczz.22-156

Ludlow, K., Smyth, S., & Falck-Zepeda, J. (2021). Socio-economic considerations and potential implications for gene-edited crops. *Journal of Regional Science*, *9*(2), 1–11. DOI: 10.21423/JRS-V09I2LUDLOW

Lukasiewicz, J. M., van de Wiel, C. C. M., Lotz, L. A. P., & Smulders, M. J. M. (2024a). Consumer transparency in the production chain for plant varieties produced using new genomic techniques. *aBIOTECH*, *5*(2), 239–246. DOI: 10.1007/s42994-024-00142-y PMID: 38974855

Lukasiewicz, J. M., van de Wiel, C. C. M., Lotz, L. A. P., & Smulders, M. J. M. (2024b). Clemens CM van de Wiel, Lambertus AP Lotz, and Marinus JM Smulders. "Intellectual property rights and plants made by new genomic techniques: Access to technology and gene-edited traits in plant breeding.". *Outlook on Agriculture*, *53*(3), 205–215. DOI: 10.1177/00307270241277219

Matsuo, M., & Tachikawa, M. (2022). Implications and lessons from the introduction of genome-edited food products in Japan. *Frontiers in Genome Editing*, *4*, 899154. DOI: 10.3389/fgeed.2022.899154 PMID: 35813972

Mbaya, H., Lillico, S., Kemp, S., Simm, G., & Raybould, A. (2022). Regulatory frameworks can facilitate or hinder the potential for genome editing to contribute to sustainable agricultural development. *Frontiers in Bioengineering and Biotechnology*, *10*, 959236. DOI: 10.3389/fbioe.2022.959236 PMID: 36246373

Menz, J., Modrzejewski, D., Hartung, F., Wilhelm, R., & Sprink, T. (2020). Genome edited crops touch the market: A view on the global development and regulatory environment. *Frontiers in Plant Science*, *11*, 586027. DOI: 10.3389/fpls.2020.586027 PMID: 33163013

Milne, R., Aidid, U., Atutornu, J., Bircan, T., Boraschi, D., Costa, A., Henriques, S., Patch, C., & Middleton, A. (2023). What Difference Can Public Engagement in Genome Editing Make, and for Whom? *The American Journal of Bioethics : AJOB*, *23*(7), 58–60. DOI: 10.1080/15265161.2023.2207545 PMID: 37339292

Milne, R., Aidid, U., Atutornu, J., Bircan, T., Boraschi, D., Costa, A., Henriques, S., Patch, C., & Middleton, A. (2023). What Difference Can Public Engagement in Genome Editing Make, and for Whom? *The American Journal of Bioethics : AJOB*, *23*(7), 58–60. DOI: 10.1080/15265161.2023.2207545 PMID: 37339292

Mishra, R., & Zhao, K. (2018). Genome editing technologies and their applications in crop improvement. *Plant Biotechnology Reports*, *12*(2), 57–68. DOI: 10.1007/s11816-018-0472-0

Myskja, B. K., & Myhr, A. I. (2020). Non-safety assessments of genome-edited organisms: Should they be included in regulation? *Science and Engineering Ethics*, *26*(5), 2601–2627. DOI: 10.1007/s11948-020-00222-4 PMID: 32424723

Oliva, R., Ji, C., Atienza-Grande, G., Huguet-Tapia, J. C., Perez-Quintero, A., Li, T., Eom, J.-S., Li, C., Nguyen, H., Liu, B., Auguy, F., Sciallano, C., Luu, V. T., Dossa, G. S., Cunnac, S., Schmidt, S. M., Slamet-Loedin, I. H., Vera Cruz, C., Szurek, B., & Yang, B. (2019). Broad-spectrum resistance to bacterial blight in rice using genome editing. *Nature Biotechnology*, *37*(11), 1344–1350. DOI: 10.1038/s41587-019-0267-z PMID: 31659337

Pan, L., Yu, Q., Wang, J., Han, H., Mao, L., Nyporko, A., Maguza, A., Fan, L., Bai, L., & Powles, S. (2021). An ABCC-type transporter endowing glyphosate resistance in plants. *Proceedings of the National Academy of Sciences of the United States of America, 118*(16), e2100136118. DOI: 10.1073/pnas.2100136118 PMID: 33846264

Pandey, S., Divakar, S., & Singh, A. (2024). S. Divakar, and Ashutosh Singh. "Genome editing prospects for heat stress tolerance in cereal crops.". *Plant Physiology and Biochemistry : PPB, 108989*, 108989. Advance online publication. DOI: 10.1016/j.plaphy.2024.108989 PMID: 39094478

Prado, G. S., Rocha, D. C., Nascimento dos Santos, L., Contiliani, D. F., Nobile, P. M., Martinati-Schenk, J. C., & Padilha, L.. (2024). CRISPR technology towards genome editing of the perennial and semi-perennial crops citrus, coffee and sugarcane. *Frontiers in Plant Science, 14*, 1331258. DOI: 10.3389/fpls.2023.1331258 PMID: 38259920

Prasad, A., & Chimata, M. K. (2023). Regulation of Genome-edited Technologies: Indian Policy Perspective. *Current Indian Science, 1*(1), e230223213951. DOI: 10 .2174/2210299X01666230223105053

Puchta, H. (2024). Regulation of gene-edited plants in Europe: From the valley of tears into the shining sun? *aBIOTECH, 5*(2), 231–238. DOI: 10.1007/s42994-023-00130-8 PMID: 38974871

Puchta, H., Jiang, J., Wang, K., & Zhao, Y. (2022). Updates on gene editing and its applications. *Plant Physiology, 188*(4), 1725–1730. DOI: 10.1093/plphys/kiac032 PMID: 35225345

Rabuma, T., Moronta-Barrios, F., & Craig, W. (2024). Navigating biosafety regulatory frameworks for genetic engineering in Africa: A focus on genome editing and gene drive technologies. *Frontiers in Bioengineering and Biotechnology, 12*, 1483279. DOI: 10.3389/fbioe.2024.1483279 PMID: 39512657

Ribarits, A., Eckerstorfer, M., Simon, S., & Stepanek, W. (2021). Genome-edited plants: Opportunities and challenges for an anticipatory detection and identification framework. *Foods, 10*(2), 430. DOI: 10.3390/foods10020430 PMID: 33669278

Rogo, U., Simoni, S., Fambrini, M., Giordani, T., Pugliesi, C., & Mascagni, F. (2024). Future-proofing agriculture: De Novo domestication for sustainable and resilient crops. *International Journal of Molecular Sciences, 25*(4), 2374. DOI: 10.3390/ijms25042374 PMID: 38397047

San-Epifanio, E., Leire, I. F., López, A. L., Puigdomènech, P., & Larrabe, J. U. (2023). Possible EU futures for CRISPR-edited plants: Little margin for optimism? *Frontiers in Plant Science*, *14*, 1141455. DOI: 10.3389/fpls.2023.1141455 PMID: 37008488

Sánchez, M. A. "The Global Advance of Genome-Edited Plants to the Market: The Key Role of Chile in Its Development." *Plants* 13, no. 24 (2024): 3597. https://doi.org/DOI: 10.3390/plants13243597

Sarangi, D., Tyre, A. J., Patterson, E. L., Gaines, T. A., Irmak, S., Knezevic, S. Z., Lindquist, J. L., & Jhala, A. J. (2017). Pollen-mediated gene flow from glyphosate-resistant common waterhemp (Amaranthus rudis Sauer): Consequences for the dispersal of resistance genes. *Scientific Reports*, *7*(1), 44913. DOI: 10.1038/srep44913 PMID: 28327669

Schiemann, J., Hartung, F., Menz, J., Sprink, T., & Wilhelm, R. (2021). Policies and governance for plant genome editing. In *Plant Biotechnology* (pp. 259–275). Springer., DOI: 10.1007/978-3-030-68345-0_18

Shew, A. M., Nalley, L. L., Snell, H. A., Nayga, R. M.Jr, & Dixon, B. L. (2018). CRISPR versus GMOs: Public acceptance and valuation. *Global Food Security*, *19*, 71–80. DOI: 10.1016/j.gfs.2018.10.005

Smyth, Stuart J. "Regulation of genome editing in plant biotechnology: Canada." *Regulation of genome editing in plant biotechnology: A comparative analysis of regulatory frameworks of selected countries and the EU* (2019): 111-135. https://doi.org/.DOI: 10.1007/978-3-030-17119-3_4

Smyth, S. J., Falck-Zepeda, J., & Ludlow, K. (2016). The costs of regulatory delays for genetically modified crops. *Estey Journal of International Law and Trade Policy*, *17*(2), 173–195. DOI: 10.22004/ag.econ.253074

Sony, S. K., Kaul, T., Abdel Motelb, K. F., Thangaraj, A., Bharti, J., Kaul, R., Verma, R., & Nehra, M. (2023). CRISPR/Cas9-mediated homology donor repair base editing confers glyphosate resistance to rice (Oryza sativa L.). *Frontiers in Plant Science*, *14*, 1122926. DOI: 10.3389/fpls.2023.1122926 PMID: 36959937

Sprink, T., Eriksson, D., Schiemann, J., & Hartung, F. (2016). Regulatory hurdles for genome editing: Process-vs. product-based approaches in different regulatory contexts. *Plant Cell Reports*, *35*(7), 1493–1506. DOI: 10.1007/s00299-016-1990-2 PMID: 27142995

Strobbe, S., Wesana, J., Van Der Straeten, D., & De Steur, H. (2023). Public acceptance and stakeholder views of gene edited foods: A global overview. *Trends in Biotechnology*, *41*(6), 736–740. DOI: 10.1016/j.tibtech.2022.12.011 PMID: 36658005

Tan, Y., Du, H., Wu, X., Liu, Y., Jiang, M., Song, S., Wu, L., & Shu, Q. (2020). Tan, Yuan-yuan, Hao Du, Xia Wu, Yan-hua Liu, Meng Jiang, Shi-yong Song, Liang Wu, and Qing-yao Shu. "Gene editing: An instrument for practical application of gene biology to plant breeding.". *Journal of Zhejiang University. Science. B*, *21*(6), 460–473. DOI: 10.1631/jzus.B1900633 PMID: 32478492

Tsuda, M., Watanabe, K. N., & Ohsawa, R. (2019). Regulatory status of genome-edited organisms under the Japanese Cartagena Act. *Frontiers in Bioengineering and Biotechnology*, *7*, 387. DOI: 10.3389/fbioe.2019.00387 PMID: 31867318

Turnbull, C., Lillemo, M., & Trine, A. K. (2021). Hvoslef-Eide. "Global regulation of genetically modified crops amid the gene edited crop boom–a review.". *Frontiers in Plant Science*, *12*, 630396. DOI: 10.3389/fpls.2021.630396 PMID: 33719302

Vishnu, L. (2024). *A Comprehensive Review of Genetically Modified Horticultural Crops: Benefits, Risks, and Public Perceptions*. AJGHAS-Australian Journal of Global Horticulture and Agronomy Science.

Voigt, B. (2023). EU regulation of gene-edited plants—A reform proposal. *Frontiers in Genome Editing*, *5*, 1119442. DOI: 10.3389/fgeed.2023.1119442 PMID: 36865566

Waghmode, Prajyot, Chetariya Chana Pitha, and Vaishnavi Banka. "The Role of Genome Editing Technologies in Sustainable Agriculture." (2024) https://doi.org/ DOI: 10.9734/jabb/2024/v27i81168

Wen, W., & Zhang, X.-B. (2022). CRISPR–Cas9 gene editing induced complex on-target outcomes in human cells. *Experimental Hematology*, *110*, 13–19. DOI: 10.1016/j.exphem.2022.03.002 PMID: 35304271

Whelan, A. I., & Lema, M. A. (2015). Regulatory framework for gene editing and other new breeding techniques (NBTs) in Argentina. *GM Crops & Food*, *6*(4), 253–265. DOI: 10.1080/21645698.2015.1114698 PMID: 26552666

Wilson, F. M., & Harrison, R. J. (2021). CRISPR/Cas9 mediated editing of the Quorn fungus Fusarium venenatum A3/5 by transient expression of Cas9 and sgRNAs targeting endogenous marker gene PKS12. *Fungal Biology and Biotechnology*, *8*(1), 1–8. DOI: 10.1186/s40694-021-00121-8 PMID: 34789333

Wolt, J. D., Yang, B., Wang, K., & Spalding, M. H. (2016). Regulatory aspects of genome-edited crops. *In Vitro Cellular & Developmental Biology. Plant : Journal of the Tissue Culture Association*, *52*(4), 349–353. DOI: 10.1007/s11627-016-9784-3

Yamaguchi, T., Ezaki, K., & Ito, K. (2024). Exploring the landscape of public attitudes towards gene-edited foods in Japan. *Breeding Science*, *74*(1), 11–21. DOI: 10.1270/jsbbs.23047 PMID: 39246435

Yang, F., Zheng, K., & Yao, Y. (2024). China's regulatory change toward genome-edited crops. *Trends in Biotechnology*, *42*(7), 801–806. DOI: 10.1016/j.tibtech.2023.12.008 PMID: 38160083

Yang, Y., & Hobbs, J. E. (2020). Supporters or opponents: Will cultural values shape consumer acceptance of gene editing? *Journal of Food Products Marketing*, *26*(1), 17–37. DOI: 10.1080/10454446.2020.1715316

Yao, D., Zhou, J., Zhang, A., Wang, J., Liu, Y., Wang, L., Pi, W., Li, Z., Yue, W., Cai, J., Liu, H., Hao, W., & Qu, X. (2023). Advances in CRISPR/Cas9-based research related to soybean [Glycine max (Linn.) Merr] molecular breeding. *Frontiers in Plant Science*, *14*, 1247707. DOI: 10.3389/fpls.2023.1247707 PMID: 37711287

Yu, C., Mo, J., Zhao, X., Li, G., & Zhang, X. (2021). "CRISPR/Cas-mediated DNA base editing technology and its application in biomedicine and agriculture." *Sheng wu Gong Cheng xue bao=. Chinese Journal of Biotechnology*, *37*(9), 3071–3087. DOI: 10.13345/j.cjb.200693 PMID: 34622618

Yu, X., Huo, G., Yu, J., Li, H., & Li, J. (2023). Prime editing: Its systematic optimization and current applications in disease treatment and agricultural breeding. *International Journal of Biological Macromolecules*, *253*, 127025. DOI: 10.1016/j.ijbiomac.2023.127025 PMID: 37769783

Yuan, P., Usman, M., Liu, W., Adhikari, A., Zhang, C., Njiti, V., & Xia, Y. (2024). Advancements in Plant Gene Editing Technology: From Construct Design to Enhanced Transformation Efficiency. *Biotechnology Journal*, *19*(12), e202400457. DOI: 10.1002/biot.202400457 PMID: 39692063

Zanatta, C. B., Benevenuto, R. F., Nodari, R. O., & Agapito-Tenfen, S. Z. (2020). Stacked genetically modified soybean harboring herbicide resistance and insecticide rCry1Ac shows strong defense and redox homeostasis disturbance after glyphosate-based herbicide application. *Environmental Sciences Europe*, *32*(1), 1–17. DOI: 10.1186/s12302-020-00379-6

Zarate, S., Cimadori, I., Jones, M. S., Roca, M. M., & Kathleen Barnhill-Dilling, S. (2023). Assessing agricultural gene editing regulation in Latin America: An analysis of how policy windows and policy entrepreneurs shape agricultural gene editing regulatory regimes. *Frontiers in Bioengineering and Biotechnology*, *11*, 1209308. DOI: 10.3389/fbioe.2023.1209308 PMID: 37362213

Zhang, C., Liu, S., Li, X., Zhang, R., & Li, J. (2022). Virus-induced gene editing and its applications in plants. *International Journal of Molecular Sciences*, *23*(18), 10202. DOI: 10.3390/ijms231810202 PMID: 36142116

Zhang, R., Liu, J., Chai, Z., Chen, S., Bai, Y., Zong, Y., Chen, K., Li, J., Jiang, L., & Gao, C. (2019). Generation of herbicide tolerance traits and a new selectable marker in wheat using base editing. *Nature Plants*, *5*(5), 480–485. DOI: 10.1038/s41477-019-0405-0 PMID: 30988404

Zhang, S., & Zhu, H. (2024). Development and prospect of gene-edited fruits and vegetables. *Food Quality and Safety*, *8*, fyad045. Advance online publication. DOI: 10.1093/fqsafe/fyad045

Zheng, Y., Guo, T., Xia, T., Guo, S., Chen, M., Ye, S., Pan, T., Xu, X., Gan, Y., Zhan, Y., Zheng, T., & Zheng, Z. (2024). Utility of Arabidopsis KASII Promoter in Development of an Effective CRISPR/Cas9 System for Soybean Genome Editing and Its Application in Engineering of Soybean Seeds Producing Super-High Oleic and Low Saturated Oils. *Journal of Agricultural and Food Chemistry*, *72*(39), 21720–21730. DOI: 10.1021/acs.jafc.4c05840 PMID: 39288439

Zhu, J.-K. (2022). future of gene-edited crops in China. *National Science Review*, *9*(4), nwac063. Advance online publication. DOI: 10.1093/nsr/nwac063 PMID: 35505661

Chapter 8
Public Perception and Acceptance of Gene-Edited Crops

Sameel Hassan
http://orcid.org/0009-0001-7559-5550
Department of Agriculture and Forest Science, University of Tuscia, Viterbo, Italy

Muhammad Azhar Sultan
Department of Plant Sciences, Faculty of Biological Sciences, Quaid-I-Azam University, Ilsmabad, Pakistan

Fakhar Javed
Department of Crop Science, College of Agriculture and Life Sciences, Chungnam National University, South Korea

Muhammad Waseem
http://orcid.org/0000-0001-7947-7722
Hainan University, Haikou, China

ABSTRACT

Public acceptance of gene-edited crops depends on scientific literacy, media framing, institutional trust, and ethical values. This chapter contrasts gene-edited crops with earlier GMOs in terms of regulatory treatment and public reception, identifying drivers of consumer attitudes. Misinformation and sensational reporting can amplify perceived risks, while cultural and ethical concerns—such as religious views and "naturalness"—shape regional differences in acceptance. Survey evidence and market trends suggest that transparent labeling and confidence in regulators strongly influence purchasing decisions. The chapter reviews case studies of public engagement to identify effective practices, including clear science communication,

DOI: 10.4018/979-8-2600-3575-7.ch008

educational outreach, and policy transparency. Overall, it argues for a balanced strategy that integrates scientific evidence with ethical reflection and sustained public dialogue, as improved transparency and misconception reduction are essential for wider adoption in support of food security and sustainable agriculture.

INTRODUCTION

Brief Overview of Gene Edited Crops

The development of gene-edited crops has supported agricultural biotechnology since it allows exact genome modifications without depending on foreign DNA integration that GMOs require (Lassoued et al., 2019). Organisms receive targeted DNA changes through CRISPR-Cas9 and TALENs and zinc-finger nucleases instead of conventional breeding selection and multiple generations of crossbreeding and genetic modification. Gene editing modifies plants to add drought-resistance along with pest protection capabilities while upgrading nutritional value and stretching product lifetime thus presenting an effective farming strategy against worldwide agricultural difficulties (Rahman et al., 2023). Gene editing functions as an advantageous tool because it allows for precise biological material modifications. Gene editing produces genome changes through specific modifications which differentiate it from traditional breeding that results in unforeseen genetic alterations. This technology appeals to scientists and agriculturalists because it provides an important method to boost farm production while reducing harmful pesticide use while ensuring safe food. The distinction between gene-edited crops and GMOs exists because regulatory agencies consider editing techniques as natural processes which duplicate traditional methods of breeding biological materials through time. The regulatory difference has brought discussions regarding their safety and information labeling while impacting public attitude. The wide acceptance of gene-edited crops remains limited by various implementation barriers. The main difficulty arises from how the public sees DNA-edited crops. Societal understanding and perspectives regarding genetically altered crops directly determines their adoption in farming operations and industry support and their effectiveness for sustainable food production (Ervin et al., 2011).

Importance of Public Perception in Technology Adoption

Emerging technologies achieve success mainly through how the public receives them because biotechnological advancements link scientific growth to public confidence level and moral issues and governmental regulations. The collective perceptions of gene-edited crops among public groups including the public, policymakers and farmers alongside food producers determine what direction gene-edited crops will take in agriculture (Lucht, 2015). Gene-edited crops will have a better chance of success with public approval because they will help boost food security through sustainable farming which decreases chemical use. Resistance toward gene-edited crops will face regulatory barriers together with market rejection while promoting a reduction in research and development. Public perception mostly depends on citizens'scientific knowledge. The public tends to accept the agricultural applications of gene-editing technology when they have fundamental knowledge of it and its potential advantages (Delhove et al., 2020; Hill, 2020). The absence of knowledge, combined with incorrect information, leads people to develop doubts about the subject. The amount of misconception has an impact on consumers' perceptions of bioscience food products and policies and regulations on biotechnology, as well as the public discourse on the technology.

Ultimately, the information that is remembered depends on how the media covers the subject. This occurs through media outlets, including traditional and social media platforms, which provide specific formats concerning gene editing. The approach to presenting media information regarding matters such as agricultural uplift and environmental sustainability leads to the enhancement of people's support for gene-edited crops. When news coverage presents technology as dangerous or irregular, people may develop concerns that lead to overproduction of resistance towards progress and more regulatory measures. This shows that the technology in gene-edited crops gains acceptance or faces rejection depending on the way in which the news articles, documentaries, or online discussions appear (Yang & Hobbs, 2020). Various issues are perceived from the level of confidence that people have in their governing institutions. It depends on the actions of regulatory agencies and scientific associations, together with agricultural companies and governments, to gain or lose the public's confidence and faith in GM crops, in that they are safe for human consumption and very useful in the nutritional needs of the people. This is because people have low trust when they feel that the biotech firm prioritizes profit over public health safety. Therefore, it is possible to achieve public discourse on work with informed stakeholders when stakeholders keep research education public and engage in open discussions.

Essentially, the viewpoints of both culture and ethics influence acceptance in some way. They are perceptions that a few citizens prefer organic foods to genetically modified foods, implying that they have more opposition to agricultural biotechnology (Shanahan et al., 2001). Religion-based beliefs form specific attitudes toward gene editing because certain communities struggle with moral questions regarding the alteration of living creatures. The successful acceptance of new technologies depends on recognizing and adapting to multicultural approaches that drive acceptance among different cultural groups. Market conditions and economic elements define how people establish their perceptions. The purchase readiness of consumers depends on their assessment of food labeling, safety concerns, and evaluation of the price-point gap versus non-gene-edited crops. The market acceptance of gene-edited products becomes stronger when consumers view them as both money-saving alternatives and sustainable and healthy choices (Ortega et al., 2022; Paudel et al., 2023). Consumer reluctance to adopt these products is likely to increase when they perceive unknown safety and ethical problems (Fig 1).

Figure 1. Factors shaping the public perception of gene editing

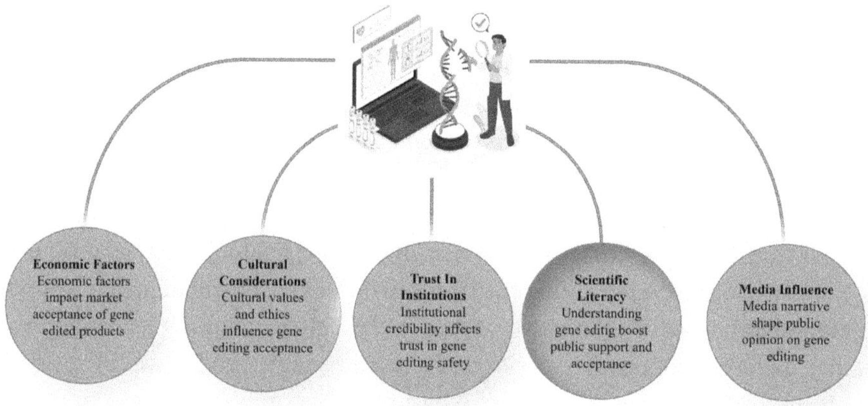

Key Factors Influencing Acceptance

The acceptance of gene-edited crops is determined by several interrelated factors. Among these factors, scientific matters go beyond and exclude other psychological, social, and regulatory factors. Therefore, understanding these influences is important for policymakers, researchers, and industry leaders engaged in introducing gene-edited crops (Henderson et al., 2024).

Scientific Literacy and Awareness

Acceptance is strongly dependent on the public's understanding of gene editing (Delhove et al., 2020). To support this, people tend to understand the technology and science behind technology, such as safety and benefits. However, resistance can come in the form of misconceptions, such as gene editing somehow equating to genetic modification or that every genetic change is bad. Knowledge gaps can be bridged and informed decision-making encouraged through educational initiatives, science communication efforts, and public engagement campaigns.

Media Representation and Framing

Public opinion is formed based on the way the media presents gene-edited crops. Skepticism or fear can be fed through sensationalist headlines, misinformation, and overly simplified explanations. On the other hand, fair reporting of the science and highlighting the benefits along with addressing any issues transparently can lead people to understand and accept it (Hill, 2020; Kuzma, 2018). Scientists and people involved in policymaking must get involved in media debates to counter misinformation and put forward the truth.

Institutional/Regulatory Bodies (Trust in)

Acceptance relies on public trust in the institutions overseeing gene-edited crops (government agencies, scientific bodies, and biotech companies). Belief that these institutions work transparently and for public health and environmental safety will undoubtedly lead to their support for gene-edited products (Lindberg et al., 2023). So far, skepticism has stemmed from past controversies regarding GMOs. To build trust, there are clear regulations and open dialogue, and agencies are required to conduct third-party-sponsored research to verify these claims.

Ethical and Cultural Considerations

Opinions on gene editing are influenced by moral, religious, and cultural beliefs. What is also in question is the ethical issue surrounding 'tampering with nature.' However, some advocate gene editing to combat food insecurity and promote sustainability (Critchley et al., 2019). Learning about these perspectives and having respectful conversations can contribute to relieving worries and acceptance.

Market and Consumer Behavior

Willingness to purchase gene-edited products will hinge on their pricing, perceived health benefits, and labeling transparency (Lindberg et al., 2023). More widely accepted foods could be gene-edited if they are marketed as being more nutritious, environmentally friendly, and affordable. However, if consumers view them as unnatural or unsafe, they may be reluctant to purchase them. Building this confidence can be achieved through clear labeling and education.

Historical Context and Comparisons

The history of the acceptance of agricultural biotechnology has been marked by change, as science, regulatory frameworks, and public perception have evolved. Debates regarding genetically modified organisms (GMOs) have figured in discussions of acceptance for newer technologies, such as gene-edited crops (Hanssen et al., 2018). The historical context of GMOs, lessons learned from past crop advances, and how the dissimilarity in regulation affects public perception offer useful signals of what might come next for gene-edited crops.

PUBLIC REACTIONS TO GMOS VS. GENE-EDITED CROPS

The Emergence of GMOs and Public Controversy

The first genetically modified organisms (GMOs) emerged in the 1990s and are roughly synonymous with the rise of agricultural biotechnology (Barrows et al. 2014). Scientists have been able to produce plants with desirable traits, such as pest resistance, herbicide tolerance, and increased nutritional content, by modifying crops at the genetic level. By 1994, the Flavr Savr tomato was the first commercial GMO approved for sale in the US, with genetically modified soybeans, corn, and cotton rapidly adopted (Schneider et al., 2002). There was a mixed public reaction to the technological promise of GMOs, and in many cases, a very negative one. Groups advocating on behalf of health risks, environmental impact, or corporate control of the food supply expressed opposition. Resistance to GMOs was so strong in Europe that GMOs were strictly regulated and sometimes even completely banned from cultivation (Lynch & Vogel, 2001). The objection was primarily based on the precautionary principle, arguing that the absence of definite evidence of safety warranted an error on the side of caution by the regulatory agencies. Misinformation coupled with media portrayal also contributed to public perception. This has been taken up by activist groups and sections of the media that portray GMOs as 'unnatural' and harmful, resulting in broad skepticism. However, although scientific opinion has generally been in favor of GMOs, people are still skeptical of technology around the world.

The Introduction of Gene-Edited Crops

The most exciting thing, however, involves gene-editing technologies, particularly CRISPR-Cas9, which offer a far closer and more precise way to improve crops than traditional genetic modification (Furrow, 2017). Whereas GMOs generally involve the placement of foreign DNA from another species into an organism's genome, gene editing is highly precise in making changes to what is already in the organism's genome (Rozas et al., 2022). This distinction has caused many scientists to contend that gene-edited crops should be subject to different regulations and an alternate frame of public perception. Public reactions to gene-edited crops have generally been more positive than those to GMOs. Several factors have been attributed to this. Perceived as Natural Since gene editing does not necessitate introducing foreign genes, it is perceived as a more 'natural,' albeit more precise, process compared to traditional breeding. Scientific Communication and Outreach: The scientific community has been much more proactive in bringing the public along with gene editing technologies, getting to them first before misinformation has a chance to take root. Differentiating from GMOs – Many advocates of gene editing stress its separate identity from GMOs to sever it from GMOs' controversies (Singh et al., 2006). However, groups that advocate organic and natural farming methods still have some skepticism. While gene editing of crops remains part of the discussion regarding gene-edited crops, concerns about unintended genetic consequences, corporate control, and long-term environmental effects continue.

Lessons from Past Agricultural Innovations

Importantly, the history of agricultural technological advances can offer valuable lessons regarding the public acceptance of new technologies (Feder & Umali, 1993). There is a wealth of learning in past food production innovations and crop development from which we can draw a number of key takeaways.

The Green Revolution and Initial Skepticism

In the mid-20th century, the Green Revolution coined high-yield crop varieties, the use of synthetic fertilizers, and modern irrigation techniques, which greatly increased food production globally (Borlaug, 2002). There was initial skepticism and worry about soil degradation, water utilization, and the inequalities that could exist between farmers. While these technologies were initially met with resistance, however, as increased food security gained an appetite from the world over, this opposition waned, and the technologies became a standard 'form' of agricultural practice. This pattern suggests that public skepticism of agricultural innovations is a common phenomenon, and over time, the benefits become clear, the risks are managed, and the projects become acceptable. If the advantages of gene-edited crops are clearly proven, a similar path might be expected for gene-edited crops (Lassoued et al., 2019).

GMO Backlash and the Importance of Transparency

This is one of the most important reasons why GMOs have met with backlash: they were developed and introduced without transparency. For many consumers, they were not given the option to not consume developing GMOs under the terms of whatever testing and public discussion has or has not happened (Westerman, 2018). As a result, this lack of engagement caused an incredibly high amount of mistrust, meaning that biotech companies struggled to change the public's perception. Unlike these tools, the emergence of gene editing has involved more public engagement, scientific outreach, and regulatory discussions. The debate has become more open and informed as scientists and policymakers have become more open to the potential benefits and risks. This strategy may help stop gene-edited crops from facing the same degree of resistance that GMO crops have had to suffer (Kumar et al., 2020).

The Role of Regulation in Shaping Public Perception

Another important lesson is that regulatory decisions also play a significant role as an important determinant of public trust. It appears that when regulatory agencies impose stringent safety assessments for new technologies and require transparency in terms of their approval, consumers have more faith in the safety of these new technologies. This contrasts with when regulations are perceived to be too lenient or subject to strong corporate influence. Gene-edited crop regulations should seek a middle ground between safety and unnecessary hurdles for innovation. Science-based policies must be implemented to maintain public trust (Macnaghten & Habets, 2020).

Regulatory Differences and Their Impact on Perception

The level of public acceptance of both GMOs and gene-edited crops is strongly influenced by the variability in the regulation of agricultural biotechnology in different parts of the world (Turnbull et al., 2021).

The United States: A Favorable Regulatory Environment

In general, the United States has been more permissive in both permitting GMOs and gene-edited crops. In a statement the U.S. Department of Agriculture (USDA) stated that it would not regulate gene edits that do not add foreign genetic material to a crop and would not consider such crops as GMOs. There is a regulatory stance based on the argument that gene editing does no more than mimic natural mutations or other traditional breeding processes. This approach will probably make gene-edited crops less controversial in the U.S. than in other regions with stricter regulations.

Nevertheless, there are contentious elements regarding labelling policies, with consumer advocacy groups calling for greater transparency (Wolt & Wolf, 2018).

The European Union: Precautionary Regulations

In the European Union (EU), regulations became stricter in 2018, deciding that gene-edited crops ought to be regulated under a framework similar that to of GMOs. Many scientists have criticized this decision as stifling innovation and giving European farmers an unfair disadvantage. This is because the EU has strict regulations on genetic modification, which reflects broader public unease over GMOs, mainly due to previous opposition against it. However, to obtain such crops on the European market, policymakers may have to reassess how these technologies should be categorized and regulated (Levidow et al., 2000).

Asia and Other Regions: A Mixed Approach

In Asia, the regulatory approaches vary. Japan has been progressive, allowing certain gene-edited crops to be commercialized as non-GMO products. Gene editing technology has sparked interest in China cautious with GMOs, particularly if it can be used to bolster food security. There are still a number of factors that determine the acceptance of gene-edited crops in developing countries, including the need for food security, economic incentives, and international trade agreements (Jones et al., 2022).

FACTORS INFLUENCING PUBLIC PERCEPTION OF GENE-EDITED CROPS

The peculiarities of gene-edited crop perception are determined by social, cultural, and informational factors. Moreover, the scientific world is in full support of techniques such as CRISPR and TALENs being implemented in agriculture, despite the high level of skepticism among the public. The issue of misinformation, the ways media are presented by different sources, the reliability of such media, concerns about users' trust, and ethical issues influence how people look at this technology. This is knowledge one needs to have to be able to undertake a balanced discussion and to avoid the stigmatization of gene-edited crops based on the mere transmission of genes (Gao et al., 2024).

Literacy & Awareness

There are two major factors that, when combined, have been observed to have the largest contribution to opinions regarding genetically modified foods: the first is scientific literacy. Therefore, there is a clear relationship between awareness and acceptance or skepticism regarding gene editing. However, this issue is not well understood because of the numerous scientific processes involved (Jasanoff et al., 2015).

Understanding of Gene-Editing Technologies (CRISPR, TALENs, etc.)

By its broad definition, gene editing can be described as a process of making changes to an organism's genetic makeup to augment certain characteristics or rid the organism of undesirable traits. Unlike GMOS, where new genes are inserted, a gene editor alters the set gene sequence. Some of the widely applied methodologies for altering genes to enhance proficiency include (Fig 2):

Figure 2. Gene-editing tools

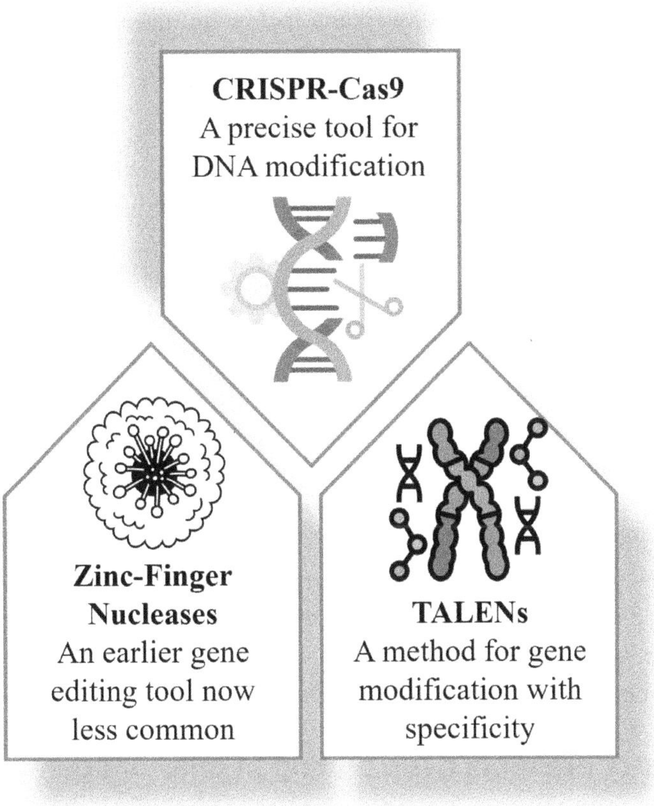

However, information regarding these technologies and their differences is still rather limited among the public, even though many of the benefits are obvious. There is nothing wrong with gene editing, and the public needs to be informed that gene editing works as a natural mutation so that the process is more effective and accurate.

Misinformation and Its Role in Shaping Opinions

The public debate surrounding gene editing faces the major problem of receiving incorrect information. Social media rapidly shares incorrect information about biotechnology, leading people to doubt its value. Common misinformation includes people wrongly linking gene-edited crops to GMOs because they think the products contain foreign DNA, which makes them worry about safety. Some people believe that changing the natural DNA structure is contrary to nature and results in unknown health effects. People believe that gene-edited foods are unsafe because of expert claims, despite laboratory findings showing their safety. Public trust in gene-editing technology depends on real science promotion and community listening, as well as sharing how this technology works well and its few limitations (Krimsky, 2019).

Media and Communication

The media strongly shapes public understanding of gene-edited farming products. People develop opinions about news content both in legitimate news channels and online across social platforms.

Role of Traditional and Social Media in Spreading Information

Necessary scientific updates from newspapers, television, and online news websites reach the public first. Newspapers and media outlets sometimes provide inaccurate or exaggerated information about their topics. News items that stress risk more than scientific evidence trigger concern and create opposition to scientific progress. Social media makes public understanding harder because it spreads news quickly, both right and wrong. The sharing platforms Twitter, Facebook, and YouTube spread incorrect information, especially when the content uses emotional triggers for impact. Rumor-based rumors and objections to biotechnology move across society before accurate scientific research does so.

News industry presents gene editing in different perspectives

The manner in which news media presents gene-editing technologies determines how people understand these technologies. Some common framing patterns include the following: Journalists sometimes explain gene editing as a powerful solution, yet other articles discuss its safety and moral implications (Rademan 2020). People become less trusting of gene editing when news outlets show that major biotech corporations' own gene-editing patents. The population response shifts from negative when public research institutions create edited crops to positive when corporate research is conducted. The public opposes food production interference and genetic engineering because media reports link them to the unnatural manipulation of natural systems. Professional reporting that uses evidence establishes a foundation for public discussions on gene editing, replacing ineffective decision-making based on fear.

Trust in Institutions

People need to trust their governing bodies and dietary research organizations before accepting gene-edited food products. People will accept agricultural biotechnology if institutions demonstrate open and honest management.

Perception of Biotech Companies, Government Agencies, and Scientists

People typically do not trust biotech companies because they oppose these businesses' corporate dominance and profit orientation while also distrusting past GMO scandals. The public questioned Bayer following Monsanto's acquisition because people thought the company used herbicide-resistant seeds to profit from farmers' expenses. The US Department of Agriculture and Food and Drug Administration shaped biotech crop rules, but their regulatory abilities affect people's beliefs in these bodies (Sciences et al., 2017). Citizens in one nation fear that official regulators allow companies to control them, while people in other nations focus on the authorities' strict oversight. Scientists and researchers who work at universities receive more trust from the public than companies, but their positive messages are blocked by louder activism against biotechnology. Everyday people should have more ways to learn from regular scientists so that public beliefs better match scientific facts.

The Role of Independent Research and NGOs

Organizations beyond government control and independent research facilities help people develop strong trust. Groups that work on food protection, public wellness, and environmental preservation can create positive or negative public opinions. The benefits of gene editing that pro-biotech organizations highlight include production boost, lower pesticide usage, and carbon emission reduction (Theresa et al., 2021). Skeptical NGOs examine possible future dangers, corporate influence, and ethical perils in modern food creation. Good choices depend on independent studies alongside many different stakeholder views.

ETHICAL AND CULTURAL CONCERNS

Religious and Philosophical Views on Modifying Nature

People base their views on gene editing on multiple religious and philosophical beliefs. Certain religious organizations reject genetic changes because they alter the natural DNA structures created by God. Gene improvements are supported when they assist in ending hunger and enhancing lives (Peters et al., 2008). Christian religious faith groups have different views on gene-editing practices. People hold different opinions about changing DNA with genetic modification because they believe these methods help life but also create unnatural risks. Several Buddhist and Hindu followers see all living things as connected, so they doubt the ethical value of altering genes.

Ethical Debates on Food Security vs. Natural Preservation

Experts debate which aim of technology enhancement in agriculture should receive priority: food supply versus preserving the biological diversity of natural ecosystems. Champions of GMO technology claim that it can fix world hunger problems while making farms greener and diminishing dependency on chemical pesticides (Taheri et al., 2017). To make the public approve of this technology the ethical values of producing crops and keeping nature alive need to be kept in proper balance

Consumer Attitudes and Market Trends

Consumer attitudes toward gene-edited crops are important in determining their future market success and adoption. The role of attitudes toward gene editing stems from surveys and studies and broader cultural contexts on how gene-edited products are accepted or rejected in different parts of the world (Yang & Hobbs, 2020). However, consumer perceptions are influenced by several cultural, economic, and regulatory factors, making it more difficult to drive the global acceptance of these innovations, even at the regional level. Furthermore, labeling and transparency are important for consumer choice experiences. Biotech companies, policymakers, and other agricultural stakeholders must understand these factors when navigating the maze of consumer preferences.

Surveys and Studies on Public Opinion

Surveys and studies on public opinion are useful in this regard. They provide insights into what people think about gene-edited crops and other biotechnology work. While consumers are open to the potential benefits of gene editing, they are also worried about its side effects, ethics, and how it could affect the environment, according to these studies (Ipsos, 2021).

General Public Opinion

There is very strong research showing that public opinion regarding gene-edited crops is divided, and opinion is squishy at the margin (Cummings & Peters, 2022). A research Center survey found that more than 60 percent of Americans are in favor of genetically modified crops, although there is a huge gulf between those who want technological advances and those who are wary. Support was somewhat muted, specifically when the question of gene-edited crops was raised (Hallman et al., 2003). Approximately half of the respondents were unafraid of technology, particularly if it could contribute to higher food security, decreased pesticides, increased nutrition, or any combination of these. However, the other half were neutral or opposed, citing environmental impact and long-term safety concerns. Attitudes toward gene-edited crops are known to differ between Europe and Asia, although similar surveys show that Europeans are the most skeptical compared to Americans or Asians (Kato-Nitta et al., 2023). This indicates that consumer attitudes toward biotechnology are inextricably bound to scientific understanding, culture, and history.

Perceived Benefits and Concerns

Studies indicate that due to improved agricultural sustainability, disease resistance, and nutritional profiles, these are considered some of the most cited benefits that could sway public opinion in favor of gene-edited crops. Many consumers are interested in the possibility that gene-edited crops will assist in addressing problems such as climate change by creating more climate change-resistant crops that can survive in harsher conditions. However, there are common worries about the environmental impacts of gene-edited crops, such as how the crops could be contaminated with non-GMO crops, lose biodiversity, and other unforeseen ecological impacts. Public opinion is affected by health-related concerns, including the long-term effects of eating gene-edited foods (Medani et al., 2024). Additionally, another factor that troubles a large group of individuals is the ethical issue of 'patenting life' and the dependence of corporations on the production of our food supply.

Regional Differences in Acceptance

Attitudes toward gene-edited crops are mixed according to regions, depending on a range of regulatory approaches, cultural values, and historical experiences with biotechnology, ranging from paranoia to blind optimism. Such regional differences determine the perception of these crops and their market potential (Ferrari 2021).

The United States

Public opinion toward gene-edited crops in parts of the United States has generally been more positive than in other regions, partly because the country has historically adopted biotechnology and GMOs. Public opinion is split, but surveys have repeatedly found that many Americans are willing to accept gene-edited crops if the technology is characterized as addressing urgent agricultural problems, such as food security or climate change (Sciences et al., 2016). Furthermore, the regulatory environment in the US is more lenient than that in other countries. Finally, the U.S. Department of Agriculture (USDA) and the Food and Drug Administration (FDA) have supported gene-editing technology as public acceptance has increased. However, the battle over labeling has not ended in the U.S., with consumers also divided on whether gene-edited crops should be labeled, which could undermine market acceptance. Consumers who think that they have the right to know what they eat may react negatively if the company does not label its products. As a result, there is a public opinion balancing act in the U.S. as to whether gene-edited plants will be useful and what level of transparency and information the U.S. public wants (Selfa et al., 2021).

The European Union

EU public opinion is much more cautious about gene-edited crops. Historically, GMOs have been held back in the EU, where strict regulations have been imposed to limit their use. Although technology is more precise than traditional methods of genetic modification, many Europeans still see gene-edited crops as 'unnatural' and potentially dangerous. These concerns are reflected in the EU's regulatory framework, which has decided to place gene-edited crops under the same stringent guidelines applicable to GMOs. Repeated surveys in Europe have shown that the majority of the population does not want such crops. A recent study from 2018 by the European Commission found that 60 per cent of Europeans thought that gene-edited crops should be banned or tightly controlled. Many Europeans also see large biotech companies as putting profit before public health and environmental concerns, and little trust is extended towards them. In addition, the EU's history of food safety scandals has led consumers to be highly skeptical of new food industry technologies (Ichim, 2023).

Asia

In Asia, people's attitudes towards gene-edited crops are more favorable than in Europe but not as enthusiastic as in the U.S. While the EU has been restrictive in allowing any gene-editing technology, countries such as China and Japan are more permissive and even these nations have an interest in using such technology to address food security (Kato-Nitta et al., 2023). China has been heavily investing in gene-editing research because it is seen as a way to address the problem of feeding a large population. Although government support exists, the Chinese public remains skeptical and has a habit of doubting the safety of genetically modified products. According to a 2020 report by the Chinese Academy of Social Sciences, 37 percent of respondents supported using gene-editing technologies in agriculture (Kang et al., 2022), while 52% were against it. In Japan, however, consumers are more tolerant than their European counterparts of gene-edited products; after all, Japan is a technologically innovative country with a history of openness to scientific progress. However, public attitudes remain mixed, and gene-edited crops are associated with health risks and ethical issues. Unless these concerns are better communicated and addressed through regulation, gene-edited crops will face significant resistance.

Influence of Labeling and Transparency on Consumer Choice

In this regard, labeling and transparency are the decisive factors that determine consumer choice for gene-edited crops. As consumers increasingly want to know more about the food they eat, many want to know that the crops from which they eat are not genetically modified or gene-edited. The way this kind of information is presented can be crucial to letting consumers feel comfortable purchasing a gene-edited product (McFadden & Lusk, 2016).

The Debate on Labeling

A major focus of the ongoing debate in the U.S. and other countries is whether gene-edited crops should be labeled. People advocating for labeling assert that consumers should have the opportunity to know what they eat, especially if the products contain genetically modified or gene-edited ingredients (Negowetti, 2020). Consumer trust in brands and the empowerment of individuals to make clean and informed decisions regarding food consumption are built when labels are transparent. However, opponents say that labeling gene-edited crops would unfairly tarnish them with a stigma that they are inherently unsafe or different from other conventionally grown products (Sattell, 2006). In the U.S., labels have proved to be a complicated battleground in the discussion about informing consumers when they are eating food derived from GMOs, thanks to federal rules on food companies to indicate whether their goods contain GMOs (Keane, 2006). However, some say there should be more transparency because gene-edited crops are frequently excluded from such rules. This is different in the EU, where labeling requirements are stricter and where most countries propose, as an alternative to a ban, the mandatory labeling of gene-edited products (Cheftel, 2005).

Impact of Labeling on Consumer Behavior

Transparency in labeling studies can greatly affect consumer behavior (McFadden et al., 2024). Research from the U.S. showed that Americans given clear information on gene-editing technology were more supportive of its use in agriculture. However, when the labeling was vague or misleading, public trust was lost. A study in Europe also discovered that, with the right labelling of genetically modified products, consumers became more accepting of GMOs.

Strategies to Improve Public Acceptance

One of the critical factors for the widespread use of gene-edited crops and their incorporation into farming practices worldwide is public approval. Opposition to GMOs products and gene-edited crops is a result of ignorance, myths, ethical issues, and culture. Thus, to increase popularity, it is necessary to solve these issues by providing detailed information, clear rules, and by stimulating cooperation. Therefore, this section discusses some of the approaches that can be taken to enhance the acceptability of gene-edited crops (Wynne, 2001).

Effective Science Communication and Education

Education and science communication is a very effective measure towards overcoming the woes that plague the acceptance of gene-edited crops by the public. This is because technologies, including CRISPR and TALENs, are usually complex, posing a big challenge in explaining them to the public (Bhardwaj & Nain, 2021; Mahfouz et al., 2014). Some people consider gene editing as something bad or unnatural due to a lack of proper knowledge on how it works and how it differs from genetic modification. Science communication must be enlightening and requires no expertise to understand to be able to understand what gene editing entails and the benefits as well as the repercussions of the same. In this regard, it is essential to reduce the mystery of science. The following are some important measures that can be taken to improve science communication. When presenting information about gene-editing technologies, it is best to describe them in basic, unambiguous language. For this reason, it is suggested that communicators should leave behind complex language, scientific diagrams and instead use analogies, charts, and videos that help the public, understand the concept of gene editing and one important thing – it is not the introduction of foreign DNA. Essential aspects of media resolution also include areas of major concern that people may have towards genetic editing, including safety and environmental influences (Sciences et al., 2016). The proper dissemination of literature on the positive outcomes of studies on the application of this technology in crops to prove safety and efficacy should be enhanced by examples of its successful implementation in farming. These debates can also be conducted through video conferences, webinars, and other cases where specialists can directly interact with the population to decrease fears and uncertainties.

Learners tend to engage better when real-life situations and examples are used rather than theories. This means that portraying gene-edited crops as real-life success stories of increased yields, reduced pesticide use, or solving issues such as climate change or pests is the way to go for science communicators seeking to increase the public understanding of gene-editing technologies (Shelake, 2025). These stories can assist in building an understanding that gene editing is the future and a way to improve the human condition and not just a science fiction rogue method. Therefore, educating the respective society's populace in scientific knowledge at school is of paramount importance in the long run. People should be educated about the next generation of consumers and farmers regarding gene-editing technologies and their advantages, which will leave a small concern (Basinskiene & Seinauskiene, 2021). The use of prior topics, such as gene editing, as part of the curriculum, shall help young individuals become more familiar with such innovations when they grow up.

Policies for Transparency and Consumer Engagement

The prerequisites for regulating production and, particularly, for the use of genetically modified crops include state interference in the sphere, transparency, and beneficial effects on consumers' intentions at the policy level. Governments, regulators, and companies should implement measures that will help shape consumer awareness, facilitate their right decisions, and encourage the correct use of these technologies. One of the most effective ways to encourage consumers to accept gene-edited products is through labeling (Lindberg et al., 2023). Today, there is an increased demand for information on whether the food that they are consuming is genetically modified or gene-edited (Ishii & Araki, 2016). This is why labeling gene-edited crops and food is beneficial for customers: it gives them the opportunity to choose in favor of or against them. Although some critics would label gene-edited foods as potentially causing more harm than good, it is evident that consumers would know exactly what is in a product in a better way knowing that it has been labeled appropriately (Rose et al., 2020).

Governments are required to work towards passing legislation that would create guidelines for the use of gene-edited crops with regard to their safety and accountability, among other aspects. Regulatory agencies should be able to assess gene-edited crops based on the evidence collected and develop categories and criteria that must be met for approval (Niazi, 2023). Such regulations should be drafted primarily to safeguard people's interests and support technological advancement in agricultural practices. Government agencies and scientific institutions should organize focus group meetings or construct public forums where citizens can speak out and solicit information and/or voice their concerns regarding genetically modified crops. Public participation in decision-making and meeting its needs and concerns are essential for trust within the sphere of technology. Independent fourth party: Third parties or other parties comprising NGOs and universities that are not affiliated with government agencies can be entrusted with the responsibility of monitoring gene-editing practices. Having independent scientific review boards or committees consisting of scientific and other related professional audiences, such as scientists, ethicists, environmentalists, consumers, and public health, may offer balanced risk and benefit determination of GE crops (Singh et al., 2006).

Role of Advocacy Groups and Partnerships

Non-governmental advocacy groups, industrial and public interest alike, may play a huge role in changing public perception with the help of true scientific support of the gene-edited crop benefits by discussing the key issues and concerns (Shukla-Jones et al., 2018). The involvement of all stakeholders, such as biotech companies, government bodies, academic workers, and civil society organizations, will foster the confidence of the public to accept technological inventions. It is crucial for any biotechnology firm selecting gene-edited crops as a production focus to be sensitive to public relations (Ruder & Kandlikar, 2023). Supporters can develop brochures, flyers, TV programs, and other forms of media that will enlighten the public on the perceived benefits of gene-editing technology. Clients can also engage other stakeholders, such as researchers and consumer interest organizations, to show that gene-edited plants are safe and have been proven to deliver as expected. Biotechnology firms and NGOS should join forces to ensure a link between the social needs of society and the available technology. Civil society organizations, especially in the fields of food and nutrition, environment, and health, can thus use their position to demand the just and moral use of gene-editing technology (Helliwell et al., 2019). It gives a bonus to technology; therefore, these groups are known to be neutral and always work for the benefit of society. Both engagements ensure that they gather the necessary information and data to come up with factual arguments and points in their advocacy (Fig 3). It is still recommended that public health organizations, environmental groups, and food security advocates engage with scientists to ensure that the current information on gene-edited crops is made easy for the public to understand. They may also show the best ways that technology can be used to address some of the challenging problems facing humanity, such as world hunger, the effects of climate change, and pests' resistance to the chemicals used in farms (Nah & Chau, 2010).

Figure 3. Stakeholder influence and engagement in the acceptance of gene-edited crops.

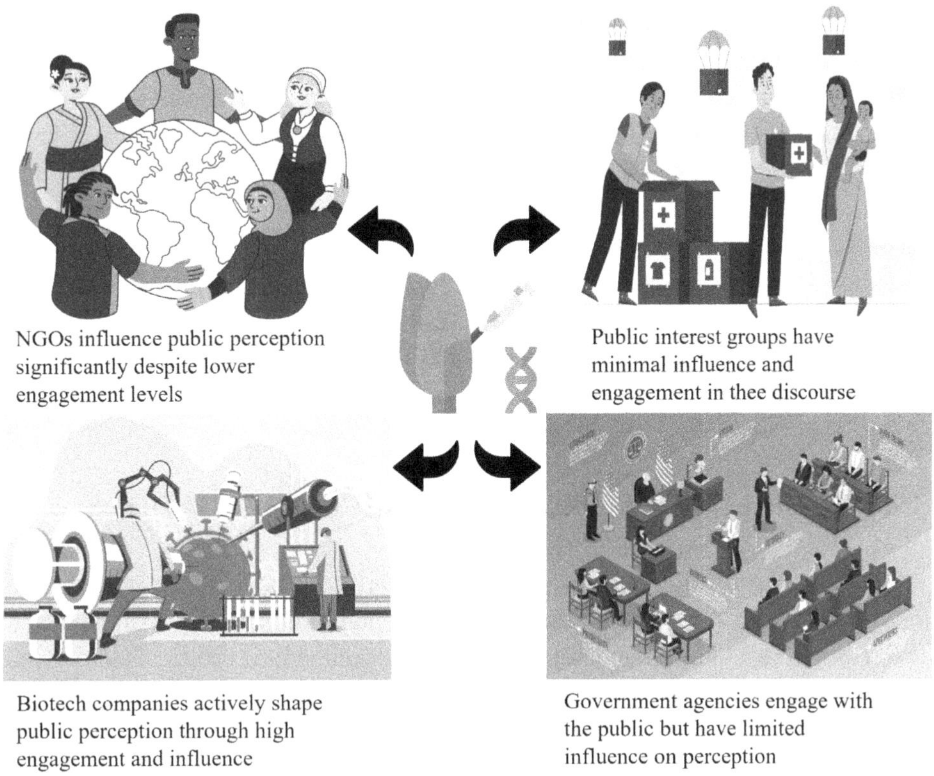

CASE STUDIES

Examples of Successful and Failed Public Engagement Efforts

Examining both successful and unsuccessful case studies for public engagement can help us understand how public acceptance of gene-edited crops can be developed (Molinari et al., 2024). For example, gene-edited crops, such as non-browning apples and Innate potatoes, have been relatively successful in the public acceptance in the U.S. The marketing of these products has been clear on what those benefits are, such as reducing food waste and helping the environment due to less need for chemical treatments. The inclusion of transparent labeling and strategies to address such concerns through public education campaigns has boosted consumer acceptance. In contrast, the European Union has proven to be a tougher market for gene-edited crops, as there is widespread suspicion of GMOs and biotechnology. Public mistrust is high, despite attempts to present gene editing as a more precise and less invasive approach than traditional GMOs (Kumar et al., 2018; Nawaz et al., 2023). This is due to the public not being sufficiently engaged and a highly restrictive regulatory environment that does not allow for adoption. Resistance in Europe has been a function of anti-GMO campaigns and the perception of the influence of large biotech companies on food systems. Gene-edited crops are more well-received in Argentina, a leader in GMO crop adoption (Whelan & Lema, 2019). The country has implemented science-based policies that emphasize a high level of clarity when communicating the advantages of gene editing in agriculture, such as increased crop production and low pesticide usage. Furthermore, there is strong evidence that Argentina is dedicated to maintaining transparency in gene-editing regulations, resulting in confidence among consumers and farmers (Friedrichs et al., 2022).

FUTURE DIRECTIONS AND POLICY RECOMMENDATIONS

Several policy recommendations moving forward will encourage the responsible use of gene-edited crops and improve public acceptance of such use. Policymakers should aim to set Global Regulatory Standards for gene-edited crops, leading to the harmonization of approval processes and uniform standards of safety evaluation. Governments and industry groups should prioritize education and transparency by investing in educational campaigns to demystify gene-editing technologies and increase the general knowledge and understanding of science in the population. A Public Involvement In: Consult the public, involve stakeholders, and employ transparent regulatory processes to enable the public to trust in gene-editing technologies and take a more informed approach when making decisions. It is vital to create ethical frameworks that consider public concerns stemming from gene-editing technologies regarding environmental sustainability, corporate control, and food sovereignty. Thus, the ultimate acceptance of gene-edited crops relies on continued efforts to engage consumers, increase transparency, and communicate the actual value of technology.

CONCLUSION

Gene-edited crop acceptance requires solutions in three main areas: science education, transparent policy development, and strategic alliances with consumer groups. Gene editing technology remains a mystery to most people; thus, public opinion about gene-edited crops continues to base its judgments on false information that circulates about risks and benefits. Public acceptance of gene-edited crops requires governmental agencies to work with biotech businesses and advocacy groups to deliver clear label information and create protective standards and educational materials about safety standards. Other key stakeholders in gene-editing technology should embrace this new science through public participation programs while working to establish trust to bring about its rewards.

REFERENCES

Barrows, G., Sexton, S., & Zilberman, D. (2014). Agricultural biotechnology: The promise and prospects of genetically modified crops. *The Journal of Economic Perspectives*, *28*(1), 99–120. DOI: 10.1257/jep.28.1.99

Basinskiene, L., & Seinauskiene, B. (2021). Gene editing versus gene modification: Awareness, attitudes and behavioral intentions of lithuanian consumers, producers, and farmers. *Chemical Engineering Transactions*, *87*, 433–438.

Bhardwaj, A., & Nain, V. (2021). TALENs—an indispensable tool in the era of CRISPR: A mini review. *Journal of Genetic Engineering and Biotechnology*, *19*(1), 125. DOI: 10.1186/s43141-021-00225-z PMID: 34420096

Borlaug, N. E. (2002). The green revolution revisited and the road ahead. Nobel-prize. org Stockholm.

Cheftel, J. C. (2005). Food and nutrition labelling in the European Union. *Food Chemistry*, *93*(3), 531–550. DOI: 10.1016/j.foodchem.2004.11.041

Critchley, C., Nicol, D., Bruce, G., Walshe, J., Treleaven, T., & Tuch, B. (2019). Predicting public attitudes toward gene editing of germlines: The impact of moral and hereditary concern in human and animal applications. *Frontiers in Genetics*, *9*, 704. DOI: 10.3389/fgene.2018.00704 PMID: 30687386

Cummings, C., & Peters, D. J. (2022). Who trusts in gene-edited foods? Analysis of a representative survey study predicting willingness to eat-and purposeful avoidance of gene edited foods in the United States. *Frontiers in Food Science and Technology*, *2*, 858277. DOI: 10.3389/frfst.2022.858277

Delhove, J., Osenk, I., Prichard, I., & Donnelley, M. (2020). Public acceptability of gene therapy and gene editing for human use: A systematic review. *Human Gene Therapy*, *31*(1-2), 20–46. DOI: 10.1089/hum.2019.197 PMID: 31802714

Ervin, D. E., Glenna, L. L., & Jussaume, R. A.Jr. (2011). The theory and practice of genetically engineered crops and agricultural sustainability. *Sustainability (Basel)*, *3*(6), 847–874. DOI: 10.3390/su3060847

Feder, G., & Umali, D. L. (1993). The adoption of agricultural innovations: A review. *Technological Forecasting and Social Change*, *43*(3-4), 215–239. DOI: 10.1016/0040-1625(93)90053-A

Ferrari, L. (2021). Attitude toward gene-edited food: and economic analysis from diverse stakeholders' viewpoint.

Friedrichs, S.. (2022). Regulatory and Policy Considerations Around Genome Editing in Agriculture. In *Applications of genome modulation and editing* (pp. 327–366). Springer. DOI: 10.1007/978-1-0716-2301-5_17

Furrow, B. R. (2017). The CRISPR-Cas9 tool of gene editing: Cheaper, faster, riskier. *Annals of Health Law*, *26*, 33.

Gao, S., ... (2024). Understanding the Factors Driving Consumers' Willingness to Pay for Gene-Edited Foods in China. Foods, 13(15), 2348.

Hallman, W. K., et al. (2003). Public perceptions of genetically modified foods: A national study of American knowledge and opinion.

Hanssen, L., Dijkstra, A., Sleenhoff, S., Frewer, L., & Gutteling, J. M. (2018). Revisiting public debate on genetic modification and genetically modified organisms. Explanations for contemporary Dutch public attitudes. *JCOM, Journal of Science Communication*, *17*(4), A01. DOI: 10.22323/2.17040201

Helliwell, R., Hartley, S., & Pearce, W. (2019). NGO perspectives on the social and ethical dimensions of plant genome-editing. *Agriculture and Human Values*, *36*(4), 779–791. DOI: 10.1007/s10460-019-09956-9

Henderson, K., Lang, B., Kemper, J., & Conroy, D. (2024). Exploring diverse food system actor perspectives on gene editing: A systematic review of socio-cultural factors influencing acceptability. *Agriculture and Human Values*, *41*(2), 883–907. DOI: 10.1007/s10460-023-10523-6

Hill, N. (2020). Public opinion of gene-editing in agriculture: A mixed-method study of online media and metaphors

Ichim, M. C. (2023). The Citizens' Awareness and Concerns During the Transition from Genetically Modified to Genome Edited Plants in Europe about Their Use in Agriculture and Food Production. A Roadmap for Plant Genome Editing, 519.

Ipsos, M. (2021). *Consumer perceptions of genome edited food*. Food Standards Agency UK. DOI: 10.46756/sci.fsa.aya629

Ishii, T., & Araki, M. (2016). Consumer acceptance of food crops developed by genome editing. *Plant Cell Reports*, *35*(7), 1507–1518. DOI: 10.1007/s00299-016-1974-2 PMID: 27038939

Jasanoff, S.. (2015). CRISPR democracy: Gene editing and the need for inclusive deliberation. *Issues in Science and Technology*, *32*(1), 25–32.

Jones, M. G., Fosu-Nyarko, J., Iqbal, S., Adeel, M., Romero-Aldemita, R., Arujanan, M., Kasai, M., Wei, X., Prasetya, B., Nugroho, S., Mewett, O., Mansoor, S., Awan, M. J. A., Ordonio, R. L., Rao, S. R., Poddar, A., Hundleby, P., Iamsupasit, N., & Khoo, K. (2022). Enabling trade in gene-edited produce in Asia and Australasia: The developing regulatory landscape and future perspectives. *Plants*, *11*(19), 2538. DOI: 10.3390/plants11192538 PMID: 36235403

Kang, Y., Deng, H., Pray, C., & Hu, R. (2022). Managers' attitudes toward gene-editing technology and companies' R&D investment in gene-editing: The case of Chinese seed companies. *GM Crops & Food*, *13*(1), 309–326. DOI: 10.1080/21645698.2022.2140567 PMID: 36382611

Kato-Nitta, N., Tachikawa, M., Inagaki, Y., & Maeda, T. (2023). Public perceptions of risks and benefits of gene-edited food crops: An international comparative study between the US, Japan, and Germany. *Science, Technology & Human Values*, *48*(6), 1360–1392. DOI: 10.1177/01622439221123830

Keane, S. (2006). Can a Consumer's Right to Know Survive the WTO: The Case of Food Labeling. *Transnat'l L. & Contemp. Probs.*, *16*, 291.

Krimsky, S. (2019). *GMOs decoded: A Skeptic's view of genetically modified foods*. MIT Press. DOI: 10.7551/mitpress/10824.001.0001

Kumar, A.. (2018). Genetic engineering and public perception. In *Biotechnology for sustainable agriculture* (pp. 335–373). Elsevier. DOI: 10.1016/B978-0-12-812160-3.00012-X

Kumar, K., Gambhir, G., Dass, A., Tripathi, A. K., Singh, A., Jha, A. K., Yadava, P., Choudhary, M., & Rakshit, S. (2020). Genetically modified crops: Current status and future prospects. *Planta*, *251*(4), 91. DOI: 10.1007/s00425-020-03372-8 PMID: 32236850

Kuzma, J. (2018). Regulating gene-edited crops. *Issues in Science and Technology*, *35*(1), 80–85.

Lassoued, R., Macall, D. M., Hesseln, H., Phillips, P. W. B., & Smyth, S. J. (2019). Benefits of genome-edited crops: Expert opinion. *Transgenic Research*, *28*(2), 247–256. DOI: 10.1007/s11248-019-00118-5 PMID: 30830581

Levidow, L., Carr, S., & Wield, D. (2000). Genetically modified crops in the European Union: Regulatory conflicts as precautionary opportunities. *Journal of Risk Research*, *3*(3), 189–208. DOI: 10.1080/13669870050043062

Lindberg, S. A., ... (2023). Gene-Edited Food Adoption Intentions and Institutional Trust in the United States: Benefits, Acceptance, and Labeling☆. Rural Sociology, 88(2), 392-425.

Lucht, J. M. (2015). Public acceptance of plant biotechnology and GM crops. *Viruses*, *7*(8), 4254–4281. DOI: 10.3390/v7082819 PMID: 26264020

Lynch, D., & Vogel, D. (2001). *The regulation of GMOs in Europe and the United States: A case-study of contemporary European regulatory politics*. Council on Foreign Relations.

Macnaghten, P., & Habets, M. G. (2020). Breaking the impasse: Towards a forward-looking governance framework for gene editing with plants. *Plants, People, Planet*, *2*(4), 353–365. DOI: 10.1002/ppp3.10107

Mahfouz, M. M., Piatek, A., & Stewart, C. N.Jr. (2014). Genome engineering via TALENs and CRISPR/Cas9 systems: Challenges and perspectives. *Plant Biotechnology Journal*, *12*(8), 1006–1014. DOI: 10.1111/pbi.12256 PMID: 25250853

McFadden, B. R., & Lusk, J. L. (2016). What consumers don't know about genetically modified food, and how that affects beliefs. *FASEB Journal : Official Publication of the Federation of American Societies for Experimental Biology*, *30*(9), 3091–3096. DOI: 10.1096/fj.201600598 PMID: 27199295

McFadden, B. R., Rumble, J. N., Stofer, K. A., & Folta, K. M. (2024). US public opinion about the safety of gene editing in the agriculture and medical fields and the amount of evidence needed to improve opinions. *Frontiers in Bioengineering and Biotechnology*, *12*, 1340398. DOI: 10.3389/fbioe.2024.1340398 PMID: 38433825

Medani, K. R., Neill, A., Garrod, G., Ojo, M., & Hubbard, C. (2024). Societal perceptions and attitudes towards genetically modified (GM) crops, feed, and food products in the Middle East, North Africa, and Turkey (MENAT) region: A systematic literature review. *Food Quality and Preference*, *117*, 105148. DOI: 10.1016/j.foodqual.2024.105148

Molinari, M. D. C., et al. (2024). Navigating the Path from Lab to Market: Regulatory Challenges and Opportunities for Genome Editing Technologies for Agriculture. Plant Genome Editing Technologies: Speed Breeding, Crop Improvement and Sustainable Agriculture, 25-63.

Nah, S.-L., & Chau, C.-F. (2010). Issues and challenges in defeating world hunger. *Trends in Food Science & Technology*, *21*(11), 544–557. DOI: 10.1016/j.tifs.2010.07.013

Nawaz, S., Satterfield, T., & Phurisamban, R. (2023). Does "Precision" matter? AQ study of public interpretations of gene editing in agriculture. *Science, Technology & Human Values*, *48*(6), 1421–1461. DOI: 10.1177/01622439221112460

Negowetti, N. E. (2020). Taking (animal-based) meat and ethics off the table: Food labeling and the role of consumers as agents of food systems change. Or. *Law Review*, *99*, 91.

Niazi, S. K. (2023). Gene Editing: The Regulatory Perspective. *Encyclopedia*, *3*(4), 1345–1357. DOI: 10.3390/encyclopedia3040096

Ortega, D. L., Lin, W., & Ward, P. S. (2022). Consumer acceptance of gene-edited food products in China. *Food Quality and Preference*, *95*, 104374. DOI: 10.1016/j.foodqual.2021.104374

Paudel, B., Kolady, D. E., Just, D., & Van der Sluis, E. (2023). Determinants of consumer acceptance of gene-edited foods and its implications for innovators and policymakers. *Agribusiness*, *39*(3), 623–645. DOI: 10.1002/agr.21799

Peters, T., et al. (2008). Religious traditions and genetic enhancement. Altering Nature: Volume Two: Religion, Biotechnology, and Public Policy, 109-159.

Rademan, A. E. (2020). *Creating expectations: the bioethics of genetics as reflected in selected South African media with specific reference to CRISPR-Cas9 Stellenbosch*. Stellenbosch University.

Rahman, S. U., McCoy, E., Raza, G., Ali, Z., Mansoor, S., & Amin, I. (2023). Improvement of soybean; A way forward transition from genetic engineering to new plant breeding technologies. *Molecular Biotechnology*, *65*(2), 162–180. DOI: 10.1007/s12033-022-00456-6 PMID: 35119645

Rose, K. M., Brossard, D., & Scheufele, D. A. (2020). Of society, nature, and health: How perceptions of specific risks and benefits of genetically engineered foods shape public rejection. *Environmental Communication*, *14*(7), 1017–1031. DOI: 10.1080/17524032.2019.1710227

Rozas, P., Kessi-Pérez, E. I., & Martínez, C. (2022). Genetically modified organisms: Adapting regulatory frameworks for evolving genome editing technologies. *Biological Research*, *55*(1), 55. DOI: 10.1186/s40659-022-00399-x PMID: 36266673

Ruder, S.-L., & Kandlikar, M. (2023). Governing gene-edited crops: Risks, regulations, and responsibilities as perceived by agricultural genomics experts in Canada. *Journal of Responsible Innovation*, *10*(1), 2167572. DOI: 10.1080/23299460.2023.2167572

Sattell, S. S. (2006). The rhetoric of risk: Institutional controversies over agricultural biotechnology Northwestern University].

Schneider, K. R.. (2002). *Genetically modified food.* Food Science and Human Nutrition Department, Florida Cooperative Extension Service, Institute of Food and Agricultural Sciences, University of Florida.

Sciences, N. A. o., et al. (2016). Genetically engineered crops: experiences and prospects. National Academies Press.

Sciences, N. A. o., et al. (2017). Preparing for future products of biotechnology. National Academies Press.

Selfa, T., Lindberg, S., & Bain, C. (2021). Governing gene editing in agriculture and food in the United States: Tensions, contestations, and realignments. *Elementa, 9*(1), 00153. DOI: 10.1525/elementa.2020.00153

Shanahan, J., Scheufele, D., & Lee, E. (2001). Trends: Attitudes about agricultural biotechnology and genetically modified organisms. *Public Opinion Quarterly, 65*(2), 267–281. DOI: 10.1086/322201 PMID: 11420759

Shelake, M. S. K. (2025). Gene Editing: Hope and Risks. SCIENCE, SOCIETY AND SUSTAINABILITY, 111.

Shukla-Jones, A., et al. (2018). Gene editing in an international context: Scientific, economic and social issues across sectors. OECD science, technology and industry working papers, 2018(4), 0_1-51.

Singh, O. V., Ghai, S., Paul, D., & Jain, R. K. (2006). Genetically modified crops: Success, safety assessment, and public concern. *Applied Microbiology and Biotechnology, 71*(5), 598–607. DOI: 10.1007/s00253-006-0449-8 PMID: 16639559

Taheri, F., Azadi, H., & D'Haese, M. (2017). A world without hunger: Organic or GM crops? *Sustainability (Basel), 9*(4), 580. DOI: 10.3390/su9040580

Theresa, S.. (2021). Governing gene editing in agriculture and food in the United States. *Elementa, 9*(1).

Turnbull, C., Lillemo, M., & Hvoslef-Eide, T. A. K. (2021). Global regulation of genetically modified crops amid the gene edited crop boom–a review. *Frontiers in Plant Science, 12*, 630396. DOI: 10.3389/fpls.2021.630396 PMID: 33719302

Westerman, L. A. (2018). Consumer choice or confusion: That GMO label doesn't mean what you think it means. *The Drake Journal of Agricultural Law, 23*, 199.

Whelan, A. I., & Lema, M. A. (2019). Regulation of genome editing in plant biotechnology: Argentina. In Regulation of Genome Editing in Plant Biotechnology: A Comparative Analysis of Regulatory Frameworks of Selected Countries and the EU (pp. 19-62). Springer.

Wolt, J. D., & Wolf, C. (2018). Policy and governance perspectives for regulation of genome edited crops in the United States. *Frontiers in Plant Science*, *9*, 1606. DOI: 10.3389/fpls.2018.01606 PMID: 30467510

Wynne, B. (2001). Creating public alienation: Expert cultures of risk and ethics on GMOs. *Science as Culture*, *10*(4), 445–481. DOI: 10.1080/09505430120093586 PMID: 15971363

Yang, Y., & Hobbs, J. E. (2020). Supporters or opponents: Will cultural values shape consumer acceptance of gene editing? *Journal of Food Products Marketing*, *26*(1), 17–37. DOI: 10.1080/10454446.2020.1715316

Chapter 9
Challenges and Future Directions in Plant Gene Editing:
Challenges and Future Directions

Muhammad Rizwan Khan
Quaid-i-Azam University, Islamabad, Pakistan

Sameel Hassan
http://orcid.org/0009-0001-7559-5550
Department of Agriculture and Forest Sciences, University of Tuscia, Viterbo, Italy

Tauqeer Ahmaed Qadri
http://orcid.org/0009-0000-9617-4512
Department of Biosciences, COMSATS University, Islamabad, Pakistan

Muhammad Waseem
http://orcid.org/0000-0001-7947-7722
Hainan University, Haikou, China

ABSTRACT

Site-specific plant gene editing is transforming agricultural biotechnology. Programmable nucleases, especially CRISPR/Cas systems, enable crops with higher yield, enhanced stress tolerance, and improved nutritional quality. However, broad deployment remains constrained by technical and translational barriers. Key limitations include off-target mutations, inefficient delivery and transformation, and low homology-directed repair (HDR) rates, which reduce reproducibility and scalability. Adoption is further slowed by complex tissue-culture requirements,

DOI: 10.4018/979-8-2600-3575-7.ch009

specialized infrastructure, and skilled personnel, increasing development costs. Regulatory uncertainty, ethical debate, and divergent public attitudes also create region-specific hurdles. This chapter traces the evolution of plant gene editing from early tools to base and prime editors, reviews current constraints, and highlights emerging solutions such as AI-guided target design, nanoparticle-enabled delivery, and in planta transformation. Case studies in wheat and leafy greens illustrate practical applications.

INTRODUCTION

Plant breeding technologies have evolved in pursuit of crop productivity and sustainability under the regimes of climate change, population growth, and limited resources. Modern gene editing, site-specific nucleases, including Zinc Finger Nucleases (ZFNs), transcription activator-like effector nucleases (TALENs), and the CRISPR-Cas system, builds on the foundation of classical breeding or transgenic approaches by delivering unprecedented precision (Bhagtaney & Sundarrajan, 2023). Gene editing differs from traditional genetic modification because it involves targeted genomic alterations, such as gene knockouts or base substitutions, without necessarily introducing foreign genetic material. This is a technological leap that is transformative for crop improvement. This technology accelerates breeding timelines and enables the development of traits that were either impossible or too long to develop. For example, CRISPR-based edits have led to the creation of crops with enhanced nutrient profiles, increased drought resistance, and pest tolerance (Khalil, 2020).

However, gene editing still faces some challenges, including technical limitations, especially in terms of delivery and regeneration in recalcitrant species. In addition, the regulatory frame for gene edited crops is where there is a lot of fragmentation and there is a lot of contention (Goralogia, Redick, & Strauss, 2021). Some countries have relaxed regulations if a gene-edited organism does not bring foreign DNA, while others, including those in the European Union, are careful about regulations. Therefore, this variation gives rise to uncertainty for scientists, developers, and stakeholders (Bachtarzi & Farries, 2019). In addition, ethical concerns and public impressions are related to the adoption trajectory of this technology. Studies show only a relatively favorable view of gene editing compared to other GMOs, and public awareness is low, with acceptance based on a lack of transparency and benefits, and the intended use (Bearth, Otten, & Cohen, 2024).

This chapter integrates a historical perspective of plant gene-editing methods that have been developed to date and the present state of the art. It then systematically goes through the technical, regulatory, and societal barriers to adoption and how one may be able to remove the obstacles from the path to realizing the full potential of gene editing. To achieve this goal, we draw on case studies and interdisciplinary perspectives to provide a holistic understanding of the opportunities and constraints faced in the future of crop genome engineering.

EVOLUTION OF GENE EDITING TECHNOLOGIES IN PLANTS

Although empirical selection and modern gene editing represent significant intellectual advances in the precise genetic improvement of plants, they have allowed the journey to trace the path from traditional breeding to molecular precision. These steps mirror the never-ending trend towards greater efficiency, accuracy, and versatility in crop improvement (Sun et al., 2024). This section traces the historical and technical development of gene-editing tools, highlighting key innovations and their implications for plant biotechnology.

Traditional and Transgenic Approaches

For centuries, plant breeding has relied on phenotypic selection, crossing, and mutagenesis to develop improved varieties. These approaches have been proven to be effective in many cases, but are by their very nature, labor- and time-intensive and restricted to what is genetically re-combinable and environmentally possible (Acquaah, 2015). The introduction of transgenic technologies in the 1980s opened the door for the direct insertion of foreign genes into the plant genome using Agrobacterium tumefaciens-mediated transformation or biolistic particle delivery. These methods have drastically changed the way in which we can develop and introduce traits, such as herbicide resistance and insect tolerance. Conversely, although transgenic approaches commonly tend to experience random integration of transgenes, as well as variable expression levels and potential gene silencing, viral expression can produce consistent gene expression and can additionally include the ability to target infected cells. Moreover, GM trees are also subject to the possibility of encountering regulatory hurdles and public resistance, such as crop pollination by genetically engineered plants, introduction of allergens, and disruption of food sovereignty (Acquaah, 2015; Tripathi & Shukla, 2024).

Modern Gene Editing Tools

Gene editing in the modern era is a paradigm shift in plant biotechnology, where gene insertion is now specific and targeted (as opposed to random). Three major classes of site-specific nucleases have driven the evolution of this field, with each class contributing to the development of this field. One of the first programmable nucleases to be developed for the targeted cleavage of DNA was the zinc-finger nuclease (ZFNs). They are fusions of a FokI nuclease domain with DNA-binding zinc-finger proteins. Although ZFNs have shown promise for genome editing, their use is limited by the fact that traditional engineering of custom zinc finger arrays on a case-by-case basis is complex and time-consuming (Tripathi & Shukla, 2024). The development of Transcription Activator-Like Effector Nucleases (TALENs) has offered greater specificity and design flexibility, partly addressing this challenge. Both DNA-binding domains are customizable, with TALENs utilizing customizable DNA-binding domains derived from *Xanthomonas* effectors being more flexible than ZFNs, but also requiring individual protein engineering for every single target site to be feasible, making them unsuitable for high-throughput usage (Jankele & Svoboda, 2014). CRISPR-Cas systems, most notably CRISPR-Cas9, followed by CRISPR-Cas12a (Cpf1), have become the most significant breakthrough because they are simple, inexpensive, and provide high multiplexing capacity. CRISPR systems utilize RNA-guided nucleases engineered to bind easily with programmable single-guide RNAs (sgRNAs) to target specific genomic loci, which is a significant advantage over ZFNs and TALENs in terms of experimental design and execution (Z. Cui et al., 2021; Hassan, Mohamed, Hassan, & El-Sayed, 2025). Recently, CRISPR innovation has quickly made CRISPR the leading platform for genome editing in plants and has accelerated functional genomics research and the development of crops with specific traits in a broad range of crop species. Base editors have been developed to enable single-nucleotide conversions on the CRISPR foundation without causing DSBs, thus minimizing the probability of inadvertent genomic rearrangements (Eid, Alshareef, & Mahfouz, 2018). However, the prime editing toolkit is further expanded by the ability to insert, delete, or perform any of the 12 possible base-to-base conversions using a reverse transcriptase fused to a Cas nickase. Next-generation technologies provide great accuracy and promise gene correction in large and polyploid plant genomes with great accuracy (Ochoa-Sanchez et al., 2021).

Key Milestones and Applications

Over the past decade, gene editing has transformed crop science, and CRISPR-based interventions have allowed for the creation of a slew of revolutionary agricultural breakthroughs. The creation of drought-tolerant rice varieties is one of the biggest achievements. Researchers have been able to enhance the capacity of rice plants to save water under drought conditions by disrupting genes that govern the control of stomatal density and water-use efficiency (Sampath et al., 2023). Similarly, tomato varieties resistant to both bacterial and viral pathogens have been developed using

CRISPR technology. Scientists have knocked out these particular susceptibility genes to reduce plant susceptibility to these diseases and improve crop resilience and yield stability in affected regions (Wang, Zhang, & Zhu, 2019). Thus, there have been numerous instances of gene editing to enhance micronutrient content, for instance, in wheat, where genes responsible for accumulating essential elements (zinc as well as iron) in the grain were targeted. The potential of this advancement to improve the nutritional value of staple crops in many parts of the world can address hidden hunger. Banana plants have also been edited to make them more resistant to Panama disease, a fungal infection that devastates banana production globally and serves as a protective measure for a crucial agricultural commodity (Roy, Kumar, Ranjan, Kumhar, & Govindan, 2022).

Cell Research Corporation is a great example of a powerful gene-editing technology that has achieved remarkable success. However, the application of gene-editing technology is challenging. The successful deployment of gene-editing tools is not yet evenly spread across different crop species (Abdallah, Prakash, & McHughen, 2015). Gene editing of many crops, especially recalcitrant genotypes, polyploidy, and long generation time crops, is challenging. However, these challenges typically require extensive optimization of the transformation method, delivery system, and regeneration protocol, making the process more labor-intensive and time-consuming. Indeed, continuous innovation in gene-editing platform technologies is key to increasing the range of crops that can be efficiently edited (Berthiaume, Maguire, & Yarmush, 2011). In addition, there is a clear need to democratize access to efforts to develop these tools so that researchers and farmers worldwide (including in developing regions) may reap the rewards of the advances in plant gene editing. Overcoming these barriers would enable genes to be edited in full, realizing their potential in food security and sustainability (de Roij, 2020).

CURRENT CHALLENGES IN PLANT GENE EDITING

Plant gene editing has high potential for transformation, but it faces a complex array of challenges. These include technical bottlenecks that restrict editing precision and rate, regulatory uncertainties that prohibit product creation and commerce and play globally, and societal concerns that form public views and approaches (Anders et al., 2021). Using case studies, this section critically analyzes the principal barriers to realizing the true potential of gene editing in agriculture (Figure 1).

Technical Limitations

Off-Target Effects

An important technical challenge in CRISPR-based gene editing is the production of off-target mutations. These are unintended genetic modifications that occur at genomic loci with partial sequence homology to the intended target, leading to changes at sites other than the ones planned to be edited (Troadec, Pagès, & Scientific Committee of the High Council for biotechnology, 2019). The frequency and impact of off-target effects vary from case to case and species to species, depending on the design of the guide RNA and the particular Cas variant used, and can also include factors such as species off-target effects, which can be very infrequent; however, their potential consequences can be very significant, even for low-frequency off-target mutations (Manghwar et al., 2020). In addition to depriving the desired phenotypic outcomes, these unintended changes can have biosafety issues, especially in genetically modified organisms (GMOs), which may require regulatory approval and public perception. In other cases, some of these off-target effects could result in undesired traits or unintended ecological consequences if the edited organism is released into the environment (Myhr, 2010).

In response to these concerns, researchers have developed high-fidelity Cas9 variants, such as SpCas9-HF1, eSpCas9, and HypaCas9, which exhibit significantly reduced binding to genomic sites. These variants have enhanced specificity and a lower chance of off-target effects, allowing for more precise gene editing and minimizing the potential for mishaps (Huang, Yang, Zhang, Xu, & Chen, 2022). In addition, computational tools have greatly improved to mitigate the risks of off-targets. Currently, in silico prediction tools for improving guide RNAs are routinely applied, which usually involve employing machine learning algorithms (Grešová, Alexiou, & Giassa, 2022). They enable the design of guide RNAs with higher specificity by predicting potential off-target sites and determining those based on genomic sequences that include partial homology to the target sequence. Although great progress has been made with these tools, it has become critical to validate CRISPR edits through whole-genome sequencing to ensure that the edits are accurate (Listgarten et al., 2018). However, whole-genome sequencing can be costly and resource-intensive; hence, it is not realistic for all breeding programs and certainly not for large-scale agricultural production. However, the development of high-fidelity CRISPR variants, predictive tools, and validation methods, which continue to improve the precision of gene editing, are gradually making gene editing a safer and more predictable technique for the genetic modification of plant and animal genomes (Y. Xu et al., 2012).

Inefficient Delivery Methods

One of the main challenges in the field of plant biotechnology today is the efficient delivery of gene-editing components to plant cells, and more specifically, to plant species with low transformation competence. For gene editing to be successful, systems such as CRISPR-Cas or other gene editing tools need to be effectively delivered into the plant genome (Kuluev et al., 2019). However, significant limitations are associated with the use of traditional delivery methods, such as Agrobacterium-mediated transformation and biolistic particle delivery. Centrifugation and treatment with pH conditions that would otherwise denature the virus are away with Agrobacterium-mediated transformation, which effectively uses the natural ability of Agrobacterium to transfer DNA into the cells of plants and is generally less efficient for monocots, such as rice and maize (Tripathi & Shukla, 2024). However, this restricted applicability creates a disincentive for monocot species, which are integral monocrops, to be expanded for thorough gene-editing. In contrast, biolistic particle delivery (also known as the gene gun method) is used for cereals and other recalcitrant crops. This technique involves shooting microscopic particles coated with DNA into plant cells; however, it causes substantial tissue damage, and a variable number of different transgenes usually makes it difficult for scientists to obtain stable, genetically edited plants (Schmidt, 2014).

The demand for more efficient and reliable gene-editing methods is growing, and emerging alternatives provide promising solutions. Active research exists in the area of nanoparticle-mediated delivery of gene editing components into plant cells via the use of materials such as carbon nanotubes or liposomes (Lv, Jiang, Chen, & Chen, 2020). These nanoparticles have the potential to enhance the delivery of genes and their functionality, as well as decrease some of the pitfalls related to classic gene transfer techniques, such as somatic clonal variation or genetic instability resulting from tissue culture and regeneration. Furthermore, viral vectors in which the viral genetic material is replaced by the 'foreign' genetic material and electroporation of protoplasts (plant cells without a cell wall) are being investigated as possible alternatives (Kujur, Senthil-Kumar, & Kumar, 2021). Although these methods can increase the efficiency of gene delivery, each has its own set of problems. For example, viral vectors and electroporation are often quite effective in one plant, but much less so in another plant species (Nayerossadat, Maedeh, & Ali, 2012). They can also differ substantially between the organs or tissues within the same plant. Moreover, although these emerging technologies promise much, their scalability for large-scale commercial applications is being investigated. Overcoming these challenges will require continued research and development on delivery methods to make gene editing a more efficient part of agricultural production for a larger number of plant species and systems of production (Breuninger & Lenhard, 2010).

Low Homology-Directed Repair (HDR) Efficiency

Precise gene insertions or allele replacements in a plant genome most often require the use of the homology-directed repair (HDR) pathway, a highly accurate DNA repair pathway used to incorporate specific genetic sequences. Nonetheless, most plant cells prefer to use the non-homologous end joining (NHEJ) pathway to repair double-strand breaks (DSBs); hence, HDR is inherently inefficient in most plant cells. NHEJ is a quick and error-prone repair method that results in random insertions or deletions (indels) that hinder precise editing, which may be required for applications such as custom genetic modifications or specific mutation correction. However, the degree of this imbalance between the two repair pathways greatly limits the ability to introduce site–specific genetic changes in many plant species (Molla, Shih, Wheatley, & Yang, 2022).

However, the efficiency of HDR in plant cells is still low, and several strategies have been developed to improve HDR efficiency, although they remain experimental and have not been widely adopted in other species. The synchronization of the plant cell cycle so that HDR active phases (S and G2 phases) occur during the time window of the desired frequency has the potential to be a practical strategy (Zhan, Lu, Zhu, & Botella, 2021). Researchers have attempted to ensure that plant cells are in the right stage of the cell cycle so that they are predisposed to perform HDR when they are given a DNA break. Alternatively, geminivirus replicons may be used to increase the availability of donor DNA templates for HDR by providing more material for the cell to work with. Moreover, chemical agents, such as DNA-PK inhibitors, which inhibit key factors of the NHEJ pathway, have been shown to be capable of reorienting repair hacia HDR (Smirnikhina, Zaynitdinova, Sergeeva, & Lavrov, 2022). These inhibitors can inhibit some of the NHEJ activity, making HDR preferable and increasing the chances of precise gene editing. Additionally, the combination of HDR-promoting proteins with editing reagents, such as the Cas9 protein or base editors, has also been tried to increase the probability of HDR integration (Shams et al., 2022).

However, challenges remain with these strategies. These approaches have been shown to have variable success in different plant species, and their applicability to a broad array of crops remains to be determined. Furthermore, the methods must be frequently fine-tuned for maximum efficiency, and there is still the possibility of unintended effects, including off-target mutations or incomplete repair (Vicente, Chaves-Ferreira, Jorge, Proença, & Barreto, 2021). Such an enhancement of HDR in plant systems has great promise for facilitating more fully controlled genetic modifications, but much work is required before these strategies can be widely utilized in agricultural biotechnology (Sedeek, Mahas, & Mahfouz, 2019).

Regeneration and Tissue Culture Bottlenecks

One of the most impactful bottlenecks in gene-editing workflows is the regeneration of whole plants from transformed cells or tissues in crops that lack efficient regeneration pathways in the first place. In the case of many economically important crops, such as legumes, woody perennials, and some cereals, the capacity to regenerate plants from cultured cells is dominated by the genetic background, and many varieties are recalcitrant to regeneration in vitro (Pathirana & Carimi, 2022). This limitation can be a roadblock to the application of gene-editing technologies and the development of genetically edited varieties; therefore, it is an important improvement target in plant biotechnology (Mohan, Easterling, & Yau, 2022).

Several innovative strategies have been considered to address this challenge. In one case, morphogenic regulators, including WUSCHEL, BABY BOOM, and PLT5, were overexpressed to stimulate plant cell differentiation and regeneration (P. Xu et al., 2024). Although the use of these regulators can greatly boost the regenerative capability of transformed cells to produce whole plants, their use normally comes at a price. Overexpression of these genes may confer unintended but potentially harmful changes in other traits (or pleiotropic effects) that may disturb the stability of the transformed plants or cause undesired phenotypic changes (Bidabadi & Jain, 2020). Furthermore, advances have been made in in planta transformation methods, enabling gene editing in the whole plant without using tissue culture. This method has great promise, especially in terms of application to species with poor tissue culture responsiveness, as it allows for more efficient and labor-intensive transformation processes (Altpeter et al., 2016).

Other approaches to this problem entail using specific explants with a higher regenerative potential, such as meristems or immature embryos. These tissues are often more receptive to signals needed in the process of regeneration, making them attractive for transformation and gene editing approaches (Bonga, 2017). Although these strategies have resulted in some improvements, the current regeneration efficiency remains a significant limitation to routine gene editing, especially in a broad germplasm. Although new methods can improve regeneration success, variation between species and genotypes prevents predictable outcomes. As a result, overcoming the regeneration bottleneck is critical for the full exploitation of these technologies for crop improvement and their wide dissemination to other plant species (Li et al., 2021).

Regulatory Hurdles

There are different regulations worldwide for gene-edited crops, which make it challenging for collaboration and trade at the international level. This rapidly evolving landscape brings with it a key issue regarding how to classify gene-edited organisms and, in turn, how such organisms should be subjected to regulatory oversight, biosafety assessment, and approval (Turnbull, Lillemo, & Hvoslef-Eide, 2021). In 2018, a landmark ruling by the European Court of Justice ruled that gene-edited organisms should be treated in the same strict manner as other genetically modified organisms (GMOs) in the European Union (EU) (Holman, 2019). Therefore, while this ruling has not rendered gene editing illegal in Europe, it has made a considerable difference to the speed with which gene-edited crops can reach the field and ultimately the shelves, because the lengthy and costly approval processes required to bring a GMO product onto the market can hold up commercialization (Marone, Mastrangelo, & Borrelli, 2023). Discussions are ongoing, and policy adjustments are being made; however, the regulatory burden is proving to be a major obstacle for the adoption of gene-editing technologies in European agriculture.

In contrast, the US has been more permissive toward gene-edited crops. Because they are not subject to foreign DNA, certain gene-edited crops have been exempted from regulation by the U.S. Department of Agriculture (USDA) to facilitate faster commercialization (Menz, Modrzejewski, Hartung, Wilhelm, & Sprink, 2020). This regulatory stance has allowed CRISPR-edited soybeans and mustard greens to be produced and sold without the rigorous oversight of GMOs (Arif, Jamil, Munir, Ahmad, & Munawar, 2024). However, the more relaxed regulatory environment has allowed for innovation, which has become more accessible to commercial enterprises in the US, while other countries such as China, Brazil, and Argentina have adopted case-by-case or product-based approaches to regulation, which allow for more flexibility to advance innovation with oversight. In these nations, a balance has been achieved, allowing technological progress in agriculture, as well as necessary safety assessments to protect public health and the environment (Ahmad et al., 2021). However, the lack of regulatory harmonization across regions hinders collaborative global efforts, trade, and market access in this field. In addition, the uncertainty in the regulatory classification of gene-edited crops may prevent investment in research and development by small companies or public sector organizations that may not be equipped to operate in an intricate and unpredictable regulatory setting. Such a pressing need arises as technology advances, and there is a need for global consensus and alignment on regulations to promote innovation, simplify the process of approval, and provide equitable access to the benefits of gene editing in agriculture (Brookes & Smyth, 2024).

Ethical, Social, and Public Perception Issues

Acceptance of gene-edited crops varies greatly across different areas, depending on a combination of cultural values, scientific literacy, and acceptance of the institutions that regulate agricultural technology. Studies also indicate that while consumers tend to have more positive views of gene editing than conventional genetic modification, this is especially true when no foreign DNA is inserted and the edits represent natural mutations. Moreover, awareness of gene editing is relatively low in the general population. Public education on the matter is both poor and minimal, leading to a number of misconceptions about the technology's safety, intentions, and possible environmental impact spreading far and wide (Lucht, 2015). However, there are some challenges associated with this technique, but an increasing understanding is shared among farmers and the populace that gene-edited crops, if regulated appropriately, can have many advantages, including increased crop resilience, raised nutritional content, and lowered requirement of chemical inputs, which can indeed help promote food security and environmental sustainability (Turnbull et al., 2021).

Effective strategies are needed to engage the public so that the community will accept more EV charging stations. One way to get around skepticism is through transparent, easily understandable communication of scientific evidence and the actual benefits of gene editing (Delhove, Osenk, Prichard, & Donnelley, 2020). This entails addressing safety issues, environmental hazards, and potential ethical problems associated with the technology. In addition, it is important that the policies and regulations on gene editing be discussed, including key stakeholders such as farmers, consumers and civil society. Incorporating these groups in decision-making processes would establish a more inclusive and democratic approach to ensure that technology fits social values and priorities (Dhirani, Mukhtiar, Chowdhry, & Newe, 2023). Additionally, messaging on gene editing needs to be context-sensitive such that it speaks to the import of the issue with respect to the specific concerns and values of each community and is presented as legitimate and credible from the perspective of accepted cultural norms and local norms towards biotechnology (Shukla-Jones, Friedrichs, & Winickoff, 2018).

Similarly, wider issues, such as the ethical debates surrounding intellectual property rights, control over seed systems, and the potential to cause socioeconomic disparities due to unequal access to this technology, are also being discussed (Paul & Mukhopadhyay, 2010). In particular, in developing countries, these concerns are most relevant, as the commercialization of gene-edited crops may further exacerbate inequalities if the commercial availability of seeds and technologies is restricted or monopolized. As such, robust governance frameworks must be created to ensure that gene-editing technologies are accessible to all, whether through small-scale farmers or corporates (Bain, Lindberg, & Selfa, 2020). Furthermore, embracing inclusive dialogue wherein the numerous viewpoints of all stakeholders, such as policymakers and grassroots organizations, are contemplated, will be fundamental in building a balanced and ethically sustainable way to deal with gene-altered harvests.

Environmental and Biodiversity Concerns

As gene-edited food crops are being deployed in agriculture, both opportunities and uncertainties exist and are still a major area of ongoing research regarding the ecological implications of such developments. Gene editing is often viewed as a more precise and possibly safer approach than the traditional methods of transgenesis when no foreign DNA is introduced, but there are also genuine fears about the way in which edited traits will play out in the environment (Sabat & Tripathy, 2024). A case in point is the unintended gene flow from a gene-edited crop to wild relatives or from a gene-edited cultivar to a non-edited cultivar. The alteration of natural plant populations due to this gene transfer might impact the stability of ecosystems and biodiversity (Schaart, van de Wiel, & Smulders, 2021). Traits such as herbicide resistance or drought tolerance that spread uncontrollably could provide a selective advantage that could alter current ecological balances or deprive the genetic diversity needed for species adaptation. Second, agroecosystems can disrupt the complex biological interactions that occur within an ecosystem (Kreiner, Stinchcombe, & Wright, 2018). The modified genes, which may provide gene-edited crop traits such as pest or disease resistance, might also impact non-target organisms such as beneficial insects and pollinators as well as natural pest predators (Han, Velasco-Hernández, Ramirez-Romero, & Desneux, 2016). Moreover, changes in plant traits can alter root exudates and nutrient cycling, affecting the composition and function of soil microbial communities. These would be unintended shifts in ecological dynamics that compromise ecosystem services essential for sustainable agriculture (Zhao et al., 2021). Moreover, the long persistence of gene-edited traits within the environment in the event that no post-release monitoring strategies are utilized is problematic. Some traits provide a selective advantage, and over time, these traits persist and proliferate, making containment and even risk management strategies difficult (Weeks et al., 2011).

Figure 1. Key challenges hindering the effective application of gene editing in plants include technical, regulatory, societal, and environmental barriers

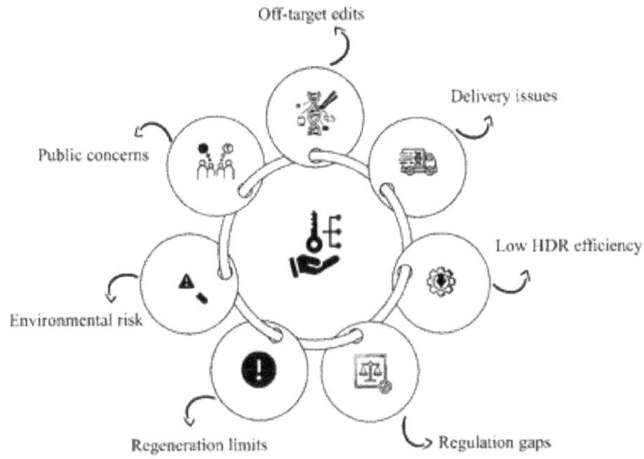

Despite these concerns, genes can be edited to improve crops so that they utilize fewer chemical inputs and are more resilient to the effects of climate change. However, for these benefits to be realized without jeopardizing ecological integrity, there must be complete, evidence-justification-based environmental risk assessments. The climate and severity of these assessments should be adapted to the crop species, the nature of the edited trait, and the local agroecosystem in which it is to be grown (KhokharVoytas et al., 2023). Before widespread deployment, rigorous evaluation of the factors affecting gene flow potential, interaction with local biota, and long-term ecological consequences is required. Ultimately, implementing ecological science into the research and regulation of gene editing will be essential to using gene editing for the good and preserving the good (Pascual, Rives, Schunter, & Macpherson, 2017).

FUTURE DIRECTIONS IN PLANT GENE EDITING

The technological landscape is being reshaped as the field of plant gene editing matures and expands. The goal, as with these other developments, is to raise precision and societal integration and lower regulatory and technical bottlenecks. This section investigates new and strategic directions for the future development of plant breeding, which are capable of changing the latter in the coming decades (Clapp & Ruder, 2020)(Figure 2).

Advancing Precision: Next-Generation Editing Tools

CRISPR–Cas9 systems have radically transformed plant biotechnology by allowing scientists to manipulate genes in an almost gene-specific (versus random) manner, but have a liability associated with their requirement of DSBs. Non-homologous end joining (NHEJ), the main cellular mechanism for repairing DSBs, is error-prone, with DSBs being repaired with insertions, deletions, or other undesirable mutations near the target site (Mahfouz, Piatek, & Stewart Jr, 2014). This can impair the predictability of genetic modifications, particularly when changing a single base is necessary to improve a trait or perform a functional genomics study. Next-generation gene-editing technologies have now emerged that make genetic alterations with high precision without DSBs, reducing the likelihood of off-target effects and improving editing specificity (Farooq, Hussain, Nazir, Javed, & Masood, 2018).

Turning a single nucleotide base to another without cleaving the DNA strands is one of the most transformative technologies, and base editing is one such technology (Eid et al., 2018). Base editors are generated by fusing a catalytically deficient Cas protein (often Cas9 or Cas12) with cytidine or adenosine deaminases, which enables defined A•T to G•C or C•G to T•A conversions (Anzalone, Koblan, & Liu, 2020). This method is particularly useful for correcting point mutations or adding specific changes to coding sequences and regulatory elements. Base editing does not rely on DSBs or donor templates, thus representing a safer and more predictable mechanism for fine-tuning genetic information, especially in species where HDR is inefficient or impossible (Y. Zheng et al., 2024).

Building on this innovation, prime editing technology presents even greater flexibility and elicitedness. Prime editors use a Cas9 nickase co-linked to reverse transcriptase and prime editing guide RNA (pegRNA), which specifies both the target site and the genetic alteration to be made. With fewer unintended effects than traditional CRISPR – Cas9 or base editing, this system can generate small insertions, deletions, and all twelve base-to-base conversions (Petrova & Smirnikhina, 2023). Nucleases are promising tools for correcting disease-associated mutations or adding complex trait changes in plants, especially those with polyploid genomes, as they can edit a range of complex and extensive edits without the need for DSBs or large donor templates (Cardi et al., 2023).

However, RNA and epigenome editing are new frontiers of gene regulation beyond DNA sequence alteration. By fusing RNA-targeting systems, including Cas13 variants or dCas9 fused to RNA-modifying enzymes, to transient and reversible RNA changes without perturbation of the underlying DNA, transient and reversible changes in transcript sequences are enabled (Yang & Patel, 2024). For example, parallel epigenetic editing tools that link catalytically dead CAS proteins (dCas9 or dCas12) to histone modifiers, DNA methyltransferases, or demethylases permit the precise modulation of gene expression without disrupting the genome (O'Geen et al., 2017). In particular, these approaches are appropriate for the regulation of genes controlling development, stress responses, or complex traits without inducing permanent genomic changes, and thus may offer regulatory relief and assuage public concerns (De Nadal, Ammerer, & Posas, 2011). Finally, transposon-based systems, such as CRISPR-associated transposases, are redefining the possibilities of gene insertion. These systems can insert large DNA sequences site-specifically without DSBs or homology-directed repair. This could be a revolutionary technology for trait stacking or for the stable integration of complex genetic circuits, allowing new approaches to the engineering of traits such as yield improvement, tolerance to abiotic stress, or metabolic enhancement in recalcitrant or genetically complex plant species (Cain et al., 2020).

These advanced platforms for gene editing represent a major advancement in precision plant biotechnology. However, it can override many of the traditional CRISPR–Cas system limitations, as well as extend the use cases possible for genome engineering in a safer, much more efficient, and flexible manner (Crauciuc et al., 2017). The growing maturity of these technologies offers the potential to increase the repertoire of tools for crop development conducive to nutritionally enhanced, climate-resilient, and sustainable food production (Gowdhaman, 2024).

Improving Delivery and Transformation Techniques

One of the major bottlenecks in plant genome engineering is the delivery of gene-editing reagents, especially in crop species that are not easily amenable to conventional transformation protocols (Atkins & Voytas, 2020). A major problem is that tissue culture is often genotype-dependent, labor-intensive, and can result in somaclonal variation. Accordingly, future goals in plant gene editing are increasingly focused on the invention of more environmentally friendly, less invasive, and regeneration-impartial supply methods that allow for focused and heritable edits in a wider variety of species, including those considered recalcitrant (Freitas-Alves et al., 2024).

Nanoparticle-based delivery systems are promising strategies. Different materials, such as carbon nanotubes, mesoporous silica particles, and DNA nanostructures, have been found to regulate the uptake of gene-editing complexes into plant cells in the absence of DNA integration or extensive genomic disturbance (Chandrasekaran, Rajiv, & Abd-Elsalam, 2020). This property makes these nanomaterials especially attractive for crops with complex or unresponsive genomes, which can be engineered to have high biocompatibility and targeted delivery, reducing their effect on the environment. This is a technological leap for precision agriculture, as these properties allow them to penetrate the plant cell wall and membrane while individually minimizing off-target effects (Ahmar et al., 2021).

An innovative approach is the use of plant viral vectors, such as Gemini viruses or tobacco rattle virus (TRV) (Shi et al., 2021). These vectors also function as transient expression platforms for the systemic delivery of CRISPR-Cas systems to plant tissues, including meristems, where heritable edits can be generated. Here, this approach of transient transformation eliminates the need to regenerate plants from transformed tissues and thus saves the time and resources associated with the generation of edited lines (Chuang et al., 2021). Additionally, because viral vector expression is transient, the issue of stable transgene integration is not a problem.

Floral dip and still-developing methods, such as pollen magnetoception, have the potential to undergo in vitro regeneration to be leveraged in planta transformation techniques. These methods aim to directly introduce gene-editing reagents into reproductive tissues for the direct production of edited progeny. These techniques are being actively adapted for other difficult-to-transform plant groups, including cereals, legumes, and woody perennials (Bélanger, Copley, Hoyos-Villegas, Charron, & O'Donoughue, 2024). Finally, protoplast editing and regeneration-free editing strategies are being investigated to limit the risk of culture-induced mutations and simplify the editing procedure. The use of these methods generally entails the transient introduction of gene-editing ribonucleoprotein (RNP) complexes into isolated plant protoplasts and subsequent selection of edited cells using fluorescent or luminescent reporters (Reed & Bargmann, 2021). However, regeneration from protoplasts remains a challenge for many species, and progress in direct regeneration and screening is making these methods increasingly practical.

These advances have led to a change in the direction towards easier and more efficient gene-editing pipelines in plants. Their potential to overcome all of these delivery and regeneration limitations is that these plants have the ability to accelerate the development of improved crop varieties, particularly high-value or those of critical importance to the food security of the area's most at risk of climate change (Vats et al., 2019).

Expanding the Range of Target Crops and Traits

To date, most agricultural gene-editing research has focused on globally important staple crops, such as rice, maize, and wheat. Although this focus has resulted in a number of important gains in yield, pest resistance, and tolerance to stress in these crops, it has also caused the neglect of a number of underutilized and orphan crops that are important for regional food security and livelihoods in developing countries (Vats et al., 2019). As many of these species are neglected, it is important to expand gene-editing efforts to these species to support climate resilience, nutritional diversity, and local conditions of sustainable agricultural systems (El Chami, Daccache, & El Moujabber, 2020).

Millions of people in arid and semi-arid regions rely on underutilized crops such as teff, millet, sorghum, cassava, and yam in their diets and economies. However, little formal genetic improvement has occurred in these species compared to that in major cereals (Mijena, Alamerew, Assefa, & Nigusse, 2024). Gene editing allows for a powerful and time-effective enhancement of economically desirable traits in such crops, including drought and heat tolerance, extended shelf life, higher protein content, and pest and disease resistance in region-specific ecosystems. Gene editing can help realize the full potential of these crops to thrive in marginal environments, helping to overcome key agronomic constraints and contributing to resilience in the food system (Kumar et al., 2021).

Gene editing promises to be of even greater application for the improvement of forest and horticultural species, such as a wide variety of fruit trees and ornamental plants, beyond staple and 'neglected' crops. Because these species typically have long generation times and are difficult to manipulate using traditional breeding methods, monitoring is insufficient (Ramirez-Torres et al., 2021). Gene editing can enable the introduction or adjustment of traits related to the resistance of fungal and viral pathogens, improved fruit quality, enhanced nutritional properties, increased shelf life, and even changes in aesthetic properties, such as color or fragrance, in ornamentals. This would revolutionize crop cultivation and the commercial value of many perennial crops and reduce post-harvest losses and pesticide use (Van Esse, Reuber, & van der Does, 2020).

Alternatively, C4 pathway engineering is an emerging cutting-edge direction of crop improvement for improving the photosynthetic efficiency of C3 plants, such as rice. Crops such as maize and sorghum, which naturally have the C4 photosynthetic pathway, have the advantages of water-use efficiency and carbon fixation under high temperatures and light intensity (Majumder, Datta, & Datta, 2023). This would offer the possibility of introducing key regulatory elements of the C4 pathway through precise gene editing into C3 crops, and therefore potentially increase productivity under climate-stressed conditions. However, a successful venture in this area would be technically demanding but potentially transformative for more sustainable and higher-yielding agriculture (H. Cui, 2021).

By expanding gene-editing applications beyond a selected group of crops and qualities, this innovation will have a broader and more significant impact on worldwide agriculture. By improving the genetics of neglected species of local and regional importance, biodiversity conservation is embraced, and technological advances are made with the actual needs of many agroecological zones and communities in mind (Qaim, 2020). Given global challenges such as global environmental and socio-economic challenges, building equitable and climate-resilient food systems requires an inclusive approach.

Multi-Omics Integration and AI-Driven Target Design

The future of precision agriculture is characterized by the seamless integration of genomic data with advanced computational technologies that can be utilized in real time. This convergence is radically changing the understanding of plant biology, identifying target genes, and generating efficient gene-editing strategies (Kumlehn, Pietralla, Hensel, Pacher, & Puchta, 2018). This innovation is based on the idea of multi-omics integration of data from biological layers, such as genomics, transcriptomics, proteomics, metabolomics, and phenomics, to present a systems-level view of gene function and trait architecture. Collectively analyzing these complex data allows scientists to identify causal genetic loci linked to favorable traits, understand how genes and the environment interact, and avoid the effects of undesirable traits that can occur from genetic redesign (Fan, Chen, Chen, & Guo, 2025).

Figure 2. Emerging directions in plant gene editing have focused on improving precision, delivery, crop range, data integration, and global accessibility.

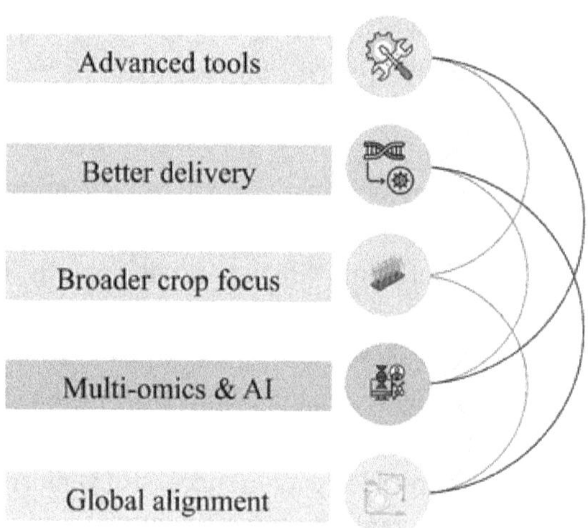

AI and ML are increasingly assumed to play an important role in the implementation of gene editing workflows, similar to their emergence in other application areas. Predicting optimal guide RNA sequences with high specificity and low off-target potential reduces the likelihood of unintended genomic modifications, and these tools are utilized for this purpose (Listgarten et al., 2018). More recently, ML algorithms have been able to simulate phenotypic outcomes from genetic modifications and identify which edit(s) are most likely to be successful in obtaining favorable agronomic traits. In addition, AI is used for the optimal design of regulatory elements, including promoters and enhancers, to fine-tune gene expression at the desired level measured in performance under certain environmental conditions (Chen, Hao, & Sun, 2025).

Advanced bioinformatic platforms such as CRISPR P, Deep CRISPR, and many ML-based editing outcome predictors are becoming inevitable components of modern genome editing. These tools improve the overall success rates of gene editing projects and facilitate the entire process from target selection to outcome prediction by reducing the need for trial and error in gene editing projects through experimental efforts. This is particularly important when dealing with complex traits whose expression is controlled by polygenic networks and complex regulatory pathways, where traditional approaches fail (Dixit, Kumar, Srinivasan, Vincent, & Ramu Krishnan, 2024).

However, with the biological depth of multi-omics tied together with the richness of AI and ML predictive analysis, researchers are ushering in a new scientific age of highly targeted data-driven plant breeding. This integrated approach not only facilitates the faster development of superior crop varieties, but also provides a broader goal of sustainable agriculture by improving productivity, increasing resilience to climate change, and reducing resource wastage. However, these technologies will mature and become a prerequisite for the advent of next-generation agricultural innovation (Cembrowska-Lech et al., 2023).

Regulatory Harmonization, Education, and Global Collaboration

To ensure that plant gene-editing technologies are a long-term success and are applied equitably, more than just scientific breakthroughs are required: robust governance, inclusive capacity building, and proactive public engagement are also essential (Selin, Friedman, Lambert, Farooque, & Guston). As cutting-edge research continues to improve the technological frontiers of gene editing, the broader acceptance and accessibility of these innovations globally will depend on how well regulatory systems, education frameworks, and institutional collaborations are aligned with the goal of responsible and inclusive uptake (Ahlawat et al., 2024).

Harmonization of the regulatory framework across countries is a critical component. International bodies such as the Food and Agriculture Organization (FAO), the Organization for Economic Cooperation and Development (OECD), and the

Convention on Biological Diversity (CBD) are affecting a meeting of minds and developing consistent guidelines for assessing and classifying gene-edited products (Keiper & Atanassova, 2020). An increasing number of nations are moving away from regulating a certain method to creating a product and figureuring out the nature and impact of that product. This approach is less restrictive than the others and encourages innovation, but will provide a level of safety that is commensurate with current safety standards and is critical for enabling cross-border trade and collaboration in agricultural biotechnology (Tizard et al., 2016).

Widespread educational and outreach efforts are equally important and involve all key stakeholders, policy practitioners, farmers, educators, students, and consumers. Confusion with earlier generations of genetically modified organisms (GMOs) and misinformation contribute to the public perception of gene editing and may result in resistance to the technology, even though science is able to present a clear benefit (Lassalle & Jarocki, 2016). Thus, clear and sustainable communication is needed to differentiate between gene editing and conventional GMO, discuss risks and benefits, and address ethical and ecological implications. Continuous, inclusive two-way communication can enhance dialogue with the public and build public trust in gene-editing technologies (Agapito-Tenfen, Okoli, Bernstein, Wikmark, & Myhr, 2018).

To democratize access to gene-editing tools, including guiding the democratic distribution of innovations to address local agricultural challenges, capacity building in the Global South is essential. However, inadequate research infrastructure, limited funding, and an inadequate number of skilled personnel present barriers to many low- and middle-income countries (Montenegro de Wit, 2020). Investments in such facilities can enable local institutions to take the lead in their own gene-editing initiatives using local expertise, and local institutions will have trained personnel and facilities that will create regionally important germplasm for their regions (Kofler et al., 2018). It helps make science self-sufficient by using locally driven priorities and enables solutions that are applicable and drive solutions according to local food security needs.

Public-private partnerships (PPPs) have also emerged as efficient and powerful mechanisms for accelerating product development in line with the principles of transparency, inclusivity, and accessibility. Gene-edited crops could be brought to market more efficiently by combining the innovation capacity of the private sector with the public sector's commitment to equity and long-term societal benefits provided through PPPs (Batan, 2025; Marx, 2019). Shared innovation models on open-source platforms, collaborative breeding networks, and pre-competitive research alliances already show how shared innovation can serve as responsible technological advancement that does not leave smallholder farmers and underrepresented regions behind (Yuen, 2017).

These non-scientific dimensions of regulation, education, capacity building, and collaboration build the basis from which the responsible deployment of gene editing in agriculture can be undertaken together (Macnaghten & Habets, 2020). Finally, gene editing is destined to play a role in ensuring that technology contributes to sustainable food systems, climate resilience, and equitable development (Table 1), but in doing so, it will depend on its integration into global innovation strategies.

Table 1. Overview of Challenges, Innovations, and Future Directions in Plant Gene Editing

Category	Key Aspects	Current Challenges	Emerging Innovations & Solutions	Future Directions	References
Technological Tools	CRISPR-Cas, Base Editing, Prime Editing	Off-target mutations Limited precision Low HDR efficiency	AI-guided sgRNA design Use of base/ prime editors with higher fidelity	Development of multi-tool editing platforms integrating editing precision with real-time feedback	(Anzalone et al., 2020)
Delivery & Transformation	Methods to deliver gene editing components (e.g., via Agrobacterium, biolistics, nanoparticles)	Low transformation efficiency Limited host range - Cellular damage	Nanoparticle-mediated delivery Electroporation Virus-based vectors	Development of in planta transformation systems for recalcitrant species	(Gad, Li, Ahmed, & Almoammar, 2020)
Regulatory Landscape	Regional and international biosafety and approval frameworks	Lack of harmonized guidelines, Delayed approval processes	Case-by-case evaluation frameworks (e.g., SDN1, SDN2, SDN3)	Creation of globally aligned policies recognizing gene-edited crops separately from GMOs	(Hock, Kian, & Wah, 2020)
Ethical and Social Concerns	Public perception, ethical debates, and societal acceptance	Misinformation and fear - Resistance due to ethical and ecological concerns	Transparent communication strategies Stakeholder engagement programs	Inclusive policy dialogues and global education initiatives	(Sadikiel Mmbando, 2024)

continued on following page

Table 1. Continued

Category	Key Aspects	Current Challenges	Emerging Innovations & Solutions	Future Directions	References
Applications in Agriculture	Use in crop improvement, disease resistance, abiotic stress tolerance, and nutritional enhancement	Limited field validation Genotype dependency of protocols	Case studies on crops (e.g., wheat, rice, lettuce) Multi-trait engineering	Large-scale field trials and scaling for commercial agriculture	(Gonzalez Guzman, Cellini, Fotopoulos, Balestrini, & Arbona, 2022)
Capacity Building & Access	Training, infrastructure, and equitable access, especially in Global South	Research disparity between developed and developing regions	International collaborations Open-source platforms	Investment in local biotech infrastructure, training centers, and knowledge sharing	(Chan & Costa, 2005)

CASE STUDIES

However, all recent laboratory breakthroughs with CRISPR and other gene-editing technologies still need to be validated for the field and enter the market. In this section, detailed case studies are presented that demonstrate the translation of gene editing into viable crop-improvement strategies. These examples illustrate the technical, regulatory, and socioeconomic challenges of taking gene-edited crops from the laboratory to the field.

CRISPR-Edited Wheat Trials in Australia: Toward Climate-Resilient Cereal Product

Climate variability, including prolonged droughts and rising temperature extremes, is increasingly challenging Australia's agricultural sector. Cereal crops are significantly affected by these environmental stressors, which have a negative impact on cereal crop productivity, posing a threat to national food security and export revenue (Bakala, Singh, & Srivastava, 2020). As a result of these challenges, a collaboration between the leading Australian cereal breeding company InterGrain and the U.S.-based biotech firm Inari Agriculture was launched. In 2024, this CRISPR-edited wheat variety field trial partnership was launched to carry out testing with the aim of improving drought resistance and increasing grain yield. The initiative also aimed to test the use of this latter class of molecular scissors as a sustainable and rapid approach to breed climate-resilient wheat cultivars (Kaur, Urban, Hammond-Kosack, & Halford, 2024).

Using a multiplexed CRISPR-Cas9 genome editing approach, specific YS1 and resistance gene regulatory regions were edited to precisely change the expression of key genes responsible for yield and stress resilience (Rai et al., 2023). TaGW2 and TaCKX were target genes, the former having been reported to affect grain weight and the latter being related to cytokinin metabolism and spikelet development. The team was able to create non-transgenic wheat lines by fine-tuning the expression levels of these genes without the addition of foreign DNA (Geng et al., 2017). The edits were precise, stable, and heritable, meaning that desired traits could be maintained over many generations. Simultaneous improvement of multiple yield components using multiplex editing has greatly accelerated the breeding process compared with traditional methods. Field trial results from the initial trial proved very promising, being carried out across diverse environments in Western Australia and New South Wales (Prober & Smith, 2009; Richards, Hunt, Kirkegaard, & Passioura, 2014). CRISPR-edited wheat lines produced grain quality traits similar to those of the controls, including protein content and milling performance, and an increased yield of 10–12% over the controls. Furthermore, these lines had higher water-use efficiency and the ability to maintain yield performance during mild drought stress, which is a necessary trait for production in increasingly arid areas of Australia. This provides practical evidence of the capacity of gene editing to address agronomic complexity in the field (Ross, Alami, Harvey, Achouak, & Ryder, 2000).

As a critical advantage of this approach, the lines could be classified as non-genetically modified organisms (non-GMO) lines under Australia's Office of the Gene Technology Regulator (OGTR) regulatory framework, given the absence of transgene insertion (Bertheau, 2012). This classification accelerated the approval process for field testing and permitted more agile commercial planning. However, while there are such regulatory benefits within Australia, they are not enough to transform the domestic acceptability of gene-edited crops; uncertainty persists regarding the global acceptability of these crops, especially in the European Union, whose regulatory stance is more restrictive. Thus, this could represent a potential barrier to international trade and emphasizes that more harmonized international regulations across the globe are needed that are in line with the scientific consensus on gene editing (Spök, Sprink, Allan, Yamaguchi, & Dayé, 2022).

This case study highlights the potential of genome editing using CRISPR-based methods to solve many of today's agricultural challenges related to climate change. Agronomic improvements in a relatively short timeframe have been achieved through non-transgenic, precise targeting of complex traits, such as yield and drought tolerance. In addition to technical success, the initiative reflects the broader consequences of regulatory policy on innovation and market access. The project itself is a model for how science-driven policy frameworks, cross-sector collaboration, and international cooperation are necessary for realizing the full potential of gene editing to benefit global food systems.

Commercialization of CRISPR-Edited Mustard Greens by Pairwise: A Consumer-Centric Breakthrough

A small agricultural biotechnology company called Pairwise, based in the U.S., embarked on a consumer-oriented mission to improve the taste of nutrient-dense vegetables to make healthy eating more appealing (Hess, 2009). The company realized that one reason flavor often stands in the way of healthier diets becoming widespread is adopted, so they concentrated their sights on sorrel mustard greens (Fubini, 2024). Although mustard greens are highly nutritious, they are under-consumed because their naturally bitter taste results from the essentially bitter components produced by the glucosinolate biosynthesis pathway. With this in mind, Pairwise aimed to use gene editing to counteract the bitterness trait in the plant without compromising its aggregate benefits (health, visual appeal, and other agronomic performance) (Wieczorek, Walczak, Skrzypczak-Zielińska, & Jeleń, 2018).

For this, the pairwise employed CRISPR Cxas12a genome editing system, which has the capacity to perform very high precision and efficient targeted gene knock-outs (Gier et al., 2020). In particular, the company aimed to target genes, including MYB28 and AOP2, which are important nodes in the glucosinolate biosynthesis pathway that contribute to the accumulation of bitter-tasting compounds. Pairwise successfully suppressed the biosynthesis of glucosinolates, which impart the sharp, pungent taste of mustard greens, by knocking out these genes. Significantly, gene editing did not require the addition of any foreign DNA into the resulting plant; therefore, these plants were non-transgenic (Burow et al., 2015; Yatusevich, 2008). Therefore, the company was able to maintain the desirable characteristics of the crop, such as growth vigor, color, and nutritional content, and massively improve its sensory characteristics.

This milestone in consumer food innovation was the gene – edited product branded as Conscious Greens. First launched in choice U.S. dining establishments in 2023, the greens provided chefs and diners with a new leafy plant with a moderate, pleasant preference profile (Toschka, 2025). Following positive reception in the food service industry, Pairwise expanded its distribution to grocery stores in the Pacific Northwest by 2025 (Ryan et al., 2025). The greens were also non-GMO, thanks to the absence of foreign genetic material, and could be certified under U.S. Department of Agriculture guidelines, making them easier to navigate through regulatory channels and put consumers' minds at ease over genetic modification (Marden, 2002). Production-oriented traits to consumer preference were no longer a matter of one at a time, and if gene editing is to be used here, then the rollout strategy demonstrated how gene editing could be used to target consumer preferences directly (Hu, House, & Gao, 2022).

Additionally, companies ran educational campaigns to explain the unique nature of CRISPR technology compared to traditional genetic modification and to explain why there is no foreign DNA in their products (Gostimskaya, 2022). Although initial consumer surveys indicated high consumer satisfaction with the flavor and texture of Conscious Greens, the brand was unable to find substantial retail outlets willing to stock the product (Barrett, Beaulieu, & Shewfelt, 2010). Intriguingly, as the highest rate of acceptance and curiosity about gene-edited foods was reported from younger demographics, especially health-conscious millennial and Gen Z consumers, this generation could prove crucial in normalizing these innovations.

Gene Editing in Cassava and Yam: Toward Food Security in Sub-Saharan Africa

Staple crops, cassava and yam, are critical staples upon which the food security of millions in Sub-Sahara Africa is reliant. Both crops are very important but are affected by some long-standing issues that hinder their productivity and value (Kavhiza et al., 2022). While cassava is highly susceptible to viral infections (e.g., cassava mosaic disease and cassava brown streak virus), yam is prone to post-harvest physiological deterioration and has a short shelf life, and banana is susceptible to black Sigatoka, Panama wilt, banana bunchy top virus, and several other diseases. Similar to flour crops, both flour and potatoes contain relatively low amounts of protein, thereby affecting the nutritional value of the crops (Chikoti & Tembo, 2022). These crops often have biological complexities, including long reproductive cycles, polyploidy, and high heterozygosity, which constrain traditional breeding approaches using these crops. However, the application of gene editing, particularly CRISPR-based technologies, has emerged as a promising tool to overcome these barriers, offering a faster and more precise alternative to conventional methods, as well as rapid improvement of traits that are of great importance to food security and nutrition (Begna, 2021).

In this regard, underutilized crops have lagged far behind better-studied crops, but recent advancements stemming from international collaborations have begun to open the possibilities for gene editing in these crops (Hossain et al., 2021). Cassava is one of the three orphan crops that the African Orphan Crops Consortium and the International Institute of Tropical Agriculture are focusing on using CRISPR-Cas9 to knock out specific susceptibility genes. Targeted edits have made cassava more resistant to CBSV, one of the most devastating viral diseases of cassava production in East and Central Africa. In parallel, gene editing work on yam has been conducted to mitigate postharvest deterioration of the yam by editing the starch synthesis pathway (Tadele, 2019). Researchers have modified genes involved in enzymatic browning and tissue breakdown to prolong the shelf life of yams and minimize post-harvest losses, an important issue for marketing yams where cold storage is limited in rural markets. These constitute some of the earliest examples of how precision breeding can be used to address crop-specific constraints that have limited productivity and sustainability in the region in the past (Raju, 2021).

Seeing that advances in technology cannot exist without local expertise, these gene-editing initiatives have included comprehensive capacity-building components. IITA and its partners are leading programs that have trained African scientists, technicians, and regulators in genome editing techniques, bioinformatics, and biosafety assessment. However, locally targeted training is essential for gene editing to be implemented and managed sustainably within Africa's agricultural systems. However, uncertainty remains regarding the regulatory bottleneck for widespread adoption. Even though some African countries are still putting together their frameworks for genome-edited crops, a regional leader has arisen in the form of Nigeria, which has allowed gene-edited crops to be taken into field testing under simplified guidelines that distinguish them from transgenic GMOs. This regulatory clarity has hastened the review of edited cassava and possibly influenced other countries in the region (Komen et al., 2020).

Although promising technical outcomes have been demonstrated in laboratory and early field trials, there are a number of interrelated hurdles for the wider adoption of gene-edited cassava and yam. Second, there is a low level of public awareness and understanding of gene editing, which may also impact trust and consumer acceptance (Norman, Dzidzienyo, Karim, & Beah, 2021). Second, many national research programs do not have easy access to affordable, scalable transformation, and regeneration technologies to independently develop and deploy edited varieties (Ricroch & Hénard-Damave, 2016). Third, the harmonization of regional policies regarding health research, movement of seeds, and biosafety oversight is crucial to enable cross-border health research collaborations and coordination in biosafety oversight and import policies (Maehira & Spencer, 2019). However, innovations may stall at the national level when there are no cohesive policies at the level of the African Union or regional economic unions.

The case of modern biotechnology for gene editing of cassava and yam presents a compelling potential for addressing the needs of crops that have otherwise been marginalized in the global research agenda. Not only do These crops represent regional food sovereignty and are also significant to African communities in terms of their cultural and economic value. If successful, gene editing would be a major milestone in achieving equitable agricultural innovation for these crops. This shows that with the right combination of technical expertise, policy support, and public engagement, biotechnology can be a powerful tool for achieving food security and resilience in the developing world.

Conscious Greens represents a significant shift in the application and reception of this type of gene editing in the food industry. This product is different from most gene-edited crops, which have usually been tailored for agronomic traits, such as pest resistance or yield, but was inspired by the consumer, improving the taste directly. Moreover, this approach does not only add value at the point of retail, but it can also contribute to a change in public discourse related to genetic technologies in food, as indirect diffusion can perpetuate the aura of science in media and everyday life (Ishii & Araki, 2016). If addressed in a way that meets tangible consumer needs and preferences, CRISPR-based innovations of this nature can help biotechnology overcome the divide with the public (van den Boomgaard, 2021). Additionally, the case of Pairwise offers a new model for biotech commercialization outside commodity chains, and potentially bypassing consumers directly will be a model for a more personalized and sustainable approach to food solutions will be followed in the future.

Tomato Engineering for Disease Resistance: A Case of Rapid Trait Deployment

As one of the world's most cultivated and consumed horticultural crops, tomatoes are threatened by a variety of devastating infectious diseases, such as bacterial spot, late blight, and tomato yellow leaf curl virus (TYLCV). In addition to reducing the yield and quality of fruits, these diseases also increase chemical pesticide dependence and raise environmental and economic problems. Given climate variability and globalized pathogen spread, there is an urgent need to develop disease-resistant tomato varieties (Panno et al., 2021). In this respect, gene editing, especially CRISPR-based technologies, has proven to be quick and powerful in responding to the challenges that are changing rapidly. An avenue of interest for durable resistance to pathogens in tomatoes is the targeted modification of genes that, when disrupted, confer susceptibility (Mushtaq et al., 2021).

In what is thought to be the first-time, scientists at the University of Florida and the Boyce Thompson Institute have used CRISPR-Cas9 to engineer defense genes in tomato plants (Tariq, Gul, Negri, & Mujtaba, 2024). The SlMlo1 gene is a key target, and a susceptibility (S) gene disrupted in which conferred resistance to powdery mildew, a serious fungal disease infecting field and greenhouse tomato crops (Z. Zheng et al., 2016). The heritably edited lines displayed strong resistance to multiple pathogens, thus showing a rare example of the development of multi-pathogen resistance by a single suite of gene edits. Most importantly, these modifications did not give rise to any unintended or deleterious phenotypic effects in terms of growth, flowering, or fruit set, demonstrating the accuracy and safety of the gene-editing strategy (Bate, Dardick, de Maagd, & Williams, 2021).

CRISPR-edited tomato lines were robustly and consistently resistant to controlled greenhouse and initial open-field testing. These results suggest the possibility of CRISPR-based breeding generating intense reductions in crop loss and chemical inputs. However, in moving towards the large-scale deployment of CIS-edited tomato plants, researchers are simultaneously engineering CIS-genic and marker-free CIS-edited tomato plants (Liu et al., 2025). This strategy entails removing selection markers and guaranteeing that only endogenous tomato sequences are retained, which could greatly relieve regulatory issues in areas with strict GMO legislation. These teams tailor the final product to work with all regulatory frameworks so that the product can be distributed on a global scale while maintaining safety standards for consumers and the environment (Bernauer, 2003).

Conventional breeding takes approximately 8–10 years to develop a disease-resistant variety, as it requires repeated crossings and selection based on phenotype. This is compressed into two to three years with gene-editing. Such an acceleration holds great significance for food security, particularly in parts of the world where tomatoes are the staple diet or a significant source of income for smallholder farmers (Khan, Hassan, & Khan, 2020). These tomato gene-editing projects lay the basis for looking forward to the extension of this model of plant breeding to other horticultural crops, marking the beginning of a new era of precision plant breeding for a changing world (Buch et al., 2023).

CONCLUSION

Plant gene editing is one of the most transformative tools in the field of agricultural biotechnology and a precision form of crop improvement. Technologies such as CRISPR-Cas systems, base editing, and prime editing could take us a long way toward solving some of the most critical challenges of our century, which range from climate-induced yield instability and strain of pests to hidden hunger and degradation of the environment. In this chapter, we have described the evolution and use of gene-editing tools and the diverse technical, regulatory, ethical, and environmental challenges that continue to steer their path. Although significant progress has been made, there are still quite a few bottlenecks. Efficient transformation and regeneration protocols, lack of homology–directed repair rates at high frequency, and off-target activity for some of these systems limit the application of these platforms to a broad set of crop species. Regulatory disparities and sociopolitical skepticism impede global harmonization and equitable access to gene-edited technologies. Things are still looking somewhat optimistic for the future, if not for the present day. Until recently, the regeneration hurdle was overcome by advances in delivery systems, including particle-based carriers and in planta transformation techniques. Much like the merging of AI, machine learning, and multi-omics, the acceleration of designing precise and efficient editing schemes will allow scientists to forecast and optimize targets for editing with unprecedented speed.

However, this scientific advancement must be accompanied by the evolution of the regulatory and ethical landscape at the same time. Regulations regarding biosafety need to be clear, science-based, harmonized, transparently communicated to the public and stakeholders, and able to establish public trust. Capacity building in regions that lack resources, especially in the Global South, can also play a role in avoiding a widening technological divide, as well as to support local-driven innovation to promote food security. Wheat trials in Australia and commercial products in the form of new mustard greens and disease-resistant tomatoes fully demonstrate the potential and realities of implementing CRISPR in the real world. This chapter demonstrates how gene editing can be used to achieve both higher agricultural yields and crop suitability for society's changing demands.

Therefore, the future of plant gene editing depends not only on improving the science but also on promoting collaboration between agricultural scientists (molecular biologists and agronomists) and natural scientists (ecologists, public policy, and the public). At this moment, we are at the crossroads of innovation and responsibility; collectively, we must ensure that these tools are used for purposes that are not only technological but also ethically, ecologically, and geographically inclusive.

REFERENCES

Abdallah, N. A., Prakash, C. S., & McHughen, A. G. (2015). Genome editing for crop improvement: Challenges and opportunities. *GM Crops & Food*, *6*(4), 183–205. DOI: 10.1080/21645698.2015.1129937 PMID: 26930114

Acquaah, G. (2015). Conventional plant breeding principles and techniques. *Advances in plant breeding strategies: Breeding, biotechnology and molecular tools*, 115-158.

Agapito-Tenfen, S. Z., Okoli, A. S., Bernstein, M. J., Wikmark, O.-G., & Myhr, A. I. (2018). Revisiting risk governance of GM plants: The need to consider new and emerging gene-editing techniques. *Frontiers in Plant Science*, *9*, 1874. DOI: 10.3389/fpls.2018.01874 PMID: 30622546

Ahlawat, U., Naruka, A., Changdeo, W. B., Rehsawla, R., Sansanwal, R., Mishra, R., & Singh, V. (2024). A review of cutting-edge Biotechnological solutions for next-generation farming. *J Exp Agric Int*, *46*(7), 1187–1204. DOI: 10.9734/jeai/2024/v46i72671

Ahmad, S., Shahzad, R., Jamil, S., Tabassum, J., Chaudhary, M. A. M., Atif, R. M., & Sheng, Z. (2021). Regulatory aspects, risk assessment, and toxicity associated with RNAi and CRISPR methods. In *CRISPR and RNAi systems* (pp. 687–721). Elsevier. DOI: 10.1016/B978-0-12-821910-2.00013-8

Ahmar, S., Mahmood, T., Fiaz, S., Mora-Poblete, F., Shafique, M. S., Chattha, M. S., & Jung, K.-H. (2021). Advantage of nanotechnology-based genome editing system and its application in crop improvement. *Frontiers in Plant Science*, *12*, 663849. DOI: 10.3389/fpls.2021.663849 PMID: 34122485

Altpeter, F., Springer, N. M., Bartley, L. E., Blechl, A. E., Brutnell, T. P., Citovsky, V., & Kausch, A. P. (2016). Advancing crop transformation in the era of genome editing. *The Plant Cell*, *28*(7), 1510–1520. DOI: 10.1105/tpc.16.00196 PMID: 27335450

Anders, S., Cowling, W., Pareek, A., Gupta, K. J., Singla-Pareek, S. L., & Foyer, C. H. (2021). Gaining Acceptance of Novel Plant Breeding Technologies. *Trends in Plant Science*, *26*(6), 575–587. DOI: 10.1016/j.tplants.2021.03.004 PMID: 33893048

Anzalone, A. V., Koblan, L. W., & Liu, D. R. (2020). Genome editing with CRISPR–Cas nucleases, base editors, transposases and prime editors. *Nature Biotechnology*, *38*(7), 824–844. DOI: 10.1038/s41587-020-0561-9 PMID: 32572269

Arif, A., Jamil, A., Munir, A., Ahmad, A., & Munawar, N. (2024). Regulatory, ethical, and social aspects of CRISPR crops. In *CRISPRized Horticulture Crops* (pp. 439–470). Elsevier.

Atkins, P. A., & Voytas, D. F. (2020). Overcoming bottlenecks in plant gene editing. *Current Opinion in Plant Biology*, *54*, 79–84. DOI: 10.1016/j.pbi.2020.01.002 PMID: 32143167

Bachtarzi, H., & Farries, T. (2019). The genetically modified organism medicinal framework in Europe, United States, and Japan: Underlying scientific principles and considerations toward the development of gene therapy and genetically modified cell-based products. *Human Gene Therapy. Clinical Development*, *30*(3), 114–128. DOI: 10.1089/humc.2019.042 PMID: 31111736

Bain, C., Lindberg, S., & Selfa, T. (2020). Emerging sociotechnical imaginaries for gene edited crops for foods in the United States: Implications for governance. *Agriculture and Human Values*, *37*(2), 265–279. DOI: 10.1007/s10460-019-09980-9

Bakala, H. S., Singh, G., & Srivastava, P. (2020). Smart breeding for climate resilient agriculture. In *Plant breeding-current and future views*. IntechOpen.

Barrett, D. M., Beaulieu, J. C., & Shewfelt, R. (2010). Color, flavor, texture, and nutritional quality of fresh-cut fruits and vegetables: Desirable levels, instrumental and sensory measurement, and the effects of processing. *Critical Reviews in Food Science and Nutrition*, *50*(5), 369–389. DOI: 10.1080/10408391003626322 PMID: 20373184

Batan, A. (2025). Assessing the Role of Public-Private Partnerships in Advancing Sustainable Development Goals Through Agri-Tech and Renewable Energy Solutions. *Transactions on Social Innovation, Digital Inclusion, and Ethical AI*, *15*(2), 1–11.

Bate, N. J., Dardick, C. D., de Maagd, R. A., & Williams, R. W. (2021). Opportunities and challenges applying gene editing to specialty crops. *In Vitro Cellular & Developmental Biology. Plant : Journal of the Tissue Culture Association*, *57*(4), 709–719. DOI: 10.1007/s11627-021-10208-x

Bearth, A., Otten, C. D., & Cohen, A. S. (2024). Consumers' perceptions and acceptance of genome editing in agriculture: Insights from the United States of America and Switzerland. *Food Research International*, *178*, 113982. DOI: 10.1016/j.foodres.2024.113982 PMID: 38309884

Begna, T. (2021). Conventional breeding methods widely used to improve self-pollinated crops. *International Journal of Research*, *7*(1), 1–16.

Bélanger, J. G., Copley, T. R., Hoyos-Villegas, V., Charron, J.-B., & O'Donoughue, L. (2024). A comprehensive review of in planta stable transformation strategies. *Plant Methods*, *20*(1), 79. DOI: 10.1186/s13007-024-01200-8 PMID: 38822403

Bernauer, T. (2003). *Genes, trade, and regulation: The seeds of conflict in food biotechnology*. Princeton University Press.

Bertheau, Y. (2012). *Genetically modified and non-genetically modified food supply chains: co-existence and traceability*: John Wiley & Sons.

Berthiaume, F., Maguire, T. J., & Yarmush, M. L. (2011). Tissue engineering and regenerative medicine: History, progress, and challenges. *Annual Review of Chemical and Biomolecular Engineering*, *2*(1), 403–430. DOI: 10.1146/annurev-chembioeng-061010-114257 PMID: 22432625

Bhagtaney, L., & Sundarrajan, P. (2023). An overview of tools for genome editing: ZFNs, mega nucleases, and TALENs. *CRISPR/Cas-mediated genome editing in plants*, 37-64.

Bidabadi, S. S., & Jain, S. M. (2020). Cellular, molecular, and physiological aspects of in vitro plant regeneration. *Plants*, *9*(6), 702. DOI: 10.3390/plants9060702 PMID: 32492786

Bonga, J. (2017). Can explant choice help resolve recalcitrance problems in in vitro propagation, a problem still acute especially for adult conifers? *Trees (Berlin)*, *31*(3), 781–789. DOI: 10.1007/s00468-016-1509-z

Breuninger, H., & Lenhard, M. (2010). Control of tissue and organ growth in plants. *Current Topics in Developmental Biology*, *91*, 185–220. DOI: 10.1016/S0070-2153(10)91007-7 PMID: 20705183

Brookes, G., & Smyth, S. J. (2024). Risk-appropriate regulations for gene-editing technologies. *GM Crops & Food*, *15*(1), 1–14. DOI: 10.1080/21645698.2023.2293510 PMID: 38215017

Buch, K., Kaushik, A., Mishra, U., Beese, S., Samanta, S., Singh, R., & Mubeen, . (2023). Unravelling the complexity of plant breeding through modern genetic techniques and tools: A review. *International Journal of Plant and Soil Science*, *35*(21), 97–105. DOI: 10.9734/ijpss/2023/v35i213950

Burow, M., Atwell, S., Francisco, M., Kerwin, R. E., Halkier, B. A., & Kliebenstein, D. J. (2015). The glucosinolate biosynthetic gene AOP2 mediates feed-back regulation of jasmonic acid signaling in Arabidopsis. *Molecular Plant*, *8*(8), 1201–1212. DOI: 10.1016/j.molp.2015.03.001 PMID: 25758208

Cain, A. K., Barquist, L., Goodman, A. L., Paulsen, I. T., Parkhill, J., & van Opijnen, T. (2020). A decade of advances in transposon-insertion sequencing. *Nature Reviews. Genetics*, *21*(9), 526–540. DOI: 10.1038/s41576-020-0244-x PMID: 32533119

Cardi, T., Murovec, J., Bakhsh, A., Boniecka, J., Bruegmann, T., Bull, S. E., Eeckhaut, T., Fladung, M., Galovic, V., Linkiewicz, A., Lukan, T., Mafra, I., Michalski, K., Kavas, M., Nicolia, A., Nowakowska, J., Sági, L., Sarmiento, C., Yıldırım, K., & Van Laere, K. (2023). CRISPR/Cas-mediated plant genome editing: Outstanding challenges a decade after implementation. *Trends in Plant Science, 28*(10), 1144–1165. DOI: 10.1016/j.tplants.2023.05.012 PMID: 37331842

Cembrowska-Lech, D., Krzemińska, A., Miller, T., Nowakowska, A., Adamski, C., Radaczyńska, M., Mikiciuk, G., & Mikiciuk, M. (2023). An integrated multi-omics and artificial intelligence framework for advance plant phenotyping in horticulture. *Biology (Basel), 12*(10), 1298. DOI: 10.3390/biology12101298 PMID: 37887008

Chan, L., & Costa, S. (2005). Participation in the global knowledge commons: Challenges and opportunities for research dissemination in developing countries. *New Library World, 106*(3/4), 141–163. DOI: 10.1108/03074800510587354

Chandrasekaran, R., Rajiv, P., & Abd-Elsalam, K. A. (2020). Carbon nanotubes: Plant gene delivery and genome editing. In *Carbon nanomaterials for agri-food and environmental applications* (pp. 279–296). Elsevier. DOI: 10.1016/B978-0-12-819786-8.00014-1

Chen, G., Hao, F., & Sun, X. (2025). Artificial intelligence-driven gene editing and crop breeding: Technological innovations and application prospects. *Advances in Resources Research, 5*(1), 235–254.

Chikoti, P. C., & Tembo, M. (2022). Expansion and impact of cassava brown streak and cassava mosaic diseases in Africa: A review. *Frontiers in Sustainable Food Systems, 6*, 1076364. DOI: 10.3389/fsufs.2022.1076364

Chuang, Y.-F., Phipps, A. J., Lin, F.-L., Hecht, V., Hewitt, A. W., Wang, P.-Y., & Liu, G.-S. (2021). Approach for in vivo delivery of CRISPR/Cas system: A recent update and future prospect. *Cellular and Molecular Life Sciences : CMLS, 78*(6), 2683–2708. DOI: 10.1007/s00018-020-03725-2 PMID: 33388855

Clapp, J., & Ruder, S.-L. (2020). Precision technologies for agriculture: Digital farming, gene-edited crops, and the politics of sustainability. *Global Environmental Politics, 20*(3), 49–69. DOI: 10.1162/glep_a_00566

Crauciuc, A., Tripon, F., Gheorghiu, A., Nemes, G., Boglis, A., & Banescu, C. (2017). Development, Applications, Benefits, Challenges and Limitations of the New Genome Engineering Technique. An Update Study. *Acta Medica Marisiensis, 63*(1), 4–9. DOI: 10.1515/amma-2017-0007

Cui, H. (2021). Challenges and approaches to crop improvement through C3-to-C4 engineering. *Frontiers in Plant Science, 12*, 715391. DOI: 10.3389/fpls.2021.715391 PMID: 34594351

Cui, Z., Liu, H., Zhang, H., Huang, Z., Tian, R., Li, L., & Zhang, S. (2021). The comparison of ZFNs, TALENs, and SpCas9 by GUIDE-seq in HPV-targeted gene therapy. *Molecular Therapy. Nucleic Acids, 26*, 1466–1478. DOI: 10.1016/j.omtn.2021.08.008 PMID: 34938601

De Nadal, E., Ammerer, G., & Posas, F. (2011). Controlling gene expression in response to stress. *Nature Reviews. Genetics, 12*(12), 833–845. DOI: 10.1038/nrg3055 PMID: 22048664

de Roij, S. (2020). *Gene-editing crops to benefit smallholder farmers in developing countries: Realised potential or empty claims of inclusive innovation?*,

Delhove, J., Osenk, I., Prichard, I., & Donnelley, M. (2020). Public acceptability of gene therapy and gene editing for human use: A systematic review. *Human Gene Therapy, 31*(1-2), 20–46. DOI: 10.1089/hum.2019.197 PMID: 31802714

Dhirani, L. L., Mukhtiar, N., Chowdhry, B. S., & Newe, T. (2023). Ethical dilemmas and privacy issues in emerging technologies: A review. *Sensors (Basel), 23*(3), 1151. DOI: 10.3390/s23031151 PMID: 36772190

Dixit, S., Kumar, A., Srinivasan, K., Vincent, P. D. R., & Ramu Krishnan, N. (2024). Advancing genome editing with artificial intelligence: Opportunities, challenges, and future directions. *Frontiers in Bioengineering and Biotechnology, 11*, 1335901. DOI: 10.3389/fbioe.2023.1335901 PMID: 38260726

Eid, A., Alshareef, S., & Mahfouz, M. M. (2018). CRISPR base editors: genome editing without double-stranded breaks. *Biochemical Journal, 475*(11), 1955-1964.

El Chami, D., Daccache, A., & El Moujabber, M. (2020). How can sustainable agriculture increase climate resilience? A systematic review. *Sustainability (Basel), 12*(8), 3119. DOI: 10.3390/su12083119

Fan, B.-L., Chen, L.-H., Chen, L.-L., & Guo, H. (2025). Integrative Multi-Omics Approaches for Identifying and Characterizing Biological Elements in Crop Traits: Current Progress and Future Prospects. *International Journal of Molecular Sciences, 26*(4), 1466. DOI: 10.3390/ijms26041466 PMID: 40003933

Farooq, R., Hussain, K., Nazir, S., Javed, M. R., & Masood, N. (2018). CRISPR/Cas9; A robust technology for producing genetically engineered plants. *Cellular and Molecular Biology, 64*(14), 31–38. DOI: 10.14715/cmb/2018.64.14.6 PMID: 30511631

Freitas-Alves, N. S., Moreira-Pinto, C. E., Távora, F. T., Paes-de-Melo, B., Arraes, F. B., Lourenço-Tessutti, I. T., & Qi, Y. (2024). CRISPR/Cas genome editing in soybean: Challenges and new insights to overcome existing bottlenecks. *Journal of Advanced Research.* PMID: 39163906

Fubini, F. (2024). *In Search of the Perfect Peach: Why flavour holds the answer to fixing our food system.* Chelsea Green Publishing.

Gad, M. A., Li, M.-j., Ahmed, F. K., & Almoammar, H. (2020). Nanomaterials for gene delivery and editing in plants: Challenges and future perspective. In *Multifunctional Hybrid Nanomaterials for Sustainable Agri-Food and Ecosystems* (pp. 135–153). Elsevier. DOI: 10.1016/B978-0-12-821354-4.00006-6

Geng, J., Li, L., Lv, Q., Zhao, Y., Liu, Y., Zhang, L., & Li, X. (2017). TaGW2-6A allelic variation contributes to grain size possibly by regulating the expression of cytokinins and starch-related genes in wheat. *Planta, 246*(6), 1153–1163. DOI: 10.1007/s00425-017-2759-8 PMID: 28825220

Gier, R. A., Budinich, K. A., Evitt, N. H., Cao, Z., Freilich, E. S., Chen, Q., Qi, J., Lan, Y., Kohli, R. M., & Shi, J. (2020). High-performance CRISPR-Cas12a genome editing for combinatorial genetic screening. *Nature Communications, 11*(1), 3455. DOI: 10.1038/s41467-020-17209-1 PMID: 32661245

Gonzalez Guzman, M., Cellini, F., Fotopoulos, V., Balestrini, R., & Arbona, V. (2022). New approaches to improve crop tolerance to biotic and abiotic stresses. *Physiologia Plantarum, 174*(1), e13547. DOI: 10.1111/ppl.13547 PMID: 34480798

Goralogia, G. S., Redick, T. P., & Strauss, S. H. (2021). Gene editing in tree and clonal crops: Progress and challenges. *In Vitro Cellular & Developmental Biology. Plant : Journal of the Tissue Culture Association, 57*(4), 683–699. DOI: 10.1007/s11627-021-10197-x

Gostimskaya, I. (2022). CRISPR–cas9: A history of its discovery and ethical considerations of its use in genome editing. *Biochemistry. Biokhimiia, 87*(8), 777–788. DOI: 10.1134/S0006297922080090 PMID: 36171658

Gowdhaman, P. (2024). Adaptation to Progressive Climate Change, Technology Development, Agricultural Practices. *Dr. KL Chaudhary*, 207.

Grešová, K., Alexiou, P., & Giassa, I.-C. (2022). Small RNA targets: Advances in prediction tools and high-throughput profiling. *Biology (Basel), 11*(12), 1798. DOI: 10.3390/biology11121798 PMID: 36552307

Han, P., Velasco-Hernández, M. C., Ramirez-Romero, R., & Desneux, N. (2016). Behavioral effects of insect-resistant genetically modified crops on phytophagous and beneficial arthropods: A review. *Journal of Pest Science*, *89*(4), 859–883. DOI: 10.1007/s10340-016-0791-2

Hassan, Y. M., Mohamed, A. S., Hassan, Y. M., & El-Sayed, W. M. (2025). Recent developments and future directions in point-of-care next-generation CRISPR-based rapid diagnosis. *Clinical and Experimental Medicine*, *25*(1), 1–15. DOI: 10.1007/s10238-024-01540-8 PMID: 39789283

Hess, D. J. (2009). *Localist movements in a global economy: Sustainability, justice, and urban development in the United States*. MIT Press. DOI: 10.7551/mitpress/9780262012645.001.0001

Hock, S. C., Kian, S. M., & Wah, C. L. (2020). Global challenges in the manufacture, regulation and international harmonization of GMP and quality standards for biopharmaceuticals. *GaBi Journal*, *9*(2), 52–64. DOI: 10.5639/gabij.2020.0902.010

Holman, C. M. (2019). A fractured international response to CRISPR-enabled gene editing of agricultural products. *Biotechnology Law Report*, *38*(1), 3–23. DOI: 10.1089/blr.2019.29100.cmh

Hossain, A., Islam, M. T., Maitra, S., Majumder, D., Garai, S., Mondal, M., . . . Brestic, M. (2021). Neglected and underutilized crop species: are they future smart crops in fighting poverty, hunger and malnutrition under changing climate? *Neglected and underutilized crops-towards nutritional security and sustainability*, 1-50.

Hu, Y., House, L. A., & Gao, Z. (2022). How do consumers respond to labels for crispr (gene-editing)? *Food Policy*, *112*, 102366. DOI: 10.1016/j.foodpol.2022.102366

Huang, X., Yang, D., Zhang, J., Xu, J., & Chen, Y. E. (2022). Recent advances in improving gene-editing specificity through CRISPR–Cas9 nuclease engineering. *Cells*, *11*(14), 2186. DOI: 10.3390/cells11142186 PMID: 35883629

Ishii, T., & Araki, M. (2016). Consumer acceptance of food crops developed by genome editing. *Plant Cell Reports*, *35*(7), 1507–1518. DOI: 10.1007/s00299-016-1974-2 PMID: 27038939

Jankele, R., & Svoboda, P. (2014). TAL effectors: Tools for DNA targeting. *Briefings in Functional Genomics*, *13*(5), 409–419. DOI: 10.1093/bfgp/elu013 PMID: 24907364

Kaur, N., Urban, M., Hammond-Kosack, K. E., & Halford, N. G. (2024). *Field trialling of CRISPR low-asparagine wheat and investigation of the increase in asparagine in response to pathogen infection in wheat.* Genetics and Agronomy for Low Acrylamide-Forming Potential in Cereals.

Kavhiza, N. J., Zargar, M., Prikhodko, S. I., Pakina, E. N., Murtazova, K. M.-S., & Nakhaev, M. R. (2022). Improving crop productivity and ensuring food security through the adoption of genetically modified crops in sub-Saharan Africa. *Agronomy (Basel), 12*(2), 439. DOI: 10.3390/agronomy12020439

Keiper, F., & Atanassova, A. (2020). Regulation of synthetic biology: Developments under the convention on biological diversity and its protocols. *Frontiers in Bioengineering and Biotechnology, 8*, 310. DOI: 10.3389/fbioe.2020.00310 PMID: 32328486

Khalil, A. M. (2020). The genome editing revolution. *Journal of Genetic Engineering and Biotechnology, 18*(1), 68. DOI: 10.1186/s43141-020-00078-y PMID: 33123803

Khan, A. H., Hassan, M., & Khan, M. N. (2020). Conventional plant breeding program for disease resistance. *Plant disease management strategies for sustainable agriculture through traditional and modern approaches*, 27-51.

KhokharVoytas, A., Shahbaz, M., Maqsood, M. F., Zulfiqar, U., Naz, N., Iqbal, U. Z., Sara, M., Aqeel, M., Khalid, N., Noman, A., Zulfiqar, F., Al Syaad, K. M., & AlShaqhaa, M. A.KhokharVoytas. (2023). Genetic modification strategies for enhancing plant resilience to abiotic stresses in the context of climate change. *Functional & Integrative Genomics, 23*(3), 283. DOI: 10.1007/s10142-023-01202-0 PMID: 37642792

Kofler, N., Collins, J. P., Kuzma, J., Marris, E., Esvelt, K., Nelson, M. P., & Semenov, M. (2018). Editing nature: Local roots of global governance. *Science, 362*(6414), 527–529. DOI: 10.1126/science.aat4612 PMID: 30385564

Komen, J., Tripathi, L., Mkoko, B., Ofosu, D. O., Oloka, H., & Wangari, D. (2020). Biosafety regulatory reviews and leeway to operate: Case studies from sub-Sahara Africa. *Frontiers in Plant Science, 11*, 130. DOI: 10.3389/fpls.2020.00130 PMID: 32210981

Kreiner, J. M., Stinchcombe, J. R., & Wright, S. I. (2018). Population genomics of herbicide resistance: Adaptation via evolutionary rescue. *Annual Review of Plant Biology, 69*(1), 611–635. DOI: 10.1146/annurev-arplant-042817-040038 PMID: 29140727

Kujur, S., Senthil-Kumar, M., & Kumar, R. (2021). Plant viral vectors: Expanding the possibilities of precise gene editing in plant genomes. *Plant Cell Reports*, *40*(6), 931–934. DOI: 10.1007/s00299-021-02697-2 PMID: 33864508

Kuluev, B., Gumerova, G., Mikhaylova, E., Gerashchenkov, G., Rozhnova, N., Vershinina, Z., & Baymiev, A. K. (2019). Delivery of CRISPR/Cas components into higher plant cells for genome editing. *Russian Journal of Plant Physiology: a Comprehensive Russian Journal on Modern Phytophysiology*, *66*(5), 694–706. DOI: 10.1134/S102144371905011X

Kumar, A., Anju, T., Kumar, S., Chhapekar, S. S., Sreedharan, S., Singh, S., Choi, S. R., Ramchiary, N., & Lim, Y. P. (2021). Integrating omics and gene editing tools for rapid improvement of traditional food plants for diversified and sustainable food security. *International Journal of Molecular Sciences*, *22*(15), 8093. DOI: 10.3390/ijms22158093 PMID: 34360856

Kumlehn, J., Pietralla, J., Hensel, G., Pacher, M., & Puchta, H. (2018). The CRISPR/Cas revolution continues: From efficient gene editing for crop breeding to plant synthetic biology. *Journal of Integrative Plant Biology*, *60*(12), 1127–1153. DOI: 10.1111/jipb.12734 PMID: 30387552

Lassalle, M. W., & Jarocki, V. (2016). Public misunderstanding of genetically modified organisms: How science and society are interconnected. *Int. J. Life. Sci. Scienti. Res*, *2*(5), 636–643. DOI: 10.21276/ijlssr.2016.2.5.22

Li, S., Zhang, C., Li, J., Yan, L., Wang, N., & Xia, L. (2021). Present and future prospects for wheat improvement through genome editing and advanced technologies. *Plant Communications*, *2*(4), 100211. DOI: 10.1016/j.xplc.2021.100211 PMID: 34327324

Listgarten, J., Weinstein, M., Kleinstiver, B. P., Sousa, A. A., Joung, J. K., Crawford, J., & Doench, J. G. (2018). Prediction of off-target activities for the end-to-end design of CRISPR guide RNAs. *Nature Biomedical Engineering*, *2*(1), 38–47. DOI: 10.1038/s41551-017-0178-6 PMID: 29998038

Liu, X., Huang, L., Li, M., Fu, Y., Zhang, W., Zhang, S., Liang, X., & Shen, Q. (2025). An Efficient and Cost-Effective Novel Strategy for Identifying CRISPR-Cas-Mediated Mutants in Plant Offspring. *The CRISPR Journal*, *8*(1), 26–36. DOI: 10.1089/crispr.2024.0057 PMID: 39804665

Lucht, J. M. (2015). Public acceptance of plant biotechnology and GM crops. *Viruses*, *7*(8), 4254–4281. DOI: 10.3390/v7082819 PMID: 26264020

Lv, Z., Jiang, R., Chen, J., & Chen, W. (2020). Nanoparticle-mediated gene transformation strategies for plant genetic engineering. *The Plant Journal : For Cell and Molecular Biology*, *104*(4), 880–891. DOI: 10.1111/tpj.14973 PMID: 32860436

Macnaghten, P., & Habets, M. G. (2020). Breaking the impasse: Towards a forward-looking governance framework for gene editing with plants. *Plants, People, Planet*, *2*(4), 353–365. DOI: 10.1002/ppp3.10107

Maehira, Y., & Spencer, R. C. (2019). Harmonization of biosafety and biosecurity standards for high-containment facilities in low-and middle-income countries: An approach from the perspective of occupational safety and health. *Frontiers in Public Health*, *7*, 249. DOI: 10.3389/fpubh.2019.00249 PMID: 31572701

Mahfouz, M. M., Piatek, A., & Stewart, C. N.Jr. (2014). Genome engineering via TALENs and CRISPR/Cas9 systems: Challenges and perspectives. *Plant Biotechnology Journal*, *12*(8), 1006–1014. DOI: 10.1111/pbi.12256 PMID: 25250853

Majumder, S., Datta, K., & Datta, S. K. (2023). Perspective of Evolution of the C 4 Plants to Develop Climate Designer C 4 Rice as a Strategy for Abiotic Stress Management. *Global Climate Change and Plant Stress Management*, 113-124.

Manghwar, H., Li, B., Ding, X., Hussain, A., Lindsey, K., Zhang, X., & Jin, S. (2020). CRISPR/Cas systems in genome editing: Methodologies and tools for sgRNA design, off-target evaluation, and strategies to mitigate off-target effects. *Advanced Science (Weinheim, Baden-Württemberg, Germany)*, *7*(6), 1902312. DOI: 10.1002/advs.201902312 PMID: 32195078

Marden, E. (2002). Risk and regulation: US regulatory policy on genetically modified food and agriculture. *BCL Rev.*, *44*, 733.

Marone, D., Mastrangelo, A. M., & Borrelli, G. M. (2023). From transgenesis to genome editing in crop improvement: Applications, marketing, and legal issues. *International Journal of Molecular Sciences*, *24*(8), 7122. DOI: 10.3390/ijms24087122 PMID: 37108285

Marx, A. (2019). *Public-private partnerships for sustainable development: Exploring their design and its impact on effectiveness* (Vol. 11). Multidisciplinary Digital Publishing Institute.

Menz, J., Modrzejewski, D., Hartung, F., Wilhelm, R., & Sprink, T. (2020). Genome edited crops touch the market: A view on the global development and regulatory environment. *Frontiers in Plant Science*, *11*, 586027. DOI: 10.3389/fpls.2020.586027 PMID: 33163013

Mijena, D. F., Alamerew, S., Assefa, K., & Nigusse, M. (2024). Indigenous and Underutilized Crops Significance in Food and Nutritional Security in the face of Climate Change: The Case of Anchote. *Ethiopian Journal of Crop Science*, *12*(2), 15–38.

Mohan, C., Easterling, M., & Yau, Y.-Y. (2022). Gene editing technologies for sugarcane improvement: Opportunities and limitations. *Sugar Tech*, *24*(1), 369–385. DOI: 10.1007/s12355-021-01045-8 PMID: 34667393

Molla, K. A., Shih, J., Wheatley, M. S., & Yang, Y. (2022). Predictable NHEJ insertion and assessment of HDR editing strategies in plants. *Frontiers in Genome Editing*, *4*, 825236. DOI: 10.3389/fgeed.2022.825236 PMID: 35481279

Montenegro de Wit, M. (2020). Democratizing CRISPR? Stories, practices, and politics of science and governance on the agricultural gene editing frontier. *Elementa*, *8*, 9. DOI: 10.1525/elementa.405

Mushtaq, M., Ahmad Dar, A., Skalicky, M., Tyagi, A., Bhagat, N., Basu, U., & Dar, T.-U.-H. (2021). CRISPR-based genome editing tools: Insights into technological breakthroughs and future challenges. *Genes*, *12*(6), 797. DOI: 10.3390/genes12060797 PMID: 34073848

Myhr, A. I. (2010). A precautionary approach to genetically modified organisms: Challenges and implications for policy and science. *Journal of Agricultural & Environmental Ethics*, *23*(6), 501–525. DOI: 10.1007/s10806-010-9234-x

Nayerossadat, N., Maedeh, T., & Ali, P. A. (2012). Viral and nonviral delivery systems for gene delivery. *Advanced Biomedical Research*, *1*(1), 27. DOI: 10.4103/2277-9175.98152 PMID: 23210086

Norman, P. E., Dzidzienyo, D. K., Karim, K. Y., & Beah, A. A. (2021). *Genetic Modification and Application in Cassava*. Sweetpotato and Yams.

O'Geen, H., Ren, C., Nicolet, C. M., Perez, A. A., Halmai, J., Le, V. M., Mackay, J. P., Farnham, P. J., & Segal, D. J. (2017). dCas9-based epigenome editing suggests acquisition of histone methylation is not sufficient for target gene repression. *Nucleic Acids Research*, *45*(17), 9901–9916. DOI: 10.1093/nar/gkx578 PMID: 28973434

Ochoa-Sanchez, A., Perez-Sanchez, G., Torres-Ledesma, A. M., Valdez, J. P. R., Rinaldi, G., Moguel, B. B., & Molina-Aguilar, C. (2021). Prime Editing, a novel genome-editing tool that may surpass conventional CRISPR-Cas9. *Re:GEN Open*, *1*(1), 75–82. DOI: 10.1089/regen.2021.0016

Panno, S., Davino, S., Caruso, A. G., Bertacca, S., Crnogorac, A., Mandić, A., Noris, E., & Matić, S. (2021). A review of the most common and economically important diseases that undermine the cultivation of tomato crop in the mediterranean basin. *Agronomy (Basel), 11*(11), 2188. DOI: 10.3390/agronomy11112188

Pascual, M., Rives, B., Schunter, C., & Macpherson, E. (2017). Impact of life history traits on gene flow: A multispecies systematic review across oceanographic barriers in the Mediterranean Sea. *PLoS One*, *12*(5), e0176419. DOI: 10.1371/journal.pone.0176419 PMID: 28489878

Pathirana, R., & Carimi, F. (2022). Management and utilization of plant genetic resources for a sustainable agriculture. *Plants*, *11*(15), 2038. DOI: 10.3390/plants11152038 PMID: 35956515

Paul, P., & Mukhopadhyay, K. (2010). Growth via intellectual property rights versus gendered inequity in emerging economies: An ethical dilemma for international business. *Journal of Business Ethics*, *91*(3), 359–378. DOI: 10.1007/s10551-009-0088-y

Petrova, I. O., & Smirnikhina, S. A. (2023). The development, optimization and future of prime editing. *International Journal of Molecular Sciences*, *24*(23), 17045. DOI: 10.3390/ijms242317045 PMID: 38069367

Prober, S. M., & Smith, F. P. (2009). Enhancing biodiversity persistence in intensively used agricultural landscapes: A synthesis of 30 years of research in the Western Australian wheatbelt. *Agriculture, Ecosystems & Environment*, *132*(3-4), 173–191. DOI: 10.1016/j.agee.2009.04.005

Qaim, M. (2020). Role of new plant breeding technologies for food security and sustainable agricultural development. *Applied Economic Perspectives and Policy*, *42*(2), 129–150. DOI: 10.1002/aepp.13044

Rai, G. K., Khanday, D. M., Kumar, P., Magotra, I., Choudhary, S. M., Kosser, R., & Rouphael, Y. (2023). Enhancing crop resilience to drought stress through CRISPR-Cas9 genome editing. *Plants*, *12*(12), 2306. DOI: 10.3390/plants12122306 PMID: 37375931

Raju, S. (2021). Strategies for enhancing post-harvest quality and shelf life of tuber crops: Insights from physiological perspectives. *Journal of Root Crops*, *47*(1 & 2), 40–52.

Ramirez-Torres, F., Ghogare, R., Stowe, E., Cerdá-Bennasser, P., Lobato-Gómez, M., Williamson-Benavides, B. A., Giron-Calva, P. S., Hewitt, S., Christou, P., & Dhingra, A. (2021). Genome editing in fruit, ornamental, and industrial crops. *Transgenic Research*, *30*(4), 499–528. DOI: 10.1007/s11248-021-00240-3 PMID: 33825100

Reed, K. M., & Bargmann, B. O. (2021). Protoplast regeneration and its use in new plant breeding technologies. *Frontiers in Genome Editing*, *3*, 734951. DOI: 10.3389/fgeed.2021.734951 PMID: 34713266

Richards, R., Hunt, J., Kirkegaard, J., & Passioura, J. (2014). Yield improvement and adaptation of wheat to water-limited environments in Australia—A case study. *Crop & Pasture Science*, *65*(7), 676–689. DOI: 10.1071/CP13426

Ricroch, A. E., & Hénard-Damave, M.-C. (2016). Next biotech plants: New traits, crops, developers and technologies for addressing global challenges. *Critical Reviews in Biotechnology*, *36*(4), 675–690. DOI: 10.3109/07388551.2015.1004521 PMID: 25641327

Ross, I. L., Alami, Y., Harvey, P. R., Achouak, W., & Ryder, M. H. (2000). Genetic diversity and biological control activity of novel species of closely related pseudomonads isolated from wheat field soils in South Australia. *Applied and Environmental Microbiology*, *66*(4), 1609–1616. DOI: 10.1128/AEM.66.4.1609-1616.2000 PMID: 10742249

Roy, C., Kumar, S., Ranjan, R. D., Kumhar, S. R., & Govindan, V. (2022). Genomic approaches for improving grain zinc and iron content in wheat. *Frontiers in Genetics*, *13*, 1045955. DOI: 10.3389/fgene.2022.1045955 PMID: 36437911

Ryan, J. P., Oestreich, W. K., Benoit-Bird, K. J., Waluk, C. M., Rueda, C. A., Cline, D. E., & Fahlbusch, J. A. (2025). Audible changes in marine trophic ecology: Baleen whale song tracks foraging conditions in the eastern North Pacific. *PLoS One*, *20*(2), e0318624. DOI: 10.1371/journal.pone.0318624 PMID: 40009591

Sabat, M., & Tripathy, A. (2024). Genetically Modified and Gene-Edited Food Crops: Recent Status and Future Prospects. *Food Production, Diversity, and Safety Under Climate Change*, 211-222.

Sadikiel Mmbando, G. (2024). The adoption of genetically modified crops in Africa: The public's current perception, the regulatory obstacles, and ethical challenges. *GM Crops & Food*, *15*(1), 185–199. DOI: 10.1080/21645698.2024.2345401 PMID: 38651587

Sampath, V., Rangarajan, N., Sharanappa, C., Deori, M., Veeraragavan, M., Ghodake, B., & Kaushal, K. (2023). Advancing crop improvement through CRISPR technology in precision agriculture trends-a review. *International Journal of Environment and Climate Change*, *13*(11), 4683–4694. DOI: 10.9734/ijecc/2023/v13i113647

Schaart, J. G., van de Wiel, C. C., & Smulders, M. J. (2021). Genome editing of polyploid crops: Prospects, achievements and bottlenecks. *Transgenic Research, 30*(4), 337–351. DOI: 10.1007/s11248-021-00251-0 PMID: 33846956

Schmidt, T. (2014). *Transient biolistic transformation in maize leaves.* University of Pretoria, Sedeek, K. E., Mahas, A., & Mahfouz, M. (2019). Plant genome engineering for targeted improvement of crop traits. *Frontiers in Plant Science, 10*, 114.

Selin, C., Friedman, L. M., Lambert, L., Farooque, M., & Guston, D. Inclusive, Equitable and Just Evolution of Gene Editing Technologies: The Case for Integrating Public Engagement in Governing Efforts. *Available at SSRN* 4853681. DOI: 10.2139/ssrn.4853681

Shams, F., Bayat, H., Mohammadian, O., Mahboudi, S., Vahidnezhad, H., Soosanabadi, M., & Rahimpour, A. (2022). Advance trends in targeting homology-directed repair for accurate gene editing: An inclusive review of small molecules and modified CRISPR-Cas9 systems. *BioImpacts, 12*(4), 371–391. DOI: 10.34172/bi.2022.23871 PMID: 35975201

Shi, G., Hao, M., Tian, B., Cao, G., Wei, F., & Xie, Z. (2021). A methodological advance of tobacco rattle virus-induced gene silencing for functional genomics in plants. *Frontiers in Plant Science, 12*, 671091. DOI: 10.3389/fpls.2021.671091 PMID: 34149770

Shukla-Jones, A., Friedrichs, S., & Winickoff, D. E. (2018). Gene editing in an international context: Scientific, economic and social issues across sectors. *OECD science, technology and industry working papers, 2018*(4), 0_1-51.

Smirnikhina, S. A., Zaynitdinova, M. I., Sergeeva, V. A., & Lavrov, A. V. (2022). Improving homology-directed repair in genome editing experiments by influencing the cell cycle. *International Journal of Molecular Sciences, 23*(11), 5992. DOI: 10.3390/ijms23115992 PMID: 35682671

Spök, A., Sprink, T., Allan, A. C., Yamaguchi, T., & Dayé, C. (2022). Towards social acceptability of genome-edited plants in industrialised countries? Emerging evidence from Europe, United States, Canada, Australia, New Zealand, and Japan. *Frontiers in Genome Editing, 4*, 899331. DOI: 10.3389/fgeed.2022.899331 PMID: 36120531

Sun, L., Lai, M., Ghouri, F., Nawaz, M. A., Ali, F., Baloch, F. S., Nadeem, M. A., Aasim, M., & Shahid, M. Q. (2024). Modern plant breeding techniques in crop improvement and genetic diversity: From molecular markers and gene editing to artificial intelligence—A critical review. *Plants, 13*(19), 2676. DOI: 10.3390/plants13192676 PMID: 39409546

Tadele, Z. (2019). Orphan crops: Their importance and the urgency of improvement. *Planta, 250*(3), 677–694. DOI: 10.1007/s00425-019-03210-6 PMID: 31190115

Tariq, S., Gul, A., Negri, S., & Mujtaba, T. (2024). Genetic engineering in tomato. In *Targeted Genome Engineering via CRISPR/Cas9 in Plants* (pp. 101–133). Elsevier. DOI: 10.1016/B978-0-443-26614-0.00015-1

Tizard, M., Hallerman, E., Fahrenkrug, S., Newell-McGloughlin, M., Gibson, J., de Loos, F., & D'Occhio, M. (2016). Strategies to enable the adoption of animal biotechnology to sustainably improve global food safety and security. *Transgenic Research, 25*(5), 575–595. DOI: 10.1007/s11248-016-9965-1 PMID: 27246007

Toschka, H. Y. (2025). *New Food Product Development: Global Strategies and Practices for Successful Innovation.* CRC Press. DOI: 10.1201/9781032724591

Tripathi, A., & Shukla, S. (2024). Methods of genetic transformation: Major emphasis to crop plants. *Journal of Microbiology, Biotechnology and Food Sciences, 13*(4), e10276–e10276.

Troadec, M.-B., Pagès, J.-C., & Scientific Committee of the High Council for biotechnology, F. (2019). *Where are we with unintended effects in genome editing applications from DNA to phenotype: focus on plant applications.* Paper presented at the Transgenic Research.

Turnbull, C., Lillemo, M., & Hvoslef-Eide, T. A. (2021). Global regulation of genetically modified crops amid the gene edited crop boom–a review. *Frontiers in Plant Science, 12*, 630396. DOI: 10.3389/fpls.2021.630396 PMID: 33719302

van den Boomgaard, M. (2021). *Public Values and Modern Biotechnology in the Dutch Patent System A multiple case-study of patents on Recombinant-DNA & PCR and CRISPR technologies in the field of green biotechnology.*

Van Esse, H. P., Reuber, T. L., & van der Does, D. (2020). Genetic modification to improve disease resistance in crops. *The New Phytologist, 225*(1), 70–86. DOI: 10.1111/nph.15967 PMID: 31135961

Vats, S., Kumawat, S., Kumar, V., Patil, G. B., Joshi, T., Sonah, H., Sharma, T. R., & Deshmukh, R. (2019). Genome editing in plants: Exploration of technological advancements and challenges. *Cells, 8*(11), 1386. DOI: 10.3390/cells8111386 PMID: 31689989

Vicente, M. M., Chaves-Ferreira, M., Jorge, J. M., Proença, J. T., & Barreto, V. M. (2021). The off-targets of clustered regularly interspaced short palindromic repeats gene editing. *Frontiers in Cell and Developmental Biology, 9*, 718466. DOI: 10.3389/fcell.2021.718466 PMID: 34604217

Wang, T., Zhang, H., & Zhu, H. (2019). CRISPR technology is revolutionizing the improvement of tomato and other fruit crops. *Horticulture Research*, *6*(1), 6. DOI: 10.1038/s41438-019-0159-x PMID: 31240102

Weeks, A. R., Sgro, C. M., Young, A. G., Frankham, R., Mitchell, N. J., Miller, K. A., & Sunnucks, P. (2011). Assessing the benefits and risks of translocations in changing environments: A genetic perspective. *Evolutionary Applications*, *4*(6), 709–725. DOI: 10.1111/j.1752-4571.2011.00192.x PMID: 22287981

Wieczorek, M. N., Walczak, M., Skrzypczak-Zielińska, M., & Jeleń, H. H. (2018). Bitter taste of Brassica vegetables: The role of genetic factors, receptors, isothiocyanates, glucosinolates, and flavor context. *Critical Reviews in Food Science and Nutrition*, *58*(18), 3130–3140. DOI: 10.1080/10408398.2017.1353478 PMID: 28718657

Xu, P., Zhong, Y., Xu, A., Liu, B., Zhang, Y., Zhao, A., Yang, X., Ming, M., Cao, F., & Fu, F. (2024). Application of developmental regulators for enhancing plant regeneration and genetic transformation. *Plants*, *13*(9), 1272. DOI: 10.3390/plants13091272 PMID: 38732487

Xu, Y., Lu, Y., Xie, C., Gao, S., Wan, J., & Prasanna, B. M. (2012). Whole-genome strategies for marker-assisted plant breeding. *Molecular Breeding*, *29*(4), 833–854. DOI: 10.1007/s11032-012-9699-6

Yang, H., & Patel, D. J. (2024). Structures, mechanisms and applications of RNA-centric CRISPR–Cas13. *Nature Chemical Biology*, *20*(6), 673–688. DOI: 10.1038/s41589-024-01593-6 PMID: 38702571

Yatusevich, R. (2008). *Analysis of the MYB28, MYB29 and MYB76 transcription factors involved in the biosynthesis of aliphatic glucosinolates in Arabidopsis thaliana.* Universität zu Köln, Yuen, K. K.-C. (2017). *New sustainable models of open innovation to accelerate technology development in cellular agriculture.* Massachusetts Institute of Technology, Zhan, X., Lu, Y., Zhu, J. K., & Botella, J. R. (2021). Genome editing for plant research and crop improvement. *Journal of Integrative Plant Biology*, *63*(1), 3–33.

Zhao, M., Zhao, J., Yuan, J., Hale, L., Wen, T., Huang, Q., Vivanco, J. M., Zhou, J., Kowalchuk, G. A., & Shen, Q. (2021). Root exudates drive soil-microbe-nutrient feedbacks in response to plant growth. *Plant, Cell & Environment*, *44*(2), 613–628. DOI: 10.1111/pce.13928 PMID: 33103781

Zheng, Y., Li, Y., Zhou, K., Li, T., VanDusen, N. J., & Hua, Y. (2024). Precise genome-editing in human diseases: Mechanisms, strategies and applications. *Signal Transduction and Targeted Therapy*, *9*(1), 47. DOI: 10.1038/s41392-024-01750-2 PMID: 38409199

Zheng, Z., Appiano, M., Pavan, S., Bracuto, V., Ricciardi, L., Visser, R. G., Wolters, A.-M. A., & Bai, Y. (2016). Genome-wide study of the tomato SlMLO gene family and its functional characterization in response to the powdery mildew fungus Oidium neolycopersici. *Frontiers in Plant Science*, *7*, 380. DOI: 10.3389/fpls.2016.00380 PMID: 27579028

Chapter 10

CRISPR-Mediated Modulation of the Plant Microbiome for Sustainable Agriculture

B. Lokeshwari
http://orcid.org/0009-0005-6612-5134
Department of Microbiology, Sacred Heart College, Tirupattur, India

P. Saranraj
http://orcid.org/0000-0002-5052-6532
Department of Microbiology, Sacred Heart College, Tirupattur, India

S. Kanshiya
Department of Microbiology, Sacred Heart College, Tirupattur, India

P. Ragavi
Department of Microbiology, Sacred Heart College, Tirupattur, India

Abdel Razzaq Mohammad Al-Tawaha
National Agricultural Research Center, Baq'a, Jordan

K. Kesavardhini
http://orcid.org/0009-0005-8235-042X
Department of Microbiology, Sacred Heart College, Tirupattur, India

ABSTRACT

Recent advances in the CRISPR-Cas system of genome editing have led to a breakthrough in the ability to precisely modify the plant microbiome and interactions between plants and microorganisms. Modifying plant and microbe communities using the CRISPR technology has the potential to enhance agricultural production, thus minimizing the excessive application of chemicals as fertilizers and pesticides. This is a powerful strategy that has shown promise in increasing the ability of beneficial microorganisms to fix atmospheric nitrogen and dissolve insoluble phosphates.

DOI: 10.4018/979-8-2600-3575-7.ch010

Moreover, the integration of CRISPR technology with synthetic biology, metagenomics, and multi-omics has provided a new perspective on developing microbiome engineering for agriculture.

1. INTRODUCTION

The development of sustainable agriculture is becoming increasingly relevant worldwide because of the growing food production requirement, poor soil quality, climatic changes, and intensive utilization of chemicals for fertilizer and pest control purposes. Traditional agricultural methods have led to a considerable increase in crop yield over the last few decades. On the other hand, these farming techniques have caused various environmental, biological, and ecological issues such as soil ecosystem disturbances. Currently, many scientists direct their studies toward the exploration of the plant microbiome as a key element of sustainable agriculture (Glick, 2012). The term "plant microbiome" refers to the microbial community of bacteria, fungi, archaea, and viruses inhabiting the soil, leaf surface, and inner parts of the plants. These organisms maintain symbiotic relations with plants and perform vital processes in agricultural ecosystems such as nutrient recycling, plant development, adaptation to stress factors, and pathogen suppression.

There are many modern biotechnological tools available at present, but one of the most significant is the CRISPR-Cas system that can be considered a breakthrough in genetic engineering because of its simplicity, accuracy, efficacy, and flexibility. It was discovered as a natural adaptive immune system in bacteria and archaea, but nowadays it is used to edit genomes in plants, animals, and microbes (Jansson and Hofmockel, 2020). This tool differs from conventional breeding technologies and previously developed genome editors because it allows making highly accurate edits of particular genes that control desired agricultural properties. The development of new types of CRISPR tools, including Cas9, Cas12a, Cas13, base editing, and prime editing, made it possible to increase the accuracy of genetic modification even more. CRISPR technology has already shown great promise in agriculture.

The incorporation of CRISPR technologies in the study of plant microbiomes has created a new direction for sustainable agriculture called microbiome engineering. The regulation of plant microbiomes using CRISPR technologies helps in the improvement of various advantageous properties of plant-associated microbes, including nitrogen fixation, phosphorous mobilization, antimicrobial activity, and adaptation to environmental conditions. Moreover, the use of CRISPR technologies to edit plant genomes that control the recruitment of plant-associated microbes may be useful in improving beneficial interactions between plants and their microbiomes and establishing robust microbial ecosystem (Trivedi et al., 2017). This approach provides sustainable solutions to the use of chemical fertilizers and pesticides through the optimization of biological processes related to nutrient availability, disease resistance, and stress tolerance.

In spite of all the impressive prospects offered by CRISPR-based microbiome engineering, there are still quite a few scientific, ecological, ethical, and regulatory obstacles that need to be tackled first. Issues related to possible off-target effects, ecological stability, biosafety, and acceptability by the general public should be carefully considered prior to employing such techniques on an agricultural scale. At the same time, constant development in genome editing technologies, omics technologies, computational biology, and synthetic biology makes microbiome engineering techniques more precise and safer. As a result, CRISPR-based manipulation of the plant microbiome will become a groundbreaking strategy for sustainable agriculture and food security (Berg et al., 2017).

2. PLANT MICROBIOME AND ITS AGRICULTURAL IMPORTANCE

Plant microbiome comprises different microbial communities that are present in various parts of the plant such as rhizosphere, endosphere, and phyllosphere. Such microbes form different interactions like symbiosis, mutualism, commensalism, or even pathogenesis with the host plants and play an important role in the physiological functions of the plants. Microbes that have beneficial effects on the plants play significant roles in nutrient acquisition, nitrogen fixation, phytohormones biosynthesis, and disease resistance. The structure and complexity of the plant microbiomes are affected by environmental conditions, plant genotypes, soil characteristics, and agricultural practices (Banerjee et al., 2018). Plant microbiomes have gained much scientific interest in recent times because of their importance in agriculture. These include beneficial microbes that play an important role in the cycling of nutrients, soil structure improvement, and tolerance against abiotic stressors like drought, salinity, and heat. Furthermore, microbial inoculants and bio-fertilizers have become increasingly popular as environmentally friendly substitutes for agrochemicals. Studies into the nature and functioning of microbial communities within plants form the basis for the development of microbial agriculture practices. Additionally, the study of plant microbiomes coupled with advancements in biotechnology, especially CRISPR technology, has provided means to manipulate the community composition (Compant et al., 2010a). Figure 1 illustrates the major components of the plant microbiome, including the rhizosphere, endosphere, phyllosphere, and soil–plant–microbe interactions, along with their roles in plant growth and health.

It also highlights the agricultural importance of beneficial microorganisms and the potential application of CRISPR technology for sustainable crop improvement.

Figure 1. Plant Microbiome and Agricultural Importance

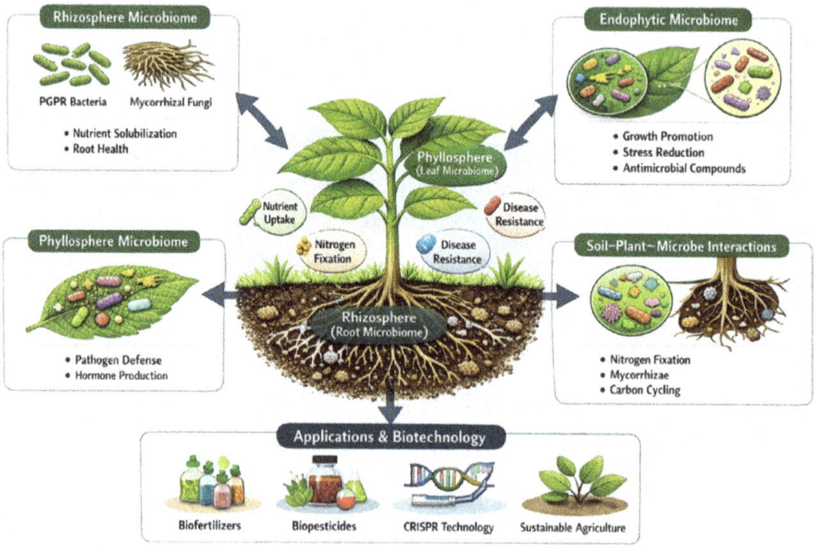

2.1 Structure and Composition of Plant Microbiomes

Microbiomes of plants refer to well-structured communities of bacteria, fungi, archaea, algae, and viruses inhabiting unique locations on plants. The presence of microbes differs based on plant species, developmental stage, soil type, environmental conditions, and farming practices. Typical bacterial phyla dominating plant habitats include Proteobacteria, Actinobacteria, Firmicutes, and Bacteroidetes. Fungal communities comprise mostly of Ascomycota and Basidiomycota species. These microbes engage in complex interactions involving communication and metabolic pathways that affect the host plant's development and defense processes. Plant-associated microbial communities play important roles in ecosystem health and resilience. Through metabolic interactions among different types of microorganisms, there is a balance of nutrients, pathogen suppression, and increased environmental adaptation in plant systems. With the use of advanced technologies like metagenomics and high-throughput sequencing, information regarding the dynamic interactions between microbial communities and gene expressions has been obtained. Knowledge of microbial community composition is crucial in the identification of beneficial microbes that can then be optimized through CRISPR technology (Raaijmakers and Mazzola, 2016).

2.2 Rhizosphere Microbiome

Rhizosphere is described as the thin layer of soil near the plant roots containing active microorganisms that respond to the presence of root exudates. Root exudates comprise sugars, amino acids, organic acids, and other secondary metabolites serving as nutrients and signaling factors for microbial populations. Microorganisms inhabiting the rhizosphere environment are responsible for nutrient solubilization, nitrogen fixation, phosphorous mineralization, and protecting plants from soil-born pathogens. Rhizobacteria, often called plant growth promoting rhizobacteria (PGPR), are beneficial for root development and plant growth due to the ability of these microorganisms to synthesize phytohormones and antimicrobial substances (Sessitsch et al.., 2018). Microbial populations of the rhizosphere play an important role in sustainable agriculture due to improvement of soil fertility and reduction of the use of chemical fertilizers. Moreover, microorganisms improve plant resistance to abiotic stresses such as salinity, water deficit, and toxicity of heavy metals. Dynamics of interactions between the root system and microbial populations is regulated by plant genetic information and environmental factors. Engineering of rhizosphere microbiota through genome modification can help to increase nutrient acquisition efficiency and pathogen resistance.

2.3 Endophytic Microbiome

Endophytes are microorganisms that colonize the internal tissues of plants without inducing any disease symptoms. Endophytic bacteria and fungi colonize the roots, stems, leaves, and seeds of host plants. They support the growth and development of host plants by secreting growth hormones, increasing nutrient absorption, and reducing stress effects. They provide defensive function through production of anti-fungal compounds which inhibit growth of pathogenic microorganisms. In contrast to rhizosphere microbes, endophytes are directly associated with host plants thus making them efficient agents for promoting healthy growth of crops. Endophyte microorganisms have become important in agriculture because of their uses as bio-fertilizers, biopesticides, and stress control strategies (Khan et al., 2019). They enable better crop performance even in adverse environmental conditions such as drought and high salt concentrations. Modern molecular biology studies help to understand more about the composition of endophytes and their mode of action. Furthermore, the use of CRISPR technology makes it easier to genetically manipulate and improve metabolism of endophytes. Such modified microorganisms could be used to deliver protective molecules to crop plants (Mendes et al., 2013).

2.4 Phyllosphere Microbiome

Phyllosphere represents the above-ground surfaces of plants such as leaves, on which various kinds of microorganisms settle. Phyllosphere microorganisms are characterized by unstable living conditions because they are subject to the action of ultraviolet radiation, temperature shifts, and poor nutrition. Nevertheless, there are many bacteria, fungi, and yeasts that settle on the leaves of different species and positively impact plant health. The major types of bacteria found in the phyllosphere are Proteobacteria, Actinobacteria, and Firmicutes. Phyllosphere microorganisms affect plant metabolism through nutrient cycling, synthesis of plant hormones, and plant pathogens reduction (Berg et al., 2014). Phyllosphere microorganisms are essential for the protection of plants from foliar pathogens and stressful environmental factors. Some microbial communities synthesize antimicrobial substances and stimulate the systemic resistance mechanisms in plants. Furthermore, phyllosphere microbes participate in nutrient transformations in the atmosphere. Studies in the field of phyllosphere microorganisms can help create new methods of disease prevention. Genetic modification of phyllosphere microbes through CRISPR technology can help improve their colonization abilities, resistance, and biocontrol features.

2.5 Soil–Plant–Microbe Interactions

The interaction between soil, plants, and microbes comprises an intricate web of biological and biochemical reactions that control nutrient cycling, plant growth, and productivity. Root exudates released by plants determine soil microbiomes, while microorganisms control nutrient cycling and plant immunity. Plant-microbe symbioses, including nitrogen fixation by rhizobium bacteria and nutrient exchange by mycorrhizal fungi, are crucial for plant growth. Soil microbe-plant interactions play a role in soil aggregate formation, decomposition of organic matter, and carbon storage. Soil microbial diversity and activity are highly dependent on environmental conditions such as pH, water content, and temperature. The excessive application of agrochemicals and climate change pose risks to the soil microbiome balance and soil quality. Understanding the dynamic relationship between soil, plants, and microbes is vital for implementing sustainable agricultural systems. CRISPR-based gene editing provides innovative opportunities to alter the activities of microbes and promote beneficial interactions with plants (Niu et al., 2017).

3. FUNDAMENTALS OF CRISPR TECHNOLOGY

CRISPR is a genome-editing tool that has emerged as a result of the immunity systems present in the microorganisms. CRISPR stands for "Clustered Regularly Interspaced Short Palindromic Repeats," which is used in relation to CRISPR-associated (Cas) proteins. CRISPR systems protect the organisms from genetic elements such as phages and plasmids by targeting and cutting foreign DNAs. Modern CRISPR systems have found applications in editing genes within living beings. In comparison to other forms of gene editing such as zinc finger nucleases and transcription activator-like effector nuclease, CRISPR systems have proved to be efficient, simple, and cheap (Jacoby et al., 2017).

The development of CRISPR technology has revolutionized agricultural and microbiological research due to the introduction of highly specific modifications to genes. It allows for introducing desired mutations, gene insertions, deletions, and gene regulation in plants as well as their associated microbiomes. CRISPR technology finds its application in improving crops, increasing resistance to diseases and environmental stresses, and engineering microbiomes. The emergence of CRISPR technologies, like base editing, prime editing, and CRISPR interference, makes it possible to modify the genome without extensive damage to DNA. Thus, CRISPR technology is considered to be a promising tool for future sustainable agriculture and microbiology-based biotechnology (Trivedi et al., 2022).

3.1 CRISPR-Cas Immune System in Bacteria

The CRISPR-Cas system serves as an adaptive immunity strategy in microbes against attacking viruses and any foreign DNAs. After being infected with foreign DNA, a short fragment of the foreign DNA gets incorporated into the CRISPR locus of the cell as spacer. Spacers are the DNA regions that allow the bacteria or archaea to remember past attacks so that they can identify the attacking organism when it re-attacks. When this happens, the crRNA targets the foreign DNA and directs Cas enzymes to make cuts at its complementary sequence, thus inactivating the foreign genetic material (Shelake et al., 2019). The CRISPR immunity response process is normally divided into three main steps: adaptation, expression, and interference. Adaptation entails integration of new spacers within the CRISPR locus, expression involves transcription and processing of CRISPR array while interference step involves cleavage of target sequences by Cas nucleases. The natural immunity system has been employed in biotechnology for genome editing purposes. Due to its ease and precision, CRISPR-Cas systems have found wide application in agriculture microbiology and genetic engineering processes.

3.2 Types of CRISPR Systems

CRISPR systems can be broadly categorized into two main groups depending on the nature and architecture of their effector complexes. The class 1 CRISPR systems employ effector complexes comprising several proteins to target and cleave DNA strands, while the class 2 CRISPR systems depend on a single Cas protein that encompasses several domains (Singh et al., 2021). Both these classes are subdivided into various types and subclasses depending on their genetic and functional attributes. Among these CRISPR classes, the class 2 systems, especially the Cas9, Cas12a, and Cas13 systems, find extensive use in genome editing. Different CRISPR systems possess unique targeting mechanisms and applications. Cas9 primarily targets double-stranded DNA and is extensively used for genome editing in plants and microorganisms. Cas12a recognizes different protospacer adjacent motif (PAM) sequences and produces staggered DNA cuts that are useful for precise insertions. Cas13 targets RNA molecules instead of DNA, enabling transient gene regulation and antiviral applications. The diversity of CRISPR systems provides flexibility for various genome engineering strategies in agricultural biotechnology and microbiome research.

3.2.1 Class 1 Systems

Characteristics of Class 1 CRISPR systems include the presence of protein complexes consisting of multiple subunits responsible for recognizing the target sequence as well as its cleavage. The Class 1 CRISPR systems include Types I, III, and IV CRISPR systems. They are mostly prevalent in bacterial and archaeal species. For example, in the case of Type I CRISPR system, the Cascade protein complex recognizes the target DNA sequence while the Cas3 nuclease degrades it. The Type III CRISPR systems target both DNA and RNA molecules and, therefore, confer a broad immune defense against any foreign genetic material. Although the Class 1 CRISPR systems are more complex than Class 2 CRISPR systems, they have varied functionalities and therefore make great study subjects for scientists to understand microbial immunology and evolution. Researchers have even begun studying the use of Class 1 CRISPR systems in the programmable regulation of genes and detecting nucleic acids (Yadav et al., 2026).

3.2.2 Class 2 Systems

The Class 2 CRISPR systems feature only one effector molecule that targets and cleaves nucleic acids, thereby offering great potential in genome editing. The representatives of this class include Types II, V, and VI, exemplified by the Cas9,

Cas12, and Cas13 molecules respectively. The simplicity of their structure coupled with the ease of reprogramming has ensured popularity of the Class 2 systems among all CRISPR-based technologies used in agriculture and biotech. The effectiveness of such systems makes them perfect candidates for modifying genomes in plants, prokaryotes, and eukaryotes, including various purposes such as knock-outs, insertions, transcription factor activity control, and epigenetic manipulations. Class 2 systems have found wide use in the field of plant microbiome studies when it comes to manipulating beneficial microbes and altering plant genes responsible for recruiting these microbes (Berg et al., 2020).

3.2.3 Cas9

Cas9 is one of the most researched and applied CRISPR associated nuclease for genome editing processes. Cas9 nuclease is extracted from Streptococcus pyogenes and acts as an RNA guided endonuclease to induce a double-stranded break in particular places in the genome. The targeting activity of Cas9 nuclease is defined by single guide RNA (sgRNA), which binds to the complementary sequence located near the protospacer adjacent motif (PAM). Cellular repair processes including non-homologous end joining and homology directed repair allow for further genome modification after the breakage induced by Cas9 nuclease. The ease, precision, and effectiveness of Cas9 technology have made it possible to apply it in agricultural and microbial biotechnology (Pour et al., 2020). Applications of Cas9 based technologies range from enhancing crop productivity, pathogen resistance, and stress tolerance to improving the efficiency of rhizobacteria, endophytes, and fungi symbionts. Nevertheless, several issues related to potential off-target activities and delivery efficacy still exist in CRISPR research area.

3.2.4 Cas12a

Cas12a, which used to be referred to as Cpf1, is a Class 2 Type V CRISPR enzyme that is different from Cas9 in terms of structural and functional aspects. First, while Cas9 needs a PAM sequence with G-rich content, Cas12a needs a T-rich PAM sequence. Second, Cas12a induces double-stranded DNA cuts that are staggered and possess sticky ends. Third, Cas12a has the capability of processing its own CRISPR arrays, which increases multiplexing capabilities during genome editing. All these features render Cas12a particularly useful in complex genetic engineering operations. Cas12a has received much interest in biotechnology for agriculture owing to its specificity and flexibility of application. Staggered DNA cuts created by Cas12a help with precise gene addition and recombination. Moreover, in the case of bacteria associated with plants, CRISPR-Cas12a could promote nutrient acquisition, stress tolerance, and antibiosis (Bulgarelli et al., 2013).

3.2.5 Cas13

Cas13 is a specific RNA-targeting Cas nuclease that belongs to the Type VI Cas system. In contrast to Cas9 and Cas12a, which are designed to recognize and cut double-stranded DNA sequences, Cas13 acts on single-stranded RNA fragments. Such feature allows achieving transient and reversible regulation of gene activity in a way that does not involve any changes in the DNA structure of an organism. Cas13 systems are very useful in research related to gene expression, virology, and RNA-based gene regulation. The RNA-targeting properties of Cas13 can be used in various areas such as the development of plant pathology and microbiome engineering. With the help of Cas13, it will be possible to fight against RNA viruses that may affect crops and regulate gene expression in microorganisms (Arif et al., 2020).

3.3 Mechanism of Gene Editing

Gene editing through CRISPR technology involves using programmable nucleases that are directed by RNA molecules to make changes within the genome. First, the creation of guide RNAs that are complementary to a particular target region takes place. When the guide RNAs guide the Cas nuclease to the target region, the Cas nuclease makes either single-stranded breaks or double-stranded breaks in the DNA. The cell then repairs the DNA through various DNA repair mechanisms. The efficiency and outcome of CRISPR editing largely depend on the DNA repair mechanisms activated after cleavage (Sharma et al., 2003). Non-homologous end joining often produces insertions or deletions that disrupt gene function, whereas homology-directed repair enables precise sequence modification using a repair template. Advances in CRISPR technology have improved editing accuracy and reduced unintended mutations. These developments have expanded the applications of CRISPR systems in crop improvement, microbial engineering, and synthetic biology.

3.3.1 DNA Cleavage and Repair Pathways

Genome editing through CRISPR technology primarily operates through the creation of breaks at targeted sites, followed by triggering the inherent repair system of the cells. After targeting the appropriate genomic site in the vicinity of the PAM sequence, cleavage of the double helix occurs by the action of the Cas enzyme. This results in either blunt-ended or staggered double stranded DNA breaks, based on the kind of Cas enzyme used. Repairing the breaks through activation of the inherent repair systems forms the crux of the process of editing and hence forms an important aspect of genome editing. DNA repair pathways play a vital role in enabling genome editing and are necessary for carrying out edits such as gene knock outs, insertions, or replacement of a specific sequence. There are two main DNA repair pathways used in genome editing via CRISPR-Cas, which are non-homologous end joining and homologous recombination. Non-homologous end joining provides fast and efficient DNA repair, but it often leads to indels, while homologous recombination provides precise gene editing using a homologous repair template (Berg and Cernava, 2022)

3.3.2 Non-Homologous End Joining (NHEJ)

The non-homologous end-joining pathway serves as the most common means of DNA repair in several species and is extensively used in CRISPR-mediated gene knockout experiments. In such a case, damaged DNA strands are connected without the presence of homologous DNA templates. Being an error-prone process, small deletions and insertions (indels) are likely to occur during DNA strand repair. Such indels are capable of disrupting the DNA sequence, which would result in inactivation of the respective genes. Thanks to its simplicity and effectiveness, the use of the non-homologous end-joining system in plant and microbe genomes modification is quite popular. NHEJ-mediated CRISPR technology has multiple applications in agricultural studies and microbiome research. The gene knockout using the NHEJ pathway can help increase the plant and microorganism resistance to various diseases and improve their stress tolerance and metabolic pathways. Additionally, the major advantage of the NHEJ pathway is that it works effectively in both mitotic and post-mitotic cells. However, the unpredictability of indel formation may produce unexpected results in the process (de la Fuente Cantó et al., 2020).

3.3.3 Homology-Directed Repair (HDR)

Homology-directed repair (HDR) is another DNA repair process with extremely high accuracy in which the damaged double strand is repaired using a homologous DNA template. The CRISPR system involves the introduction of an external donor template that has the intended DNA modification surrounded by homology (Trivedi et al., 2021). During the repair process, this donor sequence is incorporated at the target site to allow the insertion, deletion, and replacement of the gene in the genome. HDR is useful for inducing mutations and modifying complex genetic traits. Although HDR is more accurate, HDR is usually less effective than NHEJ since the DNA repair pathway only occurs during the S and G2 phases of the cell cycle. The poor efficiency of HDR is one of the limitations in the application of this technique in plants and microbial organisms. Methods for improving HDR such as cell synchronization, optimizing donor templates, and inhibition of alternative NHEJ pathways have been suggested. In agricultural biotechnology, HDR can help precisely modify genes related to nutrient metabolism, stress resistance, and microbiome attraction.

3.4 Advanced CRISPR Tools

The fast evolution of CRISPR technologies has resulted in the creation of genome engineering platforms that demonstrate an enhanced level of precision, efficacy, and flexibility. Conventional CRISPR techniques are mainly based on the formation of double-stranded breaks in the target DNA, resulting in potentially undesirable genetic alterations and genomic instability. To overcome these disadvantages, next-generation CRISPR strategies such as base editing, prime editing, CRISPR activation (CRISPRa), and CRISPR interference (CRISPRi) were created. These innovations ensure targeted manipulation of the genome without the necessity of creating significant DNA breaks, thus avoiding unwanted changes and ensuring higher accuracy in editing (Compant et al., 2010b). Advanced CRISPR systems are also increasingly used in plant and microbial genome engineering, offering opportunities for the substitution of individual nucleotides, transcriptional control, and regulation of gene activity. Agricultural applications of advanced CRISPR systems include improvements in crop yield, tolerance to stress factors, increased resistance to pathogens, and optimal conditions for beneficial microorganisms. In microbial engineering, advanced CRISPR systems can be utilized to regulate the activity of metabolic pathways and perform other desirable manipulations at the genomic level.

3.4.1 Base Editing

Base editing is a highly evolved form of CRISPR gene editing which facilitates the direct change of one nucleotide to another without making any cuts across both strands of DNA. The technique usually involves the use of a Cas protein without the ability to cut DNA along with the use of a nucleotide deaminase (Obrien et al., 2021). Cytosine base editors are responsible for changing the C-G to T-A base pair while adenine base editors change the A-T base pair to G-C. As compared to traditional CRISPR technology, base editing causes less DNA damage, resulting in low chances of insertions, deletions, and chromosome rearrangements. Base editing has vast potential in crops and microbes as it offers more accurate gene editing. Base editing can help introduce beneficial point mutations related to resistance against diseases, herbicides, and stress. In plant microbes, base editing helps edit genes encoding nutrients, antibiotics, and symbiosis. Base editing holds tremendous promise in editing complex genomes in which single nucleotide substitutions are desired.

3.4.2 Prime Editing

Prime editing is a multifunctional genome engineering tool based on the CRISPR technique that allows one to make insertions, deletions, and any base substitutions without producing double-stranded breaks and the need for donor DNAs. Prime editing consists of a modified Cas9 nickase, reverse transcriptase, and pegRNA. Prime editing is one of the latest and most innovative developments in genome engineering due to its great versatility and accuracy. The agricultural sector is one of those where prime editing can have many important uses. In particular, this technique can be used for improving the yield, nutritional value, and adaptability of crops to changing conditions (Finkel et al., 2020). Moreover, it can allow the engineering of beneficial bacteria participating in nitrogen fixing, phosphates solubilizing, and pathogen suppression. Even though the efficiency of prime editing is rather variable, it can be further optimized for its implementation into sustainable agriculture practices and microbiome engineering.

3.4.3 CRISPR Activation (CRISPRa)

CRISPR activation is a method of regulating gene transcription without changing the underlying DNA sequence. The principle of CRISPRa involves using catalytically disabled Cas9 protein called dead Cas9 (dCas9) fused with an activator domain. The dCas9 protein recognizes a sequence of interest through its complementary guide RNA molecule and activates transcription of target genes. CRISPRa is a powerful tool for activating endogenous genes in plants and microorganisms. The use of CRISPRa in metabolic engineering, stress response, and microbiome analysis holds great promise. For example, by activating genes that regulate nutrient absorption, resistance signals, and stress tolerance, researchers may increase crop yield and hardiness. For microorganisms that have positive effects on their environment, CRISPRa could be used to upregulate genes responsible for biological fertilization, biosynthesis of antimicrobials, and acclimation to adverse conditions (Saad et al., 2020). Since CRISPRa does not cause permanent changes in genomic DNA, it represents a safer alternative to other gene-editing techniques.

3.4.4 CRISPR Interference (CRISPRi)

CRISPR Interference is a system that is based on the application of non-catalytic Cas proteins to block the process of transcription. In CRISPRi systems, dCas proteins are attached to the target genes with the help of RNA molecules, which prevents the transcriptional apparatus from transcribing the genes. As a result, a gene becomes inactive. This system allows scientists to program transcription of some genes in order to regulate metabolic processes both in plants and microorganisms (Berg et al., 2023). CRISPRi systems have a lot of applications in agriculture as well, including the inhibition of genes responsible for pathogenicity, toxin synthesis, and other unwanted metabolic functions. Moreover, using CRISPRi in agricultural microbiology helps regulate microbial functions related to the process of mineralization and stimulation of plant growth. In plants, CRISPRi helps modulate stress responses and immunity-related pathways.

4. CRISPR-BASED ENGINEERING OF PLANT-ASSOCIATED MICROORGANISMS

CRISPR technology has made significant advances in genetic engineering of microbes that are associated with plants by facilitating targeted gene editing. Beneficial microorganisms associated with plants include rhizospheric bacteria, endophytes, and mycorrhizal fungi, which are key elements in the productivity of agriculture as well as ecosystem sustainability. By utilizing CRISPR systems, beneficial traits of these microorganisms may be improved while unfavorable traits are downregulated. Such engineering makes possible the development of high-quality biological inoculants and biofertilizers that are sustainable alternatives to chemicals used in agriculture (Berg and Raaijmakers, 2018). There is growing interest in beneficial microorganisms that have been engineered using CRISPR-based systems due to growing concerns regarding the impact of climate change, deterioration of soils, and the overuse of agrochemicals. With such engineering, microorganisms become capable of facilitating nutrient cycling, pathogen suppression, and tolerance to environmental stress conditions. Furthermore, engineered microorganisms have the potential to facilitate carbon sequestration.

4.1 Genome Editing in Beneficial Rhizobacteria

Beneficial rhizobacteria, also referred to as plant growth-promoting rhizobacteria (PGPR), play crucial roles within the rhizosphere microflora in stimulating plant growth and increasing soil fertility. These microbes stimulate the bioavailability of nutrients by fixing atmospheric nitrogen, dissolving phosphates, producing iron-chelating molecules, and synthesizing phytohormones. CRISPR-based genome engineering helps to target and modify genes that encode for such traits, hence enhancing the efficiency of the microbial organisms. Genome engineering of rhizobacteria utilizing CRISPR technology has led to the production of microbial strains with improved stress tolerance, root colonization, and pathogen-suppressive capabilities (Compant et al., 2019). Engineered rhizobacteria can be engineered to synthesize larger amounts of antimicrobials and growth stimulants. Additionally, CRISPR technology aids in eliminating unwanted genes that might hinder the survival of the organisms within the environment. Such engineered PGPR strains hold great promise as eco-friendly substitutes for chemical fertilizers and pesticides in contemporary agriculture.

4.2 CRISPR Editing of Endophytes

Endophytes refer to microorganisms that colonize the interior of plants without exerting any negative impacts and usually develop beneficial symbioses with the host plants. Endophytes play crucial roles in facilitating the growth of plants through synthesis of phytohormones, uptake of nutrients, and increasing the resistance of the latter against various types of abiotic and biotic stresses. CRISPR genome editing technique appears to be efficient in manipulating the genetic makeup of endophytes to increase the positive impact on plant development. By editing the genes responsible for metabolism or regulation of specific processes in endophytes, scientists can increase their capability of positively impacting plant growth and adaptation to unfavorable conditions. CRISPR editing of endophytes can be useful in several agricultural and environmental settings. It becomes possible to engineer endophytes to synthesize antibiotics, metabolites protecting against stress factors, and enzymes facilitating the mobilization of nutrients. Additionally, it is possible to alter the genes involved in the efficiency of colonization of endophytes and thus make them more persistent inside plants. CRISPR editing can also help perform research related to the interaction between plants and endophytes at genetic levels (Levy et al., 2018).

4.3 CRISPR Engineering of Mycorrhizal Fungi

Mycorrhizal fungi are fungi that have mutual relationships with plant roots. They enhance nutrient absorption, soil formation, and resistance to stress in plants. The fungi can help increase the absorption of phosphorus, nitrogen, and other micronutrients by plants and also increase their resistance to drought, salt, and infections. Gene editing using CRISPR technology on mycorrhizal fungi can be used to improve the functionality of these fungi in the agricultural ecosystem (Compant et al., 2010a). This involves modifying specific genes responsible for nutrient transport, signaling processes, and stress tolerance among others. CRISPR technology is also applied to gain knowledge on mycorrhizal biology and symbiotic processes. CRISPR technology helps modify fungi so that they become more colonized, more efficient in nutrient exchange, and more resistant to harsh environmental conditions. Moreover, the modifications can ensure that the fungi are not vulnerable to soil pollutants and other climatic stresses.

4.4 Modification of Biocontrol Agents

Biocontrol agents refer to helpful microorganisms used in limiting plant diseases to minimize damage to crops due to pathogens and pest attacks. Some of the common microbial biocontrol agents are Bacillus, Pseudomonas, Trichoderma, and Streptomyces, among others. They synthesize antimicrobials, enzymes, and volatile metabolites for inhibiting the pathogenic microorganisms. Genome editing of biocontrol agents using CRISPR technology is important for improving the pathogen inhibition and adaptation ability of such bacteria. By making modifications to various parts of the genome, certain genes responsible for pathogen suppression, tolerance to stress, and colonization of roots can be optimized. Engineering biocontrol agents using CRISPR technology offers an ecological alternative to conventional chemical pesticides (Vandenkoornhuyse et al., 2015). Such enhanced bacteria can demonstrate higher pathogen suppression and broad protection to multiple plants, in addition to increased survivability in natural settings. Additionally, CRISPR engineering allows for the deletion of genes contributing to adverse effects on the growth of the microorganism. Moreover, it is possible to create biocontrol bacteria with biofertilizer features.

5. CRISPR-MEDIATED MODULATION OF PLANT GENES AFFECTING MICROBIOME RECRUITMENT

5.1 Plant Genes Involved in Microbiome Assembly

The microbiome assembly process is controlled by a complicated gene network responsible for the root exudates production, plant immune system function, and signaling pathways. Genes impact microbial recruitment, colonization success, and the composition of the microbial community in the rhizosphere and plant tissues. The genetic differences between plants may greatly affect microbiome composition, which will lead to changes in their adaptation capacity and overall growth rate. CRISPR-based modifications of microbiome-associated genes offer a way to create crops capable of recruiting beneficial microbes (Backer et al., 2018). The advances in genomics and transcriptomics research have offered new tools to identify the genes important for interactions between plants and microbes. Experimental results based on functional analysis of selected genes using CRISPR technology showed that manipulation of certain genes may modify the composition of microbes and increase symbiotic relationships.

5.2 Root Exudate Biosynthesis Genes

The root exudates consist of complex combinations of sugars, amino acids, organic acids, flavonoids, and secondary metabolites that are released from the plant roots into the soil. These molecules act as signalling agents and nutrients that help regulate rhizosphere microbial communities. The genetic components responsible for the synthesis of root exudates play a critical role in the recruitment of plant growth-promoting microorganisms, including nitrogen-fixing bacteria and phosphate-solubilizing bacteria (Bhattacharyya et al., 2012). Using the CRISPR method, these genes can be engineered to fine-tune the process of attracting microbes. Engineering root exudate synthesis pathways could increase nutrient uptake efficiency, disease resistance, and stress tolerance in plants. An increase in flavonoid secretion could promote the establishment of rhizobacteria and increase the efficacy of nitrogen fixation processes. In a similar way, altering organic acid synthesis could lead to an increase in phosphate solubilization and micronutrients in soils (Mitter et al., 2019).

5.3 Immune Signaling Pathways

Plants' immune signaling pathway helps in sensing microorganisms and choosing if the association of those microorganisms with plants is either advantageous or harmful. Pattern recognition receptors and other immune components in plants recognize motifs in microorganisms and trigger immune responses. Although the immune system protects plants from invading pathogens, activation of the immune system may also hinder the development of beneficial microorganisms (Orozco-Mosqueda et al., 2022). By modifying immune signaling pathways through CRISPR, it will be possible to modify the defense response of plants and enable the formation of better symbiosis between plants and microorganisms while also fighting off pathogens. In addition to enhancing symbiotic relations with microorganisms, modifying the immune system would also provide plants with pest resistance. This is because the inactivation of some immune receptors in plants enables beneficial microorganisms to flourish in the environment surrounding plants, while activating certain defense-related genes allows plants to fight off invading pathogens.

5.4 Hormonal Regulation

Hormones such as auxins, cytokinins, ethylene, jasmonic acid, salicylic acid, and abscisic acid have been recognized to be involved in plant growth and microbiome interaction processes. Signaling of hormonal pathways is known to affect root morphology, plant immune system activation, and patterns of microbial colonization. Microorganisms beneficial to plants tend to regulate the levels of plant hormones to develop functional symbioses between them and plants. This means that it is possible to edit the gene responsible for the hormonal pathways using the CRISPR technology to improve plant-microbe interactions and crop yields (Gopal et al., 2013). Modulating hormonal pathways would help to promote the growth of beneficial microorganisms as well as enhancing plant resistance to environmental stresses. For example, alteration in ethylene pathways would help in improving the root colonization of growth-promoting bacteria, whereas regulation of jasmonic acid pathways would improve resistance to pathogens.

6. APPLICATIONS IN SUSTAINABLE AGRICULTURE

CRISPR-based microbiome engineering has emerged as a transformative approach for achieving sustainable agricultural development. By combining precise genome editing with beneficial plant–microbe interactions, researchers can develop crops and microbial inoculants with enhanced productivity, stress tolerance, and disease resistance. This technology provides environmentally friendly alternatives to conventional agricultural practices that rely heavily on synthetic fertilizers and pesticides. In addition, CRISPR-mediated strategies support ecological balance and resource conservation by improving nutrient use efficiency and reducing environmental pollution.

The application of CRISPR systems in sustainable agriculture extends beyond crop improvement to include soil restoration, climate resilience, and integrated pest management (Santoyo, 2026). Engineered microorganisms and optimized plant genotypes can enhance nutrient cycling, suppress pathogens, and improve tolerance to abiotic stresses such as drought and salinity. Furthermore, microbiome-based technologies contribute to reduced greenhouse gas emissions and improved soil carbon sequestration. As global food demand continues to increase, CRISPR-assisted sustainable agriculture offers promising solutions for ensuring long-term food security and environmental sustainability. Table 1 summarizes the major applications of CRISPR-mediated plant microbiome engineering in sustainable agriculture. It highlights different genome-editing strategies used to improve crop productivity, biofertilizer development, pathogen control, abiotic stress tolerance, and climate resilience, along with their associated agricultural benefits.

Table 1. CRISPR-Based Strategies and Agricultural Applications of Plant Microbiome Engineering

Applications	CRISPR-Based Strategy	Agricultural Benefits	References
Enhancement of Crop Productivity	Editing plant and microbial genes associated with nutrient uptake, stress response, and symbiosis	Improved crop yield, nutrient efficiency, disease resistance, and plant growth	(Phour et al., 2020)
Biofertilizer Development	Genetic modification of nitrogen-fixing and phosphate-solubilizing microbes	Reduced dependence on chemical fertilizers, improved soil fertility, and enhanced microbial diversity	(Olanrewaju et al., 2017)
Biocontrol of Plant Pathogens	Engineering beneficial microbes to produce antimicrobial compounds and suppress pathogens	Eco-friendly disease management and reduced pesticide usage	(Compant et al.., 2025)
Abiotic Stress Tolerance	Modification of stress-responsive genes in plants and microbes	Increased tolerance to drought, salinity, heat, and heavy metal stress	(Berendsen et al., 2012)
Climate-Resilient Agriculture	Engineering crops and microbial consortia for climate adaptation and carbon sequestration	Improved resilience to climate change, reduced greenhouse gas emissions, and sustainable crop production	(Vurukonda et al., 2016)

6.1 Enhancement of Crop Productivity

The performance of crops is highly dependent on the availability of nutrients, health of the plants, environmental factors, and microbial activities. CRISPR-based genetic modification of plants and beneficial microbes allows for the improvement of characteristics that will aid in growth enhancement and improvement of yield. CRISPR can be used to optimize pathways related to nutrient uptake, stress responses, and beneficial symbiosis, which will result in higher agricultural productivity under varied environmental conditions (Li et al., 2026). Beneficial microbes genetically engineered using CRISPR technology have the potential to enhance nitrogen fixation, solubilization of phosphate, and production of phytohormones. Edited plants that are able to recruit beneficial microbes selectively have the potential to enhance nutrient uptake and disease resistance.

6.2 Biofertilizer Development

Biofertilizers are products comprising beneficial microbes that help to make nutrients available for uptake and promote soil fertility using biological mechanisms. Nitrogen-fixing bacteria, phosphate solubilizing bacteria, cyanobacteria, and mycorrhizal fungi are examples of microbes used in biofertilizers. CRISPR technology has greatly aided the development of biofertilizers by allowing scientists to precisely edit genes responsible for nutrient acquisition, stress tolerance, and root colonization in microbes. Genetically modified microbial strains may be more efficient metabolically and better adapted to changing environmental conditions, making them more effective during field application. CRISPR-enabled biofertilizers provide more sustainable substitutes for artificial fertilizers that cause soil degradation, water contamination, and greenhouse gas emissions. Biofertilizer strains could enhance the efficiency of nutrient absorption, facilitate plant growth, and boost soil microbial diversity. In addition, biofertilizer strains modified at the genomic level could have prolonged shelf lives and better crop species compatibility (Singh et al., 2020).

6.3 Biocontrol of Plant Pathogens

Pathogens such as bacteria, fungi, viruses, and nematodes cause significant crop losses across the world. Traditional methods of disease control usually involve the use of harmful synthetic pesticides that might affect other organisms and pollute the environment. CRISPR modification of beneficial microorganisms offers novel methods to combat plant pathogens biologically. Genome-edited biocontrol agents will be able to manufacture better antimicrobial agents, lytic enzymes, and competitive metabolites that can effectively eliminate pathogens (Santoy et al., 2016). Moreover, CRISPR technology can also be used to directly target the genes responsible for the virulence of the pathogen and its resistance mechanisms. Genome-edited organisms will be able to inhibit colonization, quorum sensing, and toxins of the pathogen. Moreover, genome-editing technologies allow for the development of diagnostic techniques that can detect pathogens in plants.

6.4 Abiotic Stress Tolerance

Environmental stresses like drought, salinity, temperature extremes, heavy metals, and others adversely impact agricultural productivity worldwide, posing significant risks to food security. Plant-associated beneficial microorganisms serve as indispensable tools for increasing the tolerance levels of plants under environmental stress conditions (desouza et al., 2015). Plant and microbial genetic engineering using CRISPR provides an effective way of changing gene expression associated with osmoregulation, antioxidative response, stress signaling, and metabolism. Engineering of the beneficial soil bacteria and endophytes will facilitate improved plant water absorption, regulation of stress-related hormones, and decreased oxidative stress in plants grown under environmental stress conditions. Likewise, CRISPR editing of crops can result in improved root morphology, ion balance, and stress responses in plants grown under environmental stresses. Such advances in research allow for increasing productivity in stressed environments through enhanced tolerance mechanisms.

6.5 Climate-Resilient Agriculture

Global warming has become a major threat facing contemporary agriculture, characterized by high temperature, unpredictable precipitation, soil erosion, and climatic catastrophes. Sustainable agricultural systems need to have the ability to sustain their productivity in the face of dynamic changes in the environment. Microbiome engineering through CRISPR technology is a viable strategy to develop agriculture that is resilient to climate change by improving plant resistance, nutrient utilization, and the stability of microbial ecosystems. Genetically edited crops and microbes can help increase the resistance of plants to drought, salinity, flood, and new plant pathogens (Toju et al., 2016). The creation of climate-resilient agriculture entails minimizing greenhouse gas emissions and increasing carbon sequestration. Engineered microbial consortia can assist in enhancing soil carbon storage and minimizing nitrogen losses, which will minimize the impact of traditional agriculture on the environment. Additionally, CRISPR-enabled crop modification is faster than conventional breeding when dealing with climate change-related challenges.

7. CRISPR AND SOIL HEALTH IMPROVEMENT

7.1 Soil Microbial Diversity Restoration

Diversity of soil microflora plays an important role in the maintenance of ecological balance, nutrient cycling, and disease suppression within agricultural ecosystem systems. Farming activities and pollution tend to negatively impact soil microflora and reduce numbers of beneficial microflora. Microbial diversity can be restored via modification of microorganisms using CRISPR technology by increasing populations of beneficial bacteria and reducing pathogenic properties of some bacteria (Toju et al., 2018). Engineering will allow to restore ecological interaction between plants and microorganisms leading to more stable microbial communities. Increasing microbial diversity will contribute to the improvement of soil fertility, plant development, and increase resilience towards environmental changes. Engineered microorganisms will help to decompose organic matter, nutrients mineralization, and disease suppression within degraded soil microecosystem. Also, restoration can help in ecological restoration of soils in polluted, saline, and climatically changing regions.

7.2 Nutrient Cycling Enhancement

Nutrient cycling refers to the process through which important nutrients, including nitrogen, phosphorus, sulfur, and carbon, cycle through the ecosystem. In nutrient cycling, soil microorganisms perform critical functions by carrying out biochemical processes like nitrogen fixation, phosphate solubilization, nitrification, denitrification, and organic matter breakdown. Using CRISPR technology, manipulation of metabolic pathways of the microbes can be done to increase efficiency and improve nutrient cycling, making more nutrients available to plants. Modification of these microorganisms might result in an improvement of their ability to fix nitrogen, mobilize phosphorus, and sequester carbon, thus increasing the fertility of the soil. Moreover, using CRISPR technology, we can regulate the activity of enzymes that take part in nutrient cycling pathways (Lau et al., 2022).

7.3 Reduction of Agrochemical Dependency

Agriculture is reliant on artificial fertilizers, insecticides, and herbicides in order to achieve optimal crop production levels while protecting crops from pests and plant diseases. Excessive application of agrochemicals leads to soil degradation, water contamination, biodiversity destruction, and even negative impacts on human health. Microbiome engineering via CRISPR can provide alternative solutions based on improving biological functions related to nutrient utilization and disease prevention. Beneficial microorganisms modified using CRISPR technology can help lower agrochemical reliance by boosting nutrient availability, preventing diseases, and improving plants' stress resistance (Trivedi et al., 2020). Likewise, modification of crops via CRISPR technology can enhance their capacity for attracting microbiome thus lowering dependence on agrochemical substances. In this regard, the introduction of microbiome-oriented farming methods promotes eco-friendly cultivation systems without compromising crop yield.

8. MULTI-OMICS APPROACHES IN CRISPR–MICROBIOME RESEARCH

Multi-omics techniques have played an important part in the analysis of plant microbiome and their relationship with host plants. Multi-omics refers to techniques that analyze genomics, transcriptomics, proteomics, and metabolomics to obtain data about microbial communities, genes, proteins, and metabolic pathways (Tuncel et al., 2025). The application of multi-omics in the CRISPR-microbiome research is key to finding target genes and the effect of genome editing on different biological systems. The use of CRISPR technique with multi-omics helps in characterizing and editing plant microbiome. Researchers can identify the genes related to nutrient cycling, environmental stresses, pathogen resistance, and symbiosis. The application of omics techniques makes it easy to monitor edited microbial communities and understand the ecosystem impact of the editing.

8.1 Metagenomics

The field of metagenomics involves studying genetic information obtained directly from the environment through DNA isolation without cultivation of microbes. It allows for a complete evaluation of diversity, taxonomic identification, and gene functional profile of microbial communities in environmental samples like soils and plant microbiomes. With the help of high-throughput sequencing methods, metagenomics has made remarkable contributions to this field, providing extensive knowledge about microbial communities' structure and ecology (Kaul et al., 2016). As for the case of CRISPR technology in engineering microorganisms in agriculture, metagenomics is extensively employed for screening beneficial microorganisms as well as genes responsible for promoting plant growth and stress resistance. Metagenomic studies contribute to discovering new CRISPR technologies and microbial pathways related to agriculture, as well as allow for assessing ecological impacts on natural microbial communities.

8.2 Metatranscriptomics

The study of metatranscriptomics is the examination of all RNA transcripts expressed by microbes in certain environments. Contrary to metagenomics that provides genetic potential, metatranscriptomics gives an insight into gene expressions and functions under certain environmental conditions. The technique allows for understanding the response of microbes to the environment, the signaling from the plant and among other microbes. In CRISPR technology used for microbiome engineering, metatranscriptomics is useful for assessing the changes in gene expressions due to genetic manipulations. It aids in understanding the regulatory networks involved in microbial processes such as nutrient cycling, stress response, and colonization. Metatranscriptomics gives knowledge on how plants communicate with the beneficial microbes (de Souza et al., 2020).

8.3 Metabolomics

Metabolomics involves the analysis of all small molecules and metabolic products formed in biological systems. The study of plant-microbiome interactions, metabolomics provides in-depth information on the biochemical relationship that exists between plants and associated microbiomes (Kumar et al., 2020). Various metabolites such as organic acids, amino acids, sugars, phytohormones, and secondary metabolites contribute to microbial recruitment, nutrients transfer, and stress response mechanisms. State-of-the-art analytical platforms such as mass spectrometry and nuclear magnetic resonance are currently employed in the metabolic analysis of plant-based ecosystems. In the context of CRISPR-microbiome manipulation, metabolomics helps assess the impact of gene modification on both host plant and microbiome organisms through metabolic profiling. Metabolomics provides valuable information concerning metabolic pathways involved in nutrient mobilization, antimicrobial production, and stress response mechanisms. Moreover, the technique helps determine the effect of genetic modification on plant-microbiome interactions.

8.4 Proteomics

Proteomics is the systematic study of proteins that are produced by living cells, tissue samples, or microorganisms within defined environments. As proteins are the key functional molecules in all cellular activities, proteomics has played an indispensable role in the elucidation of microbial metabolism, plant immune responses, and metabolic pathways. With the advent of advanced proteomic technologies like two-dimensional gel electrophoresis and mass spectrometry, it has become possible to identify and quantify the proteins responsible for the biological interactions between plants and microorganisms. Within the context of CRISPR-enabled microbiome research, the use of proteomics is vital for understanding the implications of genome engineering. Proteomic profiling is instrumental in assessing the impact of genetic modifications on metabolic pathways, stress response, and microbial colonization efficiency (Zhu et al., 2020).

9. BIOSAFETY, ETHICAL, AND REGULATORY CONSIDERATIONS

The application of CRISPR technology in agriculture and microbiome engineering raises important biosafety, ethical, and regulatory concerns. Although genome editing offers significant benefits for crop improvement and sustainable agriculture, unintended genetic alterations and ecological impacts remain potential risks. Off-target mutations, horizontal gene transfer, and disruption of native microbial ecosystems are among the major biosafety considerations associated with engineered organisms.

Therefore, comprehensive risk assessment and long-term environmental monitoring are essential before large-scale agricultural deployment (Rodriguez et al., 2006).

Ethical concerns surrounding CRISPR-based technologies include issues related to biodiversity conservation, equitable access to biotechnology, intellectual property rights, and public acceptance of genetically edited organisms. Regulatory frameworks governing genome-edited crops and microorganisms vary across countries and are continuously evolving with scientific advancements. Some regulatory agencies classify CRISPR-edited organisms differently from traditional genetically modified organisms when no foreign DNA is introduced. Establishing transparent regulatory policies and promoting public awareness are important for responsible implementation of CRISPR-assisted agricultural technologies.

10. CHALLENGES AND LIMITATIONS

Despite the remarkable potential of CRISPR technology in microbiome engineering and sustainable agriculture, several scientific and technical challenges remain unresolved. One of the major limitations is the occurrence of off-target effects, where unintended genomic regions are modified, potentially affecting organismal stability and safety. Delivery of CRISPR components into plant cells and complex microbial communities also remains technically challenging, particularly in field conditions. Additionally, differences in editing efficiency among species and environmental variability can limit the reproducibility of genome engineering outcomes (Lugtenberg and Kamilova, 2009).

Another significant challenge involves the complexity of plant microbiomes and ecological interactions. Plant-associated microbial communities consist of diverse organisms that interact dynamically with environmental factors, making precise manipulation difficult. Engineered microorganisms may face competition from native microbes, limiting their survival and effectiveness in natural ecosystems. Regulatory uncertainty, public concerns, and high development costs also hinder commercialization of CRISPR-based agricultural technologies. Addressing these limitations requires interdisciplinary research integrating molecular biology, microbiology, ecology, computational biology, and agricultural sciences.

11. FUTURE PERSPECTIVES

The future of CRISPR-mediated microbiome engineering holds tremendous promise for transforming global agriculture and environmental sustainability. Continued advancements in genome editing technologies, computational biology, and systems microbiology are expected to improve the precision and efficiency of genetic manipulation in plants and microorganisms. Emerging approaches involving synthetic biology, artificial intelligence, and machine learning may facilitate predictive modelling of plant–microbe interactions and optimization of engineered microbial communities (Busby et al., 2017). These innovations could lead to the development of highly resilient agricultural systems capable of adapting to climate change and increasing food demands.

Future research is likely to focus on integrating CRISPR technologies with ecological and evolutionary principles to ensure long-term sustainability and biosafety. Precision microbiome engineering may enable the creation of customized microbial consortia designed for specific crops, soils, and environmental conditions (Kalam et al., 2020). Furthermore, advances in delivery systems and gene regulation technologies could enhance the practical implementation of CRISPR-based solutions in agriculture. As scientific understanding of microbiome dynamics continues to expand, CRISPR-assisted agricultural biotechnology is expected to play a central role in achieving sustainable food security and ecosystem resilience.

11.1 Next-Generation CRISPR Technologies

Future CRISPR technologies that are in the works include improvements in editing precision, reductions in off-target effects, and expanding genome engineering capabilities. Examples of some of these emerging CRISPR technologies include high-fidelity Cas nucleases, RNA-guided transposases, epigenome editors, and CRISPR associated transposons. Allowing increased flexibility of genetic modification, these innovative techniques can regulate gene expression, alter chromatin structure, and insert large stretches of DNA without causing any significant damage to the genome. The use of next generation CRISPR technologies in agriculture and microbiome engineering could greatly expedite developments in these sectors. Better editing tools will enable the rapid generation of crops with better nutrition, tolerance to environmental stresses, and increased resistance to disease. In much the same way, microbiome engineering will lead to improved processes for nutrient cycling, carbon sequestration, and pathogen inhibition in agricultural systems (Bashan et al., 2014).

11.2 Precision Microbiome Engineering

Precision microbiome engineering pertains to the process wherein microbial consortia are manipulated through genetic techniques to attain a certain function in plants and the overall agricultural environment. Precision microbiome engineering utilizes advancements in CRISPR technology, synthetic biology, and systems microbiology to create engineered microbial communities that are optimized for their ecological function. Contrary to traditional microbial inoculation practices, precision microbiome engineering seeks to control microbial interactions and metabolisms within the microbial community in a highly controlled manner. Some possible future applications of precision microbiome engineering will include creation of customized microbial communities for crops, climate adaptive biofertilizers, and sustainable disease management mechanisms (Finkel et al., 2017). Through precision microbiome engineering, there is potential to optimize nutrient acquisition, improve stress tolerance, and revitalize degraded soil ecosystems.

12. CONCLUSION

The CRISPR engineering of the plant microbiome is undoubtedly an impressive development in the sphere of agricultural biotechnologies that will affect future sustainable food production and environmental conservation. The combination of advanced gene editing techniques along with beneficial plant-microbe relations can result in effective strategies aimed at increasing crop productivity, optimizing nutrition processes, providing plants with additional sources of protection, and improving their resistance to different climatic conditions (Venturi et al., 2016). CRISPR modifications made both on the level of plant genes and microbes help design highly adaptable systems in agriculture that could effectively solve the issues of soil degradation, climate change, and food supply. Although the use of such biotechnologies faces certain challenges due to concerns about safety, regulatory procedures, and complexity of microbial ecosystems, ongoing improvements in CRISPR technologies and omics provide new opportunities in this sphere. It may be expected that further development of synthetic biology and computational analysis will allow researchers to design more efficient genome editing approaches.

REFERENCES

Arif, I., Batool, M., & Schenk, P. M. (2020). Plant microbiome engineering: Expected benefits for improved crop growth and resilience. *Trends in Biotechnology*, *38*(12), 1385–1396. DOI: 10.1016/j.tibtech.2020.04.015 PMID: 32451122

Backer, R., Rokem, J. S., Ilangumaran, G., Lamont, J., Praslickova, D., Ricci, E., Subramanian, S., & Smith, D. L. (2018). Plant growth-promoting rhizobacteria: Context, mechanisms of action, and roadmap to commercialization of biostimulants for sustainable agriculture. *Frontiers in Plant Science*, *9*, 1473. DOI: 10.3389/fpls.2018.01473 PMID: 30405652

Banerjee, S., Schlaeppi, K., & van der Heijden, M. G. A. (2018). Keystone taxa as drivers of microbiome structure and functioning. *Nature Reviews. Microbiology*, *16*(9), 567–576. DOI: 10.1038/s41579-018-0024-1 PMID: 29789680

Bashan, Y., de-Bashan, L. E., Prabhu, S. R., & Hernandez, J.-P. (2014). Advances in plant growth-promoting bacterial inoculant technology. *Plant and Soil*, *378*(1–2), 1–33. DOI: 10.1007/s11104-013-1956-x

Berendsen, R. L., Pieterse, C. M. J., & Bakker, P. A. H. M. (2012). The rhizosphere microbiome and plant health. *Trends in Plant Science*, *17*(8), 478–486. DOI: 10.1016/j.tplants.2012.04.001 PMID: 22564542

Berg, G., & Cernava, T. (2022). The future of plant microbiome research for sustainable agriculture. *Microbial Biotechnology*, *15*(1), 6–10. DOI: 10.1111/1751-7915.13916 PMID: 33529492

Berg, G., Cernava, T., Schloter, M., & Smalla, K. (2023). The plant microbiome in sustainable agriculture. *Annual Review of Phytopathology*, *61*, 123–145. DOI: 10.1146/annurev-phyto-021622-123456

Berg, G., Grube, M., Schloter, M., & Smalla, K. (2014). Unraveling the plant microbiome: Looking back and future perspectives. *Frontiers in Microbiology*, *5*, 148. DOI: 10.3389/fmicb.2014.00148 PMID: 24926286

Berg, G., Köberl, M., Rybakova, D., Müller, H., Grosch, R., & Smalla, K. (2017). Plant microbial diversity is suggested as the key to future biocontrol and health trends. *FEMS Microbiology Ecology*, *93*(5), fix050. Advance online publication. DOI: 10.1093/femsec/fix050 PMID: 28430944

Berg, G., & Raaijmakers, J. M. (2018). Saving seed microbiomes. *The ISME Journal*, *12*(5), 1167–1170. DOI: 10.1038/s41396-017-0028-2 PMID: 29335636

Berg, G., Rybakova, D., Fischer, D., Cernava, T., Vergès, M.-C. C., Charles, T., Chen, X., Cocolin, L., Eversole, K., Corral, G. H., Kazou, M., Kinkel, L., Lange, L., Lima, N., Loy, A., Macklin, J. A., Maguin, E., Mauchline, T., McClure, R., & Schloter, M. (2020). Microbiome definition re-visited: Old concepts and new challenges. *Microbiome*, *8*(1), 103. DOI: 10.1186/s40168-020-00875-0 PMID: 32605663

Bhattacharyya, P. N., & Jha, D. K. (2012). Plant growth-promoting rhizobacteria: Emergence in agriculture. *World Journal of Microbiology & Biotechnology*, *28*(4), 1327–1350. DOI: 10.1007/s11274-011-0979-9 PMID: 22805914

Bulgarelli, D., Schlaeppi, K., Spaepen, S., van Themaat, E. V. L., & Schulze-Lefert, P. (2013). Structure and functions of the bacterial microbiota of plants. *Annual Review of Plant Biology*, *64*(1), 807–838. DOI: 10.1146/annurev-arplant-050312-120106 PMID: 23373698

Busby, P. E., Soman, C., Wagner, M. R., Friesen, M. L., Kremer, J., Bennett, A., Morsy, M., Eisen, J. A., Leach, J. E., & Dangl, J. L. (2017). Research priorities for harnessing plant microbiomes in sustainable agriculture. *PLoS Biology*, *15*(3), e2001793. DOI: 10.1371/journal.pbio.2001793 PMID: 28350798

Compant, S., Cassan, F., Kostić, T., Johnson, L., Brader, G., Trognitz, F., & Sessitsch, A. (2025). Harnessing the plant microbiome for sustainable crop production. *Nature Reviews. Microbiology*, *23*(1), 9–23. DOI: 10.1038/s41579-024-01079-1 PMID: 39147829

Compant, S., Clément, C., & Sessitsch, A. (2010a). Plant growth-promoting bacteria in the rhizo- and endosphere of plants: Their role, colonization, mechanisms involved and prospects for utilization. *Soil Biology & Biochemistry*, *42*(5), 669–678. DOI: 10.1016/j.soilbio.2009.11.024

Compant, S., Samad, A., Faist, H., & Sessitsch, A. (2019). A review on the plant microbiome: Ecology, functions, and emerging trends in microbial application. *Journal of Advanced Research*, *19*, 29–37. DOI: 10.1016/j.jare.2019.03.004 PMID: 31341667

Compant, S., van der Heijden, M. G. A., & Sessitsch, A. (2010b). Climate change effects on beneficial plant–microorganism interactions. *FEMS Microbiology Ecology*, *73*(2), 197–214. DOI: 10.1111/j.1574-6941.2010.00900.x PMID: 20528987

de la Fuente Cantó, C., Simonin, M., King, E., Moulin, L., Bennett, M. J., Castrillo, G., & Laplaze, L. (2020). An extended root phenotype: The rhizosphere, its formation and impacts on plant fitness. *The Plant Journal : For Cell and Molecular Biology*, *103*(3), 951–964. DOI: 10.1111/tpj.14781 PMID: 32324287 de Lima Procópio, R. E., da Silva, I. R., Martins, M. K., de Azevedo, J. L., & de Araújo, J. M. (2012). Antibiotics produced by Streptomyces. *The Brazilian Journal of Infectious Diseases : an Official Publication of the Brazilian Society of Infectious Diseases*, *16*(5), 466–471. DOI: 10.1016/j.bjid.2012.08.014 PMID: 22975171

de Souza, R. S. C., Ambrosini, A., & Passaglia, L. M. P. (2015). Plant growth-promoting bacteria as inoculants in agricultural soils. *Genetics and Molecular Biology*, *38*(4), 401–419. DOI: 10.1590/S1415-475738420150053 PMID: 26537605

de Souza, R. S. C., Okura, V. K., Armanhi, J. S. L., Jorrín, B., Lozano, N., da Silva, M. J., González-Guerrero, M., de Araújo, L. M., Verza, N. C., Bagheri, H. C., & Imperial, J. (2020). Unlocking the bacterial and fungal communities assemblages of sugarcane microbiome. *Scientific Reports*, *10*(1), 19320. DOI: 10.1038/s41598-020-76221-2 PMID: 33168921

Finkel, O. M., Castrillo, G., Paredes, S. H., González, I. S., & Dangl, J. L. (2017). Understanding and exploiting plant beneficial microbes. *Current Opinion in Plant Biology*, *38*, 155–163. DOI: 10.1016/j.pbi.2017.04.018 PMID: 28622659

Finkel, O. M., Salas-González, I., Castrillo, G., Conway, J. M., Law, T. F., Teixeira, P. J. P. L., Wilson, E. D., Fitzpatrick, C. R., Jones, C. D., & Dangl, J. L. (2020). A single bacterial genus maintains root growth in a complex microbiome. *Nature*, *587*(7832), 103–108. DOI: 10.1038/s41586-020-2778-7 PMID: 32999461

Glick, B. R. (2012). Plant growth-promoting bacteria: Mechanisms and applications. *Scientifica*, *963401*, 1–15. Advance online publication. DOI: 10.6064/2012/963401 PMID: 24278762

Gopal, M., Gupta, A., & Thomas, G. V. (2013). Bespoke microbiome therapy to manage plant diseases. *Frontiers in Microbiology*, *4*, 355. DOI: 10.3389/fmicb.2013.00355 PMID: 24348466

Jacoby, R., Peukert, M., Succurro, A., Koprivova, A., & Kopriva, S. (2017). The role of soil microorganisms in plant mineral nutrition. *Current Opinion in Plant Biology*, *39*, 1–6. DOI: 10.1016/j.pbi.2017.05.007 PMID: 28974956

Jansson, J. K., & Hofmockel, K. S. (2020). Soil microbiomes and climate change. *Nature Reviews. Microbiology*, *18*(1), 35–46. DOI: 10.1038/s41579-019-0265-7 PMID: 31586158

Kalam, S., Basu, A., Ahmad, I., Sayyed, R. Z., El-Enshasy, H. A., Dailin, D. J., & Suriani, N. L. (2020). Recent understanding of soil acidobacteria and their ecological significance. *Frontiers in Microbiology*, *11*, 580024. DOI: 10.3389/fmicb.2020.580024 PMID: 33193209

Kaul, S., Sharma, T., & Dhar, M. K. (2016). "Omics" tools for better understanding the plant–endophyte interactions. *Frontiers in Plant Science*, *7*, 955. DOI: 10.3389/fpls.2016.00955 PMID: 27446181

Khan, N., Bano, A., & Babar, M. A. (2019). The stimulatory effects of plant growth-promoting rhizobacteria and plant growth regulators on wheat physiology grown in sandy soil. *Archives of Microbiology*, *201*(6), 769–785. DOI: 10.1007/s00203-019-01644-w PMID: 30843087

Kumar, A., Singh, S., Gaurav, A. K., Srivastava, S., & Verma, J. P. (2020). Plant microbiome: Trends and applications for sustainable agriculture. *Plant Gene*, *24*, 100241. DOI: 10.1016/j.plgene.2020.100241

Lau, S.-E., Teo, W. F. A., Teoh, E. Y., & Tan, B. C. (2022). Microbiome engineering and plant biostimulants for sustainable crop improvement and mitigation of biotic and abiotic stresses. *Discover Food*, *2*(1), 9. DOI: 10.1007/s44187-022-00009-5

Levy, A., Conway, J. M., Dangl, J. L., & Woyke, T. (2018). Elucidating bacterial gene functions in the plant microbiome. *Cell Host & Microbe*, *24*(4), 475–485. DOI: 10.1016/j.chom.2018.09.005 PMID: 30308154

Li, X., Zhang, Y., Wang, J., & Chen, Q. (2026). Plant microbiome regulation for sustainable agriculture. *Current Opinion in Biotechnology*, *98*, 103444. DOI: 10.1016/j.copbio.2026.103444 PMID: 41671649

Lugtenberg, B., & Kamilova, F. (2009). Plant-growth-promoting rhizobacteria. *Annual Review of Microbiology*, *63*(1), 541–556. DOI: 10.1146/annurev.micro.62.081307.162918 PMID: 19575558

Mendes, R., Garbeva, P., & Raaijmakers, J. M. (2013). The rhizosphere microbiome: Significance of plant beneficial, plant pathogenic, and human pathogenic micro-organisms. *FEMS Microbiology Reviews*, *37*(5), 634–663. DOI: 10.1111/1574-6976.12028 PMID: 23790204

Mitter, B., Brader, G., Pfaffenbichler, N., & Sessitsch, A. (2019). Next generation microbiome applications for crop production—Limitations and the need of knowledge-based solutions. *Current Opinion in Microbiology*, *49*, 59–65. DOI: 10.1016/j.mib.2019.10.006 PMID: 31731227

Niu, B., Paulson, J. N., Zheng, X., & Kolter, R. (2017). Simplified and representative bacterial community of maize roots. *Proceedings of the National Academy of Sciences of the United States of America*, *114*(12), E2450–E2459. DOI: 10.1073/pnas.1616148114 PMID: 28275097

O'Brien, F. J. M., Almaraz, M., Foster, M. A., Hill, A. F., Huber, D. P., King, E. K., Lang, J. M., Lowe, A., Mickan, B. S., Miller, S. H., & Singh, B. K. (2021). Soil salinity and the plant microbiome. *Nature Reviews. Microbiology, 19*(11), 665–681. DOI: 10.1038/s41579-021-00563-5

Olanrewaju, O. S., Glick, B. R., & Babalola, O. O. (2017). Mechanisms of action of plant growth-promoting bacteria. *World Journal of Microbiology & Biotechnology, 33*(11), 197. DOI: 10.1007/s11274-017-2364-9 PMID: 28986676

Orozco-Mosqueda, M. D. C., Santoyo, G., Glick, B. R., & de los Santos-Villalobos, S. (2022). Microbiome engineering to improve biocontrol and plant growth-promoting mechanisms. *Microbiological Research, 256*, 126938. DOI: 10.1016/j.micres.2021.126938 PMID: 29551209

Phour, M., Sindhu, S. S., Glick, B. R., & Verma, S. K. (2020). Beneficial microbes and sustainable agriculture: Recent developments and future challenges. *Current Genomics, 21*(7), 507–519. DOI: 10.2174/1389202921999200908220956

Raaijmakers, J. M., & Mazzola, M. (2016). Soil immune responses. *Science, 352*(6292), 1392–1393. DOI: 10.1126/science.aaf3252 PMID: 27313024

Rodríguez, H., Fraga, R., Gonzalez, T., & Bashan, Y. (2006). Genetics of phosphate solubilization and its potential applications for improving plant growth-promoting bacteria. *Plant and Soil, 287*(1–2), 15–21. DOI: 10.1007/s11104-006-9056-9

Saad, M. M., Eida, A. A., & Hirt, H. (2020). Tailoring plant-associated microbial inoculants in agriculture: A roadmap for successful application. *Journal of Experimental Botany, 71*(13), 3878–3901. DOI: 10.1093/jxb/eraa111 PMID: 32157287

Santoyo, G. (2026). The role of plant microbiome engineering in stressful agriculture. *Discover Applied Sciences, 8*(5), 481. DOI: 10.1007/s42452-026-08322-9

Santoyo, G., Moreno-Hagelsieb, G., Orozco-Mosqueda, M. D. C., & Glick, B. R. (2016). Plant growth-promoting bacterial endophytes. *Microbiological Research, 183*, 92–99. DOI: 10.1016/j.micres.2015.11.008 PMID: 26805622

Sessitsch, A., Brader, G., Pfaffenbichler, N., Gusenbauer, D., & Mitter, B. (2018). The contribution of plant microbiota to economy growth. *Microbial Biotechnology, 11*(5), 801–805. DOI: 10.1111/1751-7915.13290 PMID: 29926519

Sharma, A., Johri, B. N., Sharma, A. K., & Glick, B. R. (2003). Plant growth-promoting bacterium Pseudomonas sp. strain GRP3 influences iron acquisition in mung bean. *Soil Biology & Biochemistry, 35*(7), 887–894. DOI: 10.1016/S0038-0717(03)00119-6

Shelake, R. M., Pramanik, D., & Kim, J.-Y. (2019). Exploration of plant-microbe interactions for sustainable agriculture in CRISPR era. *Microorganisms*, *7*(8), 269. DOI: 10.3390/microorganisms7080269 PMID: 31426522

Singh, B. K., Delgado-Baquerizo, M., Egidi, E., Guirado, E., Leach, J. E., Liu, H., Trivedi, P., & Bardgett, R. D. (2021). Climate change impacts on plant pathogens, food security and paths forward. *Nature Reviews. Microbiology*, *19*(12), 830–843. DOI: 10.1038/s41579-021-00542-w PMID: 37131070

Singh, B. K., Trivedi, P., Egidi, E., Macdonald, C. A., & Delgado-Baquerizo, M. (2020). Crop microbiome and sustainable agriculture. *Nature Reviews. Microbiology*, *18*(11), 601–602. DOI: 10.1038/s41579-020-00446-y PMID: 33037425

Toju, H., Kishida, O., Katayama, N., & Takagi, K. (2016). Networks depicting the fine-scale co-occurrences of fungi in soil horizons. *PLoS One*, *11*(11), e0165987. DOI: 10.1371/journal.pone.0165987 PMID: 27861486

Toju, H., Peay, K. G., Yamamichi, M., Narisawa, K., Hiruma, K., Naito, K., Fukuda, S., Ushio, M., Nakaoka, S., Onoda, Y., Yoshida, K., Schlaeppi, K., Bai, Y., Sugiura, R., Ichihashi, Y., Minamisawa, K., & Kiers, E. T. (2018). Core microbiomes for sustainable agroecosystems. *Nature Plants*, *4*(5), 247–257. DOI: 10.1038/s41477-018-0139-4 PMID: 29725101

Trivedi, P., Batista, B. D., Bazany, K. E., Singh, B. K., & Leach, J. E. (2022). Plant microbiome responses under drought and climate change. *The New Phytologist*, *234*(6), 1911–1923. DOI: 10.1111/nph.18016 PMID: 35118660

Trivedi, P., Delgado-Baquerizo, M., Jeffries, T. C., Trivedi, C., Anderson, I. C., Lai, K., McNee, M., Flower, K., Singh, B. P., Minkey, D., & Singh, B. K. (2017). Soil aggregation and associated microbial communities modify the impact of agricultural management on carbon content. *Environmental Microbiology*, *19*(8), 3070–3086. DOI: 10.1111/1462-2920.13779 PMID: 28447378

Trivedi, P., Leach, J. E., Tringe, S. G., Sa, T., & Singh, B. K. (2020). Plant–microbiome interactions: From community assembly to plant health. *Nature Reviews. Microbiology*, *18*(11), 607–621. DOI: 10.1038/s41579-020-0412-1 PMID: 32788714

Trivedi, P., Singh, B. K., & Leach, J. E. (2021). Plant–microbiome interactions and their applications in sustainable agriculture. *FEMS Microbiology Ecology*, *97*(2), fiab012. Advance online publication. DOI: 10.1093/femsec/fiab012

Tuncel, A., Pan, C., Clem, J. S., Liu, D., & Qi, Y. (2025). CRISPR–Cas applications in agriculture and plant research. *Nature Reviews. Molecular Cell Biology*, *26*(6), 419–441. DOI: 10.1038/s41580-025-00834-3 PMID: 40055491

Vandenkoornhuyse, P., Quaiser, A., Duhamel, M., Le Van, A., & Dufresne, A. (2015). The importance of the microbiome of the plant holobiont. *The New Phytologist*, *206*(4), 1196–1206. DOI: 10.1111/nph.13312 PMID: 25655016

Venturi, V., & Keel, C. (2016). Signaling in the rhizosphere. *Trends in Plant Science*, *21*(3), 187–198. DOI: 10.1016/j.tplants.2016.01.005 PMID: 26832945

Vurukonda, S. S. K. P., Vardharajula, S., Shrivastava, M., & SkZ, A. (2016). Enhancement of drought stress tolerance in crops by plant growth promoting rhizobacteria. *Microbiological Research*, *184*, 13–24. DOI: 10.1016/j.micres.2015.12.003 PMID: 26856449

Yadav, A., Gehlot, P., Soni, R., & Jain, T. (2026). Plant microbiome engineering: From inoculation to genome editing. *Frontiers in Microbiology*, *17*, 1781381. DOI: 10.3389/fmicb.2026.1781381 PMID: 42100688

Zhu, H., Li, C., & Gao, C. (2020). Applications of CRISPR–Cas in agriculture and plant biotechnology. *Nature Reviews. Molecular Cell Biology*, *21*(11), 661–677. DOI: 10.1038/s41580-020-00288-9 PMID: 32973356

Compilation of References

Abdallah, N. A., Hamwieh, A., Radwan, K., Fouad, N., & Prakash, C. (2021). Genome editing techniques in plants: A comprehensive review and future prospects toward zero hunger. *GM Crops & Food*, *12*(2), 601–615. DOI: 10.1080/21645698.2021.2021724 PMID: 35135438

Abdallah, N. A., Prakash, C. S., & McHughen, A. G. (2015). Genome editing for crop improvement: Challenges and opportunities. *GM Crops & Food*, *6*(4), 183–205. DOI: 10.1080/21645698.2015.1129937 PMID: 26930114

Abdelrahman, M., Wei, Z., Rohila, J. S., & Zhao, K. (2021). Multiplex genome-editing technologies for revolutionizing plant biology and crop improvement—. *Frontiers in Plant Science*, *12*, 721203. DOI: 10.3389/fpls.2021.721203 PMID: 34691102

Acquaah, G. (2015). Conventional plant breeding principles and techniques. *Advances in plant breeding strategies: Breeding, biotechnology and molecular tools*, 115-158.

Agapito-Tenfen, S. Z., Okoli, A. S., Bernstein, M. J., Wikmark, O.-G., & Myhr, A. I. (2018). Revisiting risk governance of GM plants: The need to consider new and emerging Gene-Editing techniques. *Frontiers in Plant Science*, *9*, 1874. DOI: 10.3389/fpls.2018.01874 PMID: 30622546

Ahlawat, U., Naruka, A., Changdeo, W. B., Rehsawla, R., Sansanwal, R., Mishra, R., & Singh, V. (2024). A review of cutting-edge Biotechnological solutions for next-generation farming. *J Exp Agric Int*, *46*(7), 1187–1204. DOI: 10.9734/jeai/2024/v46i72671

Ahmad, A., Ghouri, M. Z., Munawar, N., Ismail, M., Ashraf, S., & Aftab, S. O. (2021a). Regulatory, Ethical, and Social Aspects of CRISPR Crops. In *CRISPR Crops* (pp. 261–287). DOI: 10.1007/978-981-15-7142-8_9

Ahmad, A., Munawar, N., Khan, Z., Qusmani, A. T., Khan, S. H., Jamil, A., Ashraf, S., Ghouri, M. Z., Aslam, S., Mubarik, M. S., Munir, A., Sultan, Q., Abd-Elsalam, K. A., & Qari, S. H. (2021b). An Outlook on Global Regulatory Landscape for Genome-Edited Crops. *International Journal of Molecular Sciences, 22*(21), 11753. DOI: 10.3390/ijms222111753

Ahmadikhah, A., Zarabizadeh, H., Nayeri, S., & Abbasi, M. S. (2025). Advancements in genome editing tools for genetic studies and crop improvement. *Frontiers in Plant Science*, *15*, 1370675. DOI: 10.3389/fpls.2024.1370675 PMID: 39963359

Ahmad, M. (2023). Plant breeding advancements with "CRISPR-Cas" genome editing technologies will assist future food security. *Frontiers in Plant Science*, *14*, 1133036. DOI: 10.3389/fpls.2023.1133036 PMID: 36993865

Ahmad, S., Shahzad, R., Jamil, S., Tabassum, J., Chaudhary, M. A. M., Atif, R. M., & Sheng, Z. (2021). Regulatory aspects, risk assessment, and toxicity associated with RNAi and CRISPR methods. In *CRISPR and RNAi systems* (pp. 687–721). Elsevier. DOI: 10.1016/B978-0-12-821910-2.00013-8

Ahmad, S., Tang, L., Shahzad, R., Mawia, A. M., Rao, G. S., Jamil, S., & Wei, X. (2021). CRISPR-based crop improvements: A way forward to achieve zero hunger. *Journal of Agricultural and Food Chemistry*, *69*(30), 8307–8323. DOI: 10.1021/acs.jafc.1c02653 PMID: 34288688

Ahmar, S., Gill, R. A., Jung, K.-H., Faheem, A., Qasim, M. U., Mubeen, M., & Zhou, W. (2020). Conventional and molecular techniques from simple breeding to speed breeding in crop plants: Recent advances and future outlook. *International Journal of Molecular Sciences*, *21*(7), 2590. DOI: 10.3390/ijms21072590 PMID: 32276445

Ahmar, S., Mahmood, T., Fiaz, S., Mora-Poblete, F., Shafique, M. S., Chattha, M. S., & Jung, K.-H. (2021). Advantage of nanotechnology-based genome editing system and its application in crop improvement. *Frontiers in Plant Science*, *12*, 663849. DOI: 10.3389/fpls.2021.663849 PMID: 34122485

Ahmar, S., Saeed, S., Khan, M. H. U., Ullah Khan, S., Mora-Poblete, F., Kamran, M., & Jung, K. H. (2020a). A Revolution Toward Gene-Editing Technology and Its Application to Crop Improvement. *International Journal of Molecular Sciences*, *21*(16), 5665. DOI: 10.3390/ijms21165665 PMID: 32784649

Ahmed, M., Asim, M., Ahmad, S., & Aslam, M. (2023). Climate change, agricultural productivity, and food security. In *Global agricultural production: Resilience to climate change* (pp. 31–72). Springer.

Ain, Q. U., Chung, J. Y., & Kim, Y. H. (2015). Current and future delivery systems for engineered nucleases: ZFN, TALEN and RGEN. *Journal of Controlled Release : Official Journal of the Controlled Release Society*, *205*, 120–127. DOI: 10.1016/j.jconrel.2014.12.036 PMID: 25553825

Akanmu, A. O., Asemoloye, M. D., Marchisio, M. A., & Babalola, O. O. (2024). Adoption of CRISPR-Cas for crop production: Present status and future prospects. *PeerJ*, *12*, e17402. DOI: 10.7717/peerj.17402 PMID: 38860212

Akinbo, O., Nkhabindze, B., Amedu, J., Ebegba, R., Asagbra, A., Ratemo, B. O., & Dada, S. A.. (2025). Africa and zero hunger agenda: Genome editing policy landscape, challenges and opportunities. *Frontiers in Bioengineering and Biotechnology*, *13*, 1526851. DOI: 10.3389/fbioe.2025.1526851 PMID: 40144392

Akinbo, O., Obukosia, S., Ouedraogo, J., Sinebo, W., Savadogo, M., Timpo, S., Mbabazi, R., Maredia, K., Makinde, D., & Ambali, A. (2021). Commercial release of genetically modified crops in Africa: Interface between biosafety regulatory systems and varietal release systems. *Frontiers in Plant Science*, *12*, 605937. DOI: 10.3389/fpls.2021.605937 PMID: 33828569

Akram, J., Naeem, N., Manzoor, M. T., Ali, Q., Marwat, U. A., & Zia, T. (2025). Role of modern biotechnology in improving crop protection. *Bulletin of Biological and Allied Sciences Research*, *2025*(1), 100–100. DOI: 10.54112/bbasr.v2025i1.100

Aksoy, E., Yildirim, K., Kavas, M., Kayihan, C., Yerlikaya, B. A., Çalik, I., & Demirel, U. (2022). General guidelines for CRISPR/Cas-based genome editing in plants. *Molecular Biology Reports*, *49*(12), 12151–12164. DOI: 10.1007/s11033-022-07773-8 PMID: 36107373

Alam, M. S., Kong, J., Tao, R., Ahmed, T., Alamin, M., Alotaibi, S. S., Abdelsalam, N. R., & Xu, J.-H. (2022). CRISPR/Cas9 mediated knockout of the OsbHLH024 transcription factor improves salt stress resistance in rice (Oryza sativa L.). *Plants*, *11*(9), 1184. DOI: 10.3390/plants11091184 PMID: 35567185

Aljabali, A. A. A., El-Tanani, M., & Tambuwala, M. M. (2024). Aljabali, Alaa AA, Mohamed El-Tanani, and Murtaza M. Tambuwala. "Principles of CRISPR-Cas9 technology: Advancements in genome editing and emerging trends in drug delivery.". *Journal of Drug Delivery Science and Technology*, *92*, 105338. DOI: 10.1016/j.jddst.2024.105338

Altpeter, F., Springer, N. M., Bartley, L. E., Blechl, A. E., Brutnell, T. P., Citovsky, V., & Kausch, A. P. (2016). Advancing crop transformation in the era of genome editing. *The Plant Cell*, *28*(7), 1510–1520. DOI: 10.1105/tpc.16.00196 PMID: 27335450

Anas, M., Khattak, W. A., Fahad, S., Alrawiq, N., Alrawiq, H. S., Abdelsalam, N. R., Jaremko, M., & Quraishi, U. M. (2025). Mitigating heavy metal pollution in agriculture: A multi-omics and nanotechnology approach to safeguard global wheat production. *Journal of Hazardous Materials Advances*, *17*, 100584. https://doi.org/https://doi.org/10.1016/j.hazadv.2024.100584. DOI: 10.1016/j.hazadv.2024.100584

Anders, S., Cowling, W., Pareek, A., Gupta, K. J., Singla-Pareek, S. L., & Foyer, C. H. (2021). Gaining Acceptance of Novel Plant Breeding Technologies. *Trends in Plant Science*, *26*(6), 575–587. DOI: 10.1016/j.tplants.2021.03.004 PMID: 33893048

Animasaun, D. A., Adedibu, P. A., Shkryl, Y., Emmanuel, F. O., Tekutyeva, L., & Balabanova, L. (2023). Modern plant biotechnology: An antidote against global food insecurity. *Agronomy (Basel)*, *13*(8), 2038. DOI: 10.3390/agronomy13082038

Ansari, W. A., Chandanshive, S. U., Bhatt, V., Nadaf, A. B., Vats, S., Katara, J. L., Sonah, H., & Deshmukh, R. (2020). Genome editing in cereals: Approaches, applications and challenges. *International Journal of Molecular Sciences*, *21*(11), 4040. DOI: 10.3390/ijms21114040 PMID: 32516948

Anyshchenko, A. (2022). Aligning policy design with science to achieve food security: The contribution of genome editing to sustainable agriculture. *Frontiers in Sustainable Food Systems*, *6*, 897643. DOI: 10.3389/fsufs.2022.897643

Anzalone, A. V., Koblan, L. W., & Liu, D. R. (2020). Genome editing with CRISPR–Cas nucleases, base editors, transposases and prime editors. *Nature Biotechnology*, *38*(7), 824–844. DOI: 10.1038/s41587-020-0561-9 PMID: 32572269

Anzalone, A. V., Randolph, P. B., Davis, J. R., Sousa, A. A., Koblan, L. W., Levy, J. M., Chen, P. J., Wilson, C., Newby, G. A., Raguram, A., & Liu, D. R. (2019). Search-and-replace genome editing without double-strand breaks or donor DNA. *Nature*, *576*(7785), 149–157. DOI: 10.1038/s41586-019-1711-4 PMID: 31634902

Arcidiacono, J. (2023). International Harmonization for Cell and Gene Therapy Products. In *Regulatory Aspects of Gene Therapy and Cell Therapy Products: A Global Perspective* (pp. 235–240). Springer International Publishing., DOI: 10.1007/978-3-031-34567-8_14

Arif, I., Batool, M., & Schenk, P. M. (2020). Plant microbiome engineering: Expected benefits for improved crop growth and resilience. *Trends in Biotechnology*, *38*(12), 1385–1396. DOI: 10.1016/j.tibtech.2020.04.015 PMID: 32451122

Aroge, T., Zhu, Y., Jin, D.-N., Dara, M. Z. N., Feng, J., Olajuyin, A. M., Abbas, A., & Liu, S.-Y. (2024). Omics and CRISPR-Cas9 molecular perception: A progressive review approach for powdery mildew disease management. *Physiological and Molecular Plant Pathology*, *130*, 102217. DOI: 10.1016/j.pmpp.2024.102217

Arora, L., & Narula, A. (2017). Gene Editing and Crop Improvement Using the CRISPR-Cas9 System. *Frontiers in Plant Science*, *8*, 1932. DOI: 10.3389/fpls.2017.01932 PMID: 29167680

Asmamaw, M., & Zawdie, B. (2021). Mechanism and Applications of CRISPR/Cas-9-Mediated Genome Editing. *Biologics*, *15*(1), 353–361. PMID: 34456559

Asquer, A., & Krachkovskaya, I. (2021). Uncertainty, institutions and regulatory responses to emerging technologies: CRISPR Gene editing in the US and the EU (2012–2019). *Regulation & Governance*, *15*(4), 1111–1127. DOI: 10.1111/rego.12335

Asrey, R., Barman, K., Prajapati, U., Sharma, S., & Yadav, A. (2021). Genetically modified fruit and vegetable-An overview on senescence regulation, postharvest nutraceutical quality preservation and shelf life extension. *The Journal of Horticultural Science & Biotechnology*, *96*(3), 271–287. DOI: 10.1080/14620316.2020.1845986

Atia, M., Jiang, W., Sedeek, K., Butt, H., & Mahfouz, M. (2024). Crop bioengineering via gene editing: Reshaping the future of agriculture. *Plant Cell Reports*, *43*(4), 98. DOI: 10.1007/s00299-024-03183-1 PMID: 38494539

Atimango, A. O., Wesana, J., Kalule, S. W., Verbeke, W., & De Steur, H. (2024). Genome editing in food and agriculture: From regulations to consumer perspectives. *Current Opinion in Biotechnology*, *87*, 103127. DOI: 10.1016/j.copbio.2024.103127 PMID: 38564970

Atkins, P. A., & Voytas, D. F. (2020). Overcoming bottlenecks in plant gene editing. *Current Opinion in Plant Biology*, *54*, 79–84. DOI: 10.1016/j.pbi.2020.01.002 PMID: 32143167

Avnee, S., Sood, S., Chaudhary, D. R., Jhorar, P., & Rana, R. S. (2023). Biofortification: An approach to eradicate micronutrient deficiency. *Frontiers in Nutrition*, *10*, 1233070. DOI: 10.3389/fnut.2023.1233070 PMID: 37789898

Aziz, M. A., Brini, F., Rouached, H., & Masmoudi, K. (2022). Genetically engineered crops for sustainably enhanced food production systems. *Frontiers in Plant Science*, *13*, 1027828. DOI: 10.3389/fpls.2022.1027828 PMID: 36426158

Azubuike, C. C., Couceiro, F., Robson, S. C., Piccinni, M. Z., Watts, J. E., Williams, J. B., Callaghan, A. J., & Howard, T. P. (2022). Developing biosensors for SARS-CoV-2 wastewater-based epidemiology: A systematic review of trends, limitations and future perspectives. *Sustainability (Basel), 14*(24), 16761. DOI: 10.3390/su142416761

Bachtarzi, H., & Farries, T. (2019). The genetically modified organism medicinal framework in Europe, United States, and Japan: Underlying scientific principles and considerations toward the development of gene therapy and genetically modified cell-based products. *Human Gene Therapy. Clinical Development, 30*(3), 114–128. DOI: 10.1089/humc.2019.042 PMID: 31111736

Backer, R., Rokem, J. S., Ilangumaran, G., Lamont, J., Praslickova, D., Ricci, E., Subramanian, S., & Smith, D. L. (2018). Plant growth-promoting rhizobacteria: Context, mechanisms of action, and roadmap to commercialization of biostimulants for sustainable agriculture. *Frontiers in Plant Science, 9*, 1473. DOI: 10.3389/fpls.2018.01473 PMID: 30405652

Backes, N. (2020). *CRISPR interference strategies for studies in essential gene function and live cell fluorescent imaging of DNA elements in Escherichia coli*. Iowa State University. DOI: 10.31274/etd-20200902-8

Baek, Yousoon, Lucas K. Bobadilla, Darci A. Giacomini, Jacob S. Montgomery, Brent P. Murphy, and Patrick J. Tranel. "Evolution of glyphosate-resistant weeds." *Reviews of Environmental Contamination and Toxicology Volume 255: Glyphosate* (2021): 93-128. https://doi.org/DOI: 10.1007/398_2020_55

Bain, C., Lindberg, S., & Selfa, T. (2019). Emerging sociotechnical imaginaries for gene edited crops for foods in the United States: Implications for governance. *Agriculture and Human Values, 37*(2), 265–279. DOI: 10.1007/s10460-019-09980-9

Bakala, H. S., Singh, G., & Srivastava, P. (2020). Smart breeding for climate resilient agriculture. In *Plant breeding-current and future views*. IntechOpen.

Banakar, R., Schubert, M., Kurgan, G., Rai, K. M., Beaudoin, S. F., Collingwood, M. A., & Zhang, F. (2022). Efficiency, specificity, and temperature sensitivity of Cas9 and Cas12a RNPs for DNA-free genome editing in plants. *Frontiers in Genome Editing, 3*, 760820. DOI: 10.3389/fgeed.2021.760820 PMID: 35098208

Bandyopadhyay, A., Kancharla, N., Javalkote, V. S., Dasgupta, S., & Brutnell, T. P. (2020). CRISPR-Cas12a (Cpf1): A versatile tool in the plant genome editing tool box for agricultural advancement. *Frontiers in Plant Science, 11*, 584151. DOI: 10.3389/fpls.2020.584151 PMID: 33214794

Banerjee, S., Schlaeppi, K., & van der Heijden, M. G. A. (2018). Keystone taxa as drivers of microbiome structure and functioning. *Nature Reviews. Microbiology*, *16*(9), 567–576. DOI: 10.1038/s41579-018-0024-1 PMID: 29789680

Bansal, P., & Kaur, N. (2024). Assessing risks associated with large-scale adoption of CRISPR gene-edited crops. *Journal of Crop Science and Biotechnology*, •••, 1–11.

Barbosa, S., Toe, L. P., Thizy, D., Vaz, M., & Carter, L. (2021). Engagement and social acceptance in genome editing for human benefit: Reflections on research and practice in a global context. *Wellcome Open Research*, *5*, 244. DOI: 10.12688/wellcomeopenres.16260.2 PMID: 34095505

Barragan, A. C., & Weigel, D. (2021). Plant NLR diversity: The known unknowns of pan-NLRomes. *The Plant Cell*, *33*(4), 814–831. DOI: 10.1093/plcell/koaa002 PMID: 33793812

Barrett, D. M., Beaulieu, J. C., & Shewfelt, R. (2010). Color, flavor, texture, and nutritional quality of fresh-cut fruits and vegetables: Desirable levels, instrumental and sensory measurement, and the effects of processing. *Critical Reviews in Food Science and Nutrition*, *50*(5), 369–389. DOI: 10.1080/10408391003626322 PMID: 20373184

Barrows, G., Sexton, S., & Zilberman, D. (2014). Agricultural biotechnology: The promise and prospects of genetically modified crops. *The Journal of Economic Perspectives*, *28*(1), 99–120. DOI: 10.1257/jep.28.1.99

Basak, J., & Nithin, C. (2015). Targeting Non-Coding RNAs in Plants with the CRISPR-Cas Technology Is a Challenge Yet Worth Accepting. *Frontiers in Plant Science*, *6*, 1001. DOI: 10.3389/fpls.2015.01001 PMID: 26635829

Bashan, Y., de-Bashan, L. E., Prabhu, S. R., & Hernandez, J.-P. (2014). Advances in plant growth-promoting bacterial inoculant technology. *Plant and Soil*, *378*(1–2), 1–33. DOI: 10.1007/s11104-013-1956-x

Basinskiene, L., & Seinauskiene, B. (2021). Gene editing versus gene modification: Awareness, attitudes and behavioral intentions of lithuanian consumers, producers, and farmers. *Chemical Engineering Transactions*, *87*, 433–438.

Batan, A. (2025). Assessing the Role of Public-Private Partnerships in Advancing Sustainable Development Goals Through Agri-Tech and Renewable Energy Solutions. *Transactions on Social Innovation, Digital Inclusion, and Ethical AI*, *15*(2), 1–11.

Bate, N. J., Dardick, C. D., de Maagd, R. A., & Williams, R. W. (2021). Opportunities and challenges applying gene editing to specialty crops. *In Vitro Cellular & Developmental Biology. Plant : Journal of the Tissue Culture Association, 57*(4), 709–719. DOI: 10.1007/s11627-021-10208-x

Bearth, A., Drummond Otten, C., & Segrè Cohen, A. (2024). Consumers' perceptions and acceptance of genome editing in agriculture: Insights from the United States of America and Switzerland. *Food Research International, 178*, 113982. DOI: 10.1016/j.foodres.2024.113982 PMID: 38309884

Becker, S., & Boch, J. (2021). TALE and TALEN genome editing technologies. *Gene and Genome Editing, 2*, 100007. DOI: 10.1016/j.ggedit.2021.100007

Beckie, H. J., & Martin, S. L. (2021). Monitoring herbicide resistance gene flow in weed populations. In *Gene flow: monitoring, modeling and mitigation* (pp. 86–102). CABI., DOI: 10.1079/9781789247480.0006

Begna, T. (2020). Major challenging constraints to crop production farming system and possible breeding to overcome the constraints. [IJRSAS]. *International Journal of Research Studies in Agricultural Sciences, 6*(7), 27–46.

Begna, T. (2021). Conventional breeding methods widely used to improve self-pollinated crops. *International Journal of Research, 7*(1), 1–16.

Bélanger, J. G., Copley, T. R., Hoyos-Villegas, V., Charron, J.-B., & O'Donoughue, L. (2024). A comprehensive review of in planta stable transformation strategies. *Plant Methods, 20*(1), 79. DOI: 10.1186/s13007-024-01200-8 PMID: 38822403

Bennur, P. L., O'Brien, M., Fernando, S. C., & Doblin, M. S. (2025). Improving transformation and regeneration efficiency in medicinal plants: Insights from other recalcitrant species. *Journal of Experimental Botany, 76*(1), 52–75. DOI: 10.1093/jxb/erae189 PMID: 38652155

Berendsen, R. L., Pieterse, C. M. J., & Bakker, P. A. H. M. (2012). The rhizosphere microbiome and plant health. *Trends in Plant Science, 17*(8), 478–486. DOI: 10.1016/j.tplants.2012.04.001 PMID: 22564542

Berg, G., & Cernava, T. (2022). The future of plant microbiome research for sustainable agriculture. *Microbial Biotechnology, 15*(1), 6–10. DOI: 10.1111/1751-7915.13916 PMID: 33529492

Berg, G., Cernava, T., Schloter, M., & Smalla, K. (2023). The plant microbiome in sustainable agriculture. *Annual Review of Phytopathology, 61*, 123–145. DOI: 10.1146/annurev-phyto-021622-123456

Berg, G., Grube, M., Schloter, M., & Smalla, K. (2014). Unraveling the plant microbiome: Looking back and future perspectives. *Frontiers in Microbiology*, *5*, 148. DOI: 10.3389/fmicb.2014.00148 PMID: 24926286

Berg, G., Köberl, M., Rybakova, D., Müller, H., Grosch, R., & Smalla, K. (2017). Plant microbial diversity is suggested as the key to future biocontrol and health trends. *FEMS Microbiology Ecology*, *93*(5), fix050. Advance online publication. DOI: 10.1093/femsec/fix050 PMID: 28430944

Berg, G., & Raaijmakers, J. M. (2018). Saving seed microbiomes. *The ISME Journal*, *12*(5), 1167–1170. DOI: 10.1038/s41396-017-0028-2 PMID: 29335636

Berg, G., Rybakova, D., Fischer, D., Cernava, T., Vergès, M.-C. C., Charles, T., Chen, X., Cocolin, L., Eversole, K., Corral, G. H., Kazou, M., Kinkel, L., Lange, L., Lima, N., Loy, A., Macklin, J. A., Maguin, E., Mauchline, T., McClure, R., & Schloter, M. (2020). Microbiome definition re-visited: Old concepts and new challenges. *Microbiome*, *8*(1), 103. DOI: 10.1186/s40168-020-00875-0 PMID: 32605663

Bernauer, T. (2003). *Genes, trade, and regulation: The seeds of conflict in food biotechnology*. Princeton University Press.

Bertheau, Y. (2012). *Genetically modified and non-genetically modified food supply chains: co-existence and traceability*: John Wiley & Sons.

Berthiaume, F., Maguire, T. J., & Yarmush, M. L. (2011). Tissue engineering and regenerative medicine: History, progress, and challenges. *Annual Review of Chemical and Biomolecular Engineering*, *2*(1), 403–430. DOI: 10.1146/annurev-chembioeng-061010-114257 PMID: 22432625

Bhagtaney, L., & Sundarrajan, P. (2023). An overview of tools for genome editing: ZFNs, mega nucleases, and TALENs. CRISPR/Cas-mediated genome editing in plants, 37-64.

Bhandari, R., Paudel, H., Nyaupane, S., & Poudel, M. R. (2024). Climate resilient breeding for high yields and stable wheat (Triticum aestivum L.) lines under irrigated and abiotic stress environments. *Plant Stress*, *11*, 100352. https://doi.org/https://doi.org/10.1016/j.stress.2024.100352. DOI: 10.1016/j.stress.2024.100352

Bharat, S. S., Li, S., Li, J., Yan, L., & Xia, L. (2020). Base editing in plants: Current status and challenges. *The Crop Journal*, *8*(3), 384–395. DOI: 10.1016/j.cj.2019.10.002

Bhardwaj, A., Tomar, P., & Nain, V. (2024). Machine Learning-Driven Prediction of CRISPR-Cas9 Off-Target Effects and Mechanistic Insights. *environment, 12*, 13.

Bhardwaj, A., & Nain, V. (2021). TALENs—an indispensable tool in the era of CRISPR: A mini review. *Journal of Genetic Engineering and Biotechnology*, *19*(1), 125. DOI: 10.1186/s43141-021-00225-z PMID: 34420096

Bhat, M. A., Mir, R. A., Kumar, V., Shah, A. A., Zargar, S. M., Rahman, S., & Jan, A. T. (2021). Mechanistic insights of CRISPR/Cas-mediated genome editing towards enhancing abiotic stress tolerance in plants. *Physiologia Plantarum*, *172*(2), 1255–1268. DOI: 10.1111/ppl.13359 PMID: 33576013

Bhattacharjee, S., Bhowmick, R., Kant, L., & Paul, K. (2023). Strategic transgene-free approaches of CRISPR-based genome editing in plants. *Molecular Genetics and Genomics*, *298*(3), 507–520. DOI: 10.1007/s00438-023-01998-3 PMID: 36840794

Bhattacharyya, P. N., & Jha, D. K. (2012). Plant growth-promoting rhizobacteria: Emergence in agriculture. *World Journal of Microbiology & Biotechnology*, *28*(4), 1327–1350. DOI: 10.1007/s11274-011-0979-9 PMID: 22805914

Bhatta, M., Sandro, P., Smith, M. R., Delaney, O., Voss-Fels, K. P., Gutierrez, L., & Hickey, L. T. (2021). Need for speed: Manipulating plant growth to accelerate breeding cycles. *Current Opinion in Plant Biology*, *60*, 101986. DOI: 10.1016/j.pbi.2020.101986 PMID: 33418268

Bhatt, P., Singh, S., Alfuraiji, N., & Al-Snafi, A. E. (2022). CRISPR CAS9: A new technology to modify genome-A review. *Open J. Syst. Demonstr. J*, *8*(4), 208–215.

Bidabadi, S. S., & Jain, S. M. (2020). Cellular, molecular, and physiological aspects of in vitro plant regeneration. *Plants*, *9*(6), 702. DOI: 10.3390/plants9060702 PMID: 32492786

Binnie, A., Fernandes, E., Almeida-Lousada, H., De Mello, R. A., & Castelo-Branco, P. (2021). CRISPR-based strategies in infectious disease diagnosis and therapy. *Infection*, *49*(3), 377–385. DOI: 10.1007/s15010-020-01554-w PMID: 33393066

Boch, J., Scholze, H., Schornack, S., Landgraf, A., Hahn, S., Kay, S., Lahaye, T., Nickstadt, A., & Bonas, U. (2009). Breaking the code of DNA binding specificity of TAL-type III effectors. *Science*, *326*(5959), 1509–1512. DOI: 10.1126/science.1178811 PMID: 19933107

Bode, D., Cull, A. H., Rubio-Lara, J. A., & Kent, D. G. (2021). Exploiting single-cell tools in gene and cell therapy. *Frontiers in Immunology*, *12*, 702636. DOI: 10.3389/fimmu.2021.702636 PMID: 34322133

Bonawitz, N. D., Ainley, W. M., Itaya, A., Chennareddy, S. R., Cicak, T., Effinger, K., Jiang, K., Mall, T. K., Marri, P. R., Samuel, J. P., Sardesai, N., Simpson, M., Folkerts, O., Sarria, R., Webb, S. R., Gonzalez, D. O., Simmonds, D. H., & Pareddy, D. R. (2019). Zinc finger nuclease-mediated targeting of multiple transgenes to an endogenous soybean genomic locus via non-homologous end joining. *Plant Biotechnology Journal*, *17*(4), 750–761. DOI: 10.1111/pbi.13012 PMID: 30220095

Bonga, J. (2017). Can explant choice help resolve recalcitrance problems in in vitro propagation, a problem still acute especially for adult conifers? *Trees (Berlin)*, *31*(3), 781–789. DOI: 10.1007/s00468-016-1509-z

Borah, A., Singh, S., Chattopadhyay, R., Kaur, J., & Bari, V. K. (2024). Integration of CRISPR/Cas9 with multi-omics technologies to engineer secondary metabolite production in medicinal plants: Challenges and Prospects. *Functional & Integrative Genomics*, *24*(6), 207. DOI: 10.1007/s10142-024-01486-w PMID: 39496976

Borg, E., & Policante, A. (2024). The Gene Editing Business: Rent Extraction in the Biotech Industry. *Review of Political Economy*, •••, 1–36.

Borisjuk, N., Kishchenko, O., Eliby, S., Schramm, C., Anderson, P., Jatayev, S., Kurishbayev, A., & Shavrukov, Y. (2019). Genetic modification for wheat improvement: From transgenesis to genome editing. *BioMed Research International*, *2019*(1), 6216304. DOI: 10.1155/2019/6216304 PMID: 30956982

Borlaug, N. E. (2002). The green revolution revisited and the road ahead. Nobelprize. org Stockholm.

Boros, A., Szólik, E., Desalegn, G., & Tőzsér, D. (2024). A Systematic review of opportunities and limitations of innovative practices in sustainable agriculture. *Agronomy (Basel)*, *15*(1), 76. DOI: 10.3390/agronomy15010076

Boti, M. A., Athanasopoulou, K., Adamopoulos, P. G., Sideris, D. C., & Scorilas, A. (2023). Recent Advances in Genome-Engineering Strategies. *Genes*, *14*(1), 129. DOI: 10.3390/genes14010129 PMID: 36672870

Bouché, N., & Bouchez, D. (2001). Arabidopsis gene knockout: Phenotypes wanted. *Current Opinion in Plant Biology*, *4*(2), 111–117. DOI: 10.1016/S1369-5266(00)00145-X PMID: 11228432

Bradbury, A., Clapp, O., Biacsi, A.-S., Kuo, P., Gaju, O., Hayta, S., Zhu, J.-K., & Lambing, C. (2025). Integrating genome editing with omics, artificial intelligence, and advanced farming technologies to increase crop productivity. *Plant Communications*, *6*(7), 101386. DOI: 10.1016/j.xplc.2025.101386 PMID: 40443034

Breuninger, H., & Lenhard, M. (2010). Control of tissue and organ growth in plants. *Current Topics in Developmental Biology*, *91*, 185–220. DOI: 10.1016/S0070-2153(10)91007-7 PMID: 20705183

Brock, N., Kaur, N., & Halford, N. G. (2025). Advances in genome editing in plants within an evolving regulatory landscape, with a focus on its application in wheat breeding. *Journal of Plant Biochemistry and Biotechnology*, *34*(3), 599–614. DOI: 10.1007/s13562-025-00981-w PMID: 40852733

Brokowski, C., & Adli, M. (2019). CRISPR ethics: Moral considerations for applications of a powerful tool. *Journal of Molecular Biology*, *431*(1), 88–101. DOI: 10.1016/j.jmb.2018.05.044 PMID: 29885329

Brookes, G., & Smyth, S. J. (2024). Risk-appropriate regulations for gene-editing technologies. *GM Crops & Food*, *15*(1), 1–14. DOI: 10.1080/21645698.2023.2293510 PMID: 38215017

Brooks, C., Nekrasov, V., Lippman, Z. B., & Van Eck, J. (2014). Efficient Gene Editing in Tomato in the First Generation Using the Clustered Regularly Interspaced Short Palindromic Repeats/CRISPR-Associated9 System. *Plant Physiology*, *166*(3), 1292–1297. DOI: 10.1104/pp.114.247577 PMID: 25225186

Buch, K., Kaushik, A., Mishra, U., Beese, S., Samanta, S., Singh, R., & Mubeen, . (2023). Unravelling the complexity of plant breeding through modern genetic techniques and tools: A review. *International Journal of Plant and Soil Science*, *35*(21), 97–105. DOI: 10.9734/ijpss/2023/v35i213950

Bulgarelli, D., Schlaeppi, K., Spaepen, S., van Themaat, E. V. L., & Schulze-Lefert, P. (2013). Structure and functions of the bacterial microbiota of plants. *Annual Review of Plant Biology*, *64*(1), 807–838. DOI: 10.1146/annurev-arplant-050312-120106 PMID: 23373698

Bullock, D. W., Wilson, W. W., & Neadeau, J. (2021). Gene editing versus genetic modification in the research and development of new crop traits: An economic comparison. *American Journal of Agricultural Economics*, *103*(5), 1700–1719. DOI: 10.1111/ajae.12201

Bure, I. V., Nemtsova, M. V., & Kuznetsova, E. B. (2022). Histone modifications and non-coding RNAs: Mutual epigenetic regulation and role in pathogenesis. *International Journal of Molecular Sciences*, *23*(10), 5801. DOI: 10.3390/ijms23105801 PMID: 35628612

Burow, M., Atwell, S., Francisco, M., Kerwin, R. E., Halkier, B. A., & Kliebenstein, D. J. (2015). The glucosinolate biosynthetic gene AOP2 mediates feed-back regulation of jasmonic acid signaling in Arabidopsis. *Molecular Plant*, *8*(8), 1201–1212. DOI: 10.1016/j.molp.2015.03.001 PMID: 25758208

Busby, P. E., Soman, C., Wagner, M. R., Friesen, M. L., Kremer, J., Bennett, A., Morsy, M., Eisen, J. A., Leach, J. E., & Dangl, J. L. (2017). Research priorities for harnessing plant microbiomes in sustainable agriculture. *PLoS Biology*, *15*(3), e2001793. DOI: 10.1371/journal.pbio.2001793 PMID: 28350798

Butler, N. M., Atkins, P. A., Voytas, D. F., & Douches, D. S. (2015). Generation and Inheritance of Targeted Mutations in Potato (*Solanum tuberosum L.*) Using the CRISPR/Cas System. *PLoS One*, *10*(12), e0144591. DOI: 10.1371/journal.pone.0144591 PMID: 26657719

Cain, A. K., Barquist, L., Goodman, A. L., Paulsen, I. T., Parkhill, J., & van Opijnen, T. (2020). A decade of advances in transposon-insertion sequencing. *Nature Reviews. Genetics*, *21*(9), 526–540. DOI: 10.1038/s41576-020-0244-x PMID: 32533119

Cai, R., Lv, R., Shi, X., Yang, G., & Jin, J. (2023). CRISPR/dCas9 tools: Epigenetic mechanism and application in gene transcriptional regulation. *International Journal of Molecular Sciences*, *24*(19), 14865. DOI: 10.3390/ijms241914865 PMID: 37834313

Calabrese, C., Featherstone, J. D., Robbins, M., & Barnett, G. A. (2021). Examining the relationship between gene editing knowledge, value predispositions, and general science attitudes among US farmers, scientists, policymakers, and the general public. *JCOM, Journal of Science Communication*, *20*(2), A02. DOI: 10.22323/2.20020202

Caradus, J. R. 2023. Processes for Regulating Genetically Modified and Gene-Edited Plants. *GM Crops & Food* 14 (1): 1–41. DOI: 10.1080/21645698.2023.2252947

Cardi, T., Murovec, J., Bakhsh, A., Justyna Boniecka, B., Bull, T., & Milica Zlatković, S. E. (2023). CRISPR/Cas-mediated plant genome editing: Outstanding challenges a decade after implementation. *Trends in Plant Science*, *28*(10), 1144–1165. DOI: 10.1016/j.tplants.2023.05.012 PMID: 37331842

Carlson, D. F., Fahrenkrug, S. C., & Hackett, P. B. (2012). Targeting DNA with fingers and TALENs. *Molecular Therapy. Nucleic Acids*, •••, 1. PMID: 23344620

Carroll, D. (2021). A short, idiosyncratic history of genome editing. *Gene and Genome Editing*, *1*, 100002. DOI: 10.1016/j.ggedit.2021.100002

Castel, B., Tomlinson, L., Locci, F., Yang, Y., & Jones, J. D. G. (2019). Optimization of T-DNA architecture for Cas9-mediated mutagenesis in Arabidopsis. *PLoS One*, *14*(1), e0204778. DOI: 10.1371/journal.pone.0204778 PMID: 30625150

Cathomen, T., & Joung, J. K. (2008). Zinc-finger nucleases: The next generation emerges. *Molecular Therapy : The Journal of the American Society of Gene Therapy, 16*(7), 1200–1207. DOI: 10.1038/mt.2008.114

Ceasar, S. A., Rajan, V., Prykhozhij, S. V., Berman, J. N., & Ignacimuthu, S. (2016). Insert, remove or replace: A highly advanced genome editing system using CRISPR/Cas9. Biochimica et Biophysica Acta (BBA)-. *Molecular Cell Research, 1863*(9), 2333–2344. PMID: 27350235

Cembrowska-Lech, D., Krzemińska, A., Miller, T., Nowakowska, A., Adamski, C., Radaczyńska, M., Mikiciuk, G., & Mikiciuk, M. (2023). An integrated multi-omics and artificial intelligence framework for advance plant phenotyping in horticulture. *Biology (Basel), 12*(10), 1298. DOI: 10.3390/biology12101298 PMID: 37887008

Cerdeira, A. L., & Dionsio, L. P. (2011). Gazziero, Stephen O. Duke, and Marcus B. Matallo. "Agricultural impacts of glyphosate-resistant soybean cultivation in South America.". *Journal of Agricultural and Food Chemistry, 59*(11), 5799–5807. DOI: 10.1021/jf102652y PMID: 20839871

Chandrasegaran, S., & Carroll, D. (2016). Origins of programmable nucleases for genome engineering. *Journal of Molecular Biology, 428*(5), 963–989. DOI: 10.1016/j.jmb.2015.10.014 PMID: 26506267

Chandrasekaran, J., Brumin, M., Wolf, D., Leibman, D., Klap, C., Pearlsman, M., & Gal-On, A. (2016). Development of broad virus resistance in non-transgenic cucumber using CRISPR/Cas9 technology. *Molecular Plant Pathology, 17*(7), 1140–1153. DOI: 10.1111/mpp.12375 PMID: 26808139

Chandrasekaran, R., Rajiv, P., & Abd-Elsalam, K. A. (2020). Carbon nanotubes: Plant gene delivery and genome editing. In *Carbon nanomaterials for agri-food and environmental applications* (pp. 279–296). Elsevier. DOI: 10.1016/B978-0-12-819786-8.00014-1

Chan, L., & Costa, S. (2005). Participation in the global knowledge commons: Challenges and opportunities for research dissemination in developing countries. *New Library World, 106*(3/4), 141–163. DOI: 10.1108/03074800510587354

Charpentier, E. (2017). Gene editing and genome engineering with CRISPR-Cas9. *Molecular Frontiers Journal, 1*(02), 99–107. DOI: 10.1142/S2529732517400119

Chavhan, R. L., Jaybhaye, S. G., Hinge, V. R., Deshmukh, A. S., Shaikh, U. S., Jadhav, P. K., Kadam, U. S., & Hong, J. C. (2025). Emerging applications of gene editing technologies for the development of climate-resilient crops. *Frontiers in Genome Editing, 7*, 1524767. DOI: 10.3389/fgeed.2025.1524767 PMID: 40129518

Cheftel, J. C. (2005). Food and nutrition labelling in the European Union. *Food Chemistry*, *93*(3), 531–550. DOI: 10.1016/j.foodchem.2004.11.041

Chen, M., & Qi, L. S. (2017). Repurposing CRISPR system for transcriptional activation. RNA Activation, 147-157.

Chen, F., Chen, L., Yan, Z., Xu, J., Feng, L., He, N., Guo, M., Zhao, J., Chen, Z., Chen, H., Yao, G., & Liu, C. (2024). Recent advances of CRISPR-based genome editing for enhancing staple crops. *Frontiers in Plant Science*, *15*, 1478398. Advance online publication. DOI: 10.3389/fpls.2024.1478398 PMID: 39376239

Chen, G., Hao, F., & Sun, X. (2025). Artificial intelligence-driven gene editing and crop breeding: Technological innovations and application prospects. *Advances in Resources Research*, *5*(1), 235–254.

Chen, K., Chen, J., Pi, X., Huang, L.-J., & Li, N. (2023). Isolation, purification, and application of protoplasts and transient expression systems in plants. *International Journal of Molecular Sciences*, *24*(23), 16892. DOI: 10.3390/ijms242316892 PMID: 38069215

Chen, K., Wang, Y., Zhang, R., Zhang, H., & Gao, C. (2019). CRISPR/Cas Genome Editing and Precision Plant Breeding in Agriculture. *Annual Review of Plant Biology*, *70*(1), 667–697. DOI: 10.1146/annurev-arplant-050718-100049 PMID: 30835493

Chen, L., Liu, G., & Zhang, T. (2024b). Integrating machine learning and genome editing for crop improvement. *aBIOTECH*, *5*(2), 262–277. DOI: 10.1007/s42994-023-00133-5 PMID: 38974863

Chen, Q., Washburn, J. D., Lima, D. C., Romay, M. C., Gage, J. L., Holland, J. B., Xavier, A., Murray, S. C., Ertl, D., Lopez-Cruz, M., de Los Campos, G., Aguate, F. M., Beissinger, T. M., Bohn, M. O., Buckler, E. S., Edwards, J., Flint-Garcia, S. A., Gore, M. A., Hirsch, C. N., & de Leon, N. (2026). Genomes to fields 2024 maize genotype by environment prediction competition. *BMC Research Notes*, *19*(1), 113. DOI: 10.1186/s13104-026-07629-5 PMID: 41664105

Chen, X., Zhong, Z., Tang, X., Yang, S., Zhang, Y., Wang, S., Liu, Y., Zhang, Y., Zheng, X., Zhang, Y., & Feng, X. (2024). Advancing PAM-less genome editing in soybean using CRISPR-SpRY. *Horticulture Research*, *11*(8), uhae160. Advance online publication. DOI: 10.1093/hr/uhae160 PMID: 39108580

Chen, Z. J. (2025). Empowering plant epigenetics to breed resilience of crops: From nucleolar dominance to transgenerational epigenetic inheritance. *The Plant Genome*, *18*(2), e70064. DOI: 10.1002/tpg2.70064 PMID: 40556420

Chevalier, B. S., & Stoddard, B. L. (2001). Homing endonucleases: Structural and functional insight into the catalysts of intron/intein mobility. *Nucleic acids research, 29*(18), 3757-3774. Cong, L., Ran, F. A., Cox, D., Lin, S., Barretto, R., Habib, N., Hsu, P. D., Wu, X., Jiang, W., Marraffini, L. A., & Zhang, F. (2013). Multiplex genome engineering using CRISPR/Cas systems. *Science, 339*(6121), 819–823. DOI: 10.1126/science.1231143

Chikoti, P. C., & Tembo, M. (2022). Expansion and impact of cassava brown streak and cassava mosaic diseases in Africa: A review. *Frontiers in Sustainable Food Systems, 6*, 1076364. DOI: 10.3389/fsufs.2022.1076364

Choriyeva, Z. (2024). Genetically modified ecosystems: Innovative approaches in agriculture and their environmental impact. *BIO Web of Conferences, 149*, 01009. DOI: 10.1051/bioconf/202414901009

Choudhary, A., Kumar, A., Kaur, H., Pandey, V., Singh, B., & Mehta, S. (2022). Breeding strategies for developing disease-resistant wheat: present, past, and future. In *Cereal diseases: nanobiotechnological approaches for diagnosis and management* (pp. 137–161). Springer. DOI: 10.1007/978-981-19-3120-8_8

Chuang, Y.-F., Phipps, A. J., Lin, F.-L., Hecht, V., Hewitt, A. W., Wang, P.-Y., & Liu, G.-S. (2021). Approach for in vivo delivery of CRISPR/Cas system: A recent update and future prospect. *Cellular and Molecular Life Sciences : CMLS, 78*(6), 2683–2708. DOI: 10.1007/s00018-020-03725-2 PMID: 33388855

Clapp, J., & Ruder, S.-L. (2020). Precision technologies for agriculture: Digital farming, gene-edited crops, and the politics of sustainability. *Global Environmental Politics, 20*(3), 49–69. DOI: 10.1162/glep_a_00566

Collonnier, C., Guyon-Debast, A., Maclot, F., Mara, K., Charlot, F., & Nogué, F. (2017). Towards mastering CRISPR-induced gene knock-in in plants: Survey of key features and focus on the model *Physcomitrella patens. Methods : A Companion to Methods in Enzymology, 121*, 103–117. DOI: 10.1016/j.ymeth.2017.04.024 PMID: 28478103

Compant, S., Cassan, F., Kostić, T., Johnson, L., Brader, G., Trognitz, F., & Sessitsch, A. (2025). Harnessing the plant microbiome for sustainable crop production. *Nature Reviews. Microbiology, 23*(1), 9–23. DOI: 10.1038/s41579-024-01079-1 PMID: 39147829

Compant, S., Clément, C., & Sessitsch, A. (2010a). Plant growth-promoting bacteria in the rhizo- and endosphere of plants: Their role, colonization, mechanisms involved and prospects for utilization. *Soil Biology & Biochemistry, 42*(5), 669–678. DOI: 10.1016/j.soilbio.2009.11.024

Compant, S., Samad, A., Faist, H., & Sessitsch, A. (2019). A review on the plant microbiome: Ecology, functions, and emerging trends in microbial application. *Journal of Advanced Research*, *19*, 29–37. DOI: 10.1016/j.jare.2019.03.004 PMID: 31341667

Compant, S., van der Heijden, M. G. A., & Sessitsch, A. (2010b). Climate change effects on beneficial plant–microorganism interactions. *FEMS Microbiology Ecology*, *73*(2), 197–214. DOI: 10.1111/j.1574-6941.2010.00900.x PMID: 20528987

Costa-Font, M., Gil, J. M., & Traill, W. B. (2008). Consumer acceptance, valuation of and attitudes towards genetically modified food: Review and implications for food policy. *Food Policy*, *33*(2), 99–111. https://doi.org/https://doi.org/10.1016/j .foodpol.2007.07.002. DOI: 10.1016/j.foodpol.2007.07.002

Crauciuc, A., Tripon, F., Gheorghiu, A., Nemes, G., Boglis, A., & Banescu, C. (2017). Development, Applications, Benefits, Challenges and Limitations of the New Genome Engineering Technique. An Update Study. *Acta Medica Marisiensis*, *63*(1), 4–9. DOI: 10.1515/amma-2017-0007

Critchley, C., Nicol, D., Bruce, G., Walshe, J., Treleaven, T., & Tuch, B. (2019). Predicting public attitudes toward gene editing of germlines: The impact of moral and hereditary concern in human and animal applications. *Frontiers in Genetics*, *9*, 704. DOI: 10.3389/fgene.2018.00704 PMID: 30687386

Cui, H. (2021). Challenges and approaches to crop improvement through C3-to-C4 engineering. *Frontiers in Plant Science*, *12*, 715391. DOI: 10.3389/fpls.2021.715391 PMID: 34594351

Cui, Y., Cao, Q., Li, Y., He, M., & Liu, X. (2023). Advances in cis-element-and natural variation-mediated transcriptional regulation and applications in gene editing of major crops. *Journal of Experimental Botany*, *74*(18), 5441–5457. DOI: 10.1093/jxb/erad248 PMID: 37402253

Cui, Z., Liu, H., Zhang, H., Huang, Z., Tian, R., Li, L., & Zhang, S. (2021). The comparison of ZFNs, TALENs, and SpCas9 by GUIDE-seq in HPV-targeted gene therapy. *Molecular Therapy. Nucleic Acids*, *26*, 1466–1478. DOI: 10.1016/j.omtn.2021.08.008 PMID: 34938601

Cummings, C., & Peters, D. J. (2022). Who trusts in gene-edited foods? Analysis of a representative survey study predicting willingness to eat-and purposeful avoidance of gene edited foods in the United States. *Frontiers in Food Science and Technology*, *2*, 858277. DOI: 10.3389/frfst.2022.858277

Cummings, C., Selfa, T., Lindberg, S., & Bain, C. (2024). Identifying public trust building priorities of gene editing in agriculture and food. *Agriculture and Human Values, 41*(1), 47–60. DOI: 10.1007/s10460-023-10465-z

D'Amelia, V., Docimo, T., Crocoll, C., & Rigano, M. M. (2021). Specialized metabolites and valuable molecules in crop and medicinal plants: The evolution of their use and strategies for their production. *Genes, 12*(6), 936. DOI: 10.3390/genes12060936 PMID: 34207427

De Koning, R., Daryanavard, H., Garmyn, J., Kiekens, R., Toili, M. E. M., & Angenon, G. (2023). Fine-tuning CRISPR/Cas9 gene editing in common bean (*Phaseolus vulgaris L.)* using a hairy root transformation system and in silico prediction models. *Frontiers in Plant Science, 14*, 1233418. DOI: 10.3389/fpls.2023.1233418 PMID: 37929181

de la Fuente Cantó, C., Simonin, M., King, E., Moulin, L., Bennett, M. J., Castrillo, G., & Laplaze, L. (2020). An extended root phenotype: The rhizosphere, its formation and impacts on plant fitness. *The Plant Journal : For Cell and Molecular Biology, 103*(3), 951–964. DOI: 10.1111/tpj.14781 PMID: 32324287

de Lima Procópio, R. E., da Silva, I. R., Martins, M. K., de Azevedo, J. L., & de Araújo, J. M. (2012). Antibiotics produced by Streptomyces. *The Brazilian Journal of Infectious Diseases : an Official Publication of the Brazilian Society of Infectious Diseases, 16*(5), 466–471. DOI: 10.1016/j.bjid.2012.08.014 PMID: 22975171

De Nadal, E., Ammerer, G., & Posas, F. (2011). Controlling gene expression in response to stress. *Nature Reviews. Genetics, 12*(12), 833–845. DOI: 10.1038/nrg3055 PMID: 22048664

de Roij, S. (2020). *Gene-editing crops to benefit smallholder farmers in developing countries: Realised potential or empty claims of inclusive innovation?*,

de Roij, S. (2020). *Gene-editing crops to benefit smallholder farmers in developing countries: Realised potential or empty claims of inclusive innovation?* Delhove, J., Osenk, I., Prichard, I., & Donnelley, M. (2020). Public acceptability of gene therapy and gene editing for human use: a systematic review. *Human Gene Therapy, 31*(1-2), 20–46. PMID: 31802714

de Souza, R. S. C., Ambrosini, A., & Passaglia, L. M. P. (2015). Plant growth-promoting bacteria as inoculants in agricultural soils. *Genetics and Molecular Biology, 38*(4), 401–419. DOI: 10.1590/S1415-475738420150053 PMID: 26537605 de Souza, R. S. C., Okura, V. K., Armanhi, J. S. L., Jorrín, B., Lozano, N., da Silva, M. J., González-Guerrero, M., de Araújo, L. M., Verza, N. C., Bagheri, H. C., & Imperial, J. (2020). Unlocking the bacterial and fungal communities assemblages of sugarcane microbiome. *Scientific Reports*, *10*(1), 19320. DOI: 10.1038/s41598-020-76221-2 PMID: 33168921

Deb, S., Amrita, C. B. K., & Rama, R. S. (2022). Applications of CRISPR/Cas9 technology for modification of the plant genome. *Genetica*, *150*(1), 1–12. DOI: 10.1007/s10709-021-00146-2 PMID: 35018532

Delhove, J., Osenk, I., Prichard, I., & Donnelley, M. (2020). Public acceptability of gene therapy and gene editing for human use: A systematic review. *Human Gene Therapy*, *31*(1-2), 20–46. DOI: 10.1089/hum.2019.197 PMID: 31802714

Deneault, E. (2024). Recent therapeutic gene editing applications to genetic disorders. *Current Issues in Molecular Biology*, *46*(5), 4147–4185. DOI: 10.3390/cimb46050255 PMID: 38785523

Devi, B., Kumar, M. N., Chutia, M., & Bhattacharyya, N. (2022). Abiotic and biotic stress challenges of Cassava in changing climate and strategies to overcome: A review. *Scientia Horticulturae*, *305*, 111432. DOI: 10.1016/j.scienta.2022.111432

Devi, N. B., Sen, S., & Pakshirajan, K. (2025). Artificial intelligence in synthetic biology. In *Artificial Intelligence and Biological Sciences* (pp. 278–300). CRC Press. DOI: 10.1201/9781003492726-17

Devisetty, U. K., De Neef, E., & Eric, R. L. (2024). Gordon, Valeria Velasquez-Zapata, Kenneth Narva, Laurent Mézin, Peter Mc Cahon, Kenneth W. Witwer, and Krishnakumar Sridharan. "A bioinformatics framework for human health risk assessment of externally applied dsRNA-based biopesticides.". *Computational Toxicology (Amsterdam, Netherlands)*, *100340*. Advance online publication. DOI: 10.1016/j.comtox.2024.100340

Devlet, A. (2021). Modern Agriculture and Challenges. *Frontiers in Life Sciences and Related Technologies*, *2*(1), 21–29. DOI: 10.51753/flsrt.856349

Dhaliwal, S. S., Sharma, V., Shukla, A. K., Verma, V., Kaur, M., Shivay, Y. S., Nisar, S., Gaber, A., Brestic, M., Barek, V., Skalicky, M., Ondrisik, P., & Hossain, A. (2022). Biofortification—A frontier novel approach to enrich micronutrients in field crops to encounter the nutritional security. *Molecules (Basel, Switzerland)*, *27*(4), 1340. DOI: 10.3390/molecules27041340 PMID: 35209127

Dhirani, L. L., Mukhtiar, N., Chowdhry, B. S., & Newe, T. (2023). Ethical dilemmas and privacy issues in emerging technologies: A review. *Sensors (Basel)*, *23*(3), 1151. DOI: 10.3390/s23031151 PMID: 36772190

Dixit, S., Kumar, A., Srinivasan, K., Vincent, P. D. R., & Ramu Krishnan, N. (2024). Advancing genome editing with artificial intelligence: Opportunities, challenges, and future directions. *Frontiers in Bioengineering and Biotechnology*, *11*, 1335901. DOI: 10.3389/fbioe.2023.1335901 PMID: 38260726

Doman, J. L., Pandey, S., Neugebauer, M. E., An, M., Davis, J. R., Randolph, P. B., McElroy, A., Gao, X. D., Raguram, A., Richter, M. F., Everette, K. A., Bhatt, D. L., & Liu, D. R. (2023). Phage-assisted evolution and protein engineering yield compact, efficient prime editors. *Cell*, *186*(18), 3983–4002. DOI: 10.1016/j.cell.2023.07.039 PMID: 37657419

Doudna, J. A. (2020). The promise and challenge of therapeutic genome editing. *Nature*, *578*(7794), 229–236. DOI: 10.1038/s41586-020-1978-5 PMID: 32051598

Doudna, J. A., & Charpentier, E. (2014). The new frontier of genome engineering with CRISPR-Cas9. *Science*, *346*(6213), 1258096. DOI: 10.1126/science.1258096 PMID: 25430774

Drapal, M., Enfissi, E. M., Almeida, J., Rapacz, E., Nogueira, M., & Fraser, P. D. (2023). The potential of metabolomics in assessing global compositional changes resulting from the application of CRISPR/Cas9 technologies. *Transgenic Research*, *32*(4), 265–278. DOI: 10.1007/s11248-023-00347-9 PMID: 37166587

Duensing, N., Sprink, T., Parrott, W. A., Fedorova, M., Lema, M. A., Wolt, J. D., & Bartsch, D. (2018). Novel features and considerations for ERA and regulation of crops produced by genome editing. *Frontiers in Bioengineering and Biotechnology*, *6*, 79. DOI: 10.3389/fbioe.2018.00079 PMID: 29967764

Duru, M., Therond, O., Martin, G., Martin-Clouaire, R., Magne, M.-A., Justes, E., Journet, E.-P., Aubertot, J.-N., Savary, S., Bergez, J.-E., & Sarthou, J. P. (2015). How to implement biodiversity-based agriculture to enhance ecosystem services: A review. *Agronomy for Sustainable Development*, *35*(4), 1259–1281. DOI: 10.1007/s13593-015-0306-1

Ebrahimi, V., & Hashemi, A. (2024). CRISPR-based gene editing in plants: Focus on reagents and their delivery tools. *BioImpacts*, *15*, 30019. DOI: 10.34172/bi.30019 PMID: 39963563

Eckerstorfer, M. F., Dolezel, M., Engelhard, M., Giovannelli, V., Grabowski, M., Heissenberger, A., Lener, M., Reichenbecher, W., Simon, S., Staiano, G., Wüst Saucy, A. G., Zünd, J., & Lüthi, C. (2023). Recommendations for the assessment of potential environmental effects of genome-editing applications in plants in the EU. *Plants, 12*(9), 1764. DOI: 10.3390/plants12091764 PMID: 37176822

Eckerstorfer, M. F., Grabowski, M., Lener, M., Engelhard, M., Simon, S., Dolezel, M., Heissenberger, A., & Lüthi, C. (2021). Biosafety of genome editing applications in plant breeding: Considerations for a focused case-specific risk assessment in the EU. *Biotech, 10*(3), 10. DOI: 10.3390/biotech10030010 PMID: 35822764

Eid, A., Alshareef, S., & Mahfouz, M. M. (2018). CRISPR base editors: genome editing without double-stranded breaks. *Biochemical Journal, 475*(11), 1955-1964.

El Chami, D., Daccache, A., & El Moujabber, M. (2020). How can sustainable agriculture increase climate resilience? A systematic review. *Sustainability (Basel), 12*(8), 3119. DOI: 10.3390/su12083119

Ellison, E. E., Nagalakshmi, U., Gamo, M. E., Huang, P.-J., Dinesh-Kumar, S., & Voytas, D. F. (2024). Multiplexed heritable gene editing using RNA viruses and mobile single guide RNAs. *Nature Plants, 10*, 567–578. DOI: 10.1038/s41477-024-01648-3

El-Mounadi, K., Morales-Floriano, M. L., & García-Ruiz, H. (2020). Principles, applications, and biosafety of plant genome editing using CRISPR-Cas9. *Frontiers in Plant Science, 11*(56), 1–14. DOI: 10.3389/fpls.2020.00056 PMID: 32117392

Emaru, A., & Rwotonen, I. Implications of Somaclonal variation in vitro in crop management: A review.

Endo, M., Mikami, M., & Toki, S. (2015). Biallelic Gene Targeting in Rice. *Plant Physiology, 170*(2), 667–677. DOI: 10.1104/pp.15.01663 PMID: 26668334

Enfissi, E. M., Drapal, M., Perez-Fons, L., Nogueira, M., Berry, H. M., Almeida, J., & Fraser, P. D. (2021). New plant breeding techniques and their regulatory implications: An opportunity to advance metabolomics approaches. *Journal of Plant Physiology, 258*, 153378. DOI: 10.1016/j.jplph.2021.153378 PMID: 33631493

Entine, J., Maria Sueli, S. F., Groenewald, J.-H., Kershen, D. L., Lema, M., McHughen, A., & Lima Nepomuceno, A.. (2021). Regulatory approaches for genome edited agricultural plants in select countries and jurisdictions around the world. *Transgenic Research, 30*(4), 551–584. DOI: 10.1007/s11248-021-00257-8 PMID: 33970411

Enyiukwu, D., Bassey, I., Nwaogu, G., Chukwu, L., & Maranzu, J. (2020). Post-harvest spoilage and management of fruits and vegetables: A perspective on small-holder agricultural systems of the tropics. *Greener Trends in Plant Pathology and Entomology, 3*(1), 01-17.

Erdoğan, İ., Cevher-Keskin, B., Bilir, Ö., Hong, Y., & Tör, M. (2023). Recent Developments in CRISPR/Cas9 Genome-Editing Technology Related to Plant Disease Resistance and Abiotic Stress Tolerance. *Biology (Basel), 12*(7), 1037–1037. DOI: 10.3390/biology12071037 PMID: 37508466

Ervin, D. E., Glenna, L. L., & Jussaume, R. A.Jr. (2011). The theory and practice of genetically engineered crops and agricultural sustainability. *Sustainability (Basel), 3*(6), 847–874. DOI: 10.3390/su3060847

European Parliament. (2024). Regulation on new genomic techniques for plants. *Official Journal of the European Union, L,* •••, 2024.

Faltus, T. (2023). The applicability of the European GMO legislation to epigenetically modified organisms. *Frontiers in Bioengineering and Biotechnology, 11*, 1124131. DOI: 10.3389/fbioe.2023.1124131 PMID: 36923460

Fan, B.-L., Chen, L.-H., Chen, L.-L., & Guo, H. (2025). Integrative Multi-Omics Approaches for Identifying and Characterizing Biological Elements in Crop Traits: Current Progress and Future Prospects. *International Journal of Molecular Sciences, 26*(4), 1466. DOI: 10.3390/ijms26041466 PMID: 40003933

Farooq, R., Hussain, K., Nazir, S., Javed, M. R., & Masood, N. (2018). CRISPR/Cas9; A robust technology for producing genetically engineered plants. *Cellular and Molecular Biology, 64*(14), 31–38. DOI: 10.14715/cmb/2018.64.14.6 PMID: 30511631

Feder, G., & Umali, D. L. (1993). The adoption of agricultural innovations: A review. *Technological Forecasting and Social Change, 43*(3-4), 215–239. DOI: 10.1016/0040-1625(93)90053-A

Feng, Z., Zhang, B., Ding, W., Liu, X., Yang, D.-L., Wei, P., & Zhu, J.-K. (2013). Efficient genome editing in plants using a CRISPR/Cas system. *Cell Research, 23*(10), 1229–1232. DOI: 10.1038/cr.2013.114 PMID: 23958582

Ferrari, L. (2021). Attitude toward gene-edited food: and economic analysis from diverse stakeholders' viewpoint.

Fiaz, S., Ahmar, S., Saeed, S., Riaz, A., Mora-Poblete, F., & Jung, K.-H. (2021). Evolution and application of genome editing techniques for achieving food and nutritional security. *International Journal of Molecular Sciences*, *22*(11), 5585. DOI: 10.3390/ijms22115585 PMID: 34070430

Filippova, J., Matveeva, A., Zhuravlev, E., & Stepanov, G. (2019). Guide RNA modification as a way to improve CRISPR/Cas9-based genome-editing systems. *Biochimie*, *167*, 49–60. DOI: 10.1016/j.biochi.2019.09.003 PMID: 31493470

Finkel, O. M., Castrillo, G., Paredes, S. H., González, I. S., & Dangl, J. L. (2017). Understanding and exploiting plant beneficial microbes. *Current Opinion in Plant Biology*, *38*, 155–163. DOI: 10.1016/j.pbi.2017.04.018 PMID: 28622659

Finkel, O. M., Salas-González, I., Castrillo, G., Conway, J. M., Law, T. F., Teixeira, P. J. P. L., Wilson, E. D., Fitzpatrick, C. R., Jones, C. D., & Dangl, J. L. (2020). A single bacterial genus maintains root growth in a complex microbiome. *Nature*, *587*(7832), 103–108. DOI: 10.1038/s41586-020-2778-7 PMID: 32999461

Flanagan, K. (2020). *Utilizing Science Fiction to Visualize Ethical Issues Surrounding Human Genome Editing, Influence Ethical Reasoning Skills in Education, and Guide Policy Implementation*. Drew University.

Folta, K. M. (2025). Acceptance of crop biotechnology requires a change in communication strategy. *Plant Physiology*, *198*(3), kiaf167. Advance online publication. DOI: 10.1093/plphys/kiaf167 PMID: 40314159

Food and Agriculture Organization (FAO). (2018). *Climate-Smart Agriculture: Policies, Practices, and Financing for Food Security*. FAO.

Francisco, R. P., & Camargo, R. (2020). Emerging advanced technologies to mitigate the impact of climate change in Africa. *Plants*, *9*(3), 381. DOI: 10.3390/plants9030381 PMID: 32204576

Franic, D., Dobrinic, P., & Korac, P. (2019). Key achievements in gene therapy development and its promising progress with gene editing tools (ZFN, TALEN, CRISPR/CAS9). Molecular and experimental biology in medicine, 2(1), 1-9.

Fraser, P. D., Aharoni, A., Hall, R. D., Huang, S., Giovannoni, J. J., Sonnewald, U., & Fernie, A. R. (2020). Metabolomics should be used to identify and characterize gene-edited crops. *The Plant Journal : For Cell and Molecular Biology*, *102*(5), 897–902. DOI: 10.1111/tpj.14679 PMID: 31923321

Freitas-Alves, N. S., Moreira-Pinto, C. E., Távora, F. T., Paes-de-Melo, B., Arraes, F. B., Lourenço-Tessutti, I. T., & Qi, Y. (2024). CRISPR/Cas genome editing in soybean: Challenges and new insights to overcome existing bottlenecks. *Journal of Advanced Research*. PMID: 39163906

Friedrichs, S.. (2022). Regulatory and Policy Considerations Around Genome Editing in Agriculture. In *Applications of genome modulation and editing* (pp. 327–366). Springer. DOI: 10.1007/978-1-0716-2301-5_17

Fubini, F. (2024). *In Search of the Perfect Peach: Why flavour holds the answer to fixing our food system*. Chelsea Green Publishing.

Furrow, B. R. (2017). The CRISPR-Cas9 tool of gene editing: Cheaper, faster, riskier. *Annals of Health Law*, *26*, 33.

Gad, M. A., Li, M.-j., Ahmed, F. K., & Almoammar, H. (2020). Nanomaterials for gene delivery and editing in plants: Challenges and future perspective. In *Multifunctional Hybrid Nanomaterials for Sustainable Agri-Food and Ecosystems* (pp. 135–153). Elsevier. DOI: 10.1016/B978-0-12-821354-4.00006-6

Gaj, T., Gersbach, C. A., & Barbas, C. F.III. (2013). ZFN, TALEN, and CRISPR/Cas-based methods for genome engineering. *Trends in Biotechnology*, *31*(7), 397–405. DOI: 10.1016/j.tibtech.2013.04.004 PMID: 23664777

Gallego-Bartolomé, J., Dhami, P., Haber, L., & Yao, W. (2024). Targeted demethylation of stress-response loci in wheat via dCas9-TET fusion. *The Plant Cell*, *36*(1), 145–162. DOI: 10.1093/plcell/koac340

Gallego-Bartolomé, J., Gardiner, J., Liu, W., Papikian, A., Ghoshal, B., Kuo, H. Y., Zhao, J. M.-C., Segal, D. J., & Jacobsen, S. E. (2024). Targeted DNA demethylation of the Arabidopsis genome using the human TET1 catalytic domain. *Proceedings of the National Academy of Sciences of the United States of America*, *121*(4), e2313200121. DOI: 10.1073/pnas.2313200121 PMID: 29444862

Gan, Cong, W., & P.K. Ling, A. (2022). CRISPR/Cas9 in plant biotechnology: Applications and challenges. *Biotechnologia*, *103*(1), 81–93. DOI: 10.5114/bta.2022.113919 PMID: 36605382

Gao, S., ... (2024). Understanding the Factors Driving Consumers' Willingness to Pay for Gene-Edited Foods in China. Foods, 13(15), 2348.

Gao, C. (2021). Genome engineering for crop improvement and future agriculture. *Cell*, *184*(6), 1621–1635. DOI: 10.1016/j.cell.2021.01.005 PMID: 33581057

Gao, H., Smith, J., Yang, M., Jones, S., Djukanovic, V., Nicholson, M. G., West, A., Bidney, D., Falco, S. C., Jantz, D., & Lyznik, L. A. (2010). Heritable targeted mutagenesis in maize using a designed endonuclease. *The Plant Journal : For Cell and Molecular Biology*, *61*(1), 176–187. DOI: 10.1111/j.1365-313X.2009.04041.x PMID: 19811621

Gao, J., Wang, G., Ma, S., Xie, X., Wu, X., Zhang, X., & Xia, Q. (2014). CRISPR/Cas9-mediated targeted mutagenesis in *Nicotiana tabacum*. *Plant Molecular Biology*, *87*(2), 99–110. PMID: 25344637

Gao, Y., Chen, S., Halene, S., & Tebaldi, T. (2021). Transcriptome-wide quantification of double-stranded RNAs in live mouse tissues by dsRIP-Seq. *STAR Protocols*, *2*(1), 100366. DOI: 10.1016/j.xpro.2021.100366 PMID: 33778776

Garneau, J. E., Dupuis, M. È., Villion, M., Romero, D. A., Barrangou, R., Boyaval, P., & Moineau, S. (2010). The CRISPR/Cas bacterial immune system cleaves bacteriophage and plasmid DNA. *Nature*, *468*(7320), 67–71. DOI: 10.1038/nature09523 PMID: 21048762

Gaudelli, N. M., Komor, A. C., Rees, H. A., Packer, M. S., Badran, A. H., Bryson, D. I., & Liu, D. R. (2017). Programmable base editing of A•T to G•C in genomic DNA without DNA cleavage. *Nature*, *551*(7681), 464–471. DOI: 10.1038/nature24644 PMID: 29160308

Gelvin, S. B. (2003). Agrobacterium-mediated plant transformation: The biology behind the "gene-jockeying" tool. *Microbiology and Molecular Biology Reviews : MMBR*, *67*(1), 16–37. DOI: 10.1128/MMBR.67.1.16-37.2003 PMID: 12626681

Geng, J., Li, L., Lv, Q., Zhao, Y., Liu, Y., Zhang, L., & Li, X. (2017). TaGW2-6A allelic variation contributes to grain size possibly by regulating the expression of cytokinins and starch-related genes in wheat. *Planta*, *246*(6), 1153–1163. DOI: 10.1007/s00425-017-2759-8 PMID: 28825220

Ghimire, B. K., Yu, C. Y., Kim, W.-R., Moon, H.-S., Lee, J., Kim, S. H., & Chung, I. M. (2023). Assessment of benefits and risk of genetically modified plants and products: Current controversies and perspective. *Sustainability (Basel)*, *15*(2), 1722. DOI: 10.3390/su15021722

Gier, R. A., Budinich, K. A., Evitt, N. H., Cao, Z., Freilich, E. S., Chen, Q., Qi, J., Lan, Y., Kohli, R. M., & Shi, J. (2020). High-performance CRISPR-Cas12a genome editing for combinatorial genetic screening. *Nature Communications*, *11*(1), 3455. DOI: 10.1038/s41467-020-17209-1 PMID: 32661245

Glick, B. R. (2012). Plant growth-promoting bacteria: Mechanisms and applications. *Scientifica, 963401*, 1–15. Advance online publication. DOI: 10.6064/2012/963401 PMID: 24278762

Gogoi, N., Susila, H., Leach, J., Müllner, M., Jones, B., & Pogson, B. J. (2024). Developing frameworks for nanotechnology-driven DNA-free plant genome-editing. *Trends in Plant Science*. PMID: 39477773

González Castro, N., Bjelic, J., Malhotra, G., Huang, C., & Alsaffar, S. H. (2021). Comparison of the feasibility, efficiency, and safety of genome editing technologies. *International Journal of Molecular Sciences*, *22*(19), 10355. DOI: 10.3390/ijms221910355 PMID: 34638696

Gonzalez Guzman, M., Cellini, F., Fotopoulos, V., Balestrini, R., & Arbona, V. (2022). New approaches to improve crop tolerance to biotic and abiotic stresses. *Physiologia Plantarum*, *174*(1), e13547. DOI: 10.1111/ppl.13547 PMID: 34480798

González, M. N., Massa, G. A., Andersson, M., Turesson, H., Olsson, N., Fält, A.-S., Storani, L., Décima Oneto, C. A., Hofvander, P., & Feingold, S. E. (2020). Reduced enzymatic browning in potato tubers by specific editing of a polyphenol oxidase gene via ribonucleoprotein complexes delivery of the CRISPR/Cas9 system. *Frontiers in Plant Science*, *10*, 497481. DOI: 10.3389/fpls.2019.01649 PMID: 31998338

Goodman, D. (2023). Agri-Biotechnology and the Failed Promises of the Seed-Chemical Complex, CRISPR and Gene Editing, and Regulatory Capture. In *Transforming Agriculture and Foodways* (pp. 53–70). Bristol University Press. DOI: 10.1332/policypress/9781529231465.003.0005

Gopal, M., Gupta, A., & Thomas, G. V. (2013). Bespoke microbiome therapy to manage plant diseases. *Frontiers in Microbiology*, *4*, 355. DOI: 10.3389/fmicb.2013.00355 PMID: 24348466

Goralogia, G. S., Redick, T. P., & Strauss, S. H. (2021). Gene editing in tree and clonal crops: Progress and challenges. *In Vitro Cellular & Developmental Biology. Plant : Journal of the Tissue Culture Association*, *57*(4), 683–699. DOI: 10.1007/s11627-021-10197-x

Goshisht, M. K. (2024). Machine learning and deep learning in synthetic biology: Key architectures, applications, and challenges. *ACS Omega*, *9*(9), 9921–9945. DOI: 10.1021/acsomega.3c05913 PMID: 38463314

Gostimskaya, I. (2022). CRISPR–Cas9: A History of Its Discovery and Ethical Considerations of Its Use in Genome Editing. *Biochemistry. Biokhimiia*, *87*(8), 777–788. DOI: 10.1134/S0006297922080090 PMID: 36171658

Gowdhaman, P. (2024). Adaptation to Progressive Climate Change, Technology Development, Agricultural Practices. *Dr. KL Chaudhary*, 207.

Graham, N., Patil, G. B., Bubeck, D. M., Dobert, R. C., Glenn, K. C., Gutsche, A. T., Kumar, S., Lindbo, J. A., Maas, L., May, G. D., Vega-Sanchez, M. E., Stupar, R. M., & Morrell, P. L. (2020). Plant genome editing and the relevance of Off-Target changes. *Plant Physiology*, *183*(4), 1453–1471. DOI: 10.1104/pp.19.01194 PMID: 32457089

Grešová, K., Alexiou, P., & Giassa, I.-C. (2022). Small RNA targets: Advances in prediction tools and high-throughput profiling. *Biology (Basel)*, *11*(12), 1798. DOI: 10.3390/biology11121798 PMID: 36552307

Gu, S., Bodai, Z., Cowan, Q. T., & Komor, A. C. (2021). Base editors: expanding the types of DNA damage products harnessed for genome editing. *Gene and Genome Editing, 1*, 100005.

Guertler, P., Pallarz, S., Belter, A., Eckermann, K. N., & Grohmann, L. (2023). Detection of commercialized plant products derived from new genomic techniques (NGT)-Practical examples and current perspectives. *Food Control*, *152*, 109869. DOI: 10.1016/j.foodcont.2023.109869

Guo, C., Ma, X., Gao, F., & Guo, Y. (2023). Off-target effects in CRISPR/Cas9 gene editing. *Frontiers in Bioengineering and Biotechnology*, *11*, 1143157. DOI: 10.3389/fbioe.2023.1143157 PMID: 36970624

Gupta, R. M., & Musunuru, K. (2014). Expanding the genetic editing tool kit: ZFNs, TALENs, and CRISPR-Cas9. *The Journal of Clinical Investigation*, *124*(10), 4154–4161. DOI: 10.1172/JCI72992 PMID: 25271723

Gupta, S., Kumar, A., Patel, R., & Kumar, V. (2021). Genetically modified crop regulations: Scope and opportunity using the CRISPR-Cas9 genome editing approach. *Molecular Biology Reports*, *48*(5), 4851–4863. DOI: 10.1007/s11033-021-06477-9 PMID: 34114124

Gu, X., Liu, L., & Zhang, H. (2021). Transgene-free genome editing in plants. *Frontiers in Genome Editing*, *3*, 805317. DOI: 10.3389/fgeed.2021.805317 PMID: 34927134

Hafeez, A., Ali, B., Javed, M. A., Saleem, A., Fatima, M., Fathi, A., Afridi, M. S., Aydin, V., Oral, M. A., & Soudy, F. A. (2023). Plant breeding for harmony between sustainable agriculture, the environment, and global food security: An era of genomics-assisted breeding. *Planta*, *258*(5), 97. DOI: 10.1007/s00425-023-04252-7 PMID: 37823963

Hajheidari, M., Sunyaev, S., & de Meaux, J. (2024). Are complex traits underpinned by polygenic molecular traits? A reflection on the complexity of gene expression. *Plant & Cell Physiology*, •••, pcae140. PMID: 39626022

Hallman, W. K., et al. (2003). Public perceptions of genetically modified foods: A national study of American knowledge and opinion.

Hamdan, M. F., Karlson, C. K. S., Teoh, E. Y., Lau, S.-E., & Tan, B. C. (2022). Genome editing for sustainable crop improvement and mitigation of biotic and abiotic stresses. *Plants*, *11*(19), 2625. DOI: 10.3390/plants11192625 PMID: 36235491

Han, P., Velasco-Hernández, M. C., Ramirez-Romero, R., & Desneux, N. (2016). Behavioral effects of insect-resistant genetically modified crops on phytophagous and beneficial arthropods: A review. *Journal of Pest Science*, *89*(4), 859–883. DOI: 10.1007/s10340-016-0791-2

Hanssen, L., Dijkstra, A., Sleenhoff, S., Frewer, L., & Gutteling, J. M. (2018). Revisiting public debate on genetic modification and genetically modified organisms. Explanations for contemporary Dutch public attitudes. *JCOM, Journal of Science Communication*, *17*(4), A01. DOI: 10.22323/2.17040201

Hao, L., Pu, X., & Song, J. (2021). Introduction of mutations in plants with prime editing. *Methods : A Companion to Methods in Enzymology*, *194*, 83–93. DOI: 10.1016/j.ymeth.2021.03.014 PMID: 33774158

Haque, E., Taniguchi, H., Hassan, M. M., Bhowmik, P., Karim, M. R., Śmiech, M., & Islam, T. (2018). Application of CRISPR/Cas9 genome editing technology for the improvement of crops cultivated in tropical climates: Recent progress, prospects, and challenges. *Frontiers in Plant Science*, *9*(617), 1–15. DOI: 10.3389/fpls.2018.00617 PMID: 29868073

Hartman, S., Horner, W., Jackson, C., Kovak, E., & Velan, V. (2020). Streamlining USDA Regulation of Gene Editing to Benefit US Agriculture. *Journal of Science Policy & Governance*, *17*(01). Advance online publication. DOI: 10.38126/JSPG170108

Hasani, M., Camacho-Martinez, S. V., Zai, B., Warriner, L. J., Boesch, C., Kopko, C., Marafiotti, F., & Warriner, K. (2024). Development of an online food safety toolbox based on the Codex Alimentarius General Principles of Food Hygiene: Engaging users through mapping, chunking, and learning-by-asking. *Food Quality and Safety*, *8*, fyae030. Advance online publication. DOI: 10.1093/fqsafe/fyae030

Hassan, Y. M., Mohamed, A. S., Hassan, Y. M., & El-Sayed, W. M. (2025). Recent developments and future directions in point-of-care next-generation CRISPR-based rapid diagnosis. *Clinical and Experimental Medicine*, *25*(1), 1–15. DOI: 10.1007/s10238-024-01540-8 PMID: 39789283

Hazrati, R., Asghari-Zakaria, R., & Zare, N. (2025). Crop Genome Editing for Nutrient Use Efficiency. In *Omics and Genome Editing: Revolution in Crop Improvement for Sustainable Agriculture* (pp. 87–106). Springer Nature Switzerland. DOI: 10.1007/978-3-031-81639-0_6

Helliwell, R., Hartley, S., & Pearce, W. (2019). NGO perspectives on the social and ethical dimensions of plant genome-editing. *Agriculture and Human Values*, *36*(4), 779–791. DOI: 10.1007/s10460-019-09956-9

Henderson, K., Lang, B., Kemper, J., & Conroy, D. (2024). Exploring diverse food system actor perspectives on gene editing: A systematic review of socio-cultural factors influencing acceptability. *Agriculture and Human Values*, *41*(2), 883–907. DOI: 10.1007/s10460-023-10523-6

Herman, R. A., Storer, N. P., Anderson, J. A., Amijee, F., Cnudde, F., & Raybould, A. (2021). Transparency in risk-disproportionate regulation of modern crop-breeding techniques. *GM Crops & Food*, *12*(1), 376–381. DOI: 10.1080/21645698.2021.1934353 PMID: 34107854

Hernández-Soto, A., & Gatica-Arias, A. (2024). Genome editing in Latin America: Research achievements and regulatory evolution. [PCTOC]. *Plant Cell, Tissue and Organ Culture*, *159*(3), 55. DOI: 10.1007/s11240-024-02904-4

Hertel, R., Gibhardt, J., Martienssen, M., Kuhn, R., & Commichau, F. M. (2021). Molecular mechanisms underlying glyphosate resistance in bacteria. *Environmental Microbiology*, *23*(6), 2891–2905. DOI: 10.1111/1462-2920.15534 PMID: 33876549

Hess, D. J. (2009). *Localist movements in a global economy: Sustainability, justice, and urban development in the United States*. MIT Press. DOI: 10.7551/mitpress/9780262012645.001.0001

Hibshman, G. N., Bravo, J. P. K., Hooper, M. M., Dangerfield, T. L., Zhang, H., Finkelstein, I. J., Johnson, K. A., & Taylor, D. W. (2024). Unraveling the mechanisms of PAMless DNA interrogation by SpRY-Cas9. *Nature Communications*, *15*(1), 3663. DOI: 10.1038/s41467-024-47830-3 PMID: 38688943

Hill, N. (2020). Public opinion of gene-editing in agriculture: A mixed-method study of online media and metaphors

Hillary, V. E., & Ceasar, S. A. (2023). A review on the mechanism and applications of CRISPR/Cas9/Cas12/Cas13/Cas14 proteins utilized for genome engineering. *Molecular Biotechnology*, *65*(3), 311–325. DOI: 10.1007/s12033-022-00567-0 PMID: 36163606

Hille, F., Charpentier, E. (2016). CRISPR-Cas: biology, mechanisms and relevance. Philosophical Transactions of the Royal Society. *Biological Sciences,* 371,1707. 20150496.

Hill, N., Meyers, C., Li, N., Doerfert, D. L., & Mendu, V. (2022). Persuasive effects of metaphors regarding gene-editing in agriculture. *Journal of Applied Communications*, *106*(1), 1. https://newprairiepress.org/jac/vol106/iss1/1/. DOI: 10.4148/1051-0834.2416

Hock, S. C., Kian, S. M., & Wah, C. L. (2020). Global challenges in the manufacture, regulation and international harmonization of GMP and quality standards for biopharmaceuticals. *GaBi Journal*, *9*(2), 52–64. DOI: 10.5639/gabij.2020.0902.010

Hoffman, N. E. (2021). Revisions to USDA biotechnology regulations: The SECURE rule. *Proceedings of the National Academy of Sciences of the United States of America*, *118*(22), e2004841118. DOI: 10.1073/pnas.2004841118 PMID: 34050018

Holman, C. M. (2019). A fractured international response to CRISPR-enabled gene editing of agricultural products. *Biotechnology Law Report*, *38*(1), 3–23. DOI: 10.1089/blr.2019.29100.cmh

Holmes-Davis, R., Li, G., Jamieson, A. C., Rebar, E. J., Liu, Q., Kong, Y., Case, C. C., & Gregory, P. D. (2005). Gene regulation in planta by plant-derived engineered zinc finger protein transcription factors. *Plant Molecular Biology*, *57*(3), 411–423. DOI: 10.1007/s11103-004-7820-x PMID: 15830130

Hossain, A., Islam, M. T., Maitra, S., Majumder, D., Garai, S., Mondal, M., . . . Brestic, M. (2021). Neglected and underutilized crop species: are they future smart crops in fighting poverty, hunger and malnutrition under changing climate? *Neglected and underutilized crops-towards nutritional security and sustainability*, 1-50.

Hossain, M. J., & Bakhsh, A. (2020). Development and applications of transplastomic plants; a way towards eco-friendly agriculture. *Environment, climate, plant and vegetation growth*, 285-322.

Hsu, P. D., Lander, E. S., & Zhang, F. (2014). Development and Applications of CRISPR-Cas9 for Genome Engineering. *Cell*, *157*(6), 1262–1278. DOI: 10.1016/j.cell.2014.05.010 PMID: 24906146

Hua, K., Zhang, J., Botella, J. R., Ma, C., Kong, F., Liu, B., & Zhu, J. K. (2019). Perspectives on the Application of Genome-Editing Technologies in Crop Breeding. *Molecular Plant*, *12*(8), 1047–1059. DOI: 10.1016/j.molp.2019.06.009 PMID: 31260812

Huang, T.-P., Newby, G. A., & Liu, D. R. (2023). Precision genome editing using cytosine and adenine base editors in mammalian cells. *Nature Protocols*, *16*(2), 1089–1128. DOI: 10.1038/s41596-020-00450-9 PMID: 33462442

Huang, T.-P., Zhao, K. T., Miller, S. M., Gaudelli, N. M., Oakes, B. L., Fellmann, C., Savage, D. F., & Liu, D. R. (2023). Circularly permuted and PAM-modified Cas9 variants broaden the targeting scope of base editors. *Nature Biotechnology*, *41*, 997–1008. DOI: 10.1038/s41587-022-01645-3

Huang, X., Hilscher, J., Stoger, E., Christou, P., & Zhu, C. (2021). Modification of cereal plant architecture by genome editing to improve yields. *Plant Cell Reports*, *40*(6), 953–978. DOI: 10.1007/s00299-021-02668-7 PMID: 33559722

Huang, X., Yang, D., Zhang, J., Xu, J., & Chen, Y. E. (2022). Recent advances in improving gene-editing specificity through CRISPR–Cas9 nuclease engineering. *Cells*, *11*(14), 2186. DOI: 10.3390/cells11142186 PMID: 35883629

Hüdig, M., Laibach, N., & Hein, A.-C. (2022). Genome editing in crop plant research—Alignment of expectations and current developments. *Plants*, *11*(2), 212. DOI: 10.3390/plants11020212 PMID: 35050100

Hu, H., Rincent, R., & Runcie, D. E. (2024). MegaLMM improves genomic predictions in new environments using environmental covariates. *Genetics*, *229*(1), iyae171. Advance online publication. DOI: 10.1093/genetics/iyae171 PMID: 39471330

Hu, J. H., Miller, S. M., Geurts, M. H., Tang, W., Chen, L., Sun, N., Zeina, C. M., Gao, X., Rees, H. A., Lin, Z., & Liu, D. R. (2018). Evolved Cas9 variants with broad PAM compatibility and high DNA specificity. *Nature*, *556*(7699), 57–63. DOI: 10.1038/nature26155 PMID: 29512652

Hunt, L., Lerner, F., & Ziegler, M. (2004). NAD–new roles in signaling and gene regulation in plants. *The New Phytologist*, *163*(1), 31–44. DOI: 10.1111/j.1469-8137.2004.01087.x PMID: 33873776

Husaini, A. M., & Sohail, M. (2024). Agrochemical-free genetically modified and genome-edited crops–towards a 'greener' green revolution and achieving the United Nation's Sustainable Development Goals. *Journal of Biotechnology*. Advance online publication. DOI: 10.1016/j.jbiotec.2024.04.015

Hussain, A., Ding, X., Alariqi, M., Manghwar, H., Hui, F., Li, Y., Cheng, J., Wu, C., Cao, J., & Jin, S. (2021). Herbicide resistance: Another hot agronomic trait for plant genome editing. *Plants*, *10*(4), 621. DOI: 10.3390/plants10040621 PMID: 33805182

Hu, Y., House, L. A., & Gao, Z. (2022). How do consumers respond to labels for crispr (gene-editing)? *Food Policy*, *112*, 102366. DOI: 10.1016/j.foodpol.2022.102366

Hyun, Y., Kim, J., Cho, S. W., Choi, Y., Kim, J. S., & Coupland, G. (2014). Site-directed mutagenesis in *Arabidopsis thaliana* using dividing tissue-targeted RGEN of the CRISPR/Cas system to generate heritable null alleles. *Planta*, *241*(1), 271–284. DOI: 10.1007/s00425-014-2180-5 PMID: 25269397

Ichim, M. C. (2023). The Citizens' Awareness and Concerns During the Transition from Genetically Modified to Genome Edited Plants in Europe about Their Use in Agriculture and Food Production. A Roadmap for Plant Genome Editing, 519.

Idris, S. H., Mat Jalaluddin, N. S., & Chang, L. W. (2023). Ethical and legal implications of gene editing in plant breeding: A systematic literature review. *Journal of Zhejiang University. Science. B*, *24*(12), 1093–1105. DOI: 10.1631/jzus.B2200601 PMID: 38057267

International Rice Research Institute (IRRI). (2020). *Stress-Tolerant Rice for Africa and South Asia (STRASA): Final Project Report*. IRRI.

Ipsos, M. (2021). *Consumer perceptions of genome edited food*. Food Standards Agency UK. DOI: 10.46756/sci.fsa.aya629

Ishii, T., & Araki, M. (2016). Consumer acceptance of food crops developed by genome editing. *Plant Cell Reports*, *35*(7), 1507–1518. DOI: 10.1007/s00299-016-1974-2 PMID: 27038939

Ishino, Y., Krupovic, M., & Forterre, P. (2018). History of CRISPR-Cas from Encounter with a Mysterious Repeated Sequence to Genome Editing Technology. *Journal of Bacteriology*, *200*(7), e00580–e17. DOI: 10.1128/JB.00580-17 PMID: 29358495

Jacoby, R., Peukert, M., Succurro, A., Koprivova, A., & Kopriva, S. (2017). The role of soil microorganisms in plant mineral nutrition. *Current Opinion in Plant Biology*, *39*, 1–6. DOI: 10.1016/j.pbi.2017.05.007 PMID: 28974956

Jaganathan, D., Ramasamy, K., Sellamuthu, G., Jayabalan, S., & Venkataraman, G. (2018). CRISPR for crop improvement: An update review. *Frontiers in Plant Science*, *9*, 985. DOI: 10.3389/fpls.2018.00985 PMID: 30065734

Jalaluddin, M., Sharleeza, N., Abdulah, A.-H. A. F., & Othman, R. Y. (2024). Regulatory landscape and public perception for gene-edited bananas in the Southeast Asian region. *Transgenic Research, 33*(3), 89–97. DOI: 10.1007/s11248-024-00379-9 PMID: 38600337

Janik, E., Niemcewicz, M., Ceremuga, M., Krzowski, L., Saluk-Bijak, J., & Bijak, M. (2020). Various aspects of a gene editing system—crispr–cas9. *International Journal of Molecular Sciences, 21*(24), 9604. DOI: 10.3390/ijms21249604 PMID: 33339441

Jankele, R., & Svoboda, P. (2014). TAL effectors: Tools for DNA targeting. *Briefings in Functional Genomics, 13*(5), 409–419. DOI: 10.1093/bfgp/elu013 PMID: 24907364

Jansen, R., Embden, J. D. A., & van, . (2002). Identification of genes that are associated with DNA repeats in prokaryotes. *Molecular Microbiology, 43*(6), 1565–1575. DOI: 10.1046/j.1365-2958.2002.02839.x PMID: 11952905

Jansson, J. K., & Hofmockel, K. S. (2020). Soil microbiomes and climate change. *Nature Reviews. Microbiology, 18*(1), 35–46. DOI: 10.1038/s41579-019-0265-7 PMID: 31586158

Jasanoff, S.. (2015). CRISPR democracy: Gene editing and the need for inclusive deliberation. *Issues in Science and Technology, 32*(1), 25–32.

Jenkins, D., Dobert, R., Atanassova, A., & Pavely, C. (2021). Impacts of the regulatory environment for gene editing on delivering beneficial products. *In Vitro Cellular & Developmental Biology. Plant : Journal of the Tissue Culture Association, 57*(4), 609–626. DOI: 10.1007/s11627-021-10201-4 PMID: 34429575

Jian, C., Pan, Y., Liu, S., Guo, M., Huang, Y., Cao, L., & Hou, J. (2024). The TaGW2-TaSPL14 module regulates the trade-off between tiller number and grain weight in wheat. *Journal of Integrative Plant Biology, 66*(9), 1953–1965. DOI: 10.1111/jipb.13723 PMID: 38924348

Jiang, L. (2020). Commercialization of the gene-edited crop and morality: Challenges from the liberal patent law and the strict GMO law in the EU. *New Genetics and Society, 39*(2), 191–218. DOI: 10.1080/14636778.2019.1686968

Jiang, W., Zhou, H., Bi, H., Fromm, M., Yang, B., & Weeks, D. P. (2013). Demonstration of CRISPR/Cas9/sgRNA-mediated targeted gene modification in Arabidopsis, tobacco, sorghum, and rice. *Nucleic Acids Research, 41*(20), e188. DOI: 10.1093/nar/gkt780 PMID: 23999092

Jinek, M., Chylinski, K., Fonfara, I., Hauer, M., Doudna, J. A., & Charpentier, E. (2012). A programmable dual-RNA-guided DNA endonuclease in adaptive bacterial immunity. *Science*, *337*(6096), 816–821. DOI: 10.1126/science.1225829 PMID: 22745249

Jogam, P., Sandhya, D., Alok, A., Peddaboina, V., Allini, V. R., & Zhang, B. (2022). A review on CRISPR/Cas-based epigenetic regulation in plants. *International Journal of Biological Macromolecules*, *219*(31), 1261–1271. DOI: 10.1016/j.ijbiomac.2022.08.182 PMID: 36057300

Jones, M. G., Fosu-Nyarko, J., Iqbal, S., Adeel, M., Romero-Aldemita, R., Arujanan, M., ... & Khoo, K. (2022). Enabling trade in gene-edited produce in Asia and Australasia: The developing regulatory landscape and future perspectives. Plants, 11(19), 2538.

Jones, M. G., Fosu-Nyarko, J., Iqbal, S., Adeel, M., Romero-Aldemita, R., Arujanan, M., Kasai, M., Wei, X., Prasetya, B., Nugroho, S., Mewett, O., Mansoor, S., Awan, M. J. A., Ordonio, R. L., Rao, S. R., Poddar, A., Hundleby, P., Iamsupasit, N., & Khoo, K. (2022). Enabling trade in gene-edited produce in Asia and Australasia: The developing regulatory landscape and future perspectives. *Plants*, *11*(19), 2538. DOI: 10.3390/plants11192538 PMID: 36235403

Joo, J. H., Lee, S., & Kim, K. P. (2025). Precision gene editing: The power of CRISPR-Cas in modern genetics. *Molecular Therapy. Nucleic Acids*, *36*(4), 102733. DOI: 10.1016/j.omtn.2025.102733 PMID: 41210586

Joung, J. K., & Sander, J. D. (2013). TALENs: A widely applicable technology for targeted genome editing. *Nature Reviews. Molecular Cell Biology*, *14*(1), 49–55. DOI: 10.1038/nrm3486 PMID: 23169466

Justamante, M. S., Larriba, E., Luque, A., Nicolás-Albujer, M., & Pérez-Pérez, J. M. (2025). A systematic review to identify target genes that modulate root system architecture in response to abiotic stress. *Scientific Reports*, *15*(1), 13219. DOI: 10.1038/s41598-025-97266-y PMID: 40246916

Kalaitzandonakes, N., Willig, C., & Zahringer, K. (2023). The Economics and Policy of Genome Editing in Crop Improvement. *The Plant Genome*, *16*(2), e20248. DOI: 10.1002/tpg2.20248 PMID: 36321718

Kalam, S., Basu, A., Ahmad, I., Sayyed, R. Z., El-Enshasy, H. A., Dailin, D. J., & Suriani, N. L. (2020). Recent understanding of soil acidobacteria and their ecological significance. *Frontiers in Microbiology*, *11*, 580024. DOI: 10.3389/fmicb.2020.580024 PMID: 33193209

Kalsing, A., Velini, E. D., Merotto, A.Jr, & Carbonari, C. A. (2024). The population genomics of Conyza spp. in soybean macroregions suggest the spread of herbicide resistance through intraspecific and interspecific gene flow. *Scientific Reports*, *14*(1), 19536. DOI: 10.1038/s41598-024-70153-8 PMID: 39174662

Kamyab, H., SaberiKamarposhti, M., Hashim, H., & Yusuf, M. (2023). Carbon dynamics in agricultural greenhouse gas emissions and removals: A comprehensive review. *Carbon Letters*, *34*(1), 265–289. DOI: 10.1007/s42823-023-00647-4

Kang, Y., Deng, H., Pray, C., & Hu, R. (2022). Managers' attitudes toward gene-editing technology and companies' R&D investment in gene-editing: The case of Chinese seed companies. *GM Crops & Food*, *13*(1), 309–326. DOI: 10.1080/21645698.2022.2140567 PMID: 36382611

Karavolias, N. G., Horner, W., Abugu, M. N., & Evanega, S. N. (2021). Application of gene editing for climate change in agriculture. *Frontiers in Sustainable Food Systems*, *5*, 685801. DOI: 10.3389/fsufs.2021.685801

Kar, S. R., Choudhury, S., & Chakraborty, A. (2022). CRISPR/Cas9 for soybean improvement: A review. *Asia-Pacific Journal of Molecular Biology and Biotechnology*, *30*, 40–56. DOI: 10.35118/apjmbb.2022.030.3.05

Kasi Viswanath, K., Hamid, A., Ateka, E., & Pappu, H. R. (2023). CRISPR/Cas, multiomics, and RNA interference in virus disease management. *Phytopathology*, *113*(9), 1661–1676. DOI: 10.1094/PHYTO-01-23-0002-V PMID: 37486077

Kato-Nitta, N., Maeda, T., Inagaki, Y., & Tachikawa, M. 2019. Expert and Public Perceptions of Gene-Edited Crops: Attitude Changes in Relation to Scientific Knowledge. *Palgrave Communications* 5: 1. https://doi.org/DOI: 10.1057/s41599-019-0328-4

Kato-Nitta, N., Tachikawa, M., Inagaki, Y., & Maeda, T. (2023). Public perceptions of risks and benefits of gene-edited food crops: An international comparative study between the US, Japan, and Germany. *Science, Technology & Human Values*, *48*(6), 1360–1392. DOI: 10.1177/01622439221123830

Kaul, S., Sharma, T., & Dhar, M. K. (2016). "Omics" tools for better understanding the plant–endophyte interactions. *Frontiers in Plant Science*, *7*, 955. DOI: 10.3389/fpls.2016.00955 PMID: 27446181

Kaul, T., Sony, S. K., Bharti, J., Motelb, K. F. A., Verma, R., Thangaraj, A., & Eswaran, M. (2022). CRISPR genome editing brings global food security into the first Lane: Enhancing nutrition and stress resilience in crops. In *Next-generation plant breeding approaches for stress resilience in cereal crops:285–344*. Springer Nature Singapore. DOI: 10.1007/978-981-19-1445-4_9

Kaul, T., Thangaraj, A., Jain, R., Bharti, J., Kaul, R., Verma, R., Sony, S. K., Abdel Motelb, K. F., Yadav, P., & Agrawal, P. K. (2024). CRISPR/Cas9-mediated homology donor repair base editing system to confer herbicide resistance in maize (Zea mays L.). *Plant Physiology and Biochemistry : PPB*, *207*, 108374. DOI: 10.1016/j.plaphy.2024.108374 PMID: 38310724

Kaur, N., & Pati, P. K. (2019). Harnessing the potential of brassinosteroids in abiotic stress tolerance in plants. In Brassinosteroids: plant growth and development (pp. 407-423). Singapore: Springer Singapore.

Kaur, N., Urban, M., Hammond-Kosack, K. E., & Halford, N. G. (2024). *Field trialling of CRISPR low-asparagine wheat and investigation of the increase in asparagine in response to pathogen infection in wheat.* Genetics and Agronomy for Low Acrylamide-Forming Potential in Cereals.

Kavhiza, N. J., Zargar, M., Prikhodko, S. I., Pakina, E. N., Murtazova, K. M.-S., & Nakhaev, M. R. (2022). Improving crop productivity and ensuring food security through the adoption of genetically modified crops in sub-Saharan Africa. *Agronomy (Basel)*, *12*(2), 439. DOI: 10.3390/agronomy12020439

Kavuri, N. R., Ramasamy, M., Qi, Y., & Mandadi, K. (2022). Applications of CRISPR/Cas13-based RNA editing in plants. *Cells*, *11*(17), 2665. DOI: 10.3390/cells11172665 PMID: 36078073

Kawall, K., Cotter, J., & Then, C. (2020). Broadening the GMO risk assessment in the EU for genome editing technologies in agriculture. *Environmental Sciences Europe*, *32*(1), 106. DOI: 10.1186/s12302-020-00361-2

Kaya, C. (2025). Plant Metabolic Engineering for Enhanced Nutrition and Food Security During Climate Uncertainty. *Food and Energy Security*, *14*(1), e70060. DOI: 10.1002/fes3.70060

Keane, S. (2006). Can a Consumer's Right to Know Survive the WTO: The Case of Food Labeling. *Transnat'l L. & Contemp. Probs.*, *16*, 291.

Keiper, F., & Atanassova, A. (2020). Regulation of synthetic biology: Developments under the convention on biological diversity and its protocols. *Frontiers in Bioengineering and Biotechnology*, *8*, 310. DOI: 10.3389/fbioe.2020.00310 PMID: 32328486

Ke, J., Wang, B., & Yoshikuni, Y. (2021). Microbiome engineering: Synthetic biology of plant-associated microbiomes in sustainable agriculture. *Trends in Biotechnology*, *39*(3), 244–261. DOI: 10.1016/j.tibtech.2020.07.008 PMID: 32800605

Kenzhebayeva, S., Atabayeva, S., Sarsu, F., Abekova, A., Shoinbekova, S., Omirbekova, N., Doktyrbay, G., Beisenova, A., & Shavrukov, Y. (2022). Organ-specific expression of genes involved in iron homeostasis in wheat mutant lines with increased grain iron and zinc content. *PeerJ*, *10*, e13515. DOI: 10.7717/peerj.13515 PMID: 35707120

Khalil, A. M. (2020). The genome editing revolution. *Journal of Genetic Engineering and Biotechnology*, *18*(1), 68. DOI: 10.1186/s43141-020-00078-y PMID: 33123803

Khan, A. H., Hassan, M., & Khan, M. N. (2020). Conventional plant breeding program for disease resistance. *Plant disease management strategies for sustainable agriculture through traditional and modern approaches*, 27-51.

Khan, Z., Saboor, T., Ashfaq, M., Saddique, A., & Khanum, P. (2021). Disease resistance in crops through CRISPR/Cas. *CRISPR Crops: The Future of Food Security*, 151-175.

Khan, A., Pudhuvai, B., Shrestha, A., Mishra, A. K., Shah, M. P., Koul, B., & Dey, N. (2024). CRISPR-mediated iron and folate biofortification in crops: Advances and perspectives. *Biotechnology & Genetic Engineering Reviews*, *40*(4), 4138–4168. DOI: 10.1080/02648725.2023.2205202 PMID: 37092872

Khan, N., Bano, A., & Babar, M. A. (2019). The stimulatory effects of plant growth-promoting rhizobacteria and plant growth regulators on wheat physiology grown in sandy soil. *Archives of Microbiology*, *201*(6), 769–785. DOI: 10.1007/s00203-019-01644-w PMID: 30843087

KhokharVoytas, A., Shahbaz, M., Maqsood, M. F., Zulfiqar, U., Naz, N., Iqbal, U. Z., Sara, M., Aqeel, M., Khalid, N., Noman, A., Zulfiqar, F., Al Syaad, K. M., & AlShaqhaa, M. A.KhokharVoytas. (2023). Genetic modification strategies for enhancing plant resilience to abiotic stresses in the context of climate change. *Functional & Integrative Genomics*, *23*(3), 283. DOI: 10.1007/s10142-023-01202-0 PMID: 37642792

Kishchenko, O., Zhou, Y., Jatayev, S., Shavrukov, Y., & Borisjuk, N. (2020). Gene editing applications to modulate crop flowering time and seed dormancy. *aBIOTECH*, *1*(4), 233–245. DOI: 10.1007/s42994-020-00032-z PMID: 36304127

Kleinstiver, B. P., Pattanayak, V., Prew, M. S., Tsai, S. Q., Nguyen, N. T., Zheng, Z., & Joung, J. K. (2016). High-fidelity CRISPR–Cas9 nucleases with no detectable genome-wide off-target effects. *Nature*, *529*(7587), 490–495. DOI: 10.1038/nature16526 PMID: 26735016

Kockum, I., Huang, J., & Stridh, P. (2023). Overview of genotyping technologies and methods. *Current Protocols*, *3*(4), e727. DOI: 10.1002/cpz1.727 PMID: 37026777

Kofler, N., Collins, J. P., Kuzma, J., Marris, E., Esvelt, K., Nelson, M. P., & Semenov, M. (2018). Editing nature: Local roots of global governance. *Science*, *362*(6414), 527–529. DOI: 10.1126/science.aat4612 PMID: 30385564

Komen, J., Tripathi, L., Mkoko, B., Ofosu, D. O., Oloka, H., & Wangari, D. (2020). Biosafety regulatory reviews and leeway to operate: Case studies from sub-Sahara Africa. *Frontiers in Plant Science*, *11*, 130. DOI: 10.3389/fpls.2020.00130 PMID: 32210981

Komor, A. C., Kim, Y. B., Packer, M. S., Zuris, J. A., & Liu, D. R. (2016). Programmable editing of a target base in genomic DNA without double-stranded DNA cleavage. *Nature*, *533*(7603), 420–424. DOI: 10.1038/nature17946 PMID: 27096365

Kondo, K., & Taguchi, C. (2022). Japanese regulatory framework and approach for genome-edited foods based on latest scientific findings. *Food Safety (Tokyo, Japan)*, *10*(4), 113–128. DOI: 10.14252/foodsafetyfscj.D-21-00016 PMID: 36619008

Kong, D., Liang, W., Zhou, Y., Zhu, H., & Li, H. (2024). Nanoparticle-mediated CRISPR–Cas9 ribonucleoprotein delivery enables editing in intact maize tissues without tissue culture. *Nature Plants*, *10*, 552–562. DOI: 10.1038/s41477-024-01630-z

Kong, D., Ruan, H., & Zhao, Y. (2024). Nanoparticle-mediated ribonucleoprotein delivery enables CRISPR–Cas9 editing in intact maize tissues without tissue culture. *Nature Plants*, *10*, 392–404. DOI: 10.1038/s41477-024-01620-1

Koo, D.-H., Sathishraj, R., Friebe, B., & Gill, B. S. (2021). Deciphering the mechanism of glyphosate resistance in Amaranthus palmeri by cytogenomics. *Cytogenetic and Genome Research*, *161*(12), 578–584. DOI: 10.1159/000521409 PMID: 35021177

Kordyś, M., Sen, R., & Warkocki, Z. (2022). Applications of the versatile CRISPR-Cas13 RNA targeting system. *Wiley Interdisciplinary Reviews. RNA*, *13*(3), e1694. DOI: 10.1002/wrna.1694 PMID: 34553495

Kreiner, J. M., Stinchcombe, J. R., & Wright, S. I. (2018). Population genomics of herbicide resistance: Adaptation via evolutionary rescue. *Annual Review of Plant Biology*, *69*(1), 611–635. DOI: 10.1146/annurev-arplant-042817-040038 PMID: 29140727

Krimsky, S. (2019). *GMOs decoded: A Skeptic's view of genetically modified foods*. MIT Press. DOI: 10.7551/mitpress/10824.001.0001

Kujur, S., Senthil-Kumar, M., & Kumar, R. (2021). Plant viral vectors: Expanding the possibilities of precise gene editing in plant genomes. *Plant Cell Reports*, *40*(6), 931–934. DOI: 10.1007/s00299-021-02697-2 PMID: 33864508

Kuluev, B., Gumerova, G., Mikhaylova, E., Gerashchenkov, G., Rozhnova, N., Vershinina, Z., & Baymiev, A. K. (2019). Delivery of CRISPR/Cas components into higher plant cells for genome editing. *Russian Journal of Plant Physiology: a Comprehensive Russian Journal on Modern Phytophysiology*, *66*(5), 694–706. DOI: 10.1134/S102144371905011X

Kumar, A.. (2018). Genetic engineering and public perception. In *Biotechnology for sustainable agriculture* (pp. 335–373). Elsevier. DOI: 10.1016/B978-0-12-812160-3.00012-X

Kumar, A., Anju, T., Kumar, S., Chhapekar, S. S., Sreedharan, S., Singh, S., Choi, S. R., Ramchiary, N., & Lim, Y. P. (2021). Integrating omics and gene editing tools for rapid improvement of traditional food plants for diversified and sustainable food security. *International Journal of Molecular Sciences*, *22*(15), 8093. DOI: 10.3390/ijms22158093 PMID: 34360856

Kumar, A., Kumar, R., Singh, N., & Mansoori, A. (2020). "Regulatory framework and policy decisions for genome-edited crops." *CRISPR/Cas Genome Editing: Strategies And Potential For. Crop Improvement*, •••, 193–201. DOI: 10.1007/978-3-030-42022-2_9

Kumar, A., Singh, S., Gaurav, A. K., Srivastava, S., & Verma, J. P. (2020). Plant microbiome: Trends and applications for sustainable agriculture. *Plant Gene*, *24*, 100241. DOI: 10.1016/j.plgene.2020.100241

Kumar, D., Yadav, A., Ahmad, R., Dwivedi, U. N., & Yadav, K. (2022). CRISPR-based genome editing for nutrient enrichment in crops: A promising approach toward global food security—. *Frontiers in Genetics*, *13*, 932859. DOI: 10.3389/fgene.2022.932859 PMID: 35910203

Kumar, K., Gambhir, G., Dass, A., Tripathi, A. K., Singh, A., Jha, A. K., Yadava, P., Choudhary, M., & Rakshit, S. (2020). Genetically modified crops: Current status and future prospects. *Planta*, *251*(4), 91. DOI: 10.1007/s00425-020-03372-8 PMID: 32236850

Kumar, L., Chhogyel, N., Gopalakrishnan, T., Hasan, M. K., Jayasinghe, S. L., Kariyawasam, C. S., & Ratnayake, S. (2022). Climate change and future of agri-food production. In *Future foods* (pp. 49–79). Elsevier. DOI: 10.1016/B978-0-323-91001-9.00009-8

Kumar, R., Shanker, R., Singh, P., Yadav, M. K., Chaudhary, V., & Kumar, M. (2024). Genome editing towards pests and disease management in agricultural crops: Recent developments, challenges and future prospects. *Physiological and Molecular Plant Pathology*, *134*, 102402. DOI: 10.1016/j.pmpp.2024.102402

Kumlehn, J., Pietralla, J., Hensel, G., Pacher, M., & Puchta, H. (2018). The CRISPR/Cas revolution continues: From efficient gene editing for crop breeding to plant synthetic biology. *Journal of Integrative Plant Biology*, *60*(12), 1127–1153. DOI: 10.1111/jipb.12734 PMID: 30387552

Kuzma, J. (2018). Regulating gene-edited crops. *Issues in Science and Technology*, *35*(1), 80–85.

Kuzma, J., & Grieger, K. (2020). Community-led governance for gene-edited crops. *Science*, *370*(6519), 916–918. DOI: 10.1126/science.abd1512 PMID: 33214269

LaCanne, C. E., & Lundgren, J. G. (2018). Regenerative agriculture: Merging farming and natural resource conservation profitably. *PeerJ*, *6*, e4428. DOI: 10.7717/peerj.4428 PMID: 29503771

Lassalle, M. W., & Jarocki, V. (2016). Public misunderstanding of genetically modified organisms: How science and society are interconnected. *Int. J. Life. Sci. Scienti. Res*, *2*(5), 636–643. DOI: 10.21276/ijlssr.2016.2.5.22

Lassoued, R., Smyth, S. J., & Phillips, P. W. (2024). Regulatory outlook of CRISPR-edited plants in Canada. In Global regulatory outlook for CRISPRized plants (pp. 231-250). Academic Press.

Lassoued, R., Macall, D. M., Hesseln, H., Phillips, P. W. B., & Smyth, S. J. (2019). Benefits of genome-edited crops: Expert opinion. *Transgenic Research*, *28*(2), 247–256. DOI: 10.1007/s11248-019-00118-5 PMID: 30830581

Lassoued, R., Phillips, P. W. B., Macall, D. M., Hesseln, H., & Smyth, S. J. (2021). Expert opinions on the regulation of plant genome editing. *Plant Biotechnology Journal*, *19*(6), 1104–1109. DOI: 10.1111/pbi.13597 PMID: 33834596

Lauria, M., & Rossi, V. (2011). Epigenetic control of gene regulation in plants. *Biochimica et Biophysica Acta (BBA)-. Gene Regulatory Mechanisms*, *1809*(8), 369–378.

Lau, S.-E., Teo, W. F. A., Teoh, E. Y., & Tan, B. C. (2022). Microbiome engineering and plant biostimulants for sustainable crop improvement and mitigation of biotic and abiotic stresses. *Discover Food*, *2*(1), 9. DOI: 10.1007/s44187-022-00009-5

Lawrenson, T., Shorinola, O., Stacey, N., Li, C., Østergaard, L., Patron, N., & Harwood, W. (2015). Induction of targeted, heritable mutations in barley and Brassica oleracea using RNA-guided Cas9 nuclease. *Genome Biology*, *16*(1), 258. DOI: 10.1186/s13059-015-0826-7 PMID: 26616834

Lazzarotto, C. R., Katta, V., Li, Y., Manquen, G., Wood, R. K., Chyr, J., Urbina, E., Matsubara, A., Lee, G. H., Wu, X., De Ravin, S. S., & Tsai, S. Q. (2026). Sensitive and unbiased genome-wide profiling of base-editor-induced off-target activity using CHANGE-seq-BE. *Nature Biotechnology*. Advance online publication. DOI: 10.1038/s41587-025-02948-7 PMID: 41482541

Lee, K., & Wang, K. (2023). Strategies for genotype-flexible plant transformation. *Current Opinion in Biotechnology*, *79*, 102848. DOI: 10.1016/j.copbio.2022.102848 PMID: 36463838

Lee, M. (2023). Deep learning in CRISPR-Cas systems: A review of recent studies. *Frontiers in Bioengineering and Biotechnology*, *11*, 1226182. DOI: 10.3389/fbioe.2023.1226182 PMID: 37469443

Le, H., Simmons, C. H., & Zhong, X. (2025). Functions and Mechanisms of Histone Modifications in Plants. *Annual Review of Plant Biology*, *76*(1), 1–16. DOI: 10.1146/annurev-arplant-083123-070919 PMID: 39952674

Lema, M. (2021). Regulatory assessment of off-target changes and spurious DNA insertions in gene-edited organisms for agri-food use. *Journal of Regional Science*, *9*(1), 1–15. DOI: 10.21423/JRS-V09I1LEMA

Lemmon, Z. H., Reem, N. T., Dalrymple, J., Soyk, S., Swartwood, K. E., Rodriguez-Leal, D., Van Eck, J., & Lippman, Z. B. (2018). Rapid improvement of domestication traits in an orphan crop by genome editing. *Nature Plants*, *4*(10), 766–770. DOI: 10.1038/s41477-018-0259-x PMID: 30287957

Levidow, L., Carr, S., & Wield, D. (2000). Genetically modified crops in the European Union: Regulatory conflicts as precautionary opportunities. *Journal of Risk Research*, *3*(3), 189–208. DOI: 10.1080/13669870050043062

Levy, A., Conway, J. M., Dangl, J. L., & Woyke, T. (2018). Elucidating bacterial gene functions in the plant microbiome. *Cell Host & Microbe*, *24*(4), 475–485. DOI: 10.1016/j.chom.2018.09.005 PMID: 30308154

Liang, J., Yang, X., Jiao, Y., Wang, D., Zhao, Q., Sun, Y., Li, Y., & Wu, K. (2022). The evolution of China's regulation of agricultural biotechnology. *aBIOTECH*, *3*(4), 237–249. DOI: 10.1007/s42994-022-00086-1 PMID: 36533267

Liang, Z., Zhang, K., Chen, K., & Gao, C. (2014). Targeted Mutagenesis in *Zea mays* Using TALENs and the CRISPR/Cas System. *Journal of Genetics and Genomics*, *41*(2), 63–68. DOI: 10.1016/j.jgg.2013.12.001 PMID: 24576457

Liao, W., Guo, R., Qian, K., Shi, W., Whelan, J., & Shou, H. (2024). The acyl–acyl carrier protein thioesterases GmFATA1 and GmFATA2 are essential for fatty acid accumulation and growth in soybean. *The Plant Journal : For Cell and Molecular Biology*, *118*(3), 823–838. DOI: 10.1111/tpj.16638 PMID: 38224529

Li, B., Sun, C., Li, J., & Gao, C. (2024). Targeted genome-modification tools and their advanced applications in crop breeding. *Nature Reviews. Genetics*, *25*(9), 603–622. DOI: 10.1038/s41576-024-00720-2 PMID: 38658741

Li, C., Chu, W., Gill, R. A., Sang, S., Shi, Y., Hu, X., & Zhang, B. (2022). Computational tools and resources for CRISPR/Cas genome editing. *Genomics, Proteomics & Bioinformatics*, *21*(1), 1–12. PMID: 35341983

Li, G., An, L., Yang, W., Yang, L., Wei, T., Shi, J., Wang, J., Doonan, J. H., Xie, K., Fernie, A. R., Lagudah, E. S., Wing, R. A., & Gao, C. (2025). Integrated biotechnological and AI innovations for crop improvement. *Nature*, *643*(8073), 925–937. DOI: 10.1038/s41586-025-09122-8 PMID: 40702261

Li, H., Chai, Z., Shi, X., Sun, C., Zhang, R., Zhang, Q., & Gao, C. (2026). Multiplexed, precise genome engineering in monocots with twin prime editing systems. *Nature Biotechnology*. Advance online publication. DOI: 10.1038/s41587-026-03174-5 PMID: 42249131

Li, H., Zhou, R., Liu, P., Yang, M., Xin, D., Liu, C., Zhang, Z., Wu, X., Chen, Q., & Zhao, Y. (2023). Design of high-monounsaturated fatty acid soybean seed oil using GmPDCTs knockout via a CRISPR-Cas9 system. *Plant Biotechnology Journal*, *21*(7), 1317–1319. DOI: 10.1111/pbi.14060 PMID: 37084283

Li, J. F., Norville, J. E., Aach, J., McCormack, M., Zhang, D., Bush, J., & Sheen, J. (2013). Multiplex and homologous recombination–mediated genome editing in Arabidopsis and *Nicotiana benthamiana* using guide RNA and Cas9. *Nature Biotechnology*, *31*(8), 688–691. DOI: 10.1038/nbt.2654 PMID: 23929339

Limenie, A. D., & Alehegn, M. (2025). Genetic engineering for cereal crop yield improvement and disease resistant breeding. *TheScientificWorldJournal*, *2025*(1), 6743917. DOI: 10.1155/tswj/6743917 PMID: 40225354

Lindberg, S. A., ... (2023). Gene-Edited Food Adoption Intentions and Institutional Trust in the United States: Benefits, Acceptance, and Labeling☆. Rural Sociology, 88(2), 392-425.

Lindberg, S. A., Peters, D. J., & Cummings, C. L. (2023). Gene-Edited Food Adoption Intentions and Institutional Trust in the United States: Benefits, Acceptance, and Labeling☆. *Rural Sociology, 88*(2), 392-425.

Lin, Q., Zong, Y., Xue, C., Wang, S., Jin, S., Zhu, Z., Wang, Y., Anzalone, A. V., Raguram, A., Doman, J. L., Liu, D. R., & Gao, C. (2020). Prime genome editing in rice and wheat. *Nature Biotechnology*, *38*(5), 582–585. DOI: 10.1038/s41587-020-0455-x PMID: 32393904

Li, S., Chen, K., & Grierson, D. (2021). Molecular and hormonal mechanisms regulating fleshy fruit ripening. *Cells*, *10*(5), 1136. DOI: 10.3390/cells10051136 PMID: 34066675

Li, S., Zhang, C., Li, J., Yan, L., Wang, N., & Xia, L. (2021). Present and future prospects for wheat improvement through genome editing and advanced technologies. *Plant Communications*, *2*(4), 100211. DOI: 10.1016/j.xplc.2021.100211 PMID: 34327324

Listgarten, J., Weinstein, M., Kleinstiver, B. P., Sousa, A. A., Joung, J. K., Crawford, J., & Doench, J. G. (2018). Prediction of off-target activities for the end-to-end design of CRISPR guide RNAs. *Nature Biomedical Engineering*, *2*(1), 38–47. DOI: 10.1038/s41551-017-0178-6 PMID: 29998038

Li, T., Liu, B., Spalding, M. H., Weeks, D. P., & Yang, B. (2012). High-efficiency TALEN-based gene editing produces disease-resistant rice. *Nature Biotechnology*, *30*(5), 390–392. DOI: 10.1038/nbt.2199 PMID: 22565958

Liu, A. (2022). Legal Measures Addressing Environmental Risks That May Be Caused by GM Crops in China. In *Regulating Genetically Modified Crops in View of Environmental Risks: China's Implementation of International Obligations* (pp. 143-222). Springer. DOI: 10.1007/978-981-19-2480-4_5

Liu, Yao, Xian-Hui Zhou, Shu-Hong Huang, and Xiao-Long Wang. "Prime editing: a search and replace tool with versatile base changes." *Yi Chuan= Hereditas* 44, no. 11 (2022): 993-1008. https://doi.org/.DOI: 10.16288/j.yczz.22-156

Liu, B., Dong, X., Cheng, H., Zheng, C., Chen, T., Rodríguez-Leal, D., Xu, Z., & Yan, J. (2024). Improved prime editors enable efficient and versatile large-scale editing in plants. *Nature Plants*, *10*, 296–308. DOI: 10.1038/s41477-023-01552-6

Liu, B., Dong, X., Cheng, H., Zheng, C., Chen, T., Rodríguez, T. C., Fallert, M., & Zhu, L. J. (2024). Improved prime editors enable efficient and versatile editing of human and plant genomes. *Cell, 187*(4), 938–952. DOI: 10.1016/j.cell.2023.12.001

Liu, B., Jiang, H., & Li, L. (2024). The innovative applications and prospects of artificial intelligence in agricultural research: Gene editing, crop improvement, and intelligent data analysis. *Advances in Resources Research*, *4*(4), 624–653.

Liu, G., Zhang, Y., & Zhang, T. (2020). Computational approaches for effective CRISPR guide RNA design and evaluation. *Computational and Structural Biotechnology Journal, 18*, 35–44. DOI: 10.1016/j.csbj.2019.11.006 PMID: 31890142

Liu, L., & Pei, D. S. (2022). Insights gained from RNA editing targeted by the CRISPR-Cas13 family. *International Journal of Molecular Sciences, 23*(19), 11400. DOI: 10.3390/ijms231911400 PMID: 36232699

Liu, M. S., Gong, S., Yu, H. H., Jung, K., Johnson, K. A., & Taylor, D. W. (2020). Engineered CRISPR/Cas9 enzymes improve discrimination by slowing DNA cleavage, thereby allowing the release of off-target DNA. *Nature Communications, 11*(1), 3576. DOI: 10.1038/s41467-020-17411-1 PMID: 32681021

Liu, Q., Yang, F., Zhang, J., Liu, H., Rahman, S., Islam, S., Ma, W., & She, M. (2021). Application of CRISPR/Cas9 in crop quality improvement. *International Journal of Molecular Sciences, 22*(8), 4206. DOI: 10.3390/ijms22084206 PMID: 33921600

Liu, S., He, Y., Fan, T., Zhu, M., Qi, C., Ma, Y., Yang, M., Yang, L., Tang, X., Zhou, J., Zhong, Z., An, X., Qi, Y., & Zhang, Y. (2025). PAM-relaxed and temperature-tolerant CRISPR-Mb3Cas12a single transcript unit systems for efficient singular and multiplexed genome editing in rice, maize, and tomato. *Plant Biotechnology Journal, 23*(1), 156–173. DOI: 10.1111/pbi.14486 PMID: 39387219

Liu, W., Yuan, J. S., & Stewart, C. N.Jr. (2013). Advanced genetic tools for plant biotechnology. *Nature Reviews. Genetics, 14*(11), 781–793. DOI: 10.1038/nrg3583 PMID: 24105275

Liu, X., Huang, L., Li, M., Fu, Y., Zhang, W., Zhang, S., Liang, X., & Shen, Q. (2025). An Efficient and Cost-Effective Novel Strategy for Identifying CRISPR-Cas-Mediated Mutants in Plant Offspring. *The CRISPR Journal, 8*(1), 26–36. DOI: 10.1089/crispr.2024.0057 PMID: 39804665

Li, X., Lassner, M., & Zhang, Y. (2002). Deleteagene: A fast neutron deletion mutagenesis-based gene knockout system for plants. *Comparative and Functional Genomics*, *3*(2), 158–160. DOI: 10.1002/cfg.148 PMID: 18628888

Li, X., Tian, Y., Shen, R., Pang, Y., & Tang, K. (2025). An optimized thymine base editing toolkit with various editing windows enables targeted T-to-G base conversions in rice. *Plant Biotechnology Journal*, *23*(5), 1637–1639. Advance online publication. DOI: 10.1111/pbi.14611 PMID: 39962736

Li, X., Xu, S., Fuhrmann-Aoyagi, M. B., Yuan, S., Iwama, T., Kobayashi, M., & Miura, K. (2022). CRISPR/Cas9 technique for temperature, drought, and salinity stress responses. *Current Issues in Molecular Biology*, *44*(6), 2664–2682. DOI: 10.3390/cimb44060182 PMID: 35735623

Li, X., Zhang, Y., Wang, J., & Chen, Q. (2026). Plant microbiome regulation for sustainable agriculture. *Current Opinion in Biotechnology*, *98*, 103444. DOI: 10.1016/j.copbio.2026.103444 PMID: 41671649

Li, Y., Liang, J., Deng, B., Jiang, Y., Zhu, J., Chen, L., Li, J., & Li, J. (2023). Applications and prospects of CRISPR/Cas9-mediated base editing in plant breeding. *Current Issues in Molecular Biology*, *45*(2), 918–935. DOI: 10.3390/cimb45020059 PMID: 36826004

Li, Y., Liu, Z., Gmitter, F. G.Jr, Deng, Z., Cheng, B., Duan, H., & Li, Y. (2025). Substantial enhancement of *Agrobacterium*-mediated transgene-free genome editing via short-term chemical selection using citrus as a model plant. *Horticulture Research*, *12*(9), uhaf153. Advance online publication. DOI: 10.1093/hr/uhaf153 PMID: 41019308

Li, Y., Li, W., & Li, J. (2021). The CRISPR/Cas9 revolution continues: From base editing to prime editing in plant science. *Journal of Genetics and Genomics*, *48*(8), 661–670. DOI: 10.1016/j.jgg.2021.05.001 PMID: 34362681

Li, Y., Wu, X., Zhang, Y., & Zhang, Q. (2022). CRISPR/Cas genome editing improves abiotic and biotic stress tolerance of crops. *Frontiers in Genome Editing*, *4*, 1–17. DOI: 10.3389/fgeed.2022.987817 PMID: 36188128

Li, Z., Liu, Z. B., Xing, A., Moon, B. P., Koellhoffer, J. P., Huang, L., & Cigan, A. M. (2015). Cas9-Guide RNA Directed Genome Editing in Soybean. *Plant Physiology*, *169*(2), 960–970. DOI: 10.1104/pp.15.00783 PMID: 26294043

Li, Z., Xiong, X., & Li, J.-F. (2020). The working dead: Repurposing inactive CRISPR-associated nucleases as programmable transcriptional regulators in plants. *aBIOTECH*, *1*(1), 32–40. DOI: 10.1007/s42994-019-00003-z PMID: 36305006

Lokya, V., Singh, S., Chaudhary, R., Jangra, A., & Tiwari, S. (2025). Emerging trends in transgene-free crop development: Insights into genome editing and its regulatory overview. *Plant Molecular Biology*, *115*(4), 84. DOI: 10.1007/s11103-025-01600-x PMID: 40632351

Long, Y., Wang, C., Liu, C., Li, H., Pu, A., Dong, Z., Wei, X., & Wan, X. (2023). Molecular mechanisms controlling grain size and weight and their biotechnological breeding applications in maize and other cereal crops. *Journal of Advanced Research*, *62*, 27–46. DOI: 10.1016/j.jare.2023.09.016 PMID: 37739122

Lorenzo, C. D., Debray, K., Herwegh, D., Develtere, W., Impens, L., Schaumont, D., & De Boe, Y. (2023). BREEDIT: A multiplex genome editing strategy to improve complex quantitative traits in maize. *The Plant Cell*, *35*(1), 218–238. DOI: 10.1093/plcell/koac243 PMID: 36066192

Lucht, J. M. (2015). Public acceptance of plant biotechnology and GM crops. *Viruses*, *7*(8), 4254–4281. DOI: 10.3390/v7082819 PMID: 26264020

Ludlow, K., Smyth, S., & Falck-Zepeda, J. (2021). Socio-economic considerations and potential implications for gene-edited crops. *Journal of Regional Science*, *9*(2), 1–11. DOI: 10.21423/JRS-V09I2LUDLOW

Lugtenberg, B., & Kamilova, F. (2009). Plant-growth-promoting rhizobacteria. *Annual Review of Microbiology*, *63*(1), 541–556. DOI: 10.1146/annurev.micro.62.081307.162918 PMID: 19575558

Lukasiewicz, J. M., van de Wiel, C. C. M., Lotz, L. A. P., & Smulders, M. J. M. (2024a). Consumer transparency in the production chain for plant varieties produced using new genomic techniques. *aBIOTECH*, *5*(2), 239–246. DOI: 10.1007/s42994-024-00142-y PMID: 38974855

Lukasiewicz, J. M., van de Wiel, C. C. M., Lotz, L. A. P., & Smulders, M. J. M. (2024b). Clemens CM van de Wiel, Lambertus AP Lotz, and Marinus JM Smulders. "Intellectual property rights and plants made by new genomic techniques: Access to technology and gene-edited traits in plant breeding.". *Outlook on Agriculture*, *53*(3), 205–215. DOI: 10.1177/00307270241277219

Lv, Z., Jiang, R., Chen, J., & Chen, W. (2020). Nanoparticle-mediated gene transformation strategies for plant genetic engineering. *The Plant Journal : For Cell and Molecular Biology*, *104*(4), 880–891. DOI: 10.1111/tpj.14973 PMID: 32860436

Lv, Z., Jiang, S., Kong, S., Zhang, X., Yue, J., Zhao, W., Li, L., & Lin, S. (2024). Advances in Single-Cell Transcriptome Sequencing and Spatial Transcriptome Sequencing in Plants. *Plants*, *13*(12), 1679. DOI: 10.3390/plants13121679 PMID: 38931111

Lynch, D., & Vogel, D. (2001). *The regulation of GMOs in Europe and the United States: A case-study of contemporary European regulatory politics*. Council on Foreign Relations.

Lyzenga, W. J., Pozniak, C. J., & Kagale, S. (2021). Advanced Domestication: Harnessing the Precision of Gene Editing in Crop Breeding. *Plant Biotechnology Journal*, *19*(4), 660–670. DOI: 10.1111/pbi.13576 PMID: 33657682

Maaß, O., Consmüller, N., & Kehlenbeck, H. (2019). Socioeconomic impact of genome editing on agricultural value chains: The case of Fungal-Resistant and Coeliac-Safe wheat. *Sustainability (Basel)*, *11*(22), 6421. DOI: 10.3390/su11226421

Macnaghten, P., & Habets, M. G. (2020). Breaking the impasse: Towards a forward-looking governance framework for gene editing with plants. *Plants, People, Planet*, *2*(4), 353–365. DOI: 10.1002/ppp3.10107

Maehira, Y., & Spencer, R. C. (2019). Harmonization of biosafety and biosecurity standards for high-containment facilities in low-and middle-income countries: An approach from the perspective of occupational safety and health. *Frontiers in Public Health*, *7*, 249. DOI: 10.3389/fpubh.2019.00249 PMID: 31572701

Mahfouz, M. M., Piatek, A., & Stewart, C. N.Jr. (2014). Genome engineering via TALENs and CRISPR/Cas9 systems: Challenges and perspectives. *Plant Biotechnology Journal*, *12*(8), 1006–1014. DOI: 10.1111/pbi.12256 PMID: 25250853

Mahmoudieh, M., Jariani, P., Dehghan, A. K., Karimi, M. R., Vadipour, F., Jahani, M., & Naghavi, M. R. (2025). The future of rubber production: A review of genetic regulation of biosynthetic pathways and genome editing technologies in natural rubber-producing plants. *Tree Genetics & Genomes*, *21*(2), 6. DOI: 10.1007/s11295-025-01690-0

Majumder, S., Datta, K., & Datta, S. K. (2023). Perspective of Evolution of the C 4 Plants to Develop Climate Designer C 4 Rice as a Strategy for Abiotic Stress Management. *Global Climate Change and Plant Stress Management*, 113-124.

Malhotra, N. (2025). Harnessing genomics for sustainable food systems with orphan crops. *Discov Agric*, *3*(1), 159. DOI: 10.1007/s44279-025-00354-w

Mali, P., Yang, L., Esvelt, K. M., Aach, J., Guell, M., DiCarlo, J. E., Norville, J. E., & Church, G. M. (2013). RNA-guided human genome engineering via Cas9. *Science*, *339*(6121), 823–826. DOI: 10.1126/science.1232033 PMID: 23287722

Manghwar, H., Li, B., Ding, X., Hussain, A., Lindsey, K., Zhang, X., & Jin, S. (2020). CRISPR/Cas systems in genome editing: Methodologies and tools for sgRNA design, off-target evaluation, and strategies to mitigate off-target effects. *Advanced Science (Weinheim, Baden-Württemberg, Germany)*, *7*(6), 1902312. DOI: 10.1002/advs.201902312 PMID: 32195078

Mani, V., Park, S., Kim, J. A., Lee, S. I., & Lee, K. (2021). Metabolic perturbation and synthetic biology strategies for plant terpenoid production—An updated overview. *Plants*, *10*(10), 2179. DOI: 10.3390/plants10102179 PMID: 34685985

Mao, Y., Botella, J. R., Liu, Y., & Zhu, J. K. (2019). Gene editing in plants: Progress and challenges. *National Science Review*, *6*(3), 421–437. DOI: 10.1093/nsr/nwz005 PMID: 34691892

Mao, Y., Zhang, H., Xu, N., Zhang, B., Gou, F., & Zhu, J. K. (2013). Application of the CRISPR–Cas System for Efficient Genome Engineering in Plants. *Molecular Plant*, *6*(6), 2008–2011. DOI: 10.1093/mp/sst121 PMID: 23963532

Marden, E. (2002). Risk and regulation: US regulatory policy on genetically modified food and agriculture. *BCL Rev.*, *44*, 733.

Maren, N. A., Duan, H., Da, K., Yencho, G. C., Ranney, T. G., & Liu, W. (2022). Genotype-independent plant transformation. *Horticulture Research*, *9*, uhac047. DOI: 10.1093/hr/uhac047 PMID: 35531314

Marone, D., Mastrangelo, A. M., & Borrelli, G. M. (2023). From transgenesis to genome editing in crop improvement: Applications, marketing, and legal issues. *International Journal of Molecular Sciences*, *24*(8), 7122. DOI: 10.3390/ijms24087122 PMID: 37108285

Marx, A. (2019). *Public-private partnerships for sustainable development: Exploring their design and its impact on effectiveness* (Vol. 11). Multidisciplinary Digital Publishing Institute.

Mascarenhas, M. S., Nascimento, F. S., Rocha, A. J., Ferreira, M. S., Oliveira, W. D. S., Morais Lino, L. S., Mendes, T. A. O., Ferreira, C. F., Santos-Serejo, J. A., & Amorim, E. P. (2024). Use of CRISPR Technology in Gene Editing for Tolerance to Biotic Factors in Plants: A Systematic Review. *Current Issues in Molecular Biology*, *46*(10), 11086–11123. DOI: 10.3390/cimb46100659 PMID: 39451539

Mathiazhagan, M., Chidambara, B., Hunashikatti, L. R., & Ravishankar, K. V. (2021). Genomic approaches for improvement of tropical fruits: Fruit quality, shelf life and nutrient content. *Genes*, *12*(12), 1881. DOI: 10.3390/genes12121881 PMID: 34946829

Matsuo, M., & Tachikawa, M. (2022). Implications and lessons from the introduction of genome-edited food products in Japan. *Frontiers in Genome Editing*, *4*, 899154. DOI: 10.3389/fgeed.2022.899154 PMID: 35813972

Mbaya, H., Lillico, S., Kemp, S., Simm, G., & Raybould, A. (2022). Regulatory frameworks can facilitate or hinder the potential for genome editing to contribute to sustainable agricultural development. *Frontiers in Bioengineering and Biotechnology*, *10*, 959236. DOI: 10.3389/fbioe.2022.959236 PMID: 36246373

McCarty, N. S., Graham, A. E., Studená, L., & Ledesma-Amaro, R. (2020). Multiplexed CRISPR technologies for gene editing and transcriptional regulation. *Nature Communications*, *11*(1), 1281. DOI: 10.1038/s41467-020-15053-x PMID: 32152313

McFadden, P., McColgan, M., MacLochlainn, J., Davies, H., Kirby, K., Currie, D., Schroder, H., Nicholl, P., McGrory, S., Naylor, R., Kirby, K., Mullineux, J., & Mallett, J. (2024). *Safer and Effective Staffing Research and Policy Development Older People's and Children's Social Work in Northern Ireland: Report 1- The Starting Point: Baseline Analysis*. Advance online publication.

McFadden, B. R., & Lusk, J. L. (2016). What consumers don't know about genetically modified food, and how that affects beliefs. *FASEB Journal : Official Publication of the Federation of American Societies for Experimental Biology*, *30*(9), 3091–3096. DOI: 10.1096/fj.201600598 PMID: 27199295

McFadden, B. R., Rumble, J. N., Stofer, K. A., & Folta, K. M. (2024). US public opinion about the safety of gene editing in the agriculture and medical fields and the amount of evidence needed to improve opinions. *Frontiers in Bioengineering and Biotechnology*, *12*, 1340398. DOI: 10.3389/fbioe.2024.1340398 PMID: 38433825

Medani, K. R., Neill, A., Garrod, G., Ojo, M., & Hubbard, C. (2024). Societal perceptions and attitudes towards genetically modified (GM) crops, feed, and food products in the Middle East, North Africa, and Turkey (MENAT) region: A systematic literature review. *Food Quality and Preference*, *117*, 105148. DOI: 10.1016/j.foodqual.2024.105148

Mendes, R., Garbeva, P., & Raaijmakers, J. M. (2013). The rhizosphere microbiome: Significance of plant beneficial, plant pathogenic, and human pathogenic microorganisms. *FEMS Microbiology Reviews*, *37*(5), 634–663. DOI: 10.1111/1574-6976.12028 PMID: 23790204

Mengchao, Q., Hui, T., Xuetao, S., Chongyang, Z., Feng, H., Min, W., & Shutong, W. (2023). Transcriptome Analysis of oserf922 Mutants Reveals New Insights into Rice Blast Resistance. *Rice Science*, *30*(5), 374–378. DOI: 10.1016/j.rsci.2023.05.002

Menz, J., Modrzejewski, D., Hartung, F., Wilhelm, R., & Sprink, T. (2020). Genome edited crops touch the market: A view on the global development and regulatory environment. *Frontiers in Plant Science*, *11*, 586027. DOI: 10.3389/fpls.2020.586027 PMID: 33163013

Miao, J., Guo, D., Zhang, J., Huang, Q., Qin, G., Zhang, X., & Qu, L. J. (2013). Targeted mutagenesis in rice using the CRISPR-Cas system. *Cell Research*, *23*(10), 1233–1236. DOI: 10.1038/cr.2013.123 PMID: 23999856

Michno, J. M., Wang, X., Liu, J., Curtin, S. J., Kono, T. J., & Stupar, R. M. (2015). CRISPR/Cas mutagenesis of soybean and *Medicago truncatula* using a new web tool and a modified Cas9 enzyme. *GM Crops & Food*, *6*(4), 243–252. DOI: 10.1080/21645698.2015.1106063 PMID: 26479970

Mijena, D. F., Alamerew, S., Assefa, K., & Nigusse, M. (2024). Indigenous and Underutilized Crops Significance in Food and Nutritional Security in the face of Climate Change: The Case of Anchote. *Ethiopian Journal of Crop Science*, *12*(2), 15–38.

Mikhaylova, E., Kuluev, B., Gerashchenkov, G., Chemeris, D., Garafutdinov, R., Kuluev, A., Baymiev, A. K., Baymiev, A. K., & Chemeris, A. (2024). Prime-editing methods and pegRNA design programs. *Molecular Biology*, *58*(1), 17–32. DOI: 10.1134/S0026893324010084 PMID: 39852132

Milne, R., Aidid, U., Atutornu, J., Bircan, T., Boraschi, D., Costa, A., Henriques, S., Patch, C., & Middleton, A. (2023). What Difference Can Public Engagement in Genome Editing Make, and for Whom? *The American Journal of Bioethics : AJOB*, *23*(7), 58–60. DOI: 10.1080/15265161.2023.2207545 PMID: 37339292

Minkenberg, B., Wheatley, M., & Yang, Y. (2017). CRISPR/Cas9-enabled multiplex genome editing and its application. *Progress in Molecular Biology and Translational Science*, *149*, 111–132. DOI: 10.1016/bs.pmbts.2017.05.003 PMID: 28712493

Mishra, M., Afzal, S., & Singh, N. K. (2024). CRISPR/Cas9-Mediated Genome Engineering for Production of Biofortified Crops. In *Phytoremediation and Biofortification* (pp. 303-333). Apple Academic Press.

Mishra, R., & Zhao, K. (2018). Genome editing technologies and their applications in crop improvement. *Plant Biotechnology Reports*, *12*(2), 57–68. DOI: 10.1007/s11816-018-0472-0

Mitter, B., Brader, G., Pfaffenbichler, N., & Sessitsch, A. (2019). Next generation microbiome applications for crop production—Limitations and the need of knowledge-based solutions. *Current Opinion in Microbiology*, *49*, 59–65. DOI: 10.1016/j.mib.2019.10.006 PMID: 31731227

Mittler, R., Vanderauwera, S., Gollery, M., & Van Breusegem, F. (2004). Reactive oxygen gene network of plants. *Trends in Plant Science*, *9*(10), 490–498. DOI: 10.1016/j.tplants.2004.08.009 PMID: 15465684

Modrzejewski, D., Hartung, F., Lehnert, H., Sprink, T., Kohl, C., Keilwagen, J., & Wilhelm, R. (2020). Which factors affect the occurrence of off-target effects caused by the use of CRISPR/Cas: A systematic review in plants. *Frontiers in Plant Science*, *11*, 574959. DOI: 10.3389/fpls.2020.574959 PMID: 33329634

Mohan, C., Easterling, M., & Yau, Y.-Y. (2022). Gene editing technologies for sugarcane improvement: Opportunities and limitations. *Sugar Tech*, *24*(1), 369–385. DOI: 10.1007/s12355-021-01045-8 PMID: 34667393

Mohanta, T. K., Bashir, T., & Hashem, A., Abd_Allah, E. F., and Bae, H. (. (2017). Genome Editing Tools in Plants. *Genes*, *8*(12), 399. DOI: 10.3390/genes8120399 PMID: 29257124

Moiani, A., Letort, G., Lizot, S., Chalumeau, A., Foray, C., Felix, T., Le Clerre, D., Temburni-Blake, S., Hong, P., Leduc, S., Pinard, N., Marechal, A., Seclen, E., Boyne, A., Mayer, L., Hong, R., Pulicani, S., Galetto, R., Gouble, A., & Valton, J. (2024). Non-viral DNA delivery and TALEN editing correct the sickle cell mutation in hematopoietic stem cells. *Nature Communications*, *15*(1), 4965. DOI: 10.1038/s41467-024-49353-3 PMID: 38862518

Molinari, M. D. C., et al. (2024). Navigating the Path from Lab to Market: Regulatory Challenges and Opportunities for Genome Editing Technologies for Agriculture. Plant Genome Editing Technologies: Speed Breeding, Crop Improvement and Sustainable Agriculture, 25-63.

Molinari, M. D. C., Pagliarini, R. F., Florentino, L. H., Lima, R. N., Arraes, F. B. M., Abbad, S. V.,…Nepomuceno, A. L. (2024). Navigating the Path from Lab to Market: Regulatory Challenges and Opportunities for Genome Editing Technologies for Agriculture. *Plant Genome Editing Technologies: Speed Breeding, Crop Improvement and Sustainable Agriculture*, 25-63.

Molla, K. A., Shih, J., Wheatley, M. S., & Yang, Y. (2022). Predictable NHEJ insertion and assessment of HDR editing strategies in plants. *Frontiers in Genome Editing*, *4*, 825236. DOI: 10.3389/fgeed.2022.825236 PMID: 35481279

Molla, K. A., Sretenovic, S., Bansal, K. C., & Qi, Y. (2021). Precise plant genome editing using base editors and prime editors. *Nature Plants*, *7*(9), 1166–1187. DOI: 10.1038/s41477-021-00991-1 PMID: 34518669

Montenegro de Wit, M. (2020). Democratizing CRISPR? Stories, practices, and politics of science and governance on the agricultural gene editing frontier. *Elementa*, *8*, 9. DOI: 10.1525/elementa.405

Moradpour, M., & Abdulah, S. N. (2019). CRISPR / dC as9 platforms in plants: Strategies and applications beyond genome editing. *Plant Biotechnology Journal*, *18*(1), 1–12. PMID: 31392820

Moscou, M. J., & Bogdanove, A. J. (2009). A simple cipher governs DNA recognition by TAL effectors. *Science*, *326*(5959), 1501–1501. DOI: 10.1126/science.1178817 PMID: 19933106

Mukherjee, M., Posch, K., Molina, S. J., Taymor, K., & Keller, A. (2023). Comparing regulatory processes in genome editing and autonomous vehicles: How institutional environments shape sociotechnical imaginaries. *The Review of Policy Research*, *40*(3), 433–457. DOI: 10.1111/ropr.12532

Mukherji, A., Arndt, C., Arango, J., Flintan, F. E., Derera, J., Francesconi, W.,... Mockshell, J. Y. (2023). Achieving agricultural breakthrough: A deep dive into seven technological areas.

Munawar, N., Faheem, M., Niamat, A., Munir, A., Khan, S. H., Zahoor, M. K., & Ahmad, A. (2024). Regulatory, ethical, social, and biosafety concerns in genome-edited horticultural crops. In *CRISPRized Horticulture Crops* (pp. 421–438). Elsevier. DOI: 10.1016/B978-0-443-13229-2.00026-0

Munaweera, T. I. K., Jayawardana, N. U., Rajaratnam, R., & Dissanayake, N. (2022). Modern plant biotechnology as a strategy in addressing climate change and attaining food security. *Agriculture & Food Security*, *11*(1), 1–28. DOI: 10.1186/s40066-022-00369-2

Munir, F., Azeem, S., Amir, R., Gillani, M., Gul, A., Mustafa, A., & Fatima, N. (2023). Genetic Transformation Methods in Cereal Crops. In *Cereal Crops* (pp. 273–294). CRC Press.

Murray, J., Samson, C., du Rand, A., & Sheppard, H. (2023). Unintended CRISPR-Cas9 editing outcomes: A review of the detection and prevalence of structural variants generated by gene editing in human cells. *Human Genetics*, *142*(6), 705–720. DOI: 10.1007/s00439-023-02561-1 PMID: 37093294

Mushtaq, M., Ahmad Dar, A., Skalicky, M., Tyagi, A., Bhagat, N., Basu, U., & Dar, T.-U.-H. (2021). CRISPR-based genome editing tools: Insights into technological breakthroughs and future challenges. *Genes*, *12*(6), 797. DOI: 10.3390/genes12060797 PMID: 34073848

Myhr, A. I. (2010). A precautionary approach to genetically modified organisms: Challenges and implications for policy and science. *Journal of Agricultural & Environmental Ethics*, *23*(6), 501–525. DOI: 10.1007/s10806-010-9234-x

Myskja, B. K., & Myhr, A. I. (2020). Non-safety assessments of genome-edited organisms: Should they be included in regulation? *Science and Engineering Ethics*, *26*(5), 2601–2627. DOI: 10.1007/s11948-020-00222-4 PMID: 32424723

Nadeem, M. A., Nawaz, M. A., Shahid, M. Q., Doğan, Y., Comertpay, G., Yıldız, M., & Baloch, F. S. (2018). DNA Molecular Markers in Plant Breeding: Current Status and Recent Advancements in Genomic Selection and Genome Editing. *Biotechnology, Biotechnological Equipment*, *32*(2), 261–285. DOI: 10.1080/13102818.2017.1400401

Naeem, M., Majeed, S., Hoque, M. Z., & Ahmad, I. (2020). Latest developed strategies to minimize the off-target effects in CRISPR-Cas-mediated genome editing. *Cells*, *9*(7), 1608. DOI: 10.3390/cells9071608 PMID: 32630835

Nah, S.-L., & Chau, C.-F. (2010). Issues and challenges in defeating world hunger. *Trends in Food Science & Technology*, *21*(11), 544–557. DOI: 10.1016/j.tifs.2010.07.013

Naim, F., Shand, K., Hayashi, S., O'Brien, M., McGree, J., Johnson, A. A., & Waterhouse, P. M. (2020). Are the current gRNA ranking prediction algorithms useful for genome editing in plants? *PLoS One*, *15*(1), e0227994. DOI: 10.1371/journal.pone.0227994 PMID: 31978124

Nalla, Y., & Shah, S. (2021). A new era in molecular biology clustered regularly interspaced short palindromic repeats/cas9 technology: A brief understanding. *Advances in Human Biology*, *11*(2), 152–156. DOI: 10.4103/aihb.aihb_162_20

Nangammbi, N. (2021). *Functional Testing of Xylan-Associated Genes in Arabidopsis and Application of CRISPR/Cas9 Gene Editing in Hybrid Poplar Trees*. University of Pretoria.

Nascimento, F. D. S., De Jesus Rocha, A., Da Silva Soares, J. M., Mascarenhas, M. S., Ferreira, M. D. S., Lino, L. S. M., De Souza Ramos, A. P., Diniz, L. E. C., De Oliveira Mendes, T. A., Ferreira, C. F., Santos-Serejo, J. A. D., & Amorim, E. P. (2023). Gene Editing for Plant Resistance to Abiotic Factors: A Systematic review. *Plants*, *12*(2), 305. DOI: 10.3390/plants12020305 PMID: 36679018

Nawaz, S. (2021). *Beyond naturalness?: social dimensions of gene editing in agriculture*. University of British Columbia.

Nawaz, S., & Satterfield, T. (2022). Climate solution or corporate co-optation? US and Canadian publics' views on agricultural gene editing. *PLoS One*, *17*(3), e0265635. DOI: 10.1371/journal.pone.0265635 PMID: 35313327

Nawaz, S., Satterfield, T., & Phurisamban, R. (2023). Does "Precision" matter? AQ study of public interpretations of gene editing in agriculture. *Science, Technology & Human Values*, *48*(6), 1421–1461. DOI: 10.1177/01622439221112460

Nayak, S. N., Aravind, B., Malavalli, S. S., Sukanth, B. S., Poornima, R., Bharati, P., & Puppala, N. (2021). Omics technologies to enhance plant-based functional foods: An overview. *Frontiers in Genetics*, *12*, 742095. DOI: 10.3389/fgene.2021.742095 PMID: 34858472

Nayerossadat, N., Maedeh, T., & Ali, P. A. (2012). Viral and nonviral delivery systems for gene delivery. *Advanced Biomedical Research*, *1*(1), 27. DOI: 10.4103/2277-9175.98152 PMID: 23210086

Ndour, A., Vadez, V., Pradal, C., & Lucas, M. (2017). Virtual plants need water too: Functional-structural root system models in the context of drought tolerance breeding. *Frontiers in Plant Science*, *8*, 1577. DOI: 10.3389/fpls.2017.01577 PMID: 29018456

Ndudzo, A., Makuvise, A. S., Moyo, S., & Bobo, E. D. (2024). CRISPR-Cas9 genome editing in crop breeding for climate change resilience: Implications for smallholder farmers in Africa. *Journal of Agriculture and Food Research*, *16*, 101132. DOI: 10.1016/j.jafr.2024.101132

Negowetti, N. E. (2020). Taking (animal-based) meat and ethics off the table: Food labeling and the role of consumers as agents of food systems change. Or. *Law Review*, *99*, 91.

Nekrasov, V., Staskawicz, B., Weigel, D., Jones, J. D. G., & Kamoun, S. (2013). Targeted mutagenesis in the model plant *Nicotiana benthamiana* using Cas9 RNA-guided endonuclease. *Nature Biotechnology*, *31*(8), 691–693. DOI: 10.1038/nbt.2655 PMID: 23929340

Nerkar, G., Devarumath, S., Purankar, M., Kumar, A., Valarmathi, R., Devarumath, R., & Appunu, C. (2022). Advances in Crop Breeding through Precision Genome Editing. *Frontiers in Genetics*, *13*, 880195. DOI: 10.3389/fgene.2022.880195 PMID: 35910205

NETO, R. R. D. S. (2022). Characterization of CsLIEXP1 as a new susceptibility gene to Xanthomonas citri subsp. citri and gene editing by CRISPR/Cas9 for citrus canker tolerance.

Neugebauer, M. E., Hsu, A., Arbab, M., Krasnow, N. A., McElroy, A. N., Pandey, S., & Banskota, S. (2023). Evolution of an adenine base editor into a small, efficient cytosine base editor with low off-target activity. *Nature Biotechnology*, *41*(5), 673–685. DOI: 10.1038/s41587-022-01533-6 PMID: 36357719

Niazi, S. K. (2023). Gene Editing: The Regulatory Perspective. *Encyclopedia*, *3*(4), 1345–1357. DOI: 10.3390/encyclopedia3040096

Nidhi, S., Anand, U., Oleksak, P., Tripathi, P., Lal, J. A., Thomas, G., Kuca, K., & Tripathi, V. (2021). Novel CRISPR–Cas systems: An updated review of the current achievements, applications, and future research perspectives. *International Journal of Molecular Sciences*, *22*(7), 3327. DOI: 10.3390/ijms22073327 PMID: 33805113

Nishimasu, H., Ran, F., Hsu, P. D., Konermann, S., Shehata, S. I., Dohmae, N., Ishitani, R., Zhang, F., & Nureki, O. (2014). Crystal Structure of Cas9 in Complex with Guide RNA and Target DNA. *Cell*, *156*(5), 935–949. DOI: 10.1016/j.cell.2014.02.001 PMID: 24529477

Niu, B., Paulson, J. N., Zheng, X., & Kolter, R. (2017). Simplified and representative bacterial community of maize roots. *Proceedings of the National Academy of Sciences of the United States of America*, *114*(12), E2450–E2459. DOI: 10.1073/pnas.1616148114 PMID: 28275097

Norman, P. E., Dzidzienyo, D. K., Karim, K. Y., & Beah, A. A. (2021). *Genetic Modification and Application in Cassava*. Sweetpotato and Yams.

Numan, M., Khan, A. L., Asaf, S., Salehin, M., Beyene, G., Tadele, Z., & Ligaba-Osena, A. (2021). From Traditional Breeding to Genome Editing for Boosting Productivity of the Ancient Grain Tef (*Eragrostis tef*). *Plants*, *10*(4), 628. DOI: 10.3390/plants10040628 PMID: 33806233

O'Brien, F. J. M., Almaraz, M., Foster, M. A., Hill, A. F., Huber, D. P., King, E. K., Lang, J. M., Lowe, A., Mickan, B. S., Miller, S. H., & Singh, B. K. (2021). Soil salinity and the plant microbiome. *Nature Reviews. Microbiology*, *19*(11), 665–681. DOI: 10.1038/s41579-021-00563-5

O'Geen, H., Ren, C., Nicolet, C. M., Perez, A. A., Halmai, J., Le, V. M., Mackay, J. P., Farnham, P. J., & Segal, D. J. (2017). dCas9-based epigenome editing suggests acquisition of histone methylation is not sufficient for target gene repression. *Nucleic Acids Research*, *45*(17), 9901–9916. DOI: 10.1093/nar/gkx578 PMID: 28973434

Ochiai, H., & Yamamoto, T. (2015). Genome editing using zinc-finger nucleases (ZFNs) and transcription activator-like effector nucleases (TALENs). *Targeted genome editing using site-specific nucleases: ZFNs, TALENs, and the CRISPR/Cas9 system*, 3-24.

Ochoa-Sanchez, A., Perez-Sanchez, G., Torres-Ledesma, A. M., Valdez, J. P. R., Rinaldi, G., Moguel, B. B., & Molina-Aguilar, C. (2021). Prime Editing, a novel genome-editing tool that may surpass conventional CRISPR-Cas9. Re: GEN Open, 1(1), regen-2021.

Ochoa-Sanchez, A., Perez-Sanchez, G., Torres-Ledesma, A. M., Valdez, J. P. R., Rinaldi, G., Moguel, B. B., & Molina-Aguilar, C. (2021). Prime Editing, a novel genome-editing tool that may surpass conventional CRISPR-Cas9. *Re:GEN Open*, *1*(1), 75–82. DOI: 10.1089/regen.2021.0016

Ofori, K. F., Antoniello, S., English, M. M., & Aryee, A. N. A. (2022). Improving nutrition through biofortification–A systematic review. *Frontiers in Nutrition*, *9*, 1043655. DOI: 10.3389/fnut.2022.1043655 PMID: 36570169

Olanrewaju, O. S., Glick, B. R., & Babalola, O. O. (2017). Mechanisms of action of plant growth-promoting bacteria. *World Journal of Microbiology & Biotechnology*, *33*(11), 197. DOI: 10.1007/s11274-017-2364-9 PMID: 28986676

Oliva, R., Ji, C., Atienza-Grande, G., Huguet-Tapia, J. C., Perez-Quintero, A., Li, T., Eom, J.-S., Li, C., Nguyen, H., Liu, B., Auguy, F., Sciallano, C., Luu, V. T., Dossa, G. S., Cunnac, S., Schmidt, S. M., Slamet-Loedin, I. H., Vera Cruz, C., Szurek, B., & Yang, B. (2019). Broad-spectrum resistance to bacterial blight in rice using genome editing. *Nature Biotechnology*, *37*(11), 1344–1350. DOI: 10.1038/s41587-019-0267-z PMID: 31659337

Orozco-Mosqueda, M. D. C., Santoyo, G., Glick, B. R., & de los Santos-Villalobos, S. (2022). Microbiome engineering to improve biocontrol and plant growth-promoting mechanisms. *Microbiological Research*, *256*, 126938. DOI: 10.1016/j.micres.2021.126938 PMID: 29551209

Ortega, D. L., Lin, W., & Ward, P. S. (2022). Consumer acceptance of gene-edited food products in China. *Food Quality and Preference*, *95*, 104374. DOI: 10.1016/j.foodqual.2021.104374

Ozyigit, I. I., & Yucebilgili Kurtoglu, K. (2020). Particle bombardment technology and its applications in plants. *Molecular Biology Reports*, *47*(12), 9831–9847. DOI: 10.1007/s11033-020-06001-5 PMID: 33222118

Pan, C., & Qi, Y. (2023). CRISPR-Combo–mediated orthogonal genome editing and transcriptional activation for plant breeding. *Nature Protocols*, *18*(6), 1760–1794. DOI: 10.1038/s41596-023-00823-w PMID: 37085666

Pan, C., Sretenovic, S., & Qi, Y. (2021). CRISPR/dCas-mediated transcriptional and epigenetic regulation in plants. *Current Opinion in Plant Biology*, *60*, 101980. DOI: 10.1016/j.pbi.2020.101980 PMID: 33401227

Pan, C., Ye, L., Qin, L., Liu, X., He, Y., Wang, J., & Lu, G. (2016). CRISPR/Cas9-mediated efficient and heritable targeted mutagenesis in tomato plants in the first and later generations. *Scientific Reports*, *6*(1), 24765. DOI: 10.1038/srep24765 PMID: 27097775

Pandey, S., Divakar, S., & Singh, A. (2024). S. Divakar, and Ashutosh Singh. "Genome editing prospects for heat stress tolerance in cereal crops.". *Plant Physiology and Biochemistry : PPB*, *108989*, 108989. Advance online publication. DOI: 10.1016/j.plaphy.2024.108989 PMID: 39094478

Pandian, K., Mustaffa, M. R. A. F., Mahalingam, G., Paramasivam, A., John Prince, A., Gajendiren, M., & Varanasi, S. T. (2024). Synergistic conservation approaches for nurturing soil, food security, and human health towards sustainable development goals. *Journal of Hazardous Materials Advances*, *16*, 100479. DOI: 10.1016/j.hazadv.2024.100479

Paniza, H. M. (2024). Challenges in plant breeding under climate change: a review. *Plant Quarantine Challenges under Climate Change Anxiety*, 533-556.

Pan, L., Yu, Q., Wang, J., Han, H., Mao, L., Nyporko, A., Maguza, A., Fan, L., Bai, L., & Powles, S. (2021). An ABCC-type transporter endowing glyphosate resistance in plants. *Proceedings of the National Academy of Sciences of the United States of America*, *118*(16), e2100136118. DOI: 10.1073/pnas.2100136118 PMID: 33846264

Panno, S., Davino, S., Caruso, A. G., Bertacca, S., Crnogorac, A., Mandić, A., Noris, E., & Matić, S. (2021). A review of the most common and economically important diseases that undermine the cultivation of tomato crop in the mediterranean basin. *Agronomy (Basel)*, *11*(11), 2188. DOI: 10.3390/agronomy11112188

Pascual, M., Rives, B., Schunter, C., & Macpherson, E. (2017). Impact of life history traits on gene flow: A multispecies systematic review across oceanographic barriers in the Mediterranean Sea. *PLoS One*, *12*(5), e0176419. DOI: 10.1371/journal.pone.0176419 PMID: 28489878

Pathirana, R., & Carimi, F. (2022). Management and utilization of plant genetic resources for a sustainable agriculture. *Plants, 11*(15), 2038. DOI: 10.3390/plants11152038 PMID: 35956515

Paudel, B., Kolady, D. E., Just, D., & Van der Sluis, E. (2023). Determinants of consumer acceptance of gene-edited foods and its implications for innovators and policymakers. *Agribusiness, 39*(3), 623–645. DOI: 10.1002/agr.21799

Paul, P., & Mukhopadhyay, K. (2010). Growth via intellectual property rights versus gendered inequity in emerging economies: An ethical dilemma for international business. *Journal of Business Ethics, 91*(3), 359–378. DOI: 10.1007/s10551-009-0088-y

Pavlidis, G., & Tsihrintzis, V. A. (2017). Environmental Benefits and Control of Pollution to Surface Water and Groundwater by Agroforestry Systems: A Review. *Water Resources Management, 32*(1), 1–29. DOI: 10.1007/s11269-017-1805-4

Pedrazzoli, E., Bianchi, A., Umbach, A., Amistadi, S., Brusson, M., Frati, G., & Casini, A. (2023). An optimized SpCas9 high-fidelity variant for direct protein delivery. *Molecular Therapy : The Journal of the American Society of Gene Therapy, 31*(7), 2257–2265. DOI: 10.1016/j.ymthe.2023.03.007 PMID: 36905119

Pedrozo, R., Osakina, A., Huang, Y., Nicolli, C. P., Wang, L., & Jia, Y. (2025). Status on genetic resistance to rice blast disease in the post-genomic era. *Plants, 14*(5), 807. DOI: 10.3390/plants14050807 PMID: 40094775

Peters, T., et al. (2008). Religious traditions and genetic enhancement. Altering Nature: Volume Two: Religion, Biotechnology, and Public Policy, 109-159.

Petrova, I. O., & Smirnikhina, S. A. (2023). The development, optimization and future of prime editing. *International Journal of Molecular Sciences, 24*(23), 17045. DOI: 10.3390/ijms242317045 PMID: 38069367

Phour, M., Sindhu, S. S., Glick, B. R., & Verma, S. K. (2020). Beneficial microbes and sustainable agriculture: Recent developments and future challenges. *Current Genomics, 21*(7), 507–519. DOI: 10.2174/1389202921999200908220956

Piergentili, R., Del Rio, A., Signore, F., Umani Ronchi, F., Marinelli, E., & Zaami, S. (2021). CRISPR-Cas and its wide-ranging applications: From human genome editing to environmental implications, technical limitations, hazards and bioethical issues. *Cells, 10*(5), 969. DOI: 10.3390/cells10050969 PMID: 33919194

Pietralla, J., Capdeville, N., Schindele, P., & Puchta, H. (2024). Optimizing ErCas12a for efficient gene editing in Arabidopsis thaliana. *Plant Biotechnology Journal, 22*(2), 401–412. DOI: 10.1111/pbi.14194 PMID: 37864303

Pixley, K. V., Falck-Zepeda, J. B., Paarlberg, R. L., Phillips, P. W. B., Slamet-Loedin, I. H., Dhugga, K. S., Campos, H., & Gutterson, N. (2022). Genome-edited crops for improved food security of smallholder farmers. *Nature Genetics*, *54*(4), 364–367. DOI: 10.1038/s41588-022-01046-7 PMID: 35393597

Prado, G. S., Rocha, D. C., Nascimento dos Santos, L., Contiliani, D. F., Nobile, P. M., Martinati-Schenk, J. C., & Padilha, L.. (2024). CRISPR technology towards genome editing of the perennial and semi-perennial crops citrus, coffee and sugarcane. *Frontiers in Plant Science*, *14*, 1331258. DOI: 10.3389/fpls.2023.1331258 PMID: 38259920

Pramanik, D., Shelake, R. M., Kim, M. J., & Kim, J.-Y. (2021). CRISPR-mediated engineering across the central dogma in plant biology for basic research and crop improvement. *Molecular Plant*, *14*(1), 127–150. DOI: 10.1016/j.molp.2020.11.002 PMID: 33152519

Prasad, A., & Chimata, M. K. (2023). Regulation of Genome-edited Technologies: Indian Policy Perspective. *Current Indian Science*, *1*(1), e230223213951. DOI: 10 .2174/2210299X01666230223105053

Prasad, M., & Shetty, R. (2026). Advancing crop resilience to environmental stresses: Insights from epigenetic modification and CRISPR/dCas9-based editing. *Biochimica et Biophysica Acta (BBA) -. Gene Regulatory Mechanisms*, *1869*(1), 195139. PMID: 41651287

Prober, S. M., & Smith, F. P. (2009). Enhancing biodiversity persistence in intensively used agricultural landscapes: A synthesis of 30 years of research in the Western Australian wheatbelt. *Agriculture, Ecosystems & Environment*, *132*(3-4), 173–191. DOI: 10.1016/j.agee.2009.04.005

Puchta, H., & Fauser, F. (2013). Gene targeting in plants: 25 years later. *International Journal of Developmental Biology*, *57*(6-7-8): 629–637.

Puchta, H. (2005). The repair of double-strand breaks in plants: Mechanisms and consequences for genome evolution. *Journal of Experimental Botany*, *56*(409), 1–14. DOI: 10.1093/jxb/eri025 PMID: 15557293

Puchta, H. (2024). Regulation of gene-edited plants in Europe: From the valley of tears into the shining sun? *aBIOTECH*, *5*(2), 231–238. DOI: 10.1007/s42994-023-00130-8 PMID: 38974871

Puchta, H., Dujon, B., & Hohn, B. (1996). Two different but related mechanisms are used in plants for the repair of genomic double-strand breaks by homologous recombination. *Proceedings of the National Academy of Sciences of the United States of America, 93*(10), 5055–5060. DOI: 10.1073/pnas.93.10.5055 PMID: 8643528

Puchta, H., Jiang, J., Wang, K., & Zhao, Y. (2022). Updates on gene editing and its applications. *Plant Physiology, 188*(4), 1725–1730. DOI: 10.1093/plphys/kiac032 PMID: 35225345

Qaim, M. (2020). Role of new plant breeding technologies for food security and sustainable agricultural development. *Applied Economic Perspectives and Policy, 42*(2), 129–150. DOI: 10.1002/aepp.13044

Quiroz, L. F., Khan, M., Gondalia, N., Lai, L., McKeown, P. C., Brychkova, G., & Spillane, C. (2025). Tissue culture-independent approaches to revolutionizing plant transformation and gene editing. *Horticulture Research, 12*(2), uhae292. Advance online publication. DOI: 10.1093/hr/uhae292 PMID: 39906168

Raaijmakers, J. M., & Mazzola, M. (2016). Soil immune responses. *Science, 352*(6292), 1392–1393. DOI: 10.1126/science.aaf3252 PMID: 27313024

Rabuma, T., Moronta-Barrios, F., & Craig, W. (2024). Navigating biosafety regulatory frameworks for genetic engineering in Africa: A focus on genome editing and gene drive technologies. *Frontiers in Bioengineering and Biotechnology, 12*, 1483279. DOI: 10.3389/fbioe.2024.1483279 PMID: 39512657

Rademan, A. E. (2020). *Creating expectations: the bioethics of genetics as reflected in selected South African media with specific reference to CRISPR-Cas9 Stellenbosch*. Stellenbosch University.

Rafiq, A. R. (2024). Genomic editing techniques for ensuring food security: CRISPR Cas, TALEN, ZFN, RNAi and mutagenesis. *Journal of Pure and Applied Agriculture, 9*(2).

Rahimmanesh, I., Boshtam, M., Kouhpayeh, S., Khanahmad, H., Dabiri, A., Ahangarzadeh, S., Esmaeili, Y., Bidram, E., Vaseghi, G., Javanmard, S. H., Shariati, L., Zarrabi, A., & Varma, R. S. (2022). Gene Editing-Based Technologies for beta-hemoglobinopathies treatment. *Biology (Basel), 11*(6), 862. DOI: 10.3390/biology11060862 PMID: 35741383

Rahman, M. M., & Tollefsbol, T. O. (2021). Targeting cancer epigenetics with CRISPR-dCAS9: Principles and prospects. *Methods : A Companion to Methods in Enzymology, 187*, 77–91. DOI: 10.1016/j.ymeth.2020.04.006 PMID: 32315755

Rahman, S. U., McCoy, E., Raza, G., Ali, Z., Mansoor, S., & Amin, I. (2023). Improvement of soybean; A way forward transition from genetic engineering to new plant breeding technologies. *Molecular Biotechnology*, *65*(2), 162–180. DOI: 10.1007/s12033-022-00456-6 PMID: 35119645

Rai, G. K., Khanday, D. M., Kumar, P., Magotra, I., Choudhary, S. M., Kosser, R., & Rouphael, Y. (2023). Enhancing crop resilience to drought stress through CRISPR-Cas9 genome editing. *Plants*, *12*(12), 2306. DOI: 10.3390/plants12122306 PMID: 37375931

Rajpal, V. R., Sehgal, D., Valluru, R., & Singh, S. (2023). Editorial: Current advances in genomics and gene editing tools for crop improvement in a changing climate scenario. *Frontiers in Genetics*, *14*, 1214679. DOI: 10.3389/fgene.2023.1214679 PMID: 37377598

Rajput, M., Choudhary, K., Kumar, M., Vivekanand, V., Chawade, A., Ortiz, R., & Pareek, N. (2021). RNA interference and CRISPR/Cas gene editing for crop improvement: Paradigm shift towards sustainable agriculture. *Plants*, *10*(9), 1914. DOI: 10.3390/plants10091914 PMID: 34579446

Raju, S. (2021). Strategies for enhancing post-harvest quality and shelf life of tuber crops: Insights from physiological perspectives. *Journal of Root Crops*, *47*(1 & 2), 40–52.

Ramalingam, S., Kandavelou, K., Rajenderan, R., & Chandrasegaran, S. (2011). Creating designed zinc-finger nucleases with minimal cytotoxicity. *Journal of Molecular Biology*, *405*(3), 630–641. DOI: 10.1016/j.jmb.2010.10.043 PMID: 21094162

Ramirez-Torres, F., Ghogare, R., Stowe, E., Cerdá-Bennasser, P., Lobato-Gómez, M., Williamson-Benavides, B. A., Giron-Calva, P. S., Hewitt, S., Christou, P., & Dhingra, A. (2021). Genome editing in fruit, ornamental, and industrial crops. *Transgenic Research*, *30*(4), 499–528. DOI: 10.1007/s11248-021-00240-3 PMID: 33825100

Raza, A., Anas, M., Bhardwaj, S., Mir, R. A., Charagh, S., Elahi, M., Zhang, X., Mir, R. R., Weckwerth, W., Fernie, A. R., Siddique, K. H. M., Hu, Z., & Varshney, R. K. (2025). Harnessing metabolomics for enhanced crop drought tolerance. *The Crop Journal*, *13*(2), 311–327. https://doi.org/https://doi.org/10.1016/j.cj.2025.01.001. DOI: 10.1016/j.cj.2025.01.001

Razzaq, M. K., Aleem, M., Mansoor, S., Khan, M. A., Rauf, S., Iqbal, S., & Siddique, K. H. (2021). Omics and CRISPR-Cas9 approaches for molecular insight, functional gene analysis, and stress tolerance development in crops. *International Journal of Molecular Sciences*, *22*(3), 1292. DOI: 10.3390/ijms22031292 PMID: 33525517

Reed, K. M., & Bargmann, B. O. (2021). Protoplast regeneration and its use in new plant breeding technologies. *Frontiers in Genome Editing*, *3*, 734951. DOI: 10.3389/fgeed.2021.734951 PMID: 34713266

Ribarits, A., Eckerstorfer, M., Simon, S., & Stepanek, W. (2021). Genome-edited plants: Opportunities and challenges for an anticipatory detection and identification framework. *Foods*, *10*(2), 430. DOI: 10.3390/foods10020430 PMID: 33669278

Richards, R., Hunt, J., Kirkegaard, J., & Passioura, J. (2014). Yield improvement and adaptation of wheat to water-limited environments in Australia—A case study. *Crop & Pasture Science*, *65*(7), 676–689. DOI: 10.1071/CP13426

Ricroch, A. E., Eriksson, D., Dragana Miladinović, S., J., K. van Laere, Wozniak-Gientka, E. (2023). A roadmap for plant genome editing. Europe and beyond. Springer Nature. DOI: 10.1007/978-3-031-46150-7

Ricroch, A. E., & Hénard-Damave, M.-C. (2016). Next biotech plants: New traits, crops, developers and technologies for addressing global challenges. *Critical Reviews in Biotechnology*, *36*(4), 675–690. DOI: 10.3109/07388551.2015.1004521 PMID: 25641327

Ríos, D. F., Quintana, S. A., Paniagua, P. G., Arrúa, A. A., Brozón, G. R., Hicar, M. S. B., Alegría, A. C., & Goberna, M. F. (2025). Regulatory challenges and global trade implications of genome editing in agriculture. *Frontiers in Bioengineering and Biotechnology*, *13*, 1609110. DOI: 10.3389/fbioe.2025.1609110 PMID: 40613005

Rizzo, P., Chavez, B. G., Dias, S. L., & D'Auria, J. C. (2023). Plant synthetic biology: From inspiration to augmentation. *Current Opinion in Biotechnology*, *79*, 102857. DOI: 10.1016/j.copbio.2022.102857 PMID: 36502769

Robertson, G. (2021). *Establishing the CRISPR/Cas13a genome editing system in Nicotiana benthamiana for RNA targeting applications*. Stellenbosch University.

Robertson, G., Burger, J., & Campa, M. (2022). CRISPR/Cas-based tools for the targeted control of plant viruses. *Molecular Plant Pathology*, *23*(11), 1701–1718. DOI: 10.1111/mpp.13252 PMID: 35920132

Rock, J. S., Schnurr, M. A., Kingiri, A., Glover, D., Stone, G. D., Ely, A., & Fischer, K. (2023). Beyond the Genome: Genetically modified crops in Africa and the implications for Genome Editing. *Development and Change*, *54*(1), 117–142. DOI: 10.1111/dech.12750

Rodríguez, H., Fraga, R., Gonzalez, T., & Bashan, Y. (2006). Genetics of phosphate solubilization and its potential applications for improving plant growth-promoting bacteria. *Plant and Soil*, *287*(1–2), 15–21. DOI: 10.1007/s11104-006-9056-9

Rogo, U., Simoni, S., Fambrini, M., Giordani, T., Pugliesi, C., & Mascagni, F. (2024). Future-proofing agriculture: De Novo domestication for sustainable and resilient crops. *International Journal of Molecular Sciences*, *25*(4), 2374. DOI: 10.3390/ijms25042374 PMID: 38397047

Rose, K. M., Brossard, D., & Scheufele, D. A. (2020). Of society, nature, and health: How perceptions of specific risks and benefits of genetically engineered foods shape public rejection. *Environmental Communication*, *14*(7), 1017–1031. DOI: 10.1080/17524032.2019.1710227

Ross, I. L., Alami, Y., Harvey, P. R., Achouak, W., & Ryder, M. H. (2000). Genetic diversity and biological control activity of novel species of closely related pseudomonads isolated from wheat field soils in South Australia. *Applied and Environmental Microbiology*, *66*(4), 1609–1616. DOI: 10.1128/AEM.66.4.1609-1616.2000 PMID: 10742249

Roy, C., Kumar, S., Ranjan, R. D., Kumhar, S. R., & Govindan, V. (2022). Genomic approaches for improving grain zinc and iron content in wheat. *Frontiers in Genetics*, *13*, 1045955. DOI: 10.3389/fgene.2022.1045955 PMID: 36437911

Rozas, P., Kessi-Pérez, E. I., & Martínez, C. (2022). Genetically modified organisms: Adapting regulatory frameworks for evolving genome editing technologies. *Biological Research*, *55*(1), 55. DOI: 10.1186/s40659-022-00399-x PMID: 36266673

Rozov, S. M., Permyakova, N. V., & Deineko, E. V. (2019). The problem of the low rates of CRISPR/Cas9-mediated knock-ins in plants: Approaches and solutions. *International Journal of Molecular Sciences*, *20*(13), 3371. DOI: 10.3390/ijms20133371 PMID: 31323994

Rozov, S. M., Permyakova, N. V., Sidorchuk, Y. V., & Deineko, E. V. (2022). Optimization of genome knock-in methods: Identifying the most efficient genomic regions for transgene expression in plants. *International Journal of Molecular Sciences*, *23*(8), 4416. DOI: 10.3390/ijms23084416 PMID: 35457234

Ruder, S.-L., & Kandlikar, M. (2023). Governing gene-edited crops: Risks, regulations, and responsibilities as perceived by agricultural genomics experts in Canada. *Journal of Responsible Innovation*, *10*(1), 2167572. DOI: 10.1080/23299460.2023.2167572

Rustgi, S., Zhang, H., & Oz, T. M. (2023). Editorial: Overcoming genome editing challenges in plants: new tools and nanotechnologies. *Frontiers in Genome Editing*, *5*, 1230424. DOI: 10.3389/fgeed.2023.1230424 PMID: 37324649

Ryan, J. P., Oestreich, W. K., Benoit-Bird, K. J., Waluk, C. M., Rueda, C. A., Cline, D. E., & Fahlbusch, J. A. (2025). Audible changes in marine trophic ecology: Baleen whale song tracks foraging conditions in the eastern North Pacific. *PLoS One*, *20*(2), e0318624. DOI: 10.1371/journal.pone.0318624 PMID: 40009591

Saad, M. M., Eida, A. A., & Hirt, H. (2020). Tailoring plant-associated microbial inoculants in agriculture: A roadmap for successful application. *Journal of Experimental Botany*, *71*(13), 3878–3901. DOI: 10.1093/jxb/eraa111 PMID: 32157287

Sabat, M., & Tripathy, A. (2024). Genetically Modified and Gene-Edited Food Crops: Recent Status and Future Prospects. *Food Production, Diversity, and Safety Under Climate Change*, 211-222.

Saber Sichani, A., Ranjbar, M., Baneshi, M., Torabi Zadeh, F., & Fallahi, J. (2023). A review on advanced CRISPR-based genome-editing tools: Base editing and prime editing. *Molecular Biotechnology*, *65*(6), 849–860. DOI: 10.1007/s12033-022-00639-1 PMID: 36547823

Sabzehzari, M., Zeinali, M., & Naghavi, M. R. (2020). CRISPR-based metabolic editing: Next-generation metabolic engineering in plants. *Gene*, *759*, 144993. DOI: 10.1016/j.gene.2020.144993 PMID: 32717311

Sadikiel Mmbando, G. (2024). The adoption of genetically modified crops in Africa: The public's current perception, the regulatory obstacles, and ethical challenges. *GM Crops & Food*, *15*(1), 185–199. DOI: 10.1080/21645698.2024.2345401 PMID: 38651587

Salavatipour, M. S., Poursalehi, Z., Rouzbahani, N. H., Mohammadyar, S., & Vasei, M. (2024). CRISPR-Cas9 in basic and translational aspects of cancer therapy. *BioImpacts*, *14*(6), 30087. DOI: 10.34172/bi.2024.30087 PMID: 39493894

Sami, A., Xue, Z., Tazein, S., Arshad, A., He Zhu, Z., Ping Chen, Y., Hong, Y., Tian Zhu, X., & Jin Zhou, K. (2021). CRISPR–Cas9-based genetic engineering for crop improvement under drought stress. *Bioengineered*, *12*(1), 5814–5829. DOI: 10.1080/21655979.2021.1969831 PMID: 34506262

Sampath, V., Rangarajan, N., Sharanappa, C., Deori, M., Veeraragavan, M., Ghodake, B., & Kaushal, K. (2023). Advancing crop improvement through CRISPR technology in precision agriculture trends-a review. *International Journal of Environment and Climate Change*, *13*(11), 4683–4694. DOI: 10.9734/ijecc/2023/v13i113647

Samtiya, M., Aluko, R. E., & Dhewa, T. (2020). Plant food anti-nutritional factors and their reduction strategies: An overview. *Food Production, Processing and Nutrition*, *2*(1), 6. Advance online publication. DOI: 10.1186/s43014-020-0020-5

Sánchez, M. A. "The Global Advance of Genome-Edited Plants to the Market: The Key Role of Chile in Its Development." *Plants* 13, no. 24 (2024): 3597. https://doi.org/DOI: 10.3390/plants13243597

San-Epifanio, E., Leire, I. F., López, A. L., Puigdomènech, P., & Larrabe, J. U. (2023). Possible EU futures for CRISPR-edited plants: Little margin for optimism? *Frontiers in Plant Science*, *14*, 1141455. DOI: 10.3389/fpls.2023.1141455 PMID: 37008488

Santoyo, G. (2026). The role of plant microbiome engineering in stressful agriculture. *Discover Applied Sciences*, *8*(5), 481. DOI: 10.1007/s42452-026-08322-9

Santoyo, G., Moreno-Hagelsieb, G., Orozco-Mosqueda, M. D. C., & Glick, B. R. (2016). Plant growth-promoting bacterial endophytes. *Microbiological Research*, *183*, 92–99. DOI: 10.1016/j.micres.2015.11.008 PMID: 26805622

Sarangi, D., Tyre, A. J., Patterson, E. L., Gaines, T. A., Irmak, S., Knezevic, S. Z., Lindquist, J. L., & Jhala, A. J. (2017). Pollen-mediated gene flow from glyphosate-resistant common waterhemp (Amaranthus rudis Sauer): Consequences for the dispersal of resistance genes. *Scientific Reports*, *7*(1), 44913. DOI: 10.1038/srep44913 PMID: 28327669

Sattell, S. S. (2006). The rhetoric of risk: Institutional controversies over agricultural biotechnology Northwestern University].

Saxena, R., Kumar, M., Tomar, R. S., Yadav, N., Kaur, N., Puri, P., Yadav, A., Negi, R., & Yadav, A. N. (2025). Genome editing: A new age technology for sustainable agriculture and developing abiotic stress resilience in crops. *Journal of Applied Biology and Biotechnology*. Advance online publication. DOI: 10.7324/JABB.2025.232158

Schaart, J. G., van de Wiel, C. C., & Smulders, M. J. (2021). Genome editing of polyploid crops: Prospects, achievements and bottlenecks. *Transgenic Research*, *30*(4), 337–351. DOI: 10.1007/s11248-021-00251-0 PMID: 33846956

Scharff, L. B., Saltenis, V. L. R., Jensen, P. E., Baekelandt, A., Burgess, A. J., Burow, M., Ceriotti, A., Cohan, J., Geu-Flores, F., Halkier, B. A., Haslam, R. P., Inzé, D., Lankhorst, R. K., Murchie, E. H., Napier, J. A., Nacry, P., Parry, M. J., Santino, A., Scarano, A., & Pribil, M. (2021). Prospects to improve the nutritional quality of crops. *Food and Energy Security*, *11*(1), e327. Advance online publication. DOI: 10.1002/fes3.327

Schiemann, J., Hartung, F., Menz, J., Sprink, T., & Wilhelm, R. (2021). Policies and governance for plant genome editing. In *Plant Biotechnology* (pp. 259–275). Springer., DOI: 10.1007/978-3-030-68345-0_18

Schiml, S., Fauser, F., & Puchta, H. (2014). The CRISPR/Cas system can be used as a nuclease for plant gene targeting and as paired nickases for directed mutagenesis in Arabidopsis, resulting in inheritable progeny. *The Plant Journal : For Cell and Molecular Biology*, *80*(6), 1139–1150. DOI: 10.1111/tpj.12704 PMID: 25327456

Schmidt, T. (2014). *Transient biolistic transformation in maize leaves.* University of Pretoria, Sedeek, K. E., Mahas, A., & Mahfouz, M. (2019). Plant genome engineering for targeted improvement of crop traits. *Frontiers in Plant Science*, *10*, 114.

Schneider, K. R.. (2002). *Genetically modified food.* Food Science and Human Nutrition Department, Florida Cooperative Extension Service, Institute of Food and Agricultural Sciences, University of Florida.

Schubert, M. S., Cedrone, E., Neun, B., Behlke, M. A., & Dobrovolskaia, M. A. (2024). Enhancing plant HDR: Mechanistic insights and engineering strategies. *Plant Biotechnology Journal*, *22*(3), 601–618. DOI: 10.1111/pbi.14250

Sciences, N. A. o., et al. (2016). Genetically engineered crops: experiences and prospects. National Academies Press.

Sciences, N. A. o., et al. (2017). Preparing for future products of biotechnology. National Academies Press.

Selfa, T., Cummings, C., Dahlstrom, M., Lindberg, S., Peters, D., Wolf, C., & Bain, C. (2023). *Gene editing in agriculture: Social concerns, public engagement, and governance* (Vol. 48). Sage Publications Sage CA.

Selfa, T., Lindberg, S., & Bain, C. (2021). Governing gene editing in agriculture and food in the United States: Tensions, contestations, and realignments. *Elementa*, *9*(1), 00153. DOI: 10.1525/elementa.2020.00153

Selin, C., Friedman, L. M., Lambert, L., Farooque, M., & Guston, D. Inclusive, Equitable and Just Evolution of Gene Editing Technologies: The Case for Integrating Public Engagement in Governing Efforts. *Available at SSRN* 4853681. DOI: 10.2139/ssrn.4853681

Sen, P., Chatterjee, A., Kumar, D., Bhattacharyya, S., Bandyopadhyay, S., & Sarkar, A. (2024). Genetic and functional mechanisms of yield-related genes in rice. *Acta Physiologiae Plantarum*, *46*(3), 29. DOI: 10.1007/s11738-024-03667-3

Sessitsch, A., Brader, G., Pfaffenbichler, N., Gusenbauer, D., & Mitter, B. (2018). The contribution of plant microbiota to economy growth. *Microbial Biotechnology*, *11*(5), 801–805. DOI: 10.1111/1751-7915.13290 PMID: 29926519

Shabala, S., Chen, X., Yun, P., & Zhou, M. (2025). Salinity tolerance in wheat: Rethinking the targets. *Journal of Experimental Botany*, •••, eraf152. PMID: 40201956

Shahid, M. S., Sattar, M. N., Iqbal, Z., Raza, A., & Al-Sadi, A. M. (2021). Next-generation sequencing and the CRISPR-Cas nexus: A molecular plant virology perspective. *Frontiers in Microbiology*, *11*, 609376. DOI: 10.3389/fmicb.2020.609376 PMID: 33584572

Shahzad, R., Jamil, S., Ahmad, S., Nisar, A., Khan, S., Amina, Z., Kanwal, S., Aslam, H. M. U., Gill, R. A., & Zhou, W. (2021). Biofortification of cereals and pulses using new breeding techniques: Current and future perspectives. *Frontiers in Nutrition*, *8*, 721728. DOI: 10.3389/fnut.2021.721728 PMID: 34692743

Shams, F., Bayat, H., Mohammadian, O., Mahboudi, S., Vahidnezhad, H., Soosanabadi, M., & Rahimpour, A. (2022). Advance trends in targeting homology-directed repair for accurate gene editing: An inclusive review of small molecules and modified CRISPR-Cas9 systems. *BioImpacts*, *12*(4), 371–391. DOI: 10.34172/bi.2022.23871 PMID: 35975201

Shamshirgaran, Y., Liu, J., Sumer, H., Verma, P. J., & Taheri-Ghahfarokhi, A. (2022). Tools for efficient genome editing; ZFN, TALEN, and CRISPR. *Applications of genome modulation and editing*, 29-46.

Shanahan, J., Scheufele, D., & Lee, E. (2001). Trends: Attitudes about agricultural biotechnology and genetically modified organisms. *Public Opinion Quarterly*, *65*(2), 267–281. DOI: 10.1086/322201 PMID: 11420759

Shan, Q., Wang, Y., Li, J., & Gao, C. (2014). Genome editing in rice and wheat using the CRISPR/Cas system. *Nature Protocols*, *9*(10), 2395–2410. DOI: 10.1038/nprot.2014.157 PMID: 25232936

Shan, Q., Wang, Y., Li, J., Zhang, Y., Chen, K., Liang, Z., & Gao, C. (2013). Targeted genome modification of crop plants using a CRISPR-Cas system. *Nature Biotechnology*, *31*(8), 686–688. DOI: 10.1038/nbt.2650 PMID: 23929338

Sharma, A., Johri, B. N., Sharma, A. K., & Glick, B. R. (2003). Plant growth-promoting bacterium Pseudomonas sp. strain GRP3 influences iron acquisition in mung bean. *Soil Biology & Biochemistry*, *35*(7), 887–894. DOI: 10.1016/S0038-0717(03)00119-6

Sharma, H. C., Crouch, J. H., Sharma, K. K., Seetharama, N., & Hash, C. T. (2002). Applications of biotechnology for crop improvement: Prospects and constraints. *Plant Science : An International Journal of Experimental Plant Biology*, *163*(3), 381–395. DOI: 10.1016/S0168-9452(02)00133-4

Sharma, P., Pandey, A., Malviya, R., Dey, S., Karmakar, S., & Gayen, D. (2023). Genome editing for improving nutritional quality, post-harvest shelf life and stress tolerance of fruits, vegetables, and ornamentals. *Frontiers in Genome Editing*, *5*, 1094965. DOI: 10.3389/fgeed.2023.1094965 PMID: 36911238

Sharma, S., Saroha, N. K., Sehrawat, A., Tang, G., Singh, D., & Teotia, S. (2025, December 4). Emerging tools in plant genome editing. *Frontiers in Genome Editing*, *7*, 1588089. DOI: 10.3389/fgeed.2025.1588089 PMID: 41424917

Shelake, M. S. K. (2025). Gene Editing: Hope and Risks. SCIENCE, SOCIETY AND SUSTAINABILITY, 111.

Shelake, R. M., Kadam, U. S., Kumar, R., Pramanik, D., Singh, A. K., & Kim, J.-Y. (2022). Engineering drought and salinity tolerance traits in crops through CRISPR-mediated genome editing: Targets, tools, challenges, and perspectives. *Plant Communications*, *3*(6), 100417. DOI: 10.1016/j.xplc.2022.100417 PMID: 35927945

Shelake, R. M., Pramanik, D., & Kim, J.-Y. (2019). Exploration of plant-microbe interactions for sustainable agriculture in CRISPR era. *Microorganisms*, *7*(8), 269. DOI: 10.3390/microorganisms7080269 PMID: 31426522

Shew, A. M., Nalley, L. L., Snell, H. A., Nayga, R. M.Jr, & Dixon, B. L. (2018). CRISPR versus GMOs: Public acceptance and valuation. *Global Food Security*, *19*, 71–80. DOI: 10.1016/j.gfs.2018.10.005

Shi, G., Hao, M., Tian, B., Cao, G., Wei, F., & Xie, Z. (2021). A methodological advance of tobacco rattle virus-induced gene silencing for functional genomics in plants. *Frontiers in Plant Science*, *12*, 671091. DOI: 10.3389/fpls.2021.671091 PMID: 34149770

Shimatani, Z., Kashojiya, S., Takayama, M., Terada, R., Arazoe, T., Ishii, H., Teramura, H., Yamamoto, T., Komatsu, H., Miura, K., Ezura, H., Nishida, K., & Kondo, A. (2017). Targeted base editing in rice and tomato using a CRISPR-Cas9 cytidine deaminase fusion. *Nature Biotechnology*, *35*(5), 441–443. DOI: 10.1038/nbt.3833 PMID: 28346401

Shmakov, S. A., Sitnik, V., Makarova, K. S., Wolf, Y. I., Severinov, K. V., & Koonin, E. V. (2017). The CRISPR spacer space is dominated by sequences from species-specific mobilomes. *mBio*, *8*(5), 10–1128. DOI: 10.1128/mBio.01397-17 PMID: 28928211

Shoeb, E., Venkataraman, S., Badar, U., & Hefferon, K. (2024). Genome editing for crop biofortification. *Applications of Genome Engineering in Plants*, 91-121.

Shukla-Jones, A., et al. (2018). Gene editing in an international context: Scientific, economic and social issues across sectors. OECD science, technology and industry working papers, 2018(4), 0_1-51.

Shukla-Jones, A., Friedrichs, S., & Winickoff, D. E. (2018). Gene editing in an international context: Scientific, economic and social issues across sectors. *OECD science, technology and industry working papers, 2018*(4), 0_1-51.

Silva, J. C. F., de Godoy Machado, K. L., de Souza Silva, A. F., Dias, R., Bodnar, V. R., Vieira, W. O.,...Pauli, I. (2025). Challenges and Opportunities for New Frontiers and Technologies to Guarantee Food Production: A Broad Systematic Perspective.

Singh, A. K., Mishra, P., Kashyap, S. P., Karkute, S. G., Singh, P. M., Rai, N., Bahadur, A., & Behera, T. K. (2022). Molecular insights into mechanisms underlying thermo-tolerance in tomato. *Frontiers in Plant Science, 13*, 1040532. DOI: 10.3389/fpls.2022.1040532 PMID: 36388532

Singh, A., Pandey, H., Pandey, S., Lal, D., Chauhan, D., Aparna, , Antre, S. H., B, S., & Kumar, A. (2023). Drought stress in maize: Stress perception to molecular response and strategies for its improvement. *Functional & Integrative Genomics, 23*(4), 296. DOI: 10.1007/s10142-023-01226-6 PMID: 37697159

Singh, B. K., Delgado-Baquerizo, M., Egidi, E., Guirado, E., Leach, J. E., Liu, H., Trivedi, P., & Bardgett, R. D. (2021). Climate change impacts on plant pathogens, food security and paths forward. *Nature Reviews. Microbiology, 19*(12), 830–843. DOI: 10.1038/s41579-021-00542-w PMID: 37131070

Singh, B. K., Trivedi, P., Egidi, E., Macdonald, C. A., & Delgado-Baquerizo, M. (2020). Crop microbiome and sustainable agriculture. *Nature Reviews. Microbiology, 18*(11), 601–602. DOI: 10.1038/s41579-020-00446-y PMID: 33037425

Singh, G., Raigar, O. P., Kaur, S., Bishnoi, R., Mondal, K., Abreha, K. B., Nayak, A. K., Athar, T., Sharma, V., Aziz, D., Gudi, S., Saini, P., Kumar, A., Bhardwaj, R., & Riar, A. (2025). Advances in genomics-assisted breeding strategies for enhancing nutrient uptake and use efficiency in cereals: A pathway toward sustainable agriculture. *Plant Stress, 18*, 101002. https://doi.org/https://doi.org/10.1016/j.stress.2025.101002. DOI: 10.1016/j.stress.2025.101002

Singh, J., Sharma, D., Brar, G. S., Sandhu, K. S., Wani, S. H., Kashyap, R., & Singh, S. (2022). CRISPR/Cas tool designs for multiplex genome editing and their applications in developing biotic and abiotic-stress-resistant crop plants. *Molecular Biology Reports, 49*(12), 11443–11467. DOI: 10.1007/s11033-022-07741-2 PMID: 36002653

Singh, O. V., Ghai, S., Paul, D., & Jain, R. K. (2006). Genetically modified crops: Success, safety assessment, and public concern. *Applied Microbiology and Biotechnology*, *71*(5), 598–607. DOI: 10.1007/s00253-006-0449-8 PMID: 16639559

Slaymaker, I. M., Gao, L., Zetsche, B., Scott, D. A., Yan, W. X., & Zhang, F. (2016). Rationally engineered Cas9 nucleases with improved specificity. *Science*, *351*(6268), 84–88. DOI: 10.1126/science.aad5227 PMID: 26628643

Smirnikhina, S. A., Zaynitdinova, M. I., Sergeeva, V. A., & Lavrov, A. V. (2022). Improving homology-directed repair in genome editing experiments by influencing the cell cycle. *International Journal of Molecular Sciences*, *23*(11), 5992. DOI: 10.3390/ijms23115992 PMID: 35682671

Smyth, Stuart J. "Regulation of genome editing in plant biotechnology: Canada." *Regulation of genome editing in plant biotechnology: A comparative analysis of regulatory frameworks of selected countries and the EU* (2019): 111-135. https://doi.org/.DOI: 10.1007/978-3-030-17119-3_4

Smyth, S. J., Falck-Zepeda, J., & Ludlow, K. (2016). The costs of regulatory delays for genetically modified crops. *Estey Journal of International Law and Trade Policy*, *17*(2), 173–195. DOI: 10.22004/ag.econ.253074

Sokal, K., & Kachel, M. (2025). Impact of Agriculture on Greenhouse Gas Emissions—A Review. *Energies*, *18*(9), 2272. DOI: 10.3390/en18092272

Sokolović, D., Babić, S., Petrović, M., Solís, I., Cougnon, M., Gutierrez, N., Pärssinen, P., Reheul, D., Radović, J., & Torres, A. M. (2025). Genotype by environment interactions and phenotypic traits stability of the EUCLEG faba bean collection. *Frontiers in Plant Science*, *15*, 1480110. DOI: 10.3389/fpls.2024.1480110 PMID: 39944945

Song, Q., & Chen, Z. J. (2015). Epigenetic and developmental regulation in plant polyploids. *Current Opinion in Plant Biology*, *24*(10), 101–109. DOI: 10.1016/j.pbi.2015.02.007 PMID: 25765928

Song, X., Dang, C., Wang, F., Yao, H., Song, Q., & Ye, G. (2026). CRISPR/Cas gene editing for plant resistance to pests, diseases, and herbicide: Application, risk assessment, and safety evaluation. *GM Crops & Food*, *17*(1), 2641893. DOI: 10.1080/21645698.2026.2641893

Sony, S. K., Kaul, T., Abdel Motelb, K. F., Thangaraj, A., Bharti, J., Kaul, R., Verma, R., & Nehra, M. (2023). CRISPR/Cas9-mediated homology donor repair base editing confers glyphosate resistance to rice (Oryza sativa L.). *Frontiers in Plant Science*, *14*, 1122926. DOI: 10.3389/fpls.2023.1122926 PMID: 36959937

Spök, A., Sprink, T., Allan, A. C., Yamaguchi, T., & Dayé, C. (2022). Towards social acceptability of genome-edited plants in industrialised countries? Emerging evidence from Europe, United States, Canada, Australia, New Zealand, and Japan. *Frontiers in Genome Editing*, *4*, 899331. DOI: 10.3389/fgeed.2022.899331 PMID: 36120531

Sprink, T., Eriksson, D., Schiemann, J., & Hartung, F. (2016). Regulatory hurdles for genome editing: Process-vs. product-based approaches in different regulatory contexts. *Plant Cell Reports*, *35*(7), 1493–1506. DOI: 10.1007/s00299-016-1990-2 PMID: 27142995

Sprink, T., Wilhelm, R., & Hartung, F. (2022). Genome editing around the globe: An update on policies and perceptions. *Plant Physiology*, *190*(3), 1579–1587. DOI: 10.1093/plphys/kiac359 PMID: 35976141

Squire, H. J., Tomatz, S., Voke, E., González-Grandío, E., & Landry, M. (2023). The emerging role of nanotechnology in plant genetic engineering. *Nature Reviews Bioengineering*, *1*(5), 314–328. DOI: 10.1038/s44222-023-00037-5

Sretenovic, S., Cheng, Y., Xu, Y., Yu, W., Zhang, C., & Qi, Y. (2024). Engineered ABE variants with enhanced specificity enable near-perfect A-to-G base editing in rice. *Plant Biotechnology Journal*, *22*(3), 612–625. DOI: 10.1111/pbi.14225

Sretenovic, S., Liu, S., Li, G., Cheng, Y., Fan, T., Xu, Y., Yin, D., & Qi, Y. (2024). Engineered ABE architectures achieve near-perfect specificity in rice. *Nature Plants*, *10*, 211–221. DOI: 10.1038/s41477-023-01601-4

Sternberg, S. H., Redding, S., Jinek, M., Greene, E. C., & Doudna, J. A. (2014). DNA interrogation by the CRISPR RNA-guided endonuclease Cas9. *Nature*, *507*(7490), 62–67. DOI: 10.1038/nature13011 PMID: 24476820

Strauss, S. H., Slavov, G. T., & DiFazio, S. P. (2022). Gene-editing for production traits in forest trees: Challenges to integration and gene target identification. *Forests*, *13*(11), 1887. DOI: 10.3390/f13111887

Strobbe, S., Wesana, J., Van Der Straeten, D., & De Steur, H. (2023). Public acceptance and stakeholder views of gene edited foods : a global overview. https://doi.org/DOI: 10.1016/j.tibtech.2022.12.011

Sturme, M. H., van der Berg, J. P., Bouwman, L. M., De Schrijver, A., de Maagd, R. A., Kleter, G. A., & Battaglia-de Wilde, E. (2022). Occurrence and nature of off-target modifications by CRISPR-Cas genome editing in plants. *ACS Agricultural Science & Technology*, *2*(2), 192–201. DOI: 10.1021/acsagscitech.1c00270 PMID: 35548699

Sugano, S. S., Shirakawa, M., Takagi, J., Matsuda, Y., Shimada, T., Hara-Nishimura, I., & Kohchi, T. (2014). CRISPR/Cas9-Mediated Targeted Mutagenesis in the Liverwort *Marchantia polymorpha L. Plant & Cell Physiology*, *55*(3), 475–481. DOI: 10.1093/pcp/pcu014 PMID: 24443494

Su, H., Wang, Y., Xu, J., Omar, A. A., Grosser, J. W., & Wang, N. (2024). Cas12a RNP-mediated co-transformation enables transgene-free multiplex genome editing, long deletions, and inversions in citrus chromosome. *Frontiers in Plant Science*, *15*, 1448807. DOI: 10.3389/fpls.2024.1448807 PMID: 39148610

Sun, L., Lai, M., Ghouri, F., Nawaz, M. A., Ali, F., Baloch, F. S., Nadeem, M. A., Aasim, M., & Shahid, M. Q. (2024). Modern plant breeding techniques in crop improvement and genetic diversity: From molecular markers and gene editing to artificial intelligence—A critical review. *Plants*, *13*(19), 2676. DOI: 10.3390/plants13192676 PMID: 39409546

Sun, Y., Li, J., & Xia, L. (2016). Precise Genome Modification via Sequence-Specific Nucleases-Mediated Gene Targeting for Crop Improvement. *Frontiers in Plant Science*, *7*, 1–15. DOI: 10.3389/fpls.2016.01928 PMID: 28066481

Su, Z., Hao, C., Wang, L., Dong, Y., & Zhang, X. (2011). Identification and development of a functional marker of TaGW2 associated with grain weight in bread wheat (Triticum aestivum L.). *Theoretical and Applied Genetics*, *122*(1), 211–223. DOI: 10.1007/s00122-010-1437-z PMID: 20838758

Svitashev, S., Young, J. K., Schwartz, C., Gao, H., Falco, S. C., & Cigan, A. M. (2015). Targeted Mutagenesis, Precise Gene Editing, and Site-Specific Gene Insertion in Maize Using Cas9 and Guide RNA. *Plant Physiology*, *169*(2), 931–945. DOI: 10.1104/pp.15.00793 PMID: 26269544

Tachikawa, M., & Matsuo, M. (2023). Divergence and convergence in international regulatory policies regarding genome-edited food: How to find a middle ground. *Frontiers in Plant Science*, *14*, 1105426. DOI: 10.3389/fpls.2023.1105426 PMID: 36794228

Tadele, Z. (2019). Orphan crops: Their importance and the urgency of improvement. *Planta*, *250*(3), 677–694. DOI: 10.1007/s00425-019-03210-6 PMID: 31190115

Taheri, F., Azadi, H., & D'Haese, M. (2017). A world without hunger: Organic or GM crops? *Sustainability (Basel)*, *9*(4), 580. DOI: 10.3390/su9040580

Tamburini, G., Bommarco, R., Wanger, T. C., Kremen, C., Van Der Heijden, M. G. A., Liebman, M., & Hallin, S. (2020). Agricultural diversification promotes multiple ecosystem services without compromising yield. *Science Advances*, *6*(45), eaba1715. Advance online publication. DOI: 10.1126/sciadv.aba1715 PMID: 33148637

Tang, X., Liu, G., Zhou, J., Ren, Q., You, Q., Tian, L., Xin, X., Zhong, Z., Liu, B., Zheng, X., Zhang, D., Malzahn, A. A., Gong, Z., Qi, Y., Zhang, T., & Zhang, Y. (2018). A large-scale whole-genome sequencing analysis reveals highly specific genome editing by both Cas9 and Cpf1 nucleases in rice. *Genome Biology*, *19*(1), 84. DOI: 10.1186/s13059-018-1458-5 PMID: 29973285

Tang, X., Lowder, L. G., Zhang, T., Malzahn, A. A., Zheng, X., Voytas, D. F., Zhong, Z., Chen, Y., Ren, Q., Li, Q., Kirkland, E. R., Zhang, Y., & Qi, Y. (2017). A CRISPR–Cpf1 system for efficient genome editing and transcriptional repression in plants. *Nature Plants*, *3*(3), 17018. DOI: 10.1038/nplants.2017.18 PMID: 28211909

Tang, X., & Zhang, Y. (2023). Beyond knockouts: Fine-tuning regulation of gene expression in plants with CRISPR-Cas-based promoter editing. *The New Phytologist*, *239*(3), 868–874. DOI: 10.1111/nph.19020 PMID: 37282668

Tan, Y., Du, H., Wu, X., Liu, Y., Jiang, M., Song, S., Wu, L., & Shu, Q. (2020). Tan, Yuan-yuan, Hao Du, Xia Wu, Yan-hua Liu, Meng Jiang, Shi-yong Song, Liang Wu, and Qing-yao Shu. "Gene editing: An instrument for practical application of gene biology to plant breeding.". *Journal of Zhejiang University. Science. B*, *21*(6), 460–473. DOI: 10.1631/jzus.B1900633 PMID: 32478492

Tariq, S., Gul, A., Negri, S., & Mujtaba, T. (2024). Genetic engineering in tomato. In *Targeted Genome Engineering via CRISPR/Cas9 in Plants* (pp. 101–133). Elsevier. DOI: 10.1016/B978-0-443-26614-0.00015-1

Tavakoli, K., Pour-Aboughadareh, A., Kianersi, F., Poczai, P., Etminan, A., & Shooshtari, L. (2021). Applications of CRISPR-Cas9 as an Advanced Genome Editing System in Life Sciences. *Biotech*, *10*(3), 14–27. DOI: 10.3390/biotech10030014 PMID: 35822768

Taylor, G. K., Petrucci, L. H., Lambert, A. R., Baxter, S. K., Jarjour, J., & Stoddard, B. L. (2012). LAHEDES: The LAGLIDADG homing endonuclease database and engineering server. *Nucleic Acids Research*, *40*(W1), W110–W116. DOI: 10.1093/nar/gks365 PMID: 22570419

Tembo, B., Mulenga, R. M., Sichilima, S., M'siska, K. K., Mwale, M., Chikoti, P. C., & Peterson, G. L. (2020). Detection and characterization of fungus (Magnaporthe oryzae pathotype Triticum) causing wheat blast disease on rain-fed grown wheat (Triticum aestivum L.) in Zambia. *PLoS One*, *15*(9), e0238724. DOI: 10.1371/journal.pone.0238724 PMID: 32956369

Teper, D., White, F. F., & Wang, N. (2023). The dynamic transcription activator-like effector family of Xanthomonas. *Phytopathology*, *113*(4), 651–666. DOI: 10.1094/PHYTO-10-22-0365-KD PMID: 36449529

Theresa, S.. (2021). Governing gene editing in agriculture and food in the United States. *Elementa*, *9*(1).

Thomson, G., Dickinson, L., & Jacob, Y. (2024). Genomic consequences associated with Agrobacterium-mediated transformation of plants. *The Plant Journal : For Cell and Molecular Biology*, *117*(2), 342–363. DOI: 10.1111/tpj.16496 PMID: 37831618

Thorneycroft, D., Sherson, S. M., & Smith, S. M. (2001). Using gene knockouts to investigate plant metabolism. *Journal of Experimental Botany*, *52*(361), 1593–1601. DOI: 10.1093/jexbot/52.361.1593 PMID: 11479324

Tian, Y., Li, X., Xie, J., Zheng, Z., Shen, R., Cao, X., Wang, M., Dong, C., & Zhu, J.-K. (2024). Targeted G-to-T base editing for generation of novel herbicide-resistance gene alleles in rice. *Journal of Integrative Plant Biology*, *66*(6), 1048–1051. DOI: 10.1111/jipb.13657 PMID: 38578176

Tizard, M., Hallerman, E., Fahrenkrug, S., Newell-McGloughlin, M., Gibson, J., de Loos, F., & D'Occhio, M. (2016). Strategies to enable the adoption of animal biotechnology to sustainably improve global food safety and security. *Transgenic Research*, *25*(5), 575–595. DOI: 10.1007/s11248-016-9965-1 PMID: 27246007

Tofazzal Islam, T. I. (2019). CRISPR-Cas technology in modifying food crops. CABI Reviews, (2019), 1-16.

Toju, H., Kishida, O., Katayama, N., & Takagi, K. (2016). Networks depicting the fine-scale co-occurrences of fungi in soil horizons. *PLoS One*, *11*(11), e0165987. DOI: 10.1371/journal.pone.0165987 PMID: 27861486

Toju, H., Peay, K. G., Yamamichi, M., Narisawa, K., Hiruma, K., Naito, K., Fukuda, S., Ushio, M., Nakaoka, S., Onoda, Y., Yoshida, K., Schlaeppi, K., Bai, Y., Sugiura, R., Ichihashi, Y., Minamisawa, K., & Kiers, E. T. (2018). Core microbiomes for sustainable agroecosystems. *Nature Plants*, *4*(5), 247–257. DOI: 10.1038/s41477-018-0139-4 PMID: 29725101

Toschka, H. Y. (2025). *New Food Product Development: Global Strategies and Practices for Successful Innovation*. CRC Press. DOI: 10.1201/9781032724591

Townsend, J. A., Wright, D. A., Winfrey, R. J., Fu, F., Maeder, M. L., Joung, J. K., & Voytas, D. F. (2009). High-frequency modification of plant genes using engineered zinc-finger nucleases. *Nature, 459*(7245), 442–445. DOI: 10.1038/nature07845 PMID: 19404258

Tripathi, A., & Shukla, S. (2024). Methods of genetic transformation: Major emphasis to crop plants. *Journal of Microbiology, Biotechnology and Food Sciences, 13*(4), e10276–e10276.

Tripathi, L., Dhugga, K. S., Ntui, V. O., Runo, S., Syombua, E. D., Muiruri, S., Wen, Z., & Tripathi, J. N. (2022). Genome editing for sustainable agriculture in Africa. *Frontiers in Genome Editing, 4*, 876697. DOI: 10.3389/fgeed.2022.876697 PMID: 35647578

Tripathi, S., Khatri, P., Fatima, Z., Pandey, R. P., & Hameed, S. (2022). A landscape of CRISPR/Cas technique for emerging viral disease diagnostics and therapeutics: Progress and prospects. *Pathogens (Basel, Switzerland), 12*(1), 56. DOI: 10.3390/pathogens12010056 PMID: 36678404

Trivedi, P., Batista, B. D., Bazany, K. E., Singh, B. K., & Leach, J. E. (2022). Plant microbiome responses under drought and climate change. *The New Phytologist, 234*(6), 1911–1923. DOI: 10.1111/nph.18016 PMID: 35118660

Trivedi, P., Delgado-Baquerizo, M., Jeffries, T. C., Trivedi, C., Anderson, I. C., Lai, K., McNee, M., Flower, K., Singh, B. P., Minkey, D., & Singh, B. K. (2017). Soil aggregation and associated microbial communities modify the impact of agricultural management on carbon content. *Environmental Microbiology, 19*(8), 3070–3086. DOI: 10.1111/1462-2920.13779 PMID: 28447378

Trivedi, P., Leach, J. E., Tringe, S. G., Sa, T., & Singh, B. K. (2020). Plant–microbiome interactions: From community assembly to plant health. *Nature Reviews. Microbiology, 18*(11), 607–621. DOI: 10.1038/s41579-020-0412-1 PMID: 32788714

Trivedi, P., Singh, B. K., & Leach, J. E. (2021). Plant–microbiome interactions and their applications in sustainable agriculture. *FEMS Microbiology Ecology, 97*(2), fiab012. Advance online publication. DOI: 10.1093/femsec/fiab012

Troadec, M.-B., Pagès, J.-C., & Scientific Committee of the High Council for biotechnology, F. (2019). *Where are we with unintended effects in genome editing applications from DNA to phenotype: focus on plant applications*. Paper presented at the Transgenic Research.

Tsai, S. Q., Wyvekens, N., Khayter, C., Foden, J. A., Thapar, V., Reyon, D., Goodwin, M. J., Aryee, M. J., & Joung, J. K. (2014). Dimeric CRISPR RNA-guided FokI nucleases for highly specific genome editing. *Nature Biotechnology*, *32*(6), 569–576. DOI: 10.1038/nbt.2908 PMID: 24770325

Tsuda, M., Watanabe, K. N., & Ohsawa, R. (2019). Regulatory status of genome-edited organisms under the Japanese Cartagena Act. *Frontiers in Bioengineering and Biotechnology*, *7*, 387. DOI: 10.3389/fbioe.2019.00387 PMID: 31867318

Tsutsui, H., & Higashiyama, T. (2016). pKAMA-ITACHI Vectors for Highly Efficient CRISPR/Cas9-Mediated Gene Knockout in *Arabidopsis thaliana. Plant & Cell Physiology*, *58*(1), pcw191. DOI: 10.1093/pcp/pcw191 PMID: 27856772

Tuncel, A., Pan, C., Clem, J. S., Liu, D., & Qi, Y. (2025). CRISPR–Cas applications in agriculture and plant research. *Nature Reviews. Molecular Cell Biology*, *26*(6), 419–441. DOI: 10.1038/s41580-025-00834-3 PMID: 40055491

Turnbull, C., Lillemo, M., & Hvoslef-Eide, T. A. (2021). Global regulation of genetically modified crops amid the gene edited crop boom–a review. *Frontiers in Plant Science*, *12*, 630396. DOI: 10.3389/fpls.2021.630396 PMID: 33719302

U.S. Department of Agriculture. 2023. Biotechnology and USDA's Role in Regulating Genetically Engineered Organisms. https://www.usda.gov/topics/biotechnology

Ukhatova, Y. V., Erastenkova, M. V., Korshikova, E. S., Krylova, E. A., Mikhailova, A. S., Semilet, T. V., Tikhonova, N. G., Shvachko, N. A., & Khlestkina, E. K. (2023). Improvement of crops using the CRISPR/CAS system: New target genes. *Molecular Biology*, *57*(3), 375–397. DOI: 10.1134/S0026893323030135 PMID: 37326044

Umeohia, U. E., & Olapade, A. A. (2024). Quality attributes, physiology, and Postharvest Technologies of Tomatoes (Lycopersicum esculentum)–A review. *American Journal of Food Science and Technology*, *12*(2), 42–64. DOI: 10.12691/ajfst-12-2-1

Upadhyay, S. K., Kumar, J., Alok, A., Tuli, R. (2013). RNA-Guided Genome Editing for Target Gene Mutations in Wheat. *G3: Genes|Genomes|Genetics*, 3(12): 2233–2238.

Urquidi Camacho, R. A., Lokdarshi, A., & von Arnim, A. G. (2020). Translational gene regulation in plants: A green new deal. *Wiley Interdisciplinary Reviews. RNA*, *11*(6), e1597. DOI: 10.1002/wrna.1597 PMID: 32367681

Usman, B., Nawaz, G., Zhao, N., Liao, S., Liu, Y., & Li, R. (2020). Precise editing of the OsPYL9 gene by RNA-guided Cas9 nuclease confers enhanced drought tolerance and grain yield in rice (Oryza sativa L.) by regulating circadian rhythm and abiotic stress responsive proteins. *International Journal of Molecular Sciences*, *21*(21), 7854. DOI: 10.3390/ijms21217854 PMID: 33113937

van den Boomgaard, M. (2021). *Public Values and Modern Biotechnology in the Dutch Patent System A multiple case-study of patents on Recombinant-DNA & PCR and CRISPR technologies in the field of green biotechnology.*

Van Eck, J. (2020). Applying gene editing to tailor precise genetic modifications in plants. *The Journal of Biological Chemistry*, *295*(38), 13267–13276. DOI: 10.1074/jbc.REV120.010850 PMID: 32723863

Van Esse, H. P., Reuber, T. L., & van der Does, D. (2020). Genetic modification to improve disease resistance in crops. *The New Phytologist*, *225*(1), 70–86. DOI: 10.1111/nph.15967 PMID: 31135961

Vandenkoornhuyse, P., Quaiser, A., Duhamel, M., Le Van, A., & Dufresne, A. (2015). The importance of the microbiome of the plant holobiont. *The New Phytologist*, *206*(4), 1196–1206. DOI: 10.1111/nph.13312 PMID: 25655016

Vaschetto, R., Pecere, A., Perkins, G. D., Mistry, D., Cammarota, G., Longhini, F., Ferrer, M., Pletsch-Assunção, R., Carron, M., Moretto, F., Qiu, H., Della Corte, F., Barone-Adesi, F., & Navalesi, P. (2021). Effects of early extubation followed by noninvasive ventilation versus standard extubation on the duration of invasive mechanical ventilation in hypoxemic non-hypercapnic patients: A systematic review and individual patient data meta-analysis of randomized controlled trials. *Critical Care : The Official Journal of the Critical Care Forum*, *25*(1), 189. DOI: 10.1186/s13054-021-03595-5 PMID: 34074314

Vats, S., Kumawat, S., Kumar, V., Patil, G. B., Joshi, T., Sonah, H., Sharma, T. R., & Deshmukh, R. (2019). Genome editing in plants: Exploration of technological advancements and challenges. *Cells*, *8*(11), 1386. DOI: 10.3390/cells8111386 PMID: 31689989

Vazquez-Vilar, M., Selma, S., & Orzaez, D. (2023). The design of synthetic gene circuits in plants: New components, old challenges. *Journal of Experimental Botany*, *74*(13), 3791–3805. DOI: 10.1093/jxb/erad167 PMID: 37204924

Venezia, M., & Krainer, K. M. C. (2021). Current advancements and limitations of gene editing in orphan crops. *Frontiers in Plant Science*, *12*, 742932. DOI: 10.3389/fpls.2021.742932 PMID: 34630494

Venkataraman, S., Zaruk, D., & Hefferon, K. (2025). Recent developments, challenges and opportunities in genome editing for crop science from a societal perspective. *Frontiers in Genome Editing*, *7*, 1568072. DOI: 10.3389/fgeed.2025.1568072 PMID: 40528898

Venturi, V., & Keel, C. (2016). Signaling in the rhizosphere. *Trends in Plant Science*, *21*(3), 187–198. DOI: 10.1016/j.tplants.2016.01.005 PMID: 26832945

Verma, V., Kumar, A., Partap, M., Thakur, M., & Bhargava, B. (2023). CRISPR-Cas: A robust technology for enhancing consumer-preferred commercial traits in crops. *Frontiers in Plant Science*, *14*, 1122940. DOI: 10.3389/fpls.2023.1122940 PMID: 36824195

Vicente, M. M., Chaves-Ferreira, M., Jorge, J. M., Proença, J. T., & Barreto, V. M. (2021). The off-targets of clustered regularly interspaced short palindromic repeats gene editing. *Frontiers in Cell and Developmental Biology*, *9*, 718466. DOI: 10.3389/fcell.2021.718466 PMID: 34604217

Vikas, and Ranjan, R. (2024). Agroecological approaches to sustainable development. *Frontiers in Sustainable Food Systems, 8*. DOI: 10.3389/fsufs.2024.1405409

Vishnu, L. (2024). *A Comprehensive Review of Genetically Modified Horticultural Crops: Benefits, Risks, and Public Perceptions*. AJGHAS-Australian Journal of Global Horticulture and Agronomy Science.

Voigt, B. (2023). EU regulation of gene-edited plants—A reform proposal. *Frontiers in Genome Editing*, *5*, 1119442. DOI: 10.3389/fgeed.2023.1119442 PMID: 36865566

Vurukonda, S. S. K. P., Vardharajula, S., Shrivastava, M., & SkZ, A. (2016). Enhancement of drought stress tolerance in crops by plant growth promoting rhizobacteria. *Microbiological Research, 184*, 13–24. DOI: 10.1016/j.micres.2015.12.003 PMID: 26856449

Vu, T. V., Doan, D. T. H., Tran, M. T., Sung, Y. W., Song, Y. J., & Kim, J. Y. (2021). Improvement of the LbCas12a-crRNA system for efficient gene targeting in tomato. *Frontiers in Plant Science*, *12*, 722552. DOI: 10.3389/fpls.2021.722552 PMID: 34447405

Vu, T. V., Nguyen, N. T., Kim, J., Song, Y. J., Nguyen, T. H., & Kim, J.-Y. (2024). Optimized dicot prime editing enables heritable desired edits in tomato and *Arabidopsis*. *Nature Plants*, *10*(10), 1502–1513. DOI: 10.1038/s41477-024-01786-w PMID: 39242983

Wada, N., Ueta, R., Osakabe, Y., & Osakabe, K. (2020). Precision genome editing in plants: State-of-the-art in CRISPR/Cas9-based genome engineering. *BMC Plant Biology*, *20*(1), 1–12. DOI: 10.1186/s12870-020-02385-5 PMID: 32450802

Waghmode, Prajyot, Chetariya Chana Pitha, and Vaishnavi Banka. "The Role of Genome Editing Technologies in Sustainable Agriculture." (2024) https://doi.org/ DOI: 10.9734/jabb/2024/v27i81168

Wahab, A., Abdi, G., Saleem, M. H., Ali, B., Ullah, S., Shah, W., Mumtaz, S., Yasin, G., Muresan, C. C., & Marc, R. A. (2022). Plants' Physio-Biochemical and Phyto-Hormonal Responses to Alleviate the Adverse Effects of Drought Stress: A Comprehensive Review. *Plants*, *11*(13), 1620. DOI: 10.3390/plants11131620 PMID: 35807572

Walton, R. T., Christie, K. A., Whittaker, M. N., & Kleinstiver, B. P. (2020). Unconstrained genome targeting with near-PAMless engineered CRISPR-Cas9 variants. *Science*, *368*(6488), 290–296. DOI: 10.1126/science.aba8853 PMID: 32217751

Wang, F., Xu, H., Zhang, J., Yin, K., & Liu, G. (2025). Genome-edited wheat lines show variable performance under multi-environment field trials. *The Plant Journal : For Cell and Molecular Biology*, *121*(2), 344–358. DOI: 10.1111/tpj.16835

Wang, J. Y., & Doudna, J. A. (2023). CRISPR technology: A decade of genome editing is only the beginning. *Science*, *379*(6629), eadd8643. DOI: 10.1126/science.add8643 PMID: 36656942

Wang, J. Y., Pausch, P., & Doudna, J. A. (2022). Structural biology of CRISPR–Cas immunity and genome editing enzymes. *Nature Reviews. Microbiology*, *20*(11), 641–656. DOI: 10.1038/s41579-022-00739-4 PMID: 35562427

Wang, P., Si, H., Li, C., Xu, Z., Guo, H., Jin, S., & Cheng, H. (2025). Plant genetic transformation: Achievements, current status and future prospects. *Plant Biotechnology Journal*, *23*(6), 2034–2058. DOI: 10.1111/pbi.70028 PMID: 40052992

Wang, S., Sun, S.-T., Zhang, X.-Y., Ding, H.-R., Yuan, Y., He, J.-J., Wang, M.-S., Yang, B., & Li, Y.-B. (2023). The evolution of single-cell RNA sequencing technology and application: Progress and perspectives. *International Journal of Molecular Sciences*, *24*(3), 2943. DOI: 10.3390/ijms24032943 PMID: 36769267

Wang, T., Zhang, H., & Zhu, H. (2019). CRISPR technology is revolutionizing the improvement of tomato and other fruit crops. *Horticulture Research*, *6*(1), 6. DOI: 10.1038/s41438-019-0159-x PMID: 31240102

Wang, Y., Cheng, X., Shan, Q., Zhang, Y., Liu, J., Gao, C., & Qiu, J. L. (2014). Simultaneous editing of three homoeoalleles in hexaploid bread wheat confers heritable resistance to powdery mildew. *Nature Biotechnology*, *32*(9), 947–951. DOI: 10.1038/nbt.2969 PMID: 25038773

Wang, Y., Zafar, N., Ali, Q., Manghwar, H., Wang, G., Yu, L., & Jin, S. (2022). CRISPR/Cas genome editing technologies for plant improvement against biotic and abiotic stresses: Advances, limitations, and future perspectives. *Cells*, *11*(23), 3928. DOI: 10.3390/cells11233928 PMID: 36497186

Wang, Z. P., Xing, H. L., Dong, L., Zhang, H. Y., Han, C. Y., Wang, X. C., & Chen, Q. J. (2015). Egg cell-specific promoter-controlled CRISPR/Cas9 efficiently generates homozygous mutants for multiple target genes in Arabidopsis in a single generation. *Genome Biology*, *16*(1), 144. DOI: 10.1186/s13059-015-0715-0 PMID: 26193878

Wang, Z., Wang, J., Cheng, M., Du, Y., Godwin, I., Wang, L., Lv, P., & Liu, G. (2026). Biotechnological and genetic innovations to enhance sorghum adaptation under climate change. *Frontiers in Plant Science*, *17*, 1757792. DOI: 10.3389/fpls.2026.1757792 PMID: 41809630

Watson, O., & Hayta, S. (2024). Precision breeding in agriculture and food systems in the United Kingdom. *Transgenic Research*, *33*(6), 1–6. DOI: 10.1007/s11248-024-00397-7 PMID: 39105945

Weeks, A. R., Sgro, C. M., Young, A. G., Frankham, R., Mitchell, N. J., Miller, K. A., & Sunnucks, P. (2011). Assessing the benefits and risks of translocations in changing environments: A genetic perspective. *Evolutionary Applications*, *4*(6), 709–725. DOI: 10.1111/j.1752-4571.2011.00192.x PMID: 22287981

Wei, X., Pu, A., Liu, Q., Hou, Q., Zhang, Y., An, X., & Wu, S. (2022). The bibliometric landscape of gene editing innovation and regulation in the worldwide. *Cells*, *11*(17), 2682. DOI: 10.3390/cells11172682 PMID: 36078090

Wei, Z., Arazi, T., Hod, N., Zohar, M., Isaacson, T., Doron-Faigenboim, A., & Yedidia, I. (2020). Transcriptome profiling of *Ornithogalum dubium* leaves and flowers to identify key carotenoid genes. *Plants*, *9*(4), 540. DOI: 10.3390/plants9040540 PMID: 32326260

Wen, W., & Zhang, X.-B. (2022). CRISPR–Cas9 gene editing induced complex on-target outcomes in human cells. *Experimental Hematology*, *110*, 13–19. DOI: 10.1016/j.exphem.2022.03.002 PMID: 35304271

Westerman, L. A. (2018). Consumer choice or confusion: That GMO label doesn't mean what you think it means. *The Drake Journal of Agricultural Law*, *23*, 199.

Whelan, A. I., & Lema, M. A. (2019). Regulation of genome editing in plant biotechnology: Argentina. In Regulation of Genome Editing in Plant Biotechnology: A Comparative Analysis of Regulatory Frameworks of Selected Countries and the EU (pp. 19-62). Springer.

Whelan, A. I., & Lema, M. A. (2015). Regulatory framework for gene editing and other new breeding techniques (NBTs) in Argentina. *GM Crops & Food*, *6*(4), 253–265. DOI: 10.1080/21645698.2015.1114698 PMID: 26552666

Wieczorek, M. N., Walczak, M., Skrzypczak-Zielińska, M., & Jeleń, H. H. (2018). Bitter taste of Brassica vegetables: The role of genetic factors, receptors, isothiocyanates, glucosinolates, and flavor context. *Critical Reviews in Food Science and Nutrition*, *58*(18), 3130–3140. DOI: 10.1080/10408398.2017.1353478 PMID: 28718657

Wilson, F. M., & Harrison, R. J. (2021). CRISPR/Cas9 mediated editing of the Quorn fungus Fusarium venenatum A3/5 by transient expression of Cas9 and sgRNAs targeting endogenous marker gene PKS12. *Fungal Biology and Biotechnology*, *8*(1), 1–8. DOI: 10.1186/s40694-021-00121-8 PMID: 34789333

Wolt, J. D., & Wolf, C. (2018). Policy and governance perspectives for regulation of genome edited crops in the United States. *Frontiers in Plant Science*, *9*, 1606. DOI: 10.3389/fpls.2018.01606 PMID: 30467510

Wolt, J. D., Yang, B., Wang, K., & Spalding, M. H. (2016). Regulatory aspects of genome-edited crops. *In Vitro Cellular & Developmental Biology. Plant : Journal of the Tissue Culture Association*, *52*(4), 349–353. DOI: 10.1007/s11627-016-9784-3

Wu, S., Hou, Y., Ma, Y., Wang, P., & Lang, Z. (2025). Epigenetic regulation and its applications in plants. *Epigenetics Insights*, *18*(1), e010. DOI: 10.48130/epi-0025-0009

Wynne, B. (2001). Creating public alienation: Expert cultures of risk and ethics on GMOs. *Science as Culture*, *10*(4), 445–481. DOI: 10.1080/09505430120093586 PMID: 15971363

Xie, H., Ge, X., Yang, F., Wang, B., Li, S., Duan, J., & Gu, F. (2020). High-fidelity SaCas9 identified by directional screening in human cells. *PLoS Biology*, *18*(7), e3000747. DOI: 10.1371/journal.pbio.3000747 PMID: 32644995

Xie, K., Minkenberg, B., & Yang, Y. (2015). Boosting CRISPR/Cas9 multiplex editing capability with the endogenous tRNA-processing system. *Proceedings of the National Academy of Sciences of the United States of America*, *112*(11), 3570–3575. DOI: 10.1073/pnas.1420294112 PMID: 25733849

Xie, K., Wu, C., & Xiong, L. (2023). Genomic comparison, characterization, and delineation of the OsNAC gene family and its transcriptional responses to abiotic stresses and phytohormones. *Plant Molecular Biology*, *65*(1–2), 107–124. DOI: 10.1007/s11103-007-9244-x PMID: 36219366

Xie, K., & Yang, Y. (2013). RNA-Guided Genome Editing in Plants Using a CRISPR–Cas System. *Molecular Plant*, *6*(6), 1975–1983. DOI: 10.1093/mp/sst119 PMID: 23956122

Xie, M., Zhang, S., Song, G., Wang, J., Liang, W., & Li, J. (2023). Targeted epigenetic remodeling at yield-related loci produces heritable phenotypic gains in rice. *Nature Biotechnology*, *41*(6), 801–812. DOI: 10.1038/s41587-023-01756-5

Xie, Y., Shen, Q., Li, F., Ni, S., & Yu, J. (2023). Temperature response of plants and heat tolerance in rice: A review. *Advances in Agronomy*, *179*, 135–203. DOI: 10.1016/bs.agron.2023.01.003

Xing, H. L., Dong, L., Wang, Z. P., Zhang, H. Y., Han, C. Y., Liu, B., & Chen, Q. J. (2014). A CRISPR/Cas9 toolkit for multiplex genome editing in plants. *BMC Plant Biology*, *14*(1), 327. DOI: 10.1186/s12870-014-0327-y PMID: 25432517

Xiong, W., Lin, E., Ju, H., & Xu, Y. (2007). Climate change and critical thresholds in China's food security. *Climatic Change*, *81*(2), 205–221. DOI: 10.1007/s10584-006-9123-5

Xu, L., Sun, B., Liu, S., Gao, X., Zhou, H., Li, F., & Li, Y. (2023). The evaluation of active transcriptional repressor domain for CRISPRi in plants. *Gene*, *851*, 146967. DOI: 10.1016/j.gene.2022.146967 PMID: 36261092

Xu, P., Zhong, Y., Xu, A., Liu, B., Zhang, Y., Zhao, A., Yang, X., Ming, M., Cao, F., & Fu, F. (2024). Application of developmental regulators for enhancing plant regeneration and genetic transformation. *Plants*, *13*(9), 1272. DOI: 10.3390/plants13091272 PMID: 38732487

Xu, R., Li, H., Qin, R., Wang, L., Li, L., Wei, P., & Yang, J. (2014). Gene targeting using the Agrobacterium tumefaciens-mediated CRISPR-Cas system in rice. *Rice (New York, N.Y.)*, *7*(1), 5. DOI: 10.1186/s12284-014-0005-6 PMID: 24920971

Xu, W., Zhang, C., Yang, Y., Zhao, S., Kang, G., He, X., Song, J., & Wu, Y. (2020). Versatile nucleotides substitution in plant using an improved prime editing system. *Molecular Plant*, *13*(5), 675–678. DOI: 10.1016/j.molp.2020.03.012 PMID: 32234340

Xu, Y., Lu, Y., Xie, C., Gao, S., Wan, J., & Prasanna, B. M. (2012). Whole-genome strategies for marker-assisted plant breeding. *Molecular Breeding*, *29*(4), 833–854. DOI: 10.1007/s11032-012-9699-6

Yadav, A., Gehlot, P., Soni, R., & Jain, T. (2026). Plant microbiome engineering: From inoculation to genome editing. *Frontiers in Microbiology*, *17*, 1781381. DOI: 10.3389/fmicb.2026.1781381 PMID: 42100688

Yadav, H., Roberts, P. A., & Lopez-Arredondo, D. (2025). Combating Root-Knot Nematodes (Meloidogyne spp.): From Molecular Mechanisms to Resistant Crops. *Plants*, *14*(9), 1321. DOI: 10.3390/plants14091321 PMID: 40364350

Yamaguchi, T., Ezaki, K., & Ito, K. (2024). Exploring the landscape of public attitudes towards gene-edited foods in Japan. *Breeding Science*, *74*(1), 11–21. DOI: 10.1270/jsbbs.23047 PMID: 39246435

Yang, C., Huang, L., Wang, B.-C., Zhong, Y., Ma, X., Zhang, C., Sun, Q., Wu, Y., Yao, Y., & Liu, Q. (2025a). Enhancing quality traits in staple crops: Current advances and future perspectives. *Journal of Genetics and Genomics*. *Journal of Genetics and Genomics*, *52*(12), 1438–1459. DOI: 10.1016/j.jgg.2025.05.001 PMID: 40348082

Yang, F., Zheng, K., & Yao, Y. (2024). China's regulatory change toward genome-edited crops. *Trends in Biotechnology*, *42*(7), 801–806. DOI: 10.1016/j.tibtech.2023.12.008 PMID: 38160083

Yang, H., & Patel, D. J. (2024). Structures, mechanisms and applications of RNA-centric CRISPR–Cas13. *Nature Chemical Biology*, *20*(6), 673–688. DOI: 10.1038/s41589-024-01593-6 PMID: 38702571

Yang, X., Yang, M., Deng, H., & Ding, Y. (2018). A new era in studying RNA secondary structure and its influence on gene regulation in plants. *Frontiers in Plant Science*, *9*, 671. DOI: 10.3389/fpls.2018.00671 PMID: 29872445

Yang, Y. (2017). CRISPR goes global: A snapshot of rules, policies, and attitudes. *The Bulletin*.

Yang, Y., & Hobbs, J. E. (2020). Supporters or opponents: Will cultural values shape consumer acceptance of gene editing? *Journal of Food Products Marketing*, *26*(1), 17–37. DOI: 10.1080/10454446.2020.1715316

Yang, Y., Saand, M. A., Huang, L., Abdelaal, W. B., Zhang, J., Wu, Y., & Wang, F. (2021). Applications of multi-omics technologies for crop improvement. *Frontiers in Plant Science*, *12*, 563953. DOI: 10.3389/fpls.2021.563953 PMID: 34539683

Yang, Z., Yan, H., Liu, H., Yang, L., Mi, G., & Wang, P. (2025b). Enhancing crop nitrogen efficiency: The role of mixed nitrate and ammonium supply in plant growth and development. *Biology (Basel), 14*(5), 546. DOI: 10.3390/biology14050546 PMID: 40427735

Yao, D., Zhou, J., Zhang, A., Wang, J., Liu, Y., Wang, L., Pi, W., Li, Z., Yue, W., Cai, J., Liu, H., Hao, W., & Qu, X. (2023). Advances in CRISPR/Cas9-based research related to soybean [Glycine max (Linn.) Merr] molecular breeding. *Frontiers in Plant Science, 14*, 1247707. DOI: 10.3389/fpls.2023.1247707 PMID: 37711287

Yao, Q., Ye, Y., Yu, M., Tian, Y., Liu, Q., Zheng, H., & Huang, L. (2026). Engineering prime editors in Salvia miltiorrhiza for precise genome modification. *Journal of Integrative Plant Biology, 68*(4), 855–857. DOI: 10.1111/jipb.70006 PMID: 41987526

Yaqoob, H., Tariq, A., Bhat, B. A., Bhat, K. A., Nehvi, I. B., Raza, A., Djalovic, I., Prasad, P. V., & Mir, R. A. (2023). Integrating genomics and genome editing for orphan crop improvement: A bridge between orphan crops and modern agriculture system. *GM Crops & Food, 14*(1), 1–20. DOI: 10.1080/21645698.2022.2146952 PMID: 36606637

Yaşar, S., Gehrke, F., Capdeville, N., & Puchta, H. (2026). Recent progress in plant genome engineering: From large insertions to chromosome number changes. *Current Opinion in Biotechnology, 97*, 103426. DOI: 10.1016/j.copbio.2025.103426 PMID: 41506049

Yatusevich, R. (2008). *Analysis of the MYB28, MYB29 and MYB76 transcription factors involved in the biosynthesis of aliphatic glucosinolates in Arabidopsis thaliana.* Universität zu Köln, Yuen, K. K.-C. (2017). *New sustainable models of open innovation to accelerate technology development in cellular agriculture.* Massachusetts Institute of Technology, Zhan, X., Lu, Y., Zhu, J. K., & Botella, J. R. (2021). Genome editing for plant research and crop improvement. *Journal of Integrative Plant Biology, 63*(1), 3–33.

Yuan, P., Usman, M., Liu, W., Adhikari, A., Zhang, C., Njiti, V., & Xia, Y. (2024). Advancements in Plant Gene Editing Technology: From Construct Design to Enhanced Transformation Efficiency. *Biotechnology Journal, 19*(12), e202400457. DOI: 10.1002/biot.202400457 PMID: 39692063

Yu, C., Mo, J., Zhao, X., Li, G., & Zhang, X. (2021). "CRISPR/Cas-mediated DNA base editing technology and its application in biomedicine and agriculture." *Sheng wu Gong Cheng xue bao=. Chinese Journal of Biotechnology, 37*(9), 3071–3087. DOI: 10.13345/j.cjb.200693 PMID: 34622618

Yu, X., Huo, G., Yu, J., Li, H., & Li, J. (2023). Prime editing: Its systematic optimization and current applications in disease treatment and agricultural breeding. *International Journal of Biological Macromolecules*, *253*, 127025. DOI: 10.1016/j.ijbiomac.2023.127025 PMID: 37769783

Yu, Y., Yu, J., Wang, Q., Wang, J., Zhao, G., Wu, H., Zhu, Y., Chu, C., & Fang, J. (2021). Overexpression of the rice ORANGE gene OsOR negatively regulates carotenoid accumulation, leads to higher tiller numbers and decreases stress tolerance in Nipponbare rice. *Plant Science : An International Journal of Experimental Plant Biology*, *310*, 110962. DOI: 10.1016/j.plantsci.2021.110962 PMID: 34315587

Zafar, K., Khan, M. Z., Amin, I., Mukhtar, Z., Yasmin, S., Arif, M., Ejaz, K., & Mansoor, S. (2020). Precise CRISPR-Cas9 mediated genome editing in super basmati rice for resistance against bacterial blight by targeting the major susceptibility gene. *Frontiers in Plant Science*, *11*, 575. DOI: 10.3389/fpls.2020.00575 PMID: 32595655

Zafar, S. A., Zaidi, S. S. E. A., Gaba, Y., Singla-Pareek, S. L., Dhankher, O. P., Li, X., Mansoor, S., & Pareek, A. (2020). Engineering abiotic stress tolerance via CRISPR/Cas-mediated genome editing. *Journal of Experimental Botany*, *71*(2), 470–479. DOI: 10.1093/jxb/erz476 PMID: 31644801

Zagelbaum, J., & Gautier, J. (2023). Double-strand break repair and mis-repair in 3D. *DNA Repair*, *121*, 103430. DOI: 10.1016/j.dnarep.2022.103430 PMID: 36436496

Zahoor, M. K., Ahmad, A., Zahoor, M. A., Majeed, H. N., Zulhussnain, M., & Ranian, K. (2021). CRISPR/Cas-based insect resistance in crops. *CRISPR Crops: The Future of Food Security*, 117-149.

Zanatta, C. B., Benevenuto, R. F., Nodari, R. O., & Agapito-Tenfen, S. Z. (2020). Stacked genetically modified soybean harboring herbicide resistance and insecticide rCry1Ac shows strong defense and redox homeostasis disturbance after glyphosate-based herbicide application. *Environmental Sciences Europe*, *32*(1), 1–17. DOI: 10.1186/s12302-020-00379-6

Zarate, S., Cimadori, I., Jones, M. S., Roca, M. M., & Kathleen Barnhill-Dilling, S. (2023). Assessing agricultural gene editing regulation in Latin America: An analysis of how policy windows and policy entrepreneurs shape agricultural gene editing regulatory regimes. *Frontiers in Bioengineering and Biotechnology*, *11*, 1209308. DOI: 10.3389/fbioe.2023.1209308 PMID: 37362213

Zetsche, B., Gootenberg, J. S., Abudayyeh, O. O., Slaymaker, I. M., Makarova, K. S., Essletzbichler, P., Volz, S. E., Joung, J., van der Oost, J., Regev, A., Koonin, E. V., & Zhang, F. (2015). Cpf1 is a single RNA-guided endonuclease of a class 2 CRISPR-Cas system. *Cell*, *163*(3), 759–771. DOI: 10.1016/j.cell.2015.09.038 PMID: 26422227

Zhakypbek, Y., Belkozhayev, A. M., Kerimkulova, A., Kossalbayev, B. D., Murat, T., Tursbekov, S., & Allakhverdiev, S. I. (2025). MicroRNAs in Plant Genetic Regulation of Drought Tolerance and Their Function in Enhancing Stress Adaptation. *Plants*, *14*(3), 410. DOI: 10.3390/plants14030410 PMID: 39942972

Zhang, C., Liu, S., Li, X., Zhang, R., & Li, J. (2022). Virus-induced gene editing and its applications in plants. *International Journal of Molecular Sciences*, *23*(18), 10202. DOI: 10.3390/ijms231810202 PMID: 36142116

Zhang, D., Hussain, A., Manghwar, H., Xie, K., Xie, S., Zhao, S., Larkin, R. M., Qing, P., Jin, S., & Ding, F. (2020). Genome editing with the CRISPR-Cas system: An art, ethics and global regulatory perspective. *Plant Biotechnology Journal*, *18*(8), 1651–1669. DOI: 10.1111/pbi.13383 PMID: 32271968

Zhang, D., Xu, F., Wang, F., Le, L., & Pu, L. (2024). Synthetic biology and artificial intelligence in crop improvement. *Plant Communications*, *6*(2), 101220. DOI: 10.1016/j.xplc.2024.101220 PMID: 39668563

Zhang, F. (2019). Development of CRISPR-Cas systems for genome editing and beyond. *Quarterly Reviews of Biophysics*, *52*, e6. DOI: 10.1017/S0033583519000052

Zhang, F., Yan, D., Hou, J., Yang, D., Xiong, Y., Wen, M., & Zhu, X. (2026). Writing Big in Plant Genomes: Advances, Challenges and Strategies for Targeted Large-Fragment DNA Insertion. *Plant, Cell & Environment*, *0*, 1–15. DOI: 10.1111/pce.70634 PMID: 42212714

Zhang, H. X., Zhang, Y., & Yin, H. (2019). Genome editing with mRNA encoding ZFN, TALEN, and Cas9. *Molecular Therapy : The Journal of the American Society of Gene Therapy*, *27*(4), 735–746. DOI: 10.1016/j.ymthe.2019.01.014 PMID: 30803822

Zhang, H., Qin, C., An, C., Zheng, X., Wen, S., Chen, W., Liu, X., Lv, Z., Yang, P., Xu, W., Gao, W., & Wu, Y. (2021). Application of the CRISPR/Cas9-based gene editing technique in basic research, diagnosis, and therapy of cancer. *Molecular Cancer*, *20*(1), 1–22. DOI: 10.1186/s12943-021-01431-6 PMID: 34598686

Zhang, H., Zhang, J., Lang, Z., Botella, J. R., & Zhu, J. K. (2017). Genome editing—Principles and applications for functional genomics research and crop improvement. *Critical Reviews in Plant Sciences*, *36*(4), 291–309. DOI: 10.1080/07352689.2017.1402989

Zhang, H., Zhang, J., Wei, P., Zhang, B., Gou, F., Feng, Z., Mao, Y., Yang, L., Zhang, H., Xu, N., & Zhu, J.-K. (2013). The CRISPR/Cas9 system produces specific and homozygous targeted gene editing in rice in one generation. *Plant Biotechnology Journal*, *12*(6), 797–807. DOI: 10.1111/pbi.12200 PMID: 24854982

Zhang, L., Li, G., Zhang, Y., Cheng, Y., Roberts, N., Glenn, S. E., & Qi, Y. (2023). Boosting genome editing efficiency in human cells and plants with novel LbCas12a variants. *Genome Biology*, *24*(1), 102. DOI: 10.1186/s13059-023-02929-6 PMID: 37122009

Zhang, R., He, Z., Shi, Y., Sun, X., Chen, X., Wang, G., Zhang, Y., Gao, P., Wu, Y., Lu, S., Duan, J., Sun, S., Yang, N., Fan, W., Zhao, K., Yang, B., Xia, Y., Zhang, Y., Zhang, Y., & Yin, H. (2024, July 25). Amplification editing enables efficient and precise duplication of DNA from short sequence to megabase and chromosomal scale. *Cell*, *187*(15), 3936–3952.e19. DOI: 10.1016/j.cell.2024.05.056 PMID: 38936359

Zhang, R., Liu, J., Chai, Z., Chen, S., Bai, Y., Zong, Y., Chen, K., Li, J., Jiang, L., & Gao, C. (2019). Generation of herbicide tolerance traits and a new selectable marker in wheat using base editing. *Nature Plants*, *5*(5), 480–485. DOI: 10.1038/s41477-019-0405-0 PMID: 30988404

Zhang, S., & Zhu, H. (2024). Development and prospect of gene-edited fruits and vegetables. *Food Quality and Safety*, *8*, fyad045. Advance online publication. DOI: 10.1093/fqsafe/fyad045

Zhang, X., Song, M., Wang, Y., yao, Q., Shen, R., Tian, Y., Lu, Y., & Zhu, J.-K. (2024). Programmable broad-spectrum resistance to bacterial blight using targeted insertion in rice. *Cell Discovery*, *10*(1), 100. DOI: 10.1038/s41421-024-00714-8 PMID: 39375327

Zhang, Y., Iaffaldano, B., & Qi, Y. (2021). CRISPR ribonucleoprotein-mediated genetic engineering in plants. *Plant Communications*, *2*(2), 100168. DOI: 10.1016/j.xplc.2021.100168 PMID: 33898980

Zhang, Y., Massel, K., Godwin, I. D., & Gao, C. (2018). Applications and potential of genome editing in crop improvement. *Genome Biology*, *19*(1), 210. DOI: 10.1186/s13059-018-1586-y PMID: 30501614

Zhang, Y., Ma, X., Xie, X., & Liu, Y. G. (2017). CRISPR/Cas9-Based Genome Editing in Plants. *Progress in Molecular Biology and Translational Science*, *149*, 133–150. DOI: 10.1016/bs.pmbts.2017.03.008 PMID: 28712494

Zhang, Y., Zhang, H., Tao, T., Chen, J., Li, P., Wang, Y., Liu, P., Zhu, Y., Routledge, M. N., Yang, C., & Zhang, C. (2025). Investigation and insights into the technical strategies for reducing agrochemical inputs in rice farming to enhance agroecosystem resilience and food safety, A case study in China. *Journal of Environmental Management*, *392*, 126619. DOI: 10.1016/j.jenvman.2025.126619 PMID: 40695181

Zhang, Z., Mao, Y., Ha, S., Liu, W., Botella, J. R., & Zhu, J. K. (2016). A multiplex CRISPR/Cas9 platform for fast and efficient editing of multiple genes in Arabidopsis. *Plant Cell Reports*, *35*(7), 1519–1533. DOI: 10.1007/s00299-015-1900-z PMID: 26661595

Zhao, M., Zhao, J., Yuan, J., Hale, L., Wen, T., Huang, Q., Vivanco, J. M., Zhou, J., Kowalchuk, G. A., & Shen, Q. (2021). Root exudates drive soil-microbe-nutrient feedbacks in response to plant growth. *Plant, Cell & Environment*, *44*(2), 613–628. DOI: 10.1111/pce.13928 PMID: 33103781

Zhao, Y., Huang, Z., Zhou, X., Teng, W., Liu, Z., Wang, W., Tang, S., Liu, Y., Liu, J., Wang, W., Chai, L., Zhang, N., Guo, W., Liu, J., Ni, Z., Sun, Q., Wang, Y., & Zong, Y. (2025, February). Precise deletion, replacement and inversion of large DNA fragments in plants using dual prime editing. *Nature Plants*, *11*(2), 191–205. DOI: 10.1038/s41477-024-01898-3 PMID: 39805934

Zhao, Z., Shang, P., Mohanraju, P., & Geijsen, N. (2023). Prime editing: Advances and therapeutic applications. *Trends in Biotechnology*, *41*(8), 1000–1012. DOI: 10.1016/j.tibtech.2023.03.004 PMID: 37002157

Zheng, Y., Guo, T., Xia, T., Guo, S., Chen, M., Ye, S., Pan, T., Xu, X., Gan, Y., Zhan, Y., Zheng, T., & Zheng, Z. (2024). Utility of Arabidopsis KASII Promoter in Development of an Effective CRISPR/Cas9 System for Soybean Genome Editing and Its Application in Engineering of Soybean Seeds Producing Super-High Oleic and Low Saturated Oils. *Journal of Agricultural and Food Chemistry*, *72*(39), 21720–21730. DOI: 10.1021/acs.jafc.4c05840 PMID: 39288439

Zheng, Y., Li, Y., Zhou, K., Li, T., VanDusen, N. J., & Hua, Y. (2024). Precise genome-editing in human diseases: Mechanisms, strategies and applications. *Signal Transduction and Targeted Therapy*, *9*(1), 47. DOI: 10.1038/s41392-024-01750-2 PMID: 38409199

Zheng, Z., Appiano, M., Pavan, S., Bracuto, V., Ricciardi, L., Visser, R. G., Wolters, A.-M. A., & Bai, Y. (2016). Genome-wide study of the tomato SlMLO gene family and its functional characterization in response to the powdery mildew fungus Oidium neolycopersici. *Frontiers in Plant Science*, *7*, 380. DOI: 10.3389/fpls.2016.00380 PMID: 27579028

Zhou, L., Iqbal, A., Yang, M., & Yang, Y. (2025). Research Progress on Gene Regulation of Plant Floral Organogenesis. *Genes*, *16*(1), 79. DOI: 10.3390/genes16010079 PMID: 39858626

Zhou, Q., Galindo-González, L., Hwang, S. F., & Strelkov, S. E. (2021). Application of genomics and transcriptomics to accelerate development of clubroot-resistant canola. *Canadian Journal of Plant Pathology*, *43*(2), 189–208. DOI: 10.1080/07060661.2020.1794541

Zhou, W., Yuan, Y., Zhang, Y., & Chen, D. (2021). A decade of CRISPR gene editing in China and beyond: A scientometric landscape. *The CRISPR Journal*, *4*(3), 313–320. DOI: 10.1089/crispr.2020.0148 PMID: 34152220

Zhu, C., Wang, Y., Zhu, M., Chen, G., Wang, F., Li, B., Xu, Z., Wang, G., Xu, J., Lu, X., Wang, Y., & Jin, S. (2026). High-Density CRISPR/Cas12a-Mediated Multiplex Genome Editing Reveals Genome Instability in Allotetraploid Cotton. *Genes*, *17*(6), 622. DOI: 10.3390/genes17060622 PMID: 42353780

Zhu, H., Li, C., & Gao, C. (2020). Applications of CRISPR–Cas in agriculture and plant biotechnology. *Nature Reviews. Molecular Cell Biology*, *21*(11), 661–677. DOI: 10.1038/s41580-020-00288-9 PMID: 32973356

Zhu, H., Wu, X., Zheng, R., & Zhu, Y. (2025). The impact of gene editing technology on agricultural economic efficiency Biomechanical Applications: An empirical analysis based on international CRISPR patent data. *Molecular & Cellular Biomechanics*, *22*(3), 1031. DOI: 10.62617/mcb1031

Zhu, J.-K. (2022). future of gene-edited crops in China. *National Science Review*, *9*(4), nwac063. Advance online publication. DOI: 10.1093/nsr/nwac063 PMID: 35505661

Zong, Y., Wang, Y., Li, C., Zhang, R., Chen, K., Ran, Y., Qiu, J.-L., Wang, D., & Gao, C. (2017). Precise base editing in rice, wheat and maize with a Cas9-cytidine deaminase fusion. *Nature Biotechnology*, *35*(5), 438–440. DOI: 10.1038/nbt.3811 PMID: 28244994

Zuo, E., Sun, Y., Wei, W., Yuan, T., Ying, W., Sun, H., Yuan, L., Steinmetz, L. M., Li, Y., & Yang, H. (2019). Cytosine base editor generates substantial off-target single-nucleotide variants in mouse embryos. *Science*, *364*(6437), 289–292. DOI: 10.1126/science.aav9973 PMID: 30819928

About the Contributors

Sana Basharat is working as a postdoctoral researcher at college of tropical agriculture and forestry at hainan university, haikou, hainan, China. She got her doctoral degree from Department of botany, university of agriculture faislabad, Pakistan. Her research direction involved structural and functional Strategies in Cenchrus Species to Combat Environmental Extremities Imposed by Multiple Abiotic Stresses. She had published more than 15 research articles in renowned journals including Plants, Pakistan Journal botany, Acta Physiologiae Plantarum, so on.

Muhammad Waseem is Associate Professor/Researcher at College of Tropical Agriculture and Forestry, Hainan University, Haikou, Hainan, China. He received a doctoral degree (Ph.D., specialization in Plant Molecular Biology) from the School of Life Sciences, Chongqing University, Chongqing, PR China. He has five years of research experience including postdoctoral. His interests include transcriptional regulation of flowering time regulation, fruit ripening and development, and abiotic stresses such as salinity and drought. In addition, he has hands-on experience in next-generation sequencing data analysis, transcriptomics, and data visualization. Currently, Dr. Waseem is actively in investigating flowering time regulation in Brassica crops. He has been awarded with a National Science Foundation Project, RFIS-I in 2023. In brief, he has published more than 58 research articles, 5 review articles, and 9 book chapters in various high-impact factor publications including New Phytologist, International Journal of Macromolecules, Planta, GM Crops & Food, Environmental and Experimental Botany, Physiologia Plantarum, and Frontiers in Plant Science. He has also edited books with Springer publisher.

***

Sai Ramesh A is an Assistant Professor of Bioengineering at Vels Institute of Science, Technology and Advanced Studies, Chennai, with over 14 years of academic and research experience. He holds a Ph.D. in Biotechnology and an MBA in Educational Management, and has published extensively in SCI and Scopus indexed journals with research interests spanning biotechnology, nanotechnology, computational biology, and science education. He has served as Associate Editor and reviewer for several international journals, supervised doctoral and postgraduate research, and actively contributes to faculty development, interdisciplinary education, and ethical scholarly publishing.

Shabbar Abbas is currently working as an Associate Professor at the Department of Biosciences, COMSATS University Islamabad, Pakistan. Graduated from the School of Food Science & Technology, Jiangnan University, Wuxi, Jiangsu, P R China. Area of research includes, but is not limited to, Food Nanotechnology, Encapsulation Technologies, Nutrient Delivery, Nano-ingredients, Nano-nutrition and Functional food ingredients.

Banzeer Ahsan Abbasi received her graduate degree in Botany from Quaid-i-Azam University, Islamabad, Pakistan. Dr. Banzeer is currently working as an Assistant Professor at Department of Botany, Rawalpindi Women University, Rawalpindi, Pakistan. She has published more than 60 research articles in various international journals and 10 chapters in reputed publishers (ELSIEVER, RSC and SPRINGER). Currently, she is supervising 2 Ph.D, 6 MPhil and 6 BS students. She has expertise in the synthesis of novel nanomaterials from medicinal plants and their multiple biological and envirnemental applications. Her research is also focusing on phytochemistry, natural products of medicinal plants and Phylogenetics. She has a very decent research background with research published in well-reputed journals with H-index 28, i-10 index 44 and overall citations 3317.

Haider Ali is a research scientist at the National Agricultural Research Center (NARC), Islamabad, Pakistan. His research focuses on Plant Tissue culture, plant molecular biology and crop improvement technologies, with particular emphasis on the application of CRISPR/Cas9 systems for genome editing in agricultural crops. He holds an M.Sc. in Horticulture with a specialization in Plant Biotechnology and Genetics.

Amjid Khan is an interdisciplinary scientist and nanomaterials expert currently serving as a Postdoctoral Research Fellow under the prestigious UNESCO-UNISA Africa Chair in Nanoscience and Nanotechnology, College of Graduate Studies, at the University of South Africa, South Africa. Dr. Khan is a Senior Fellow under the prestigious UNESCO-UNISA-iTLABS/NRF Africa Chair in

Nanoscience and Nanotechnology in Pretoria, Republic of South Africa. Dr. Khan's interdisciplinary research bridges advanced nanotechnology applications across biomedicine and sustainable agriculture. In biomedicine, his work focuses on the green and biomimetic synthesis of functional nanomaterials, including plasmonic nanoparticles, quantum dots, and metal-oxide nanocomposites, tailored for diagnostic nanoprobes, bio-imaging, targeted drug delivery, and precision therapeutics. In sustainable agriculture, he specializes in smart nanobiotechnology, abiotic stress mitigation, and the development of next-generation nano-biofertilizers to enhance crop resilience, soil health, and sustainable yield. A highly productive scholar, Dr. Khan has authored over 50 SCI-indexed peer-reviewed articles. Beyond his research appointments, he serves as an Advisory and Affiliate Member of the Paschim Medinipur Microbiological Innovation and Technological Advancement Society (PMMITAS). Demonstrating global editorial leadership, Dr. Khan serves as an Associate Editor for Frontiers in Systems Biology, a Reviewing Editor for the Springer Nature Reviewer Community, and a Lead Volume Editor for academic books with premier publishers, including IGI Global, John Wiley & Sons, and CRC Press/Taylor & Francis.

Waseem Ahmed Khattak is a plant scientist with research expertise spanning plant physiology, nanobiotechnology, and sustainable agriculture. His work focuses on the application of green-synthesized and biogenic nanoparticles to enhance crop productivity, strengthen antioxidant defense systems, and mitigate abiotic stresses such as heavy metal toxicity. He has conducted extensive field and pot experiments involving major crops, with particular emphasis on nutrient use efficiency, root morphology, biochemical profiling, and stress physiology. Dr. Khattak has authored and co-authored multiple research and review articles in reputable journals and has contributed chapters to academic books addressing sustainable agriculture, nanotechnology, and environmental remediation. His research integrates advanced biochemical, physiological, and nanotechnological approaches to develop eco-friendly strategies for resilient and sustainable crop production.

Saranraj P holds a Ph.D. in Microbiology from Annamalai University, Chidambaram, India, and is NET (ASRB–ICAR) qualified in Microbiology. He is currently serving as Assistant Professor and Head, Department of Microbiology, at Sacred Heart College (Autonomous), Tirupattur, Tamil Nadu, India. In addition, he is the Proprietor and Managing Director of JPS Scientific Publications, contributing actively to academic publishing and research dissemination. Dr. Saranraj has over 15 years of research experience and 10 years of teaching experience in the fields of Environmental Microbiology, Agricultural Microbiology, Industrial Microbiology, Bioremediation, Microbial Biotechnology, and Sustainable Ecosystem Management.

His research work has focused on Microbial degradation of pollutants, Effluent treatment, Soil health improvement, Plant–microbe interactions, and Applied environmental sustainability. He has an extensive publication record, with more than 240 international research articles, over 90 books (international and national), and 70+ book chapters published with reputed academic publishers.

Muhammad Rizwan is a doctoral researcher in the Department of Plant Sciences at Quaid-i-Azam University, Islamabad, Pakistan, with academic and research expertise in plant systematics, molecular biology, and contemporary crop improvement strategies. His scholarly interests integrate classical botany with advanced biotechnological approaches, including gene editing, molecular taxonomy, and functional genomics, aimed at enhancing crop resilience, productivity, and sustainability. He has contributed to interdisciplinary research addressing plant evolution, biodiversity conservation, and climate-responsive agricultural systems, with a growing focus on the applications and implications of gene-editing technologies in modern plant science. Through his academic work and authorship, he seeks to bridge foundational plant biology with next-generation tools for sustainable agriculture and food security.

Index

A

B

C

R

S

T

Z